7th International Interdisciplinary Scientific Conference SOCIETY. HEALTH. WELFARE 2018

AA003555

SHS Web of Conferences Volume 68 (2019)

Riga, Latvia
10 – 12 October 2018

Editors:

U. Berkis
L. Vilka

ISBN: 978-1-5108-9864-6

Printed from e-media with permission by:

Curran Associates, Inc.
57 Morehouse Lane
Red Hook, NY 12571

Some format issues inherent in the e-media version may also appear in this print version.

No changes have been made to the content of these proceedings. There may be changes to pagination and minor adjustments for aesthetics.

Printed with permission by Curran Associates, Inc. (2020)

For additional information, please contact EDP Sciences – Web of Conferences at the address below.

EDP Sciences – Web of Conferences
17, Avenue du Hoggar
Parc d'Activité de Courtabœuf
BP 112
F-91944 Les Ulis Cedex A
France

Phone: +33 (0) 1 69 18 75 75
Fax: +33 (0) 1 69 28 84 91

contact-edps@webofconferences.org

Additional copies of this publication are available from:

Curran Associates, Inc.
57 Morehouse Lane
Red Hook, NY 12571 USA
Phone: 845-758-0400
Fax: 845-758-2633
Email: curran@proceedings.com
Web: www.proceedings.com

TABLE OF CONTENTS

Differences in selected areas of social capital regarding residence

Angela Almasiova, *Katarina* Kohutova, and *Zuzana* Gejdosova

Catholic University in Ružomberok, Slovakia

Abstract. Introduction. In addition to a massive increase in interest in the concept of social capital in the social sciences, the research has shown its considerable impact on various areas of human life. OECD points to the correlation between social capital and health and points out that social isolation is associated with misery and disease. **Aim.** The contribution aims at pointing out the differences in selected indicators of social capital, focusing on the participation and social involvement of respondents in their place of residence. **Results.** Quantitative research was conducted using the questionnaire method, which was compiled from the Social Capital Question Bank database, covered by the UK Statistical Office. Result Cluster analysis divided respondents into those with higher and lower social capital, and comparative analysis showed significant differences in selected indicators of social capital between urban and rural respondents. **Conclusions.** Respondents from cities and rural areas differ in the level of social capital in our areas of interest: Participation or involvement in local groups, Political activity or voting, Taking positive action about a local issue, Participation in religious activity, Completed or received a practical favour, Familiarity with neighbourhood, Provide regular service, help or care for others.

Key words: social capital, participation, social involvement.

1 Selected aspects of social capital

The present paper is primarily concerned with the question of whether there are differences in social capital concerning the person's place of residence. Why is it of any relevance to look at these differences? It results from the very character of the place of residence of the person – the countryside is characterized by less social control, less access to public services, lack of employment, etc., and, on the other hand, a city with much more anonymity is associated with weak social control, a much faster way of life strongly linked to employment. However, as Giddens [1] points out, the city can sometimes take a "new human being" much friendlier than a countryside where a long-term established community and neighbourhood can be unavailable for a long time for the new individual. The difference between the countryside and the city also concerns the bonds between the people and thus also the social capital itself.

In trying to organize the definition of social capital, we should base it on its two basic forms of perception – individual and collective, depending on who benefits from the sources of capital. Some authors perceive it at an individual level [2–4], or collective level [5, 6].

One of the leading representatives of the perception of social capital at an individual level is Pierre Bourdieu, who perceives it as an "aggregate of current or potential resources that are based on ownership of a sustained network of more or less institutionalized relationships and acquaintances, in other words, group membership that equips each member with collectively owned capital, that is, with different entitlements [2]. The author argues that the baseline of any capital is the economic capital, but in order to understand the social world, it is necessary to understand all forms of capital, not only the economic one. For economic capital, Bourdieu considers capital in the traditional perception of this word – the various forms of material wealth – not just the possession of means of production. It is possible to "transform" it into real power only if it is associated with other forms of capital – social and cultural [8].

Another significant figure of the perception of an individual form of social capital is James Coleman [7], which is mainly associated with the theory of rational choice and its instrumental view of the behaviour of people in order to maximize profits. He defined social capital as ... "standards, social networks, and relationships between adults and children that are a value within childhood. Social capital exists in the family but also outside of the community and also serves to maximize its profit". Another dimension of social capital was provided by Robert Putnam [6] who also "levelled" it to the macro-social level in the public sphere – especially in the politics and functioning of democracy in general: "... social capital refers to the characteristics of a social organization, such as trust, standards, and networks that facilitate coordinated behaviour and thus contribute to the company's performance".

Since our paper is devoted to social capital in the context of the size of the human capital, we will focus on perceiving this dimension of social capital. The author mentioned above, Putnam [6] states that social capital depends on the size of the community in which a person lives. He claims that altruism, volunteering, philanthropy, participation in community projects, helping foreigners, participation in charity projects are more common in smaller communities than in big cities. In the case of church attendance and membership in church organizations, rural residents are more involved than inhabitants of the city (10%–20% difference). In the countryside there is 30–40% higher chance of people becoming members of interest or other organizations than in the city. Putnam claims that people from the city are less committed not because of who and how they are, but because of where they are. Neighbourhoods with a high level of social capital tend to be suitable places for child upbringing, public spaces are cleaner, people are more friendly, and the streets are safer. Research conducted with pregnant women has shown that women living in smaller neighbourhoods with more considerable social capital perceive their health as better compared to women living in larger neighbourhoods with smaller social capital [9].

2 The goal of the research and key hypothesis

The goal of the research is to identify the extent of social capital in selected areas and to compare it between urban and rural respondents. Based on the facts above, we assume that there are significant differences in selected areas of social capital due to the size of the residence (up to 1,000 inhabitants, from 1001 to 5000 inhabitants and over 5001 inhabitants).

3 Research sample

The research sample consisted of 381 respondents, of whom 117 were men (30.7%) and 264 women (69.3%). The median age of the research file was 40.6 years (SD 13.24 years). The youngest respondent was 18 years old, the oldest 72 years old. Secondary education with a leaving exam (N 138, 36.2%) and higher education (N 132, 34.6%) prevailed. To a lesser extent, we also counted respondents with secondary education without leaving exam (N 87, 22.8%) and respondents with elementary education, those were 24 (6.3%).

Regarding marital status, married respondents (N 222, 58.3%) prevailed, followed by single/single (N 123, 32.3%) and least respondents were divorced (N 21, 5.5%). 15 respondents marked option "other."

An important variable was also the size of the residence where the respondent lives. 114 respondents (29.9%) lived in a place up to 1,000 inhabitants, 138 respondents (36.2%) lived in a residence of 1001 to 5000 inhabitants, and 120 respondents (31.5%) lived in a residence of over 5001 inhabitants. 9 respondents did not respond.

4 Research method

Respondents were asked to submit a questionnaire consisting of 18 items devoted to different areas of the social capital. 5 items surveyed demographic characteristics of respondents. The items in the questionnaire were selected from the Social Capital Question Bank[1] database on social capital.

We have observed 6 areas of social capital related to Participation, social engagement and commitment: Participation or involvement in local groups, Political activity or voting, Taking positive action on a local issue, Participation in religious activities, Completed or received a practical favour, providing regular service, help or care for others.

The data were analyzed using IBM®SPSS (version 22), and descriptive and inductive statistics were used for statistical analysis (Mann-Whitney U-test, Kruskal-Wallis Test).

5 Results

In the result section, we are gradually presenting the analysis of the six focused areas. Each area consisted of several questions in the questionnaire, and we created a compound variable from these items. We continued to work with this variable, and we found differentiations in it concerning the size of the respondent's residence within the framework of the research questions.

5.1 Participation or involvement in local groups

The area mentioned above of social capital has been identified with three items: a) How often have you visited a group, association, club or religious institution for the last year to ask for help or just for joy, sport, relaxation and more? b) How often in the past year have you helped an organization – school, hospital, jail, charity, volunteer organization? c) Are you currently a member of one of the following organizations: political party, trading company; environmental society[2], etc.?

From the items mentioned above regarding participation in local groups, we created a compound variable[3]. It is true that if the respondent has more points in this area of the social capital, his/her participation in local groups is higher. Respondents could have reached a minimum of 0 points (if they answered the first two questions as "never" and are not members of any organization), and a maximum of 29 points (if they would answer "two times a week" and would be a member of all organizations). 39 respondents had a minimum possible value of 0 points, while the maximum was 17 points, which was reached by 3 respondents. The average level of participation in local groups was 3.74 points (SD 3.35).

---

[1] The Social Capital Question Bank is an extensive database of suggestions for making its own methodological proposals for social capital research, which is covered by the UK Office for National Statistics. The database consists of 5 areas of social capital: 1. Participation, social engagement, commitment, 2. Control, self-efficacy, 3. Social interaction, social networking, 5. Trust, reciprocity, social cohesion. In our research we deal with part 1.

[2] A brief description of answers to individual questionnaire questions is presented in the discussion.

[3] Sum of items 1, 2, 3. Scaling of question no. 1 and 2: never = 0, in the last year three-four times = 1, every second month = 2, etc., scaling of question no. 3: membership in the organization yes = 1, no = 0.

Table 1. Differentiation in participation or involvement in local groups relative to the size of the homeplace.

	N	Mean Rank	Mean	$\chi2$	df	Sig.
Up to 1000 inhab.	114	199.72	3.45			
From 1001 to 5000 inhab.	138	196.64	3.91	9.2417	2	**0.010**
Above 5001 inhab.	120	162.28	2.34			

Table 2. Differentiation in political activity or voting regarding the size of the place of residence.

	N	Mean Rank	Mean	χ^2	df	Sig.
Up to 1000 inhab.	114	200.87	9.63			
From 1001 to 5000 inhab.	135	177.93	8.69	6.256	2	**0.044**
Above 5001 inhab.	114	167.95	8.32			

RQ1: Is there a significant difference in the area of participation in local groups due to the size of the residence place?

Based on the significance values, we can say that there is a significant difference between groups (Table 1). Further testing of the differences between groups has shown that there is no significant difference between residents living in the population of up to 1,000 and between 1001 and 5,000. A significant difference was only among the population of over 5001 respondents compared to the two groups mentioned. Thus, we can say that there is a significant difference in the participation in local groups of respondents living in the residence of over 5001 inhabitants compared to the other two groups. Based on the values of the median and the average ranking, respondents living in areas of over 5001 inhabitants are significantly participating lower in local groups compared to those living in areas below 5,000 inhabitants.

5.2 Political activity or voting

Political activity and election participation were identified by three items: a) Did you take part in the elections? b) How likely are you going to participate in the next election in the future? c) Please, indicate to what extent it is crucial for you to participate in municipal elections/municipal authorities.

Just as in the previous case, we have created a composite variable from the items mentioned above on political activity and election participation.[4] It is true that when the respondent has more points in this area of the social capital, their political activity and participation in elections are higher.[5] The average domain value of political activity and participation in the election was 8.81 points (SD 3.59).

RQ2: Is there a significant difference in the area of political activity and participation in elections based on the size of the place of residence?

Based on statistical test we can say that there is a significant difference between groups (Table 2). Further difference testing has shown that there is no significant difference between

[4] Sum of items 1, 2, 3. Scaling of question no. 1 and 2: yes = 1, no = 0; Scaling Question # 3: important = 2, not important/not important = 1, not important = 0.

[5] Theoretically, respondents could reach at least 0 points, reaching a maximum of 13 points. In fact, 12 respondents had a minimum possible value of 0 points, while a maximum of 12 points was reached by 132 respondents.

Table 3. Differentiation in taking positive action about a local issue regarding the size of the place of residence.

	N	Mean Rank	Mean	χ^2	df	Sig.
Up to 1000 inhab.	111	194.65	1.16	7.93	2	**0.019**
From 1001 to 5000 inhab.	135	186.60	1.13			
Above 5001 inhab.	114	159.50	0.84			

residents living in the place with a population of 1001–5000 and over 5001 inhabitants. A significant difference was only among residents up to 1000 respondents compared to the two groups mentioned.

From median values and average ranking, respondents living in areas under 1000 inhabitants have significantly higher levels of political participation and election participation than those living in areas of over 1001 inhabitants.

5.3 Taking positive action about a local issue

The third area consisted of the following two items: a) Have you performed any of the following activities in the last 12 months to solve the problem affecting people around you? b) Have you been in contact with a member of the municipal council in the last 12 months? We have created a compound variable from these items.[6,7] The average domain value of political activity and participation in the election was 1.04 points (SD 1,086).

RQ3: Is there a significant difference in active participation in local affairs due to the size of the home place?

Based on statistical test Table 3 shows that there is a significant difference between groups. Further testing of group differences has shown that there is significant difference between residents living in the population of up to 1,000 and between 1001 and 5,000. A significant difference was only among the population of over 5001 respondents compared to the two groups mentioned.

Based on the values of the median and the average ranking we can state, that respondents living in areas over 5001 inhabitants have a significantly lower rate of active participation in local affairs than those living in areas of up to 5,000 inhabitants.

5.4 Participation in religious activity

This section concerned one item in the questionnaire, namely: How often, if any, are you attending worship or religious meetings?

RQ4: Is there a significant difference in participation in religious activities due to the size of the home place?

Table 4 shows differences. Further testing of group differences has shown that there is no significant difference between residents living in the place with a population of up to 1,000 and between 1001 and 5,000. A significant difference was only among the population of over 5001 respondents compared to the two groups mentioned. From the values of the median and

[6] Sum of items 1 and 2. Scaling of questions: yes = 1, no = 0.

[7] It is proven that if the respondent has more points in this area of social capital, his active participation in local affairs is higher. Theoretically, respondents could reach at least 0 points, reaching a maximum of 9 points. In fact, up to 147 respondents could reach a minimum of 0 points, while a maximum of 4 points was reached by 15 respondents.

Table 4. Differentiation in participation in religious activity relative to the size of the place of residence.

	N	Mean Rank	Mean	χ^2	df	Sig.
Up to 1000 inhab.	111	192.54	3.32	38.276	2	**0.000**
From 1001 to 5000 inhab.	135	207.67	3.44			
Above 5001 inhab.	114	136.61	2.53			

Table 5. Differentiation in completed or received practical favour.

	N	Mean Rank	Mean	χ^2	df	Sig.
Up to 1000 inhab.	111	174.18	3.27	15.912	2	**0.000**
From 1001 to 5000 inhab.	138	212.36	3.87			
Above 5001 inhab.	120	163.55	2.95			

the average ranking, respondents living in areas over 5001 inhabitants have a significantly lower rate of participation in religious activities than those living in areas of up to 5,000 inhabitants.

5.5 Completed or received a practical favour

The fifth area consisted of the following three items: a) Have you done any kind of favour to your neighbour in the last 6 months? b) Has any of your neighbours made a favour to you in the past 6 months? c) Select what you have been given free of charge in the last 12 months. Please exclude family members and relatives. From the items mentioned above regarding accepted or delivered kindness, we have created a composite variable.[8,9] The average value of domain received and given favour was 3.37 points (SD 1,823).

RQ5: Is there a significant difference in giving or receiving favour concerning the size of the place of residence?

Based on the result of statistical tests we can say that respondents living in areas of 1001 to 5,000 inhabitants have a significantly higher rate of providing and receiving kindness than those living in areas of up to 1,000 inhabitants and over 5001 inhabitants.

5.6 Provide regular service, help or care for others

The sixth area consisted of the following three items: a) Do you provide regular services or assistance to an ill person, a disabled person or an elderly person who does not live with you? b) Have you made any of these activities free of charge in the last 12 months for a person who is not your relative? c) Have you helped someone or provided any help for someone who is not a member of your household for the last 4 weeks? We have created a compound

[8] The sum of items a), b) c). Scaling questions: yes = 1, no = 0.

[9] It is proven that if the respondent has more points in this area of social capital, that is his activity in giving or receiving favours is higher. Theoretically, respondents could reach at least 0 points, reaching a maximum of 14 points. In fact, 36 respondents had a minimum of 1 point, while the maximum was 13 points, which was 3 respondents.

Table 6. Differentiation in provide regular service, help or care for others.

	N	Mean Rank	Mean	χ^2	df	Sig.
Up to 1000 inhab.	111	144.05	2.27	26.718	2	**0.000**
From 1001 to 5000 inhab.	138	211.54	3.39			
Above 5001 inhab.	120	192.35	2.80			

variable from these items.[10,11] Average Domain Value Regular Assistance/Service/Care for Others was 2.83 points (SD 2,025).

RQ6: Is there a significant difference in the regular provision of help/service/care relative to the size of the home place?

Based on the results of relevant statistical tests we can say that respondents living in areas of up to 1,000 inhabitants have a significantly lower rate of regular assistance/care/care compared to those living in areas above 1001 inhabitants.

6 Results evaluation and discussion

The research aimed to identify the level of social capital in selected areas and to compare it between urban and rural respondents. From the theoretical analysis and conducted research, there are significant differences in the social capital of the inhabitants of smaller and larger settlements, or between rural and urban residents [10]. Also, rural residents are characterized by a more significant number of social interactions than urban residents [11]. Our research did not investigate the impact of social capital on human health, but a study that observed the effects of social capital in 45 countries showed a significant direct relationship between the amount of social capital and human health [12]. For this reason, we conducted our research.

Based on this fact we assumed that there are significant differences in selected areas of social capital due to the size of the home place.

The first observed area was the rate of participation in local groups. Three items measured this area. The first item was focused on visit of a group, association, club or religious organization because of the support for somebody (or just for fun, relaxation, ...) last year, while the descriptive analysis showed that 19% of respondents never realized such a visit, on the contrary almost half did it three or four times in the last year. Other respondents (34%) did it several times over the last year (e.g., several times a month or a week). Furthermore, under the second item in the questionnaire, we found out that 39.2% of the respondents never realized the help to any organization (e.g., school, charity, hospital), but 42.2% realized it in the last year at least 3 to 4 times. Other respondents (18.6%) reported a more frequent frequency of assistance. The third question in this area was whether the respondent was a member of an organization. We found that most respondents are members of voluntary organizations (24.6%), followed by membership in sports clubs (21.3%), religious organizations (19.7%), retiree organization (18%), professional organization or other organization, a school parent organization and a business company (in the last four cases, 11.5% of the respondents reached the same share). Below 10% of respondents were the members of a general organization, a neighbourhood organization, a women's organization,

[10] The sum of items a), b) c). Scaling questions: yes = 1, no = 0.

[11] It is proven that if the respondent has more points in this area of social capital, his or her assistance, service or care for others is higher. Theoretically, respondents could reach at least 0 points, reaching a maximum of 14 points. In fact, 6 respondents achieved a minimum possible value of 0 points, while the maximum was 11 points, which was reached by 3 respondents.

a social club for men and a youth organization. From these items, we compiled a composite variable named Participation or involvement in local groups, and we compared the overall score in this area to the size of the respondent's residence place. We can say that **respondents living in bigger cities (over 5,000 inhabitants) are less involved in local affairs.**

The second area concerned **political action or voting** and was measured by three items. From the descriptive analysis of the items, 80.3% of the respondents participated in the parliamentary elections, 75.6% of the respondents participated in the municipality elections, 74% of the respondents took part in the election of the President, 45.7% participated in elections when voting for members of EU Parliament and 62.2% of respondents participated in the elections in self-governing regions. The second question, if the respondents are planning to participate in the next elections, was answered positively by 81.1% in the case of parliamentary elections, 77.2% in the municipal elections, the election of the President was planned to be attended by 78% of the respondents, 57.5% of the respondents for the elections to EU Parliament and for the bodies of the self-governing regions it was 66.1% of the respondents. The third item, concerning the importance of participation in the municipal elections, was answered by 70.9% of the respondents, who consider it was important, 19.7 did not consider it significant but not insignificant, and 6.3% said it was irrelevant. Overall, we have found that ***the highest rate of political participation and participation in the elections belongs to the respondents in the smallest villages to 1000 inhabitants.***

The third area was aimed at identifying **active participation in local affairs,** and two items measured it. A descriptive analysis showed that only 2.4% of respondents mentioned that they approached local newspapers, 8.7% of respondents contacted the general council, 19.7% contacted a municipal representative or mayor/responsible person, 13.4% participated in a civil meeting due to discussion, 0.8% participated in the protest meeting, 6.3% helped organize the petition. The second item, which focused on contact with a member of the municipal (village) council, was answered positively by 49.6% of respondents. By this, we can assume that the Slovak respondents are relatively active in the elections (see results for the previous area), but after the elections, they are no longer actively involved in public affairs. Based on the comparative analysis, ***the respondents in the smaller villages and towns of up to 5,000 inhabitants are more involved in this area compared to larger cities.***

Within the fourth area, focused on **participation in religious activities**, we found the frequency of visits to religious services or religious meetings. We can say that more than half of the respondents attend worship once a week or even more (55.9%), 16.5% visit them at least once a month, 13.4% only during weddings and funerals, 8.7% at least once a year and 2.4% never or almost never. We also found that ***higher rates of participation in religious activities are among respondents living in smaller villages and towns of less than 5,000 inhabitants.***

The fifth section was related to accepted or provided favour and was measured by three items in the questionnaire. The descriptive analysis showed that 85% of the respondents had done the favour for the neighbour in the last 6 months, and 70.1% of the respondents received help from the neighbour. As a free help or support, 40.9% of the respondents mentioned free assistance, 18.1% of respondents received it in the form of shopping or mail assistance, 15.7% of respondents in the form of cooking or household duties, in the form of home repairs (e.g., painting, car repair) 11.8% of the respondents, babysitting was mentioned by 13.4% of respondents, support with dressing or doing basic hygiene for 7.1% of the respondents, 40.2% of the respondents got the advice, the help with the official affairs claimed 3.9% of the respondents, representing with official instances 2.4% of respondents, transport assistance for 16.5% of respondents and 2.4% of respondents claimed that they got another form of support. Comparative analysis has shown that ***the most active and engaged in this area are respondents living in areas from 1,000 to 5001 inhabitants.***

In the sixth section, we focused on the rate of **regular service, assistance, or care for others.** Three items were measured, with a descriptive analysis showing that a regular service or assistance to a sick, disabled or older adult who does not live with the respondent was provided by 11% of the respondents. 70% of respondents assisted someone who is not a member of their household for the last 4 weeks. As a free help to a non-relatives, 59.8% of the respondents provided support in form of regular contact, in the form of shopping or mail assistance 15% of respondents, in the form of food preparation or household duties 18.1% of respondents, in the form of home repairs (painting, car repair) 9.5% of respondents, 12.6% of respondents provided assistance in the form of childcare, 9.4% of respondents provided help with washing or dressing, 41.7% gave advice, 1.6% guarded the household or house pet, official assistance (filling in forms, etc.) was provided by 9.4% of respondents, representing with official instances by 6.3% of respondents, transport assistance, accompanying somebody by 11% of respondents and 5.5% of respondents offered another type of help. Comparison analysis showed that ***social capital in this area is the lowest among respondents living in municipalities under 1000 inhabitants.***

7 Conclusion

The results of the study presented by us are aligned with the results of the authors mentioned above, who have mentioned the differences between social capital in smaller and bigger settlements, in the city and the countryside. In all of our social capital areas surveyed, respondents from smaller venues reached higher values. The reason we have identified these differences is that social capital, among other things, affects human health and its quality of life, so it is necessary to find out what can contribute to the greater well-being of a person.

References

[1] Giddens, A. Sociologie. Vyd. 1., Praha: Argo, 1049 s. ISBN 978-802-5708-071 (2013)
[2] Bourdieu, P. The forms of capital. In J. Richardson (Ed.) Handbook of Theory and Research for the Sociology of Education (New York, Greenwood), 241–258 (1986)
[3] Narayan, D., Cassidy, M.F. A Dimensional Approach to Measuring Social Capital: Development and Validation of a Social Capital Inventory. Current Sociology. 2001 **49**(2), 9–102 (2001)
[4] Bjørnskov, Ch., Sønderskov, K. Is Social Capital a Good Concept? Midwest Political Science Association 67th Annual National Conference, Chicago, April 22–25, 2010. Available at SSRN: `http://ssrn.com/abstract=1593507` (2010)
[5] Lochner, K., Kawachi, I., Brennan R.T., Buka, S.T. Social capital and neighborhood mortality rates in Chicago. Social Science & Medicine. Elsevier **56**(8), 1797–1805. DOI: `https://doi.org/10.1016/S0277-9536(02)00177-6` (2003)
[6] Putnam, R.D. Bowling alone: the collapse and revival of American community. New York: Simon & Schuster, ISBN 0-7432-0304-6 (2000)
[7] Coleman, J. Social Capital in the Creation of Human Capital. The American Journal of Sociology, 2009 **94**. Available at: `http://courseweb.lis.illinois.edu/~katewill/forchina/readings/coleman%201988%20 social%20capital.pdf` (1998)
[8] Šubrt, J., Balon, J. Soudobá sociologická teorie. Vyd. 1. Praha: Grada, 232 s. Sociologie (Grada). ISBN 978-802-4724-577 (2010)

[9] Lamarca, G., Leal, M., Sheiham, A., Vettore, M. 'The association of neighborhood and individual social capital with consistent self-rated health: a longitudinal study in Brazilian pregnant and postpartum women', BMC Pregnancy & Childbirth **13**(1), 1–17 (2013)

[10] Yang, T., Jensen, L., Haran, M. 'Social Capital and Human Mortality: Explaining the Rural Paradox with County-Level Mortality Data', Rural Sociology **76**(3), 347–374 (2011)

[11] Hofferth, S.L., Iceland, J. "Social Capital in Rural and Urban Communities". Rural Sociology **63**, 574–98 (1998)

[12] Mansyur, C., Amick, B.C., Harrist, R., Fransini, L. Social capital, income inequality, and self-rated health in 45 countries. Social Science & Medicine. Elsevier **1**(66), 43–56. DOI: 10.1016/j.socscimed.2007.08.015 (2008)

SHS Web of Conferences **68**, 01002 (2019) https://doi.org/10.1051/shsconf/20196801002
Int. Conf. SOCIETY. HEALTH. WELFARE. 2018

Relational place-making in Latvian countryside: Straupe and Kaldabruņa case

Dina Bite[1,*] and *Zenija* Kruzmetra[2]

[1] Rīga Stradiņš University, Department of Humanities, Riga, Latvia
[2] Latvia University of Life Sciences and Technologies, Institute of Social Sciences and Humanities, Jelgava, Latvia

Abstract. The paper deals with the relational place-making term describing it through the two examples of Latvian countryside. Straupe and Kaldabruņa have developed strong social networks and cooperation during more than 10 years. Both cases are built on certain values that gather similar people and organisations around them. Administrative and social borders are not as important as the purpose of coming together and maintaining of social values. Relational place-making approach can also give an alternative view of consumers' society, as social agents involved tend to assign specific meaning to their actions and interactions. Relational place-making approach is powerful enough to bring up new ideas and implement new economic and social activities, and both theoretically and practically offer a new logic for trans–territorial relations and rural development.

Key words: relational place-making, Latvian countryside, rural development.

1 Introduction

The idea of the topic emerged after several published papers about the role of culture in the revitalisation of rural territories in Latvia. Data were collected during the National Research Programme ECOSOC (2014–2017) when authors noticed relatively high activity of rural communities in order to save and develop their places. Local activities that are directed towards urging local people to take responsibilities about themselves and their place have become a natural path of endogenous and neo-endogenous development in many rural counties of Latvia. The role of human capital and social resources is appreciated more and more in rural territories. The practice of involving different social agents in their activities and networks has led to the relational place-making approach.

Relational place-making as an approach to rural development gives a new insight into activities of rural communities and developed practices. The aim of the paper is to describe two cases that have been established and developed in rural territories of Latvia from the point of view of the relational place-making approach. The sequence of the paper is the following: theoretical description of the relational place-making perspective, methodology, the description of Straupe and Kaldabruņa cases, discussion and conclusion.

* Corresponding author: Dina.Bite@rsu.lv

2 Theoretical insight

Relational place-making is one of the terms that have been related to rural development through last years. Relational place-making (also place-shaping and place-framing) concerns such a way of interactions or relations between social agents that shapes development and is a key to social innovation [1, 2]. Dutch rural sociologist Bettina B. Bock leads the discussion on rural development further and offers a term "nexogenous approach" that means "importance of reconnecting and binding together forces across space", including endogenous and exogenous resources, and necessity to apply this concept especially to marginal areas [1]. Relational approach seems to be a relatively new way of exploring and enabling rural development through emphasis on social networks much more than on static territories and relations between them [3]. Similar view describes relational places as a new "coming together" kind of places [4]. It stresses the importance of social connections. Relational approach can be analysed through a sociological point of view using agent network theory, system theory etc., as well as postmodern perspective where borders between places and structures merge, and quite flexible and unsavoury practices emerge in the foreground. As a sociologist of postmodernism Zygmunt Bauman wrote, "Culture no longer refers to a structure that imposes boundaries, but is seen as a critical process in which individuals transcend boundaries. The "cultural stance" is a questioning position in which the existence of multiple realities is recognized" [5]. Also, territorial development becomes "a variable set of interconnected places" and places are seen as "nodes in a dynamic network of interactions" [6] and successfully cross stereotypical borders of urban and rural territories. For example, Danish researcher Anne-Mette Hjalagar describes the "new trans-territorial logics" that emphasize rural enterprises as primary agents in partnership building [7]. Thus, flexible connections, networks and relations are prevailing over static perception of places.

From the geographical perspective, relational place-making is defined as a "networked, political processes of place-framing", analysing it as a conceptual part of the politics of place, networked politics and networked place [8]. This perspective highlights co-existence of different individual and collective views towards place-making, shared place understandings around these views and competition between them consequently. As the place is actively made and remade through cultural, discursive and material practices [9], these views and accordingly results of place-making are flexible and continuously changing. Further development of the geographical approach towards relational place-making exactly highlights temporary aspect [10]. According to theoretical phases of place-framing – "naming/blaming, claiming, and practicing – place-frames may go through multiple rounds of re-articulation as a result of the converging or diverging interests or understandings of places among actors" [10, 11]. It affirms that even if social agents involved have agreed on particular aims or practices, they need to "re-affirm" these agreements from time to time. It can be considered as a systemic flow of negotiations and re-articulations of the meaning of the place.

Relational place-making approach as well as cultural assets is also proved as an alternative for filling the gaps in infrastructure in resource-limited communities. Theoretically, significant components of relational approach are traditional authority, strategic leadership of non-profit sector, funds, diaspora linkages, social and cultural assemblages, and inclusion of all stakeholders [12]. Necessity to have strong leadership is confirmed by several studies [7, 13] and real needs of rural communities. Leadership brings out a cluster of important functions and facilitates the territorial development and also is a ground of relational place-making.

The importance of relational place-making is also recognised by the OECD and the European Commission, thus bringing new philosophy and style of operating [12]. There have been several attempts to implement it into policies. For example, Scotland has developed their place-based policy approaches considering following general principles of place-based development: mobilising endogenous potential, harnessing creative energy and knowledge, engaging an array of interests and social agents across diverse territories and multiple scales, supported by effective and adaptable institutions [14, 15]. In particular, they highlight the role of public sector institutions that has to involve and "re-think" how to engage communities [14]. Similarly, the UK and the Nordic countries have been developing institutional and financial support for rural social innovations including social economy, sharing of responsibilities between public, private and nongovernmental sector, and considering relational networks as a main source for implementing social changes [16].

Considering all above-mentioned theoretical material in relation to Latvia, there are some peculiarities in the area of moving towards the relational place-making approach. We cannot find specific plans and policies that support relational place-making in rural territories on national level. Local governments are starting to appreciate local human resources; however existing policies of place development are more adaptive to the rules given by central government, potential sources of financing and actually follow to a kind of inertia of previous policies. At the same time LEADER programme as an EU mechanism operates well and encourages rural communities to develop cooperation and social networks. Shrinking population and infrastructure as well as centralisation of public institutions during last decades have forced local people to look for alternative surviving ways and to almost automatically choose relational place-based strategy. This is a way how a significant part of villages and counties (communities) have surpassed formal statements and directions that come "from above". As authors have mentioned before [17], local agents like rural NGOs, entrepreneurs, sometimes local governments seem like pioneers and drivers of changes and territorial development implementing new products, new economic and social structures as well as new ways of thinking. However, there is the problem of fragmentation and instability of such structures, as they are not linked to any kind of support.

3 Methodology

During the implementation of the research project in the framework of the National Research Programme ECOSOC (2014–2017) best practices for building local communities and renewing rural territories were collected. The cases of Kaldabruņa and Straupe stood out of the context because of strong network both of local and external social agents. They have proved the thesis of endogenous and also nexogenous territorial development and can offer more or less replicable social practices.

Case study strategy was chosen in order to analyse phenomena in their context [18]. Document analysis (local and national policy documents, reviews, and statistics data) and observation were used initially during two or three years (2016–2019). Semi-structured individual and group interviews were conducted with the representatives of networks involved until the information collected was saturated enough ($n = 12$). The content of interviews has covered both ideological principles of establishing relational networks as well as practical issues and description of daily life and practical examples.

As relational place-making is not widely explored in Latvia, qualitative approach of social researches was chosen. Also, research questions are quite open: 1) what social agents are involved in the network of analysed cases? 2) What influences the dynamics of the network established? 3) What are the main consequences of the network in terms of relational place-making? Interpretative analysis of qualitative research approach was used and the following discussion in the theoretical framework developed.

The paper deals with the widening of understanding of relational place-making cases in Latvia that could help to design potential policies in order to support existing practices as well as increase knowledge about their features, development patterns and implications.

4 Straupe case

Straupe is a small parish (1341 inhabitants in July 2018) in the Pargauja district, which is located in the middle of Latvia, about 80 km northwest of Riga, with a rich cultural and historical heritage [19].

Straupe Farmers' Market is at the heart of Straupe's case, which was launched as a project idea in 2007 under the auspices of Straupe Tourism Society. In 2010, 5 activists founded the association "Healthy in City and Countryside", which continued to develop the idea of the fair. At the time of its creation, the Latvian legislation did not allow for the regular rural market, therefore, the association, in cooperation with the representatives of Slow Food movement in Riga, achieved favourable legislative changes. The market is open twice a month on Sundays. It aims to support local producers and "create the conditions for a shorter route to healthy rural products to the consumer". The motto of the store is: "*Growing ourselves. Making ourselves. Selling ourselves*" [20]. Straupe Rural Benefits Market is the only one in the Baltic that is a member of the Alliance of "Slow Food International Earth Markets" and is eligible to operate under this association's logo. This means that the products coming to the market are produced according to the Slow Food philosophy – they are local, seasonal and high quality, cultivated in good faith, respecting nature diversity and thinking about environmental sustainability [20].

As the organizers of the fair admit, the ideas have to be developed gradually, considering leaders or people who believe in the idea and are ready to represent it as the main driving force. Gradually, the association has developed contacts with local government, manufacturers, sales partners, buyers, other non-governmental organizations, the planning region, policy makers, entrepreneurs, cooperatives, international partners in the Slow Food Movement (Italy, Estonia, etc.).

In the history of its existence, the shop has created several successful companies, producers who have started their activities gradually: "*He bakes wonderful bread. You persuade him. He says it's no big deal, I don't do anything special. Then he comes back and sees that some people like his products. Then he understands that his stuff is good. Then he makes the next 10 buns. And then he registers with the Revenue Service. And so, they go forth, the entrepreneurs*".

As the chairman of the board of the association emphasizes, the organizers of the market consider this the most important function – the opportunity given to local producers to develop their own business, which has raised the self-assurance of the manufacturers and confidence in their product. Especially in rural areas, the creation of such relatively unconventional employment structures is very important: "*They start looking at their work with different eyes because they see that there are same problems all over the world, and that people everywhere work and overcome them*". On a larger scale, it is the home producers who can preserve a particular culture, retaining its original value and idea, as well as practical skills and skills of various crafts.

Their customers as another part of the network are described as a relatively prosperous part of the Latvian population seeking true values. There are quite many buyers with small children as well as buyers who support a certain lifestyle: "*But there are also those who just go to enjoy the atmosphere*". The market is mostly visited by the residents of the surrounding counties and cities, but there are many regular customers from Riga, as well as from remote locations and foreign countries.

With regard to social networking, the organisation's policy can be seen as cautious, with no intention of expanding the network and gaining popularity, but faithfully representing the declared values and philosophy. For example, domestic producers and craftsmen who meet certain criteria are invited to trade and local producers are preferred. This is in line with the idea previously stated in the theoretical material that a certain part of the local community has a certain vision of what ideas should be implemented. It should be noted, however, that these ideas are not always related to the purpose of developing a particular place. Most often, the goals stem from private initiative, perhaps for rational reasons, such as in this case, from the desire to provide their family with good products and to create a place for local producers to sell their products. Only afterwards is the scale expanding to the understanding that it is possible to influence a wider audience with their activities, for example, realizing that tourists, when coming to the market, also see the culture of the place and nation, its everyday life and traditions. As a result, community leaders must exhibit a sense of responsibility not only for the realization of a narrow idea, but also for its wider impact.

Another social function worth mentioning is the renewing of awareness of the historical and cultural monuments. In 2016 the association started cooperation with the owners of Straupe Horse Post Station [21], which is currently the venue of the market. In a broader sense, this can be described as a successful collaboration that provides buyers/visitors with a specific framework vision or idea, and includes a wide range of activities (market, tours, rental of premises, etc.) to integrate the Horse Post Stations as an old Hanz trade road stop pointing out its historical and social significance: "*This is a wonderful case when the landlord did not close up staying in his own little world but opened it to the public.*"

In general, it can be said that a stable community has been established around Straupe Market, which maintains and supports certain values that serve as basis for the network of agents. For individual projects that are limited in time and place, the range of cooperating partners is being expanded, but the core group for the operation of the market remains practically unchanged. One of the essential functions of the organizers of the market is advising entrepreneurs, non-governmental organizations, and cooperatives on how to develop similar types of activities. Leaders act as mentors for people from other regions of Latvia and also from abroad: "*We are also giving a push to some extent in all regions of Latvia, because we have large number of visitors*". At the same time, it is not possible to respond to all the calls for cooperation due to the capacity and values of the organization. Therefore, this network is considered as flexible and at the same time stable.

Referring to the theoretical material on relationship re-articulation, the following examples can also be identified in the case of Straupe, e.g., they have accepted a practice that not only organizers, but also the producers themselves share their experiences in seminars. Thus, the role of one social agent in this network can be flexible.

The case of Straupe Market confirms the theoretical concepts of the importance of social resources in the development of a place: "*There may be all kinds of money at the bottom, any projects… if not for the person who carries the idea out and makes it true, there is zero value to all research and stimulus from the outside*".

Summarizing the information about the Straupe Market, it can be concluded that in this case a certain value-based view of the development of the site is realized, which does not submit to the demands of the surrounding consumer society and thus offers an alternative view on the development of a certain area and also the place. Relational place-making, even if not called by this name, is being implemented and has a significant impact at both local and national level. Social relations are formed and strengthened over geographical constraints.

5 Kaldabruņa case

Kaldabruņa is a small village in Jēkabpils county of Rubene parish, located in the eastern part of Latvia, 153 km from Riga. The population in the village is very small. According to the data of the Statistical Office of Latvia, 118 residents lived in the village in 2000, but in 2018 – only 59 [22]. The village has a very poor infrastructure – no school, no store, no post office, no preschool etc. There are poor quality ground roads in the area, which makes the village in to a marginal area.

However, thanks to the activities of local leaders, since 2008 the place has slowly come to life and gained both the national and international recognition. The former school building is run by the Association "Watermarks". The activities of the Association are focused on the following directions: environmental protection and sustainable development, development of civil society, art and culture, education and development of children and youth [23].

The most visible activities of the Association are related to various artistic activities, such as the annual Hay Days, the exposition of the Meadow Museum, concerts, artists' plenary sessions, the glass workshop, and the provision of various social services to the local population. There is the internet available in the Association's facility, laundry can be washed, practical help is provided to people in crisis, etc.

As the population is small and the infrastructure is scarce, the Association's only option is to network with other social agents. The head of the Association also works in the local LEADER partnership. An attempt has been made to bring the experience of Kaldabruņa into the neighbouring areas. In cooperation with neighbouring parishes and counties, a popular theme route has been established in 17 small settlements "Sēlija Islands" [24] and a 28 km bicycle route "The First Journey of Jānis Rainis" opened. In this way, local associations, in cooperation with municipalities and other agents work for the sake of economic and social revitalization of rural areas. Social networks and cross-border relationships have been designed purposefully and resolutely to create more opportunities to attract funding and thus develop the territory. The agents involved in the Sēlija Islands project have created their own operational philosophy called "island emanation": "*How big is the emanation from the place you maintain. This place is sometimes geographic, sometimes it is purely spiritual, and its geographic significance is not so great*". Representatives of the Association especially emphasize the spiritual and emotional aspects to the development of the place. Otherwise, the services and products offered will quickly exhaust themselves and become one of the standard products. In other words, the social agents involved in the relationship network must constantly create and maintain certain values or the story of the place. At the same time, the "emanation approach" expresses a vision of site development, cultivates certain values, and takes care of site marketing, social and economic development. Moreover, "Sēlija Islands" is formed as a place of inspiration, good practice that can be taken over by other rural districts: "*For example, the association right here in Dviete... the girls said – if you can do it, we can as well! Now they are very strong, now they only need us as friends*".

More than 10 years of activity from Kaldabruņa have served as inspiration to several neighbouring counties and also more distant guests. Mutual cooperation allows for reciprocal complementarities, specialization and synergy. Consequently, there are other non-governmental organizations in the region and also nationally, local government and its institutions (e.g., schools), local community, artists from different places in Latvia and also in Lithuania, representatives of universities or academic environment etc. among the network partners of Kaldabruņa Association. Kaldabruņa is an example of how borders between populated areas, cities and fields are dissolved and the focus can be on ideas and people with certain values. For example, in the Church of Chervonca, which the Association is gradually trying to renew, concerts with the participation of well-known artists, classical music, etc. take place. Interestingly, the idea of restoring the Church of Chervonca was proposed by artists who came to visit. This means that within a relational network, local people have the opportunity to look at their resources from the outside and see other opportunities.

The case of Kaldabruņa is one of the rare networks in Latvia that has deliberately and purposefully set culture as the most important development resource. It is not an accidental or copied action, but a purposeful and meaningful initiative. As a result, communities in the area are actively looking for and developing a variety of local tangible and intangible cultural resources. As the Association explains, the development of a place in a cultural context can be explained by the use of traditional cultural resources and the creation of a new added value.

The existence and dynamics of the network in the case of Kaldabruņa is strongly influenced by national policies, such as how much financial resources are allocated to the non-governmental sector. As the Association reveals, in recent years the support for rural areas has been channelled more towards local governments than to civil society, which undoubtedly affects the ability of the non-governmental sector to realize its vision of local development. Likewise, the local population underlines the constant ignorance and worry about the status of small schools in the region, as the state's policy is to reduce the number of schools without thinking about the consequences of such decisions. Hence, the relationships and trust unique to small rural communities are eroding and people tend to deal with it individually, most often by moving away from so-called non-viable rural areas.

The case of Kaldabruņa reveals the dynamics of networking also in relations with the local community. In principle, the Association has in 10 years created a legal and practical framework for local people's activities. At present, however, Association's representatives disclose that the local community has a more consumer-oriented approach, namely the desire to enjoy existing opportunities, such as attending cultural events, rather than engaging themselves or starting their own business. Commenting on the theories of these situations, one of the visions of site development has become dominant in this small village, while other alternative visions are in a latent position. Undoubtedly, this is also caused by the small population and infrastructure constraints.

Evaluating social and economic consistency in the example of relational place-making in Kaldabruņa, tourism activities and their contribution are at the forefront. Tourists from Latvia and foreign countries are visiting Kaldabruņa, and "Sēlija Islands" is becoming a more and more popular tourist destination, which shows that the strategy of cooperation and networking is the right one. Local communities and the non-governmental sector in the neighbouring areas and elsewhere in Latvia have also become more active. Several local activists have raised their human capital through education, for instance, in the field of tourism. Several jobs have been created, cultural and tourism activities bring economic benefits to local entrepreneurs, etc..

6 Discussion and conclusion

In both cases, the informants admit that they have not thought about their activities in any theoretical categories, so the interviews also serve as an educational, self-reflective and cooperative factor. This aspect also confirms that rural communities have developed relational place-making as a natural response to local, national and global challenges, having arrived to the need and development of a relational network through learning and experience.

There is a certain vision and realization of ideas about the development of the place in Kaldabruņa and Straupe. For various reasons, they are dominant in given areas and do not especially compete with alternative views. Operational and development visions are based on certain values and their maintenance practices. In both cases, there is a periodic re-articulation of interests, directions of action, and social norms, but this takes place within the framework of existing moral norms and values.

Creating and realizing a vision of place development always encounter the characteristics and manifestations of the consumer society. The direction that is maintained in both Kaldabruņa and Straupe includes a deeper understanding of site development based on cultural resources. Both cases are considered good practices precisely because over time their creators and implementers have not used any quick marketing tricks that would have quickly drained their social practices. On the contrary, community leaders have fought hard for each product or each cooperation practice to represent certain values.

An important conclusion of the study is the dissolution of social borders between places, cities and rural areas, social groups and even social classes. The assertion that administrative or social borders are irrelevant in the cases investigated would be confusing, but the existence of borders was not identified as a problem. Latvia's countryside is birthing a new logic of place development and a new form of community, which has certain values and a view on the order of things at the heart. In other words, boundaries are formed around a set of values, and the specialization of cultural and economic activities is gradually taking place. Relational place-making is a tool for implementing these activities.

The study confirms the importance of social resources for place development. Attraction of new human resources, as well as the unlocking of the potential of human resources, is important in both studied places, however, in the case of Kaldabruņa, the importance of infrastructure becomes more visible. The situation in Kaldabruņa illuminates a significant problem with relational place-making. If there is not enough infrastructure and institutional support, if the human capital, social and cultural capital of the local population falls over time, the structures created by relational place-making are not as sustainable and lasting in themselves. They need new human, material and financial resources, as well as a supportive regulatory framework. In the case of Latvia, it could be argued that the official rural policy is lagging behind the established structures in the countryside. To some extent, rural people have shown a direction for further development, but the support of central power is currently insufficient.

The results of the research, to some extent, challenge the traditional notions of the countryside as a place requiring marginal and special policies. As informants say, a large part of the rural population is well-off, have opportunities to develop different ideas, and cultural and entertainment opportunities are often more accessible than in the city. For example, people deliberately choose to visit a fair or cultural activity in the countryside or in small towns, where the relative "quality level" is not lower than in the city, but where there are other benefits. Thanks to the processes of globalization and the activity of the local population, the range of goods and services in the countryside is not very different from the cities. The dependence of the rural population on cities is therefore diminishing, and the ideas offered by the countryside can compete seriously with cities. However, as stated above, the dominant (political) discourse in Latvia does not appreciate or underestimate rural resources.

Summarizing the above, there are relational place-making features in rural areas of Latvia, where both endogenous and exogenous forces are engaged in a single network, overcoming administrative and social boundaries. The research carried out generally confirms the theoretical assumptions of the importance and functions of relational place-making. This approach helps to better understanding the processes, actions and interactions in rural areas. Future studies call for analyzing sustainability issues and dynamics of existing relational networks.

Authors acknowledge the National research programme 2014–2017 ECOSOC and The National research programme "Latvian heritage and future challenges for sustainable development" 2018–2021, research project "Towards sustainable development and inclusive society in Latvia: response to demographic and migration challenges".

References

[1] B.B. Bock, SociolRuralis **56**, 552 (2016)
[2] M. Woods, Cultural Sustainability and Regional Development (2015)
[3] M. Woods, J. Heley, ROBUST D1.1 (2017)
[4] J. Anderson, Environ. Planning **30**, 570 (2012)
[5] V. Marotta, Thesis Eleven **70**, 36 (2002)
[6] B.B. Bock, TARRN Meeting (2018)
[7] A.M. Hjalager, Local Econ. **32**, 34 (2017)
[8] J. Pierce, D.G. Martin, J.T. Murphy, Trans. Inst. Br. Geogr. **36**, 54 (2011)
[9] A.L. Bain, F. Landau, Urban. Aff. Rev., 1 (2017)
[10] A.Y. Zhang, Geoforum **90**, 91 (2018)
[11] S.L. Van Neste, D.G. Martin, Trans. Inst. Brit. Geogr. **43**, 47 (2017)
[12] C.F. Che, Societies **8**, 12 (2018)
[13] O. Potluka, J. Kalman, I. Musiałkowska, P. Idczak, Local Econ. **32**, 297 (2017)
[14] J. Atterton, RESAS Strategic Research Programme Research Deliverable 3.4.2 (2017)
[15] L. Pugalis, G. Bentley, Local Econ. **29**, 273 (2014)
[16] A. Copus, L. Perjo, L. Jungsberg, L. Randall, H. Sigurjónsdóttir, Nordregio Working Paper, 2 (2017)
[17] D. Bite, Ž. Krūzmētra, Int. J. Econ. Manag. Syst. **2**, 277 (2017)
[18] C. Robson, K. McCartan, Real World Research, 4th ed., John Wiley & Sons Ltd. (2016)
[19] Straupe, world's smallest Hansa town. Available: `http://hansa.pargaujasnovads.lv/en/`
[20] Slow Food Straupe. Available: `https://straupestirdzins.lv/en/us/`
[21] Straupe Horse Post Station. Available at: `https://zirgupasts.lv/`
[22] CSP data bases. Available at: `https://data1.csb.gov.lv/pxweb/lv/iedz/iedz__riga/RIG010.px/table/tableViewLayout1/?rxid=32f317cb-2d8a-46f1-ab11-b1fa54391ab4`
[23] Association "Ūdenszīmes" ("Watermarks"). Available at: `http://www.udenszimes.lv/cms/par-mums/`
[24] Sēlijas Rural Partnership cooperation project "Sēlijas Islands". Available at: `http://www.laukutikls.lv/nozares/lauku-telpa/notikumi/2017-11-01-000000-lidz-2017-11-17-000000/selijas-lauku-partneribu`

Foreigners in the town. Experience of voluntary immigrants living in a small town

Daniel Markovic* and *Anna* Zilova

Catholic University in Ružomberok, Faculty of Education, Department of Social Work, Ružomberok, Slovakia

Abstract. The aim of the paper is to unveil the experiences of voluntary immigrants to Slovakia related to their life in a small Slovak town and to identify opportunities for intervention through social work, particularly experience in the field of interpersonal relations, employment or during studies, and in the economic field. In research, we are interested in voluntary migrants who chose small towns in Slovakia, which are culturally and linguistically homogeneous. We have chosen a qualitative research strategy. Research design is grounded theory. As a research tool, we chose an individual structured interview (In-Depth Interview). For qualitative analysis of data, we used the coding; we set descriptive and inductive codes, subcategories and categories. The subject of research is eight voluntary immigrants to Slovakia who live in small towns for at least one year at the time of research. Thus, they meet the long-term migration criterion. Participants were selected on a deliberate basis to cover the phenomenon of voluntary migration. Due to the chosen research design, theoretical sampling is the case.

Key words: human migration, voluntary immigrants, social work.

1 Introduction

Human migration is the movement of people from one place to another with the intention of settling in a new location [1]. Asylum Seekers and Supplementary Protection Statistics show that Slovakia is rarely the target country of forcibly displaced people [2]. Emigration from Slovakia and its reasons are being examined relatively well [3, 4], but we see specific gaps in the research of voluntary immigration to Slovakia. Statistics after Slovakia joined the European Union (EU) show a significant, quadruple increase in the number of foreigners from both EU and non-EU countries living in Slovakia [5]. Migrants arriving in EU countries are usually trying to settle in large cities, where there are support nets made up of their natives and job opportunities in an international environment, which confirms the migration network theory. In research, we are interested in voluntary migrants who chose small towns in Slovakia, which are culturally and linguistically homogeneous. The aim of the paper is to find out the experiences of voluntary immigrants related to their life in a small Slovak town and to identify opportunities for intervention through social work. We were particularly interested in their experience in the field of interpersonal relations, in the employment or studies, and in the economic field.

* Corresponding author: daniel.markovic@ku.sk

2 Research design and methods

We have chosen a qualitative research strategy. Research design is grounded theory. Since we have only carried out the first three steps of the grounded theory, the compilation of a new theory is not the content of this paper. As the research tool, we chose an individual structured interview (In-Depth Interview). For qualitative data analysis, we used the coding; we set descriptive and inductive codes, subcategories and categories. We have identified two research questions, looking for answers to them through 22 interview questions:

1. What is the life story of the participants before coming to Slovakia?
2. What are the participants' life experience in Slovakia in the field of interpersonal relations, own study or work, and in their economic situation?

As part of research, we used elements of biographical design. When interpreting the participants' answers to the question *who is in your social network?* we used the word cloud method to create an image consisting of words used to denote the terms, the size of each word denoting the frequency of the term in the participants' testimonials.

The subject of the research is voluntary migrants who moved to Slovakia and have been living in Slovakia at the time of research (July 2018) for at least one year or longer. There are eight participants; they were selected on a deliberate basis to cover the phenomenon of voluntary migration to Slovakia. Due to the chosen research design, the choice is in the form of theoretical sampling, a deliberate theoretical choice. In our research, it showed its typical features: changing the sample size during the research, selecting the participants with respect to the research goal and selecting special to borderline cases (differently developed countries, migration for study, work, and family reasons). All participants have experienced an immigration experience in Slovakia; they are competent to talk about the topic. Because of the research design, we can talk about them as informants. Participants come from European Union member states (Poland, United Kingdom), from other European country (Ukraine), from Asia (Turkmenistan), from North America (United States of America), from the Caribbean (Haiti). Participants belong to all four groups of countries grouped according to the HDI Human Development Index. The specificity of the informants, which was the intention of the researcher, is the education of the participants. Higher level of education develops the ability of abstract thinking, argumentation and generalization, so we expect skilled and relevant answers from participants. In the presentation of the research results, we do not give the names of the participants, their place of residence, the exact age (but the age group information), the employer or college. These data are known to the researcher but are not included in the research results as part of the anonymity and privacy of the participants.

3 Results

3.1 Life stories of participants are captured

In participants' profiles, their compact characteristics generated from answers to 18 interview questions are captured. We present shortened life stories (biographical design) of participants bound to their countries of origin for the period before coming to Slovakia.

The participant 1 belongs to the age group of 16–20 years, she is a woman, single, and her highest achieved education is secondary school – general education. She has been in Slovakia for two years; currently she is studying social work at the university. She is from Ukraine (HDI 0.743), has Ukrainian nationality. She worked there as a photographer and as a worker in Copy centre. She left Ukraine as a seventeen-year-old girl. She decided to change the country because of more information about her study field in Slovakia, comparing to Ukraine. She moved to Slovakia and not to other country because of security. Before moving to Slovakia, she looked up information from the Internet and "heard a lot from a teacher of the Slovak language". She studies here because, in her words, social work is more developed in Slovakia than in Ukraine. During her stay in Slovakia, she offers unpaid or volunteer work as a part of the study program. She often helps students from Ukraine because she knows how it feels to be alone in a foreign country.

The participant 2 belongs to the age group of 16–20 years. She is a woman and is single. Her highest achieved education is secondary school – in the field of fine arts. She is currently studying at the university in the second year and is studying at the Faculty of Fine Arts. She likes painting, likes to draw. She is from Ukraine (HDI 0.743), has Ukrainian nationality. She did not work in Ukraine before she left for Slovakia. The main reason for the change of country for P2 was "more options for the future" than in Ukraine. She moved to Slovakia and not to other country because of study opportunities in Slovakia. She helps her peers in Slovakia because "people need it".

The participant 3 is a woman who belongs to the age group of 31–40 years. She is married. Her highest achieved education is third, doctoral degree. She studied the teaching of Slovak language and literature and Religious education, and later the Theory of Teaching of Religious Education. She has been living in Slovakia with a break from 1999, first for study, and since 2007 because of marriage. She works as an assistant in a facility focusing on education. She joined the workplace because she could not find other job near her home in 2009. Now she likes her work and she is doing it with pleasure. She is from Poland (HDI 0.855), her nationality is Slovak. After studies she returned to Poland, where she worked as a teacher. She started living in Slovakia for the second time because of marriage. She moved to Slovakia and not to other country because of the university and marriage. She gained information about Slovakia before coming to Slovakia at school and in her own family. She is financially supporting the charitable and volunteer area in Slovakia.

Participant 4 is a man; he belongs to the age group of 21–30 years. He is single; his highest achieved education is baccalaureate. He studied Russian language and culture. He chose the study program because he likes the Russian language. He has been in Slovakia for 6 years. He works as a manager at the bar. He is from Turkmenistan (HDI 0.691) and has Turkmen nationality. In his country, he worked as a manager. He left Turkmenistan in 2012. The reason for the change of the country was university studies. The choice of Slovakia and no other country was the choice of his parents. He did not have much information about this country before he arrived in Slovakia, nor did he try to find them anywhere. He worked as a volunteer in Slovakia in organization Smile as a gift.

Participant 5 is a man; he belongs to the age group of 21–30 years. He is married; his highest achieved education will soon be a third, doctoral degree. He studied management and now is studying social work because his country currently needs many social workers due to its difficult situation. He has been living in Slovakia for 7 years. He is from Haiti (HDI 0.493), his nationality is Haitian. He studied in his own country and worked part-time as a "Help Manager". The main reason for moving to Slovakia was university studies. P5 chose Slovakia and no other country thanks to a Slovak friend, "with whom I met in my country and who invited me to study here in Slovakia." At the same time, this friend provided him with information on Slovakia before moving to the Slovak Republic. He helped other people in Slovakia many times, whenever they asked for it.

Participant 6 is a woman who belongs to the age group of 31–40 years. She is married. She has achieved third degree, a PhD in Theology and Biblical Sciences. She works as an assistant at a college because she can use her education and knowledge there. She has been living in Slovakia for 13 years, she is from Ukraine (HDI 0.743), is of Ukrainian nationality. Before coming to Slovakia, she did not work, but she studied. She left her native country when she was 19 years old. After leaving Ukraine she did not come to Slovakia straight away, she came "for her husband who returned to his native country after graduation abroad". She decided to move to Slovakia and not to other country because of job opportunities for her and her husband. Before the immigration, she looked up information about Slovakia in various sources: from interviews with a future husband, from newspapers, from acquaintances who were Slovaks. In Slovakia, she did not work as a volunteer because she was very busy and gradually gained family responsibilities.

Participant 7 is a man belonging to the age group of 31–40 years. He is married. He has received Bachelor's degree in Corporate Governance and Finance. He has been in Slovakia for three years. He is from the United States of America (HDI 0.920), nationality is stated as American. This is because Americans do not recognize their nationality by ethnicity but by citizenship. He was born in North Carolina, where he also graduated from college. He worked there as a regional manager of the Evangelization Association. He left for family reasons, his wife is originally from Slovakia, and she "owns and operates business in Slovakia". He decided for Slovakia and not for other country because "the wife is from Slovakia and business was in Slovakia". He jokes that he did not have a choice. He works as a private English teacher, owns and operates commercial property for rent. In Slovakia, he also does volunteer and unpaid work, especially for Christian communities and churches. He says his family has a heart for those who are not Christians, or who have been "hurt" by the church in the past.

Participant 8 is a man belonging to the age group of 31–40 years. He is single. During his studies, he achieved the Bachelor's degree in graphic design. He has been in Slovakia for 3 and half years already. He is from the United Kingdom (HDI 0.909) and is of British nationality. In his country, he graduated from college, worked as a lifeguard. The main reason for moving to Slovakia was a different lifestyle. He decided for Slovakia and not for other country because of job opportunity. He gained information about Slovakia before his arrival from his partner. He works as a customer care agent in a facility focused on computers because he has good experience with this service himself. He does not work as a volunteer.

3.2 Comparison of native country with Slovakia

Since the informants have lived in Slovakia for at least one year, we were interested in their comparison of Slovakia with their country of origin in the field of human relations, in employment or study, and in the economic sphere.

When comparing, participants tended to think in the scale as better – worse, more – less, so we offer the following breakdown of response groups based on subjective ratings of participants:

- In Slovakia, there are better: "You have less corruption than in Ukraine" (P1). "The economy is more stable in Slovakia. People are more positive and better" (than in Ukraine, author´s note) (P1). "I think there are better opportunities in Slovakia, better teachers and more opportunities for travelling and studying. From the economic point of view, it is also better" (P2). "Slovakia is a more stable country than mine in finance as well as in government areas" (P4). "There are many options for young people – more than in my country. There are many laws in Slovakia that are very good, unlike in my own country" (P5). "At work, I can deal with what I studied, I do not know if I would have the opportunity to have a similar position in the country I come from" (P6). "Economically, Slovakia is much more developed and more stable; even if I worked in a similar region and a similar workplace in the country I come from, I would not have a comparable social and economic standard" (P6). "Some of the biggest

Table 1. Codes, subcategories and categories – Comparison of native country with Slovakia.

Category	Subcategory	Inductive code	Descriptive code
In Slovakia, there are better:	Economic, social and communal sphere	Less corruption is present	Less corruption
		Better economic situation	Better conditions from the economic point of view
		Cheaper accommodation	Lower living costs
		Cheaper healthcare	Lower healthcare costs
		Lower costs related to own children	Lower costs related to children
		More job and business opportunities	Business opportunities
			More job opportunities
			Work application in the field
		More stability in Slovakia	More stable economy
			More stable financial sphere
			More stable government
			Slovakia is more stable
		More developed country (Slovakia)	Slovakia is more developed
		Better legislative	Better laws
		Better public transport	Good public transport
	Personal improvement in Slovakia	Better and more opportunities for people in fields of study and travel	Better opportunities
			Better opportunities for young
			Better study opportunities
			Better travel opportunities
	Environment	Positively perceived environment	Magnificent nature
	People and interpersonal relations	Positively perceived personal characteristics	More positive and better people
		Strong bonds between family members	Strong bonds within family
	Lifestyle	More active lifestyle of Slovaks	Slovaks are physically more active
			Healthier lifestyle of Slovaks
In Slovakia there are the same:	Social and economic situation	Interpersonal relations in Slovakia	Relations
		Work in Slovakia	Work
		Economic situation in Slovakia	Economic situation
	Environment	Small area of the country	Slovakia is also small (as Haiti, author´s note)
In Slovakia, there are worse:	Economic and social sphere	Interpersonal relations are limited to smaller social circle of close relatives	Smaller circle of close relatives
		Smaller emphasis on customer service	Smaller emphasis on customer service
		Less of a comfort	Less of a comfort in everyday life
		Lower wages	Lower wages
		Older, slower and less modern healthcare facilities	Worse healthcare facilities
	Environment	Weather (cold, because the participant is from warmer climate)	Weather

benefits: strong family relationships, low housing costs, business opportunities, health care costs, children's costs, spectacular nature and the surrounding area, good public transport" (P7). "Economically speaking, England is overpriced and unavailable for workers; more work is available in Slovakia. Slovaks complain less, are physically more active and care for a healthy lifestyle" (P8). "You do not have a war" (P1).
- In Slovakia, there are the same: "Relationships and work are the same" (P3). "Economic situation? Neither it is good in Slovakia, nor in Poland" (P3). "Slovakia is as small as my country" (P5).
- In Slovakia, there are worse: "In the area of interpersonal relations, I feel a certain limitation in Slovakia – the circle of the acquaintances is smaller" (P6). "Some of the most important negative: weather (originally from the hot climate), much lower wages, less emphasis on customer service, medical facilities are much older, slower and less modern, less comfort in everyday life" (P7).

P3 is aware of the limits of his opinion: *"I feel like I have been here for too long, I cannot comment on it"* (P3). Interestingly, P5 talks about his country of origin as of my country. Participants from countries with lower HDI positively assess the Slovak economy (and the rate of corruption), better and more positive people (and teachers), study and travel opportunities. They positively assess the stability of Slovakia, Slovak laws and possibilities of employment at work. Participants from countries with a higher HDI than Slovakia also commend business and business opportunities in Slovakia. The participant born in Poland perceives the work and relationships in Slovakia as the same and the economic situation as equally bad as it is in Poland. There are fewer negative statements about Slovakia than positive, participants view as the worst areas of human relations, salaries in the SR, and health care facilities. Participants coming from the warmer weather presented the weather in Slovakia as worse.

In this question, we were interested in the frequency of opinions on the stability changes in Slovakia. Participants perceive Slovakia as stable (more stable economy, more stable area of finance, more stable government, more stable Slovakia).

3.3 Social network of voluntary migrants

Young Immigrants, although coming to their family or friends, are getting into a new country, in a community where they start to make new contacts, new friendships and relationships. At the same time, they keep contacts with close people from the original country. That is why we were interested in who they are in touch with. When interpreting the participants' answers to the question *who is in your social network*? we used the *word cloud.* Most often, the friends and family were mentioned in the testimonies.

We have included the codes obtained from participants' testimonies into more general terms. For example, friends, fellows – to *friends*, partner, girlfriend and my boyfriend to a more general – *partner*. Interestingly, although there are four married participants, no one explicitly spoke of a spouse. It is possible that the spouse is hidden in the terms partner and family.

3.4 The feelings of an immigrant in Slovakia

Some of participants may be perceived as physically different for Slovaks, while with others, Slovaks may find out their foreign origin only during the communication (e.g. accent). With some participants, especially those who live longer in Slovakia, is difficult to notice some

Picture 1. Social networks of participants. Source: own processing in the application WordItOut.

differences. The person living in the new country as an immigrant perceives different feelings. That is why we asked the participants how they feel in Slovakia as foreigners.

During interviews, the informants talked about both positive and negative feelings, so we divided their testimonies into the following groups:

- Positive feelings: "Sometimes I can forget that I am a stranger. I am comfortable in Slovakia"(P1). "Well, because I do not feel like a stranger, I feel like a Slovak" (P2). "I am at home (smile)" (P3). "I feel good, perhaps because of my perfect Slovak language, no one even knows I am a foreign person" (P4). "All right, I do not feel bound to be limited in some sphere of life because I am a stranger. It is the experience I have practically from the start of my stay" (P6).
- Neutral feelings: "Normal (P5)".
- Ambient feelings: "The difficult question, it depends on the situation. The main daily barriers are of a cultural nature. There are positive and negative aspects of being foreign in Slovakia" (P7).
- Negative feelings: "I feel like an outsider because I do not speak a language" (P8).

It is natural that the feelings of an immigrant, even if it is a voluntary migrant, are not only positive in the new country. Of course, they are variable, depending on many factors.

3.5 Reactions of Slovaks on immigrants

We asked the participants how the Slovaks responded to them being foreign.

As in the previous question, the responses ranged on the scale positively – negative.

- Positive: "They say I am already good at speaking Slovak" (P1). "They always try to help me" (P2). "Some responses are positive ..." (P5) "My English is sometimes attractive because people want to learn more effectively" (P7).
- Neutral: "They always ask me: Where are you from? How long do you travel home? Do you like it here?" (P1). "Nobody even knows I am a foreigner" (P4). "Neutrally, sometimes I awaken curiosity" (P6). "They are usually shy and surprised that I live here" (P8).
- Negative: "Sometimes I can hear:" Slovakia for the Slovaks!. "Foreigners have no rights but have duties. Even: "go home" " (P1). "During the study, they sometimes reacted negatively, but now it is OK, they no longer hear the accent" (P3). "... and at the same time many negative reactions that I do not care about because I know who I am, so for

Table 2. Codes, subcategories and category – Reactions of Slovaks on Immigrants.

Category	Subcategory	Inductive code	Descriptive code
Reactions of Slovaks to foreigners:	Positive reactions	Positive reactions about speaking Slovak language	They say that I speak good Slovak
		Experience with helpfulness of Slovaks	They always try to help
		Positive reactions because of attractive mother language (English)	Mostly positive reactions because of English
		Non-specified positive reactions	Some of the reactions are positive
	Neutral reactions	Slovaks often react with asking the same question: where is she from, how long does it take to travel home, if he/she likes it here.	They ask repetitive, stereotypical questions
		Slovaks do not realize she is a foreigner	They do not know that I am a foreigner
		Arousing curiosity in Slovaks	Neutrally, I arouse curiosity
		Slovaks react surprised and shyly	They are shy and surprised that I live here
	Negative reactions	Xenophobic reactions emphasizing, that Slovakia belongs to Slovaks	"Slovakia for Slovaks"
		Xenophobic reactions denying the rights of foreigners and emphasizing their duties	"Foreigners do not have rights, only duties"
		In past negative reactions because of accent	They are no longer negative, they do not hear accent.
		Many non-specified negative reactions, is not concerned with them	Many negative reactions, I do not care about them
		Slovaks are avoiding him because they do not speak English well or do not speak it at all	They avoid me because of English

me these negative reactions are nothing" (P5). "But there are many people who avoid me because they either do not speak English or are unsure communicating in English language" (P7).

Since the P1, P7 and P5 statements pointed to both positive and negative aspects, we divided them into two groups of testimonies. From the P5 response, we can also capture the strategy of dealing with negative response. A group of neutral responses is specific because one might argue that testimonies have a positive or negative undertone. From the interview, it was not possible to determine this sub-tone unambiguously, in contrast to clearly positive and clearly negative shades. In this question, we have captured the widest range of negative testimonies from all the questions.

4 Discussion

While staying in a small town, voluntary immigrants consciously or unknowingly integrate into the majority society; assimilation is also taking place. There are four models of migrant integration: Assimilative model of the migrant's integration, Segregate model of the migrant's integration, Multicultural model of the migrant's integration, and Model of the individual civic integration [6]. Our participants are more likely to remain in their home in Slovakia if an assimilation model of integration or a model of individual civic integration is present in their environment. Reardon describes seven areas of addressing social needs of migrants through social work: 1. Identifying and overcoming barriers, 2. The need for culturally competent practice, 3. Circumstances of the arrival of clients in destination country, 4. Health concerns of the client, 5. Family dynamics, 6. Creating social connections, 7. The need for patience and respect [7]. Participants sometimes experience barriers in social contacts or even feelings of outsiders. From this area of social work solutions for migrants, we propose that social workers focus on creating social connections (such as prevention of solitude and isolation, pooling of clients with each other, community organizations and other entities) and identifying and overcoming barriers related to life in the country.

5 Conclusion

Voluntary immigrants identified these areas of positive experience of living in Slovakia: employment, economic and social conditions, leisure, cultural and development activities, environmental aspects. Participants perceive areas of human relations, wages and medical facilities as the worst. Participants have the advantage of quickly creating their new social network. Nevertheless, they sometimes experience barriers in social contacts or even feelings of outsiders. We propose that social workers focus on creating social connections (such as prevention of solitude and isolation, pooling of clients with each other, community organizations and other entities) and identifying and overcoming barriers related to life in the country.

References

[1] R. Perruchoud, J. Redpath-Cross (eds.) *Glossary on Migration* (IOM, Geneva, 2011)
[2] EUROSTAT, Asylum statistics (Accessed on 10/02/2019) Retrieved from `https://ec.europa.eu/eurostat/statistics-explained/index.php/Asylum_statistics#Countries_of_destination:_Germany.2C_Italy_and_France_the_main`
[3] S. Vojtovich, The Impact of Emigration on Unemployment in Slovakia InzEkon **24**(3), 207–216 (2013)
[4] A. Pavličkovâ& A. Dayanandan, Social networks and emigration in Slovakia, Migration and Development **4**(2), 220–231 (2015), doi: 10.1080/21632324.2014.976942
[5] *Štatistickýprehadlegálnej a nelegálnejmigrácie v Slovenskejrepublike*, (Prezídium Policajnéhozboru, Bratislava, 2018)

[6] IOM, Základnépojmy v oblastimigrácie a integráciecudzincov (Accessed on 10/02/2019). Retrieved from `https://www.iom.sk/sk/pre-media/zakladne-pojmy-o-migracii.html`
[7] C. Reardon, Addressing the Social Work Needs of Older Immigrants and Refugees, Social Work Today (Accessed on 10/02/2019). Retrieved from `http://www.socialworktoday.com/news/enews_0911_01.shtml`

Public opinion in Slovakia about physical punishment of children

Dominika Uhnakova and *Jana* Levicka

Trnava University, Department of Social work, 91843 Trnava, Slovakia

Abstract. Not so long ago, when the issue of physical punishment of children was not discussed at all. Progress occurred in the second half of the 20th century when children began to be perceived as authentic and legally protected beings. Over the past two decades we have seen an international shift in perspectives concerning the physical punishment of children. In 1990, research showing an association between physical punishment and negative developmental outcomes was starting to accumulate, and the Convention on the Rights of the Child had just been adopted by the General Assembly of the United Nations. The aim of our research was to find out how citizens of the Slovak Republic perceive physical punishments of children and how they apply them in the educational process. We have found out whether opinions on physical punishments are varied according to gender, education and religion of respondents. The research has been conducted using a quantitative research strategy. In order to collect the data a questionnaire was used. The survey sample consisted of 168 respondents. The data obtained was processed and analyzed using the SPSS program. Significant differences were confirmed from the point of view of education – respondents with lower education preferred physical punishment more than respondents with higher education. It is important for parents, to realize that the use of educational methods, such as explanation, persuasion, demand, or personal example is the base for achieving educational goals.

Key words: children, child's right, physical punishment, negative pedagogical methods.

1 Introduction

All children misbehave. Every parent faces the challenge of how to discipline his or her child. It can be frustrating when a child acts out or has significant behaviour problems. Children need limits and rules. There are many ways to give children rules and help change their behaviour. Examples include convention, explanation, positive reinforcement, taking away of privileges, and physical punishment. Physical punishment, sometimes called corporal punishment, is anything done to cause pain or discomfort in response to child's behaviours. There are several definitions of physical punishment. Straus and Stewart [1] define physical punishment as the use of physical force with the intention of causing a child to experience pain but not injury, for the purposes of correction or control of the child's behaviour. Lovasová, Schmidová [2] describe physical punishment as intentional or deliberate causing pain on the child's body, which interferes with his or her physical integrity, pushes the child to a primitive reaction, and thus hinders the development of his/her personality. The authors also emphasize that physical punishment increases the readiness for aggressive behaviour in the child's mind. The child perceives it as a form of social behaviour, as a method of solving difficult life situations. Children can see it as a guide for their future. The United Nations Committee on the Rights of the Child offers a more detailed definition of corporal punishment: "*Any punishment in which physical force is used and intended to cause some degree of pain or discomfort, however light. Most involve hitting ("smacking", "slapping", "spanking") children, with the hand or with an implement – whip, stick, belt, shoe, wooden spoon, etc. But it can also involve, for example, kicking, shaking or throwing children, scratching, pinching, biting, pulling hair or boxing ears, forcing children to stay in uncomfortable positions, burning, scalding or forced ingestion (for example, washing children's mouths out with soap or forcing them to swallow hot spices). In the view of the Committee, corporal punishment is invariably degrading. In addition, there are other non-physical forms of punishment which are also cruel and degrading and thus incompatible with the Convention. These include, for example, punishment which belittles, humiliates, denigrates, scapegoats, threatens, scares or ridicules the child* [3]. In general, physical punishment includes

- spanking (one of the most common methods of physical punishment),
- slapping, pinching, or pulling,
- hitting with an object, such as a paddle, belt, hairbrush, whip, or stick,
- making someone eat soap, hot sauce, hot pepper, or other unpleasant substances,
- wounds, caused by hitting with a ruler, splitting, pulling hair, dipping hands into hot or ice water [4].

Langmeier distinguishes two basic functions of physical punishments – positive and negative. The positive function of punishment makes children learn to understand the words "no," "you can't," consequently they know what is wrong. Langmeier states that if children are punished for the bad and praised for the good, not just rewarded for the good, they learn to recognize more quickly between the right and the wrong. The speed of changing undesirable behaviour using punishment is advantageous in the situations where the change of child's behaviour is urgent (for example, when a child runs out to the road to catch a ball, or if he/she physically hurts another child or animal). As a negative consequence of the use of punishment, Langmaier mentions child's strong emotional responses, such as fear and anxiety, which can lead to the development of generalized anxiety and shyness or, on the contrary, sometimes defiance and antisocial behaviour. Furthermore, he claims that punishment leads to the suppression of undesirable behaviour, but the child does not learn the desired behaviour, so it is important to combine it with the remuneration system. Another negative point of punishment is that it interferes with relationships between the child and the parent, and children can resort to lying and hiding in order to avoid the punishment [5].

2 Reasons for punishment

Parents apply punishment because they believe that punishment is a good method of education that teaches a child to behave properly and accept moral values. The reason for the use of punishment can also be a warning and guidance. Parents believe that after the punishment the child will not repeat bad behaviour. Using punishments, parents can also compensate for their own negative emotions, which result from the lack of control of the situation or their helplessness. Parents use the physical punishment when they feel completely helpless and use it as the last resort, when other methods of education have not helped [6]. Kopřiva [7] provides a systematic overview of why parents punish their child. She divides these reasons into the following categories:

Education

- sense of responsibility; grasping the right principles, observance of the rules,
- awareness of the borders ("stop" signal),
- changing behaviour, correcting what they have done,
- for lesson,
- for warning of others,
- prevention – avoiding further negative behaviour.

Tradition, adopted social model

- a model of education taken from own family; customs,
- obedience as educational value,
- social pressure (tradition of other adults to punish the child).

Emotions

- emotional failure to manage the situation; deregulation of own negative emotions, bad mood,
- anger, fear about the child's safety and health,
- ventilation of own problems.

Confirmation of power

- recognition of authority; enforcing their truth,
- feeling of superiority,
- revenge, retaliation,
- satisfaction from the suffering of the other.

Ignorance

- helplessness, ignorance of alternatives, inability to solve the situation differently,
- sense of helplessness.

Time stress

- effort to respond to inappropriate behaviour of children.

Convenience

- quick solution ("short process"),
- to act educationally without much effort,
- to have peace.

Personal issues

- solving own problems,
- unfulfilled expectations,
- moodiness, weariness, irritability,
- communication problems.

Although punishments reduce undesirable behaviour, they also generate a series of negative emotions – fear, humiliation, anger. These emotions will manifest as aggression and violence in the future and, in addition, distorted perceptions of punishment may occur. The long-term consequence of physical punishment of children is trauma. Trauma is an overly stressful impact on a person. There are two types of trauma: 1. Unexpected and unpredictable trauma; 2. Frequent, painful, expected and degrading trauma. Usually physical punishments belong to the second group. For this type of trauma, predictable, painful and degrading experience for children is characteristic. This experience is accompanied by fear, sadness, shame, guilt, apathy. The children mobilize all defence mechanisms against humiliation. This can be a denial, suppression, identification with the aggressor or self-destruction behaviour [6]. Sometimes non-physical punishments have more serious consequences than physical punishments. Prohibition of games and television, forced to unpopular activity (e.g., cleaning as a punishment), causes feelings of guilt and inferiority. Parental insults are very harmful and painful to the child [2, 6].

2.1 Approaches to the use of physical punishment

Severe [8] states that parents use punishments in three ways: impulsive, in anger and deliberately.

Impulsive approaches – this is a way of punishing when parents act almost automatically and impulsively. Impulsive punishment does not have lasting positive effect on the behaviour of the child. If parents repeatedly use a punishment, this punishment becomes insignificant. Many parents often hit children if they are doing something wrong. However, in this case, it is necessary for parents to explain what children have done wrong otherwise the children will not learn anything. If the parent is more irritated, the child will be hit for every little thing. This inconsistency is confusing for the child. Children, who are often impulsively hit, will begin to believe that physical punishment is a normal part of life. Importance of the education is lost. Frequent physical punishments do not lead to the goal of education – and that is, to teach the child a responsible decision.

Punishment in anger – this approach is the most harmful, especially because this way of punishment is impulsive, and it is likely that parents will not manage their behaviour and harm the child. This punishment creates strong negative emotions, both in parents and children. These emotions can interfere with the child's self-confidence. The child begins to be afraid of parents and loses confidence in them. In this way, parents teach the child to respond in anger with brutal force.

Premeditated punishment – this method of punishment involves warning to the child that if the child does some particular activity, a spanking will follow. For example, "if you speak unmannerly, you will get on your ass". This type of punishment is the most effective, but as the author states, there is no evidence that physical punishment would lead to a greater responsibility of children. Moreover, the planned punishment is also a pattern of negative behaviour. Many parents forget that the planned education must contain positive solutions, not only the use of physical punishment, but also the use of motivation techniques.

It is necessary to understand that all physical punishments, however "mild" and "light", carry an inbuilt risk of escalation: its effectiveness in controlling children's behaviour decreases over time, encouraging the punisher to increase the intensity of the punishment. The risk of escalation is increased by the fact that adults who inflict physical punishment are often angry: their anger can increase the level of force used beyond what was intended, and their intent may be retaliatory as well as punitive [9]. Physical punishment can reduce empathy and moral regulation. It does not teach children how to behave. Rather than helping children to develop the desire and motivation to behave well of their own accord, it teaches them that it is desirable not to get caught. Physical punishment may also decrease the likelihood of long-term compliance by damaging adult-child relationships and introducing fear [9]. Just the meaning of empathy is mainly in empathic communication and empathic understanding. The main tool of understanding and empathy is that through empathy, one person's feelings and different tensions are equipped with those of the others. This is a sense of empathy that requires the laying down of oneself [10].

3 Methodology

The main aim of our research was to find out how citizens of the Slovak Republic perceive physical punishments of children and how they apply them in the educational process. We have found out whether opinions on physical punishment differ according to gender, education and religion of respondents. The research has been conducted using a quantitative research strategy. Questionnaire consisted of 24 questions. To assess participants' views on corporal punishment, we used a modified version of the *Attitudes toward corporal punishment questionnaire*. The questionnaire consisted of 9 closed-ended questions and then participants read 9 statements and indicated their level of agreement using a 5-point Likert-type scale, with 1 = strongly agree and 5 = strongly disagree. The end of the questionnaire contained 6 demographic questions on participants' age, gender, religion, education, marital status and area of residence. In addition to the printed version the questionnaire was also conceived in an electrical form. The data obtained was processed and analyzed using the SPPS Statistic 20 – Kruskal Wallis test and Mann Whitney U test.

3.1 Research sample

The survey sample consisted of 168 respondents (75.4% women and 24.6% men). Majority of respondents were 31–40 years old (47.3%) and 18–30 years old (31.9%). In the range of 41–50 years old were 12.5% of respondents, 51–60 years old were 5.4% of respondents and over 60 years old were only 1.8% of respondents. Regarding the education, respondents were classified into six groups: PhD degree (4.2%), Master degree (55.4%), Bachelor degree (3.5%), post-secondary education (6.5%), secondary education (29.2%) and primary education (1.2%).

3.2 Results

86.9% of the respondents stated that their parents used physical punishment in their upbringing. Most often it was slapping the bottom (63.1%) and spanking (15.5%). 13.1% of respondents stated that their parents had never beaten them. Physical punishment was considered as effective for upbringing by 62.5% respondents. Total prohibition of physical punishment was supported by only 8.9% of respondents. As an alternative to physical punishment, the respondents mostly mentioned explanation (29.2%), prohibition of favourite activities (25.6%) and educational discussion of the situation (17.9%).

The first factor which we examined was gender. We wanted to find out whether there was a difference in views on physical punishment between men and women. We used the Mann Whitney U test ($U = 2311, 500, P = 0.169$). The results did not show significant difference between men and women in opinion on physical punishments. Regarding the education of the respondents, we wanted to find out whether there were differences in views on physical punishment depending on education. We used the Kruskal–Wallis test.

Table 1. The results Mean Rank for education.

Ranks			
Education		N	Mean Rank
PP	PhD	8	57.25
	Master degree	93	87.55
	Bachelor degree	6	47.17
	Postgraduate	10	120.45
	Graduate degree	49	77.04
	Primary education	2	166.50
	Total	168	

The results showed differences in opinion on physical punishment according to education ($\chi^2(3) = 18.944$, $P = 0.002$).

The table shows (*a lower score indicates a lower degree of agreement to physical punishment)* that the lowest mean score was reported by respondents with university education – that means they prefer more alternative methods in upbringing to physical punishment. It suggests that people with university education can have more information about negative influence of physical punishment on children. Another factor that we examined was whether views on physical punishment are influenced by religion of respondents. Results showed that most of the respondents stated that they were believers and practised faith regularly (54.2%). 29% said they were believers, but did not practise faith, and 16.8% were unbelievers. We used the Kruskal–Wallis test again and we did not find out significant differences in opinion on physical punishment according to religion of the respondents ($\chi^2(3) = 3.838$, $P = 0.147$).

3.3 Discussion

Physical punishments in Slovakia were often used at schools or in families as an educational technique. However, what was considered to be educational in the past is taken as a limitation of children's rights and freedom nowadays. This study showed that majority of the respondents were exposed to physical punishment during childhood and they often use it in upbringing their children, too. This confirms the findings of Global Initiative to End All Corporal Punishment of Children that have found that parents in households, where physical punishment was applied, are more likely to inflict physical punishment on their children and associate it with a belief that physical punishment is necessary in upbringing of children. From the above-mentioned findings, we can say that physical punishment in Slovak republic is still very widely used in upbringing. This is confirmed by the finding that only a small percentage of respondents agree with prohibition of physical punishment. However, studies show that in the states which have prohibited all kinds of physical punishment, the positive effects of it are becoming visible. A study in 2000, which examined the impact of the ban in Sweden, found that there had been a decrease in the number of 15 to 17-year olds involved in theft, narcotics crimes, assaults against young children and rape, and a decrease in suicide and use of alcohol and drugs by young people. Studies in Finland have found that there has been a clear reduction in all forms of corporal punishment and other parental violence against children since prohibition in 1983 and that the decline in physical punishment was associated with a similar decline in the number of children who were murdered [9]. Next our finding is that education is a significant factor in the opinion on punishment. Physical punishment can have a negative impact on children's cognitive development: associations have been found between physical punishment and lower IQ scores, smaller word stock, poorer cognitive abilities, slower cognitive development and poorer school marks. Similar associations have been found about physical punishment at school. Physical punishment at school violates children's right to education, including by creating a violent and intimidating environment in which children are less able to learn. This punishment is often a reason given by children for not attending or for dropping out of school. Even adults who experienced corporal punishment as children were less likely to graduate from college and to have high status and highly paid jobs [9].

Change from the traditional upbringing style to the collaborative one takes time, patience, and energy. Even so, collaboration does much more to help us achieve our long-term goals of facilitating the development of adults, who are driven to do good not for the sake of avoiding a punishment or getting a reward, but rather to do what is right [11]. Rewards and punishments seem, at first glance, two completely different concepts, but these are two sides of the same coin. Their similarity is that both are only an external impulse (incentive), which focuses on the external behaviour of the child. Both tend to have an immediate effect, which most people see as the primary result and never think of their long-term negative impact on personality development. The research results can be summarized in the clear findings by Čáp and Mareš: education based on rewards has better results than education based on punishments [12].

In Slovakia, no significant differences were found between men and women in the frequency of used punishment. However, in international research in seven of the nine countries, mothers reported spanking, slapping, or hitting their child significantly more frequently than fathers did in the same families. In Colombia and Kenya, this parent-gender effect was qualified by a significant interaction with child's gender. Mothers reported slapping or hitting daughters more frequently than sons, whereas fathers reported slapping, or hitting sons more frequently than daughters [13]. It can be caused by the fact that Slovakia is still a Christian country, in which both parents have the same position in upbringing.

4 Conclusion

Despite the fact that parents seek to follow humanistic principles in education, the results of our research have confirmed that in many families, negative pedagogical methods such as physical punishment are used. Significant differences were confirmed from the point of view of education – respondents with lower education preferred physical punishment more than respondents with higher education. However, these and many other negative sanctions are contrary to the child's fundamental right to respect for his personality. It is important for parents to realize, that the use of educational methods such as explanations, persuasion, demand or personal example is the base for achieving educational goals.

The paper was prepared as a part of project APVV-16-0205 – "Identification of Mechanisms of Early Diagnosis of CAN Syndrome".

References

[1] M.A. Straus, J.H. Stewart, "Corporal punishment by American parents: National data on prevalence, chronicity, severity, and duration, in relation to child and family characteristics" Clin. Child Fam. Psychology Rev. **2**(2), 55–70 (1999), https://www.researchgate.net/publication/12105415_Corporal_Punishment_by_American_Parents_National_Data_on_Prevalence_Chronicity_Severity_and_Duration_in_Relation_to_Child_and_Family_Characteristics

[2] L. Lovasová, K. Schmidová, Tělesné tresty. Available at: http://www.vyzkum-mladez.cz/zprava/1385375890.pdf
[3] Commissioner for Humans Rights, Council of Europe, Children and Corporal Punishment: "The right not to be hit, also a children's right" (2006). Available at: https://rm.coe.int/children-and-corporal-punishment-the-right-not-to-be-hit-also-a-childr/16806da87b
[4] American academy of child and adolescent psychiatry: Physical punishment (2018). Available at: https://www.aacap.org/AACAP/Families_and_Youth/Facts_for_Families/FFF-Guide/Physical-Punishment-105.aspx
[5] J. Langmeier, Z. Matějček, Psychická deprivace v dětství, Praha: Avicenum. ISBN 08-049-74 (1974)
[6] E. Vaníčková, Tělesné tresty dětí: Definice - popis – následky. Praha: Grada Publishing. ISBN 80247081409 (2004)
[7] P. Kopřiva et al., Respektovat a být respektován. ISBN 9788090403000 (2007)
[8] S. Severe, Co delat, aby sa vaše deti správne chovaly, ISBN 8071783684 (2000)
[9] Global Initiative to End All Corporal Punishment of Children, Corporal punishment of children: summary of research on its impact and associations (2015)
[10] L. Banovčinová, M. Bubeníková, Empatia vošetrovatetve **2**, 7 Ošetrovateství a porodní asistence (2011)
[11] J. Lasser, Alternatives to Punishment and Rewards (2018). Available at: https://www.psychologytoday.com/us/blog/school-and-family-matters/201802/alternatives-punishment-and-rewards
[12] J. Jakešová, S. Slezáková, Rewards and Punishments in the Education of Preschool Children. Procedia – Social and Behavioral Sciences, Vol. 217 (2016), p. 322–328
[13] J.E. Lansford et al., Corporal Punishment of Children in Nine Countries as a Function of Child Gender and Parent Gender. Int. J. Pediatrics **2010**. Available at: https://www.hindawi.com/journals/ijpedi/2010/672780/

Modern youth's value orientations and activity in a cross-cultural context

Inna Atamanova[1,*], *Sergey* Bogomaz[1], and *Tatyana* Filippova[2]

[1] National Research Tomsk State University, Tomsk, Russia
[2] Karaganda State Technical University, Karaganda, Kazakhstan

Abstract. Addressing the issue of youth's personal and professional development, taking into account their value orientations and psychological characteristics of their activity, seems to be relevant to the modern world's challenges. Cultural dimensions are believed to be the very factors that can contribute to a better understanding of the interplay between personal and environmental aspects when analysing the problem of high quality human resource development. The paper presents a cross-cultural study on cultural values characterising socio-economic and cultural settings of two countries and university students' basic values and parameters of their personal potential in the context of their activity. 307 university students from Tomsk (Russia) and 295 university students from Karaganda (Kazakhstan) participated in the study. The research tools involved a number of questionnaires aimed at exploring cultural dimensions, subjective evaluation of realisation of basic values in the urban environment and parameters of personal potential as factors of one's psychological system of activity. The study results revealed a number of statistically significant differences among the study participants. First, there was a difference in some cultural values, namely individualism, uncertainty avoidance and long-term orientation. Second, university students from Kazakhstan higher evaluated their urban environment in the context of possibility to realise their basic values. Third, the study participants from Kazakhstan showed higher scores in such parameters of their psychological system of activity as purposefulness, satisfaction with life, need for autonomy as well as the achievement and affiliation scales. Russian university students had higher scores in reflection which is a significant factor of one's self-determination. The results obtained are discussed in the context of modern youth's personal and professional development in a specific socio-economic and cultural environment.

Key words: youth, students' value orientations, cross-cultural context.

1 Introduction

The collapse of the Soviet Union in 1991 generated a unique situation when 15 countries that had been a single unit before simultaneously started their own pathways to building independent national communities [1]. As Smolentseva, Huisman and Froumin highlight, all newly-born states faced similar challenges, namely, consolidating their national societies and introducing market economies [2]. These processes were accompanied by political instability, economic decline, demographic changes and even armed conflicts in most of them, having

* Corresponding author: `iatamanova@yandex.ru`

a deep impact on their societies and economies. More than 25 years of the post-Soviet development in these countries have resulted in the emergence of specific political, economic and socio-cultural conditions. Moreover, in these countries there have already appeared young people who were even not "born in the USSR". They have never lived in a huge multinational united country speaking the same language.

Youth is known to be the most sensitive-to-changes age category, so it represents a major research interest in the context of the future trends of the society development. On the one hand, such a research focus is needed for better understanding of what is going on in the countries concerned. The issue of youth's personal and professional development is tightly connected with that of each country's high-quality human capital in a long-term perspective. On the other hand, globalisation processes are enhancing cooperation in different spheres between the post-soviet states as well. What do we know about people living in these countries nowadays, especially about young people who will be a driving force in their countries and, therefore, globally in a couple of decades? To what extent are they similar to each other? How much do they differ? Do they value the same things?

Cultural dimensions are believed to be the very factors that can contribute to a better understanding of the interplay between personal and environmental aspects when analysing the problem in question. The researchers emphasise that socio-cultural characteristics of human capital are becoming more and more important in identifying growth points of national economies nowadays [3]. Addressing the issue of youth's personal and professional development, taking into account their value orientations and psychological characteristics of activity, seems to be relevant to the modern world's challenges.

According to Vygotsky [4], "every human being represents a certain historical unit, living at a certain historical period and in certain historical circumstances" (p. 352). This highlights the role of socio-cultural factors in human development and means that one's socio-cultural environment greatly affects the process of personal and professional development. In other words, "the environment's role in the development of higher, specifically human characteristics and forms of activity is as a source of development" (p. 351). In turn, Sorokin [5] viewed socio-cultural interaction as an indivisible integrity of three interdependent modes, namely, personality, society and culture. Therefore, as emphasised in [6], "any economic, social or cultural transformations will certainly have an effect on both personal and professional development" (p. 553).

For the purpose of this study we also address Hofstede's concept of national culture [7], whose essence is "collective programming of the mind that distinguishes the members of one group or category of people from others" (p. 5). This means that values and beliefs shared by the members of the society (at the national or local level) have a significant effect on their behavioural patterns in all life spheres.

Furthermore, recent studies revealed that educational settings also had a marked effect on university students' personal and professional development [8]. It was shown, for example, that different career orientations among master's students and doctoral students were connected with their university's orientation towards classical or engineering education. Thus, university students' socio-cultural environment has its potential in the context of their personal and professional development. The conducted studies highlighted the role of their basic values as well or, more exactly, their subjective evaluation of the possibilities to realize their basic values in the local settings. Such a subjective evaluation may differ in its profiles depending on the regional factor [6, 9].

Moreover, recent studies [6] showed that the higher was young peoples' personal potential, the more possibilities they revealed in their local settings for their personal and professional development. The relationships found confirmed the idea that "not only the socio-cultural environment affects university students' personal characteristics, but also their subjective perception of this environment as a factor of their personal and professional development depends on their personal potential" (p. 556). The university youth with pronounced leadership qualities, purposefulness, inclination for self-organisation of their activity and inclination for reflection on it higher evaluated the possibilities to realize basic values in their socio-cultural environment. They were also more positive in their perception of possibilities for their development and self-realisation.

Thus, the paper presents a cross-cultural study on cultural values characterising socio-economic and cultural settings of two post-soviet countries (Russia and Kazakhstan) and modern youth's basic values and parameters of their psychological system of activity.

2 Research materials and methods

The main research focus of the study presented was to compare university students' value orientations (at the individual level and that of culture) and parameters of their psychological system of activity. This is the first part of the research project supported by the Russian Foundation for Basic Research and aimed at studying the specific interaction of modern youth's basic values and their psychological system of activity, taking into account cultural factors of the country's socio-economic development (cultural dimensions). The three-year project involves conducting a cross-cultural analysis based on the data from three post-Soviet countries. In the year of 2018 the data were collected in Russia and Kazakhstan.

2.1 Study design

The study was designed as follows. The study participants were asked to fill in several questionnaires. The paper-and-pencil forms were offered in Russian. University students from Kazakhstan who participated in the study could understand the Russian language because their educational programs were implemented in Russian. The data collected were then statistically treated applying the Statistica 6.0 package. Descriptive analysis and analysis of variance (ANOVA, the F-test criterion) were used to compare the subsamples at this stage of the research project.

2.2 Study sample

307 university students from Tomsk (Russia) and 295 university students from Karaganda (Kazakhstan) participated in the study. Their demographic characteristics were the following. The study participants from Russia: the mean age was 20.84, 47.6% were male, 51.5% were female, and three people did not indicate their sex. The study participants from Kazakhstan: the mean age was 20.87, 47.8% were male, 50.8% were female, and four people did not indicate their sex either. The study participants were majoring in various subjects, but this aspect was not in question in this study.

2.3 Research tools

There were three research foci in collecting the data. The first one was connected with *cultural dimensions* (the study participants' value orientations at the level of culture). These data were collected using the Values Survey Module (VSM 2013) by Hofstede [10] and the World Values Survey questionnaire by Inglehart [11], modified by Khabibulin [12]. The former research tool is a 24-item questionnaire developed for comparing culturally influenced values, which measures six cultural dimensions, namely, *power distance* (large versus small), *individualism* versus *collectivism*, *masculinity* versus *femininity*, *uncertainty avoidance* (strong versus weak), *long-* versus *short-term orientation*, and *indulgence* versus *restraint*. Each item is evaluated based on a 5-point Likert-like scale where 1 is *of utmost importance* and 5 is *of very little or no importance*. Each cultural dimension or index is calculated according to the formula, for more details see [10]. Inglehart's tool modified by Khabibulin is a 13-item questionnaire which allows one to evaluate two culture-specific value orientations: *traditional values* versus *secular-rational values* and *survival values* versus *self-expression values*. Each item is measured according to a 7-point Likert-like scale (1 means *absolutely disagree* and 7 means *absolutely agree*) [12].

The second research focus was on the study participants' value orientations at the individual level, i.e. their *basic values*. The research tool was developed by Bogomaz for examining one's subjective evaluation of possibilities for basic values realisation in the local settings [9]. The Basic Values Realisability technique is a 20-item instrument based on the semantic differential method originated by Osgood (see [13]) and analogous to the Noetic Orientations Test [14]. It contains 20 items corresponding to 20 basic values. Among them are *have a good job, be in good health, be financially secure, have a happy family, achieve success in a profession, be respected, achieve success in your career, love and be loved, be free, feel safe, be known and famous, achieve desired goals, live full life, find meaning of your life, know everything, be an example for others, assert yourself, be unique and original, have power*, and *be fair*. Taken together, each basic value and its opposite represent one item of the technique. The task is to evaluate each bipolar set using a 7-point Likert-like scale (3-2-1-0-1-2-3) [6].

The third research focus was on the study participants' *psychological system of activity*. The psychological parameters of interest were collected by using a number of questionnaires. We applied the Self-Organisation of Activity Questionnaire developed by Mandrikova [15] to evaluate the study participants' tendency to self-organisation of activity. This 25-item research tool allows measuring one's ability to plan and organise one's own activity. It has 6 subscales: *planning, purposefulness, persistence, fixation, self-organisation* and *orientation in time*. For the purposes of our study only two subscales (*planning* and *purposefulness*) were involved (10 items) and their items were evaluated based on a 5-point Likert-like scale with 1meaning *absolutely disagree* and 5 meaning *absolutely agree*. The Satisfaction with Life Scale by Diener (see [16]), translated into Russian and adapted by Osin and Leontiev [17], was used to measure the study participants' *satisfaction with life*. The Balanced Measure of Psychological Needs (BMPN) scale by Sheldon and Hilpert [18], an 18-item tool adapted by Leontiev, was applied to evaluate the study participants' needs for *autonomy, competence* and *relatedness* with others. In the Reflexivity Type Assessment Test by Leontiev (see [19]), a 30-item instrument developed for measuring reflective processes in a person, we used only one subscale modified for the purposes of our study (7 items) – *systemic reflection*. Finally, we used two subscales of the translated into Russian and modified version of the Personality Research Form by Jackson [20, 21], namely *achievement* and *affiliation* (6 items). To evaluate items in these research tools, a 5-point Likert-like scale was also applied.

3 Study results and discussion

As mentioned above, at this stage of the research project the data collected were statistically treated using descriptive analysis and analysis of variance (ANOVA, the F-test criterion). The paper presents statistically significant differences between the study subsamples (university students from Russia and Kazakhstan) and discusses them from the cross-cultural perspective.

There were three research foci (see the Research tools subsection), namely on the study participants' value orientations at the level of culture (*cultural dimensions*), their value

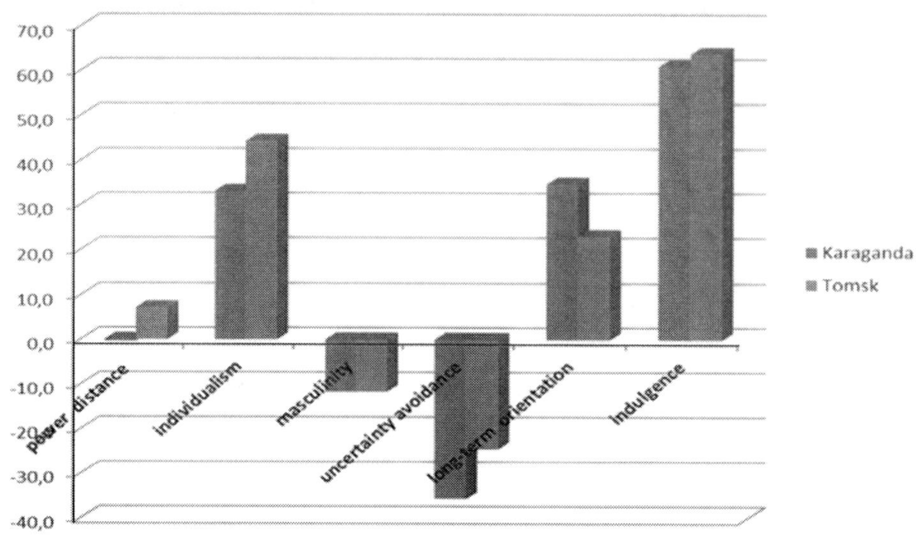

Fig. 1. Differences in Hofstede's cultural dimensions between university students from Russia (Tomsk) and Kazakhstan (Karaganda).

orientations at the individual level (*basic values*) and parameters of their *psychological system of activity*.

Cultural dimensions. Figure 1 illustrates between-group differences in cultural dimensions that were revealed using Hofstede's research tool when comparing university students from Russia (N = 307) and Kazakhstan (N = 295).

The ANOVA results showed some statistically significant differences ($p < 0.05$) between the subsamples in such cultural dimensions as individualism versus collectivism (F = 4.034, $p = 0.045$), uncertainty avoidance (F = 4.305, $p = 0.038$) and long-term versus short-term orientation (F = 5.615, $p = 0.018$). As seen in Figure 1, university students from Kazakhstan can be characterised as having more collective mindsets, feeling more threatened by unknown, uncertain or ambiguous situations and being to a greater extent oriented towards long-term perspectives [10]. In contrast, university students from Russia can be described as being more individualistic, feeling less threatened by uncertainty and having short-term orientations.

Inglehart's questionnaire also allowed us to reveal a statistically significant ($p < 0.05$) difference between the study subsamples. The study participants from Kazakhstan (M = 4.82, SD = 1.01) were found to have more adherence to traditional values (F = 32.446, $p = 0.000$) compared to those from Russia (M = 4.37, SD = 0.89). Based on [10], we can conclude that university youth from Kazakhstan more appreciate traditional parent-child relations and traditional family values and norms. In turn, this is in line with the results obtained by Hoftede's measure. In contrast, university students from Russia are to a larger extent oriented towards post-materialist values.

Basic values. Figure 2 shows between-group differences in the study participants' value orientations at the individual level. These are the scores in their subjective evaluation of basic values realisability in their local settings. As mentioned in [6], such a characteristic can describe the potential of one's socio- cultural environment in terms of one's personal and professional development. The subjective perception of one's local settings with taking into account one's basic values seems to be essential for shedding some light on the problem of initiating human activity and achieving one's meaningful goals.

Table 1. The study participants' subjective evaluation of basic values realisability in the local settings.

Basic values	**Russia (N = 307)**		**Kazakhstan (N = 295)**		**F**	***p***
	Mean	**SD**	**Mean**	**SD**		
Have a good job	5.95	1.41	6.37	1.05	16.561	0.000
Be in good health	5.65	1.60	6.26	1.29	24.776	0.000
Be financially secure	5.42	1.67	5.60	1.82	1.558	0.212
Have a happy family	5.93	1.47	6.22	1.47	5.366	0.021
Achieve success in a profession	5.96	1.35	6.19	1.34	4.394	0.036
Be respected	5.67	1.42	5.66	1.61	0.019	0.890
Achieve success in your career	5.91	1.35	6.18	1.30	5.948	0.015
Love and be loved	6.09	1.27	6.33	1.22	4.967	0.026
Be free	5.24	1.70	5.43	1.73	1.863	0.173
Feel safe	5.62	1.50	6.06	1.41	13.328	0.000
Be known and famous	4.84	1.72	4.79	1.74	0.116	0.733
Achieve desired goals	5.83	1.45	6.02	1.54	2.408	0.121
Live full life	5.93	1.45	6.19	1.45	4.726	0.030
Find meaning of your life	5.71	1.53	6.11	1.40	10.456	0.001
Know everything	5.62	1.66	5.90	1.53	4.571	0.033
Be an example for others	5.58	1.30	5.67	1.51	0.538	0.464
Assert yourself	5.88	1.22	6.13	1.15	6.222	0.013
Be unique and original	4.79	1.97	5.14	1.93	4.491	0.035
Have power	4.70	1.60	4.92	1.75	2.535	0.112
Be fair	5.89	1.25	6.19	1.28	8.084	0.005

The ANOVA results showed that there were statistically significant ($p < 0.05$) differences in the subjective evaluation of basic values realisability between the study subsamples. The study participants from Kazakhstan higher evaluated their urban environment for realising such basic values as being unique and original (F = 4.491, $p = 0.035$), knowing everything (F = 4.571, $p = 0.033$), feeling safe (F = 13.328, $p = 0.000$), finding meaning of your life (F = 10.456, $p = 0.001$), asserting yourself (F = 6.222, $p = 0.013$), achieving success in your career (F = 5.948, $p = 0.015$), being fair (F = 8.084, $p = 0.005$), living full life (F = 4.726, $p = 0.030$), achieving success in a profession (F = 4.394, $p = 0.036$), having a happy family (F = 5.366, $p = 0.021$), being in good health (F = 24.776, $p = 0.000$), loving and being loved (F = 4.967, $p = 0.026$) and having a good job (F = 16.561, $p = 0.000$). Table 1 presents basic values realisability means, standard deviations and ANOVA results for the study subsamples.

The study participants from Kazakhstan higher evaluated their local settings for realising most of the basic values in question compared to those from Russia, as seen in Table 1 and Fig. 2. On the one hand, this may be a consequence of their adherence to traditional values and norms, including traditional ways of understanding the essence of their personal and professional development as well. On the other hand, such a result may reflect their less realistic views on their local settings.

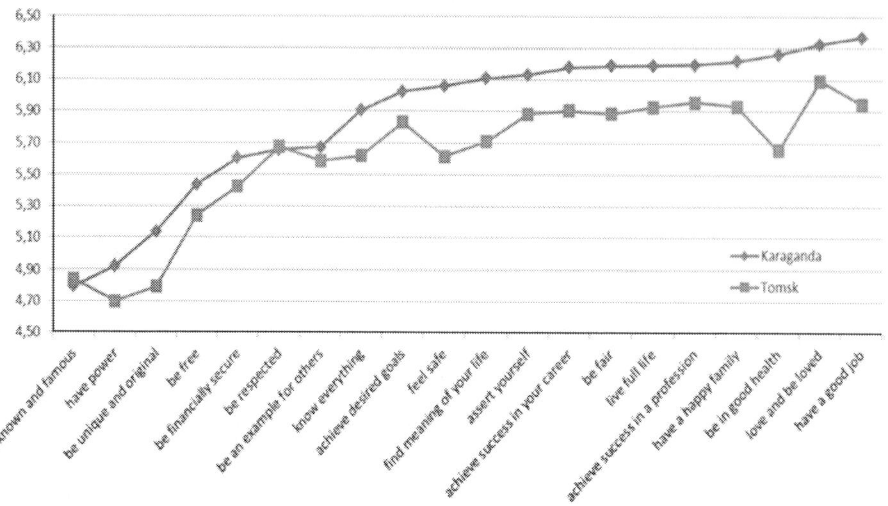

Fig. 2. Differences in the subjective evaluation of basic values realisability in the local settings between university students from Russia (Tomsk) and Kazakhstan (Karaganda).

Table 2. The study participants' psychological system of activity.

Parameters	Russia (N = 307)		Kazakhstan (N = 295)		F	p
	Mean	SD	Mean	SD		
Purposefulness	3.95	0.78	4.09	0.66	5.756	0.017
Planning	3.21	0.98	3.32	0.88	1.824	0.177
Systemic reflection	4.03	0.58	3.88	0.60	8.904	0.003
Satisfaction with life	3.34	0.75	3.50	0.81	5.969	0.015
Readiness for activity	3.64	0.51	3.70	0.54	1.771	0.184
Need for achievement	3.96	0.71	4.20	0.65	18.222	0.000
Need for affiliation	3.99	0.78	4.14	0.74	5.685	0.017

One more issue of interest was connected with the study participants' ***psychological system of activity***. As described in the Research tools subsection, there were a number of parameters in question. First, statistically significant ($p < 0.05$) between-group differences were found in such parameters as purposefulness (F = 5.756, $p = 0.017$), systemic reflection (F = 8.904, $p = 0.003$) and satisfaction with life (F = 5.969, $p = 0.015$). The study participants from Kazakhstan had higher scores in purposefulness (M = 4.09, SD = 0.66) and satisfaction with life (M = 3.50, SD = 0.81), while the ones from Russia were more disposed to systemic reflection (M = 4.03, SD = 0.58). Second, the study participants from Kazakhstan showed higher scores in their needs for achievement (M = 4.20, SD = 0.65) and affiliation (M = 4.14, SD = 0.74) compared to the ones from Russia (M = 3.96, SD = 0.71 and M = 3.99, SD = 0.78, respectively). The statistically significant between-group differences were the following: F = 18.222, $p = 0.000$ for the achievement scale and F = 5.685, $p = 0.017$ for the affiliation one. The study participants' needs for autonomy also differed at the statistically significant level ($p < 0.05$). The university students from Kazakhstan had higher scores in their need for autonomy (F = 9.279, $p = 0.002$). For more details see Table 2.

Interpreting the identified differences in parameters of the study participants' psychological system of activity, taking into account their value orientations at the individual level and that of culture, we would like to emphasise the following. Higher scores in the achievement and affiliation scales in university students from Kazakhstan may be connected with their greater adherence to traditional values compared to their counterparts from Russia. It is these traditional views that may determine young people's priorities in their personal and professional development as well as their life strategies. Meanwhile, they had higher scores in their need for autonomy; this may signal their desire to rely to a greater extent on their own.

In addition, higher scores in purposefulness and planning among university students from Kazakhstan may be also connected with their adherence to traditional values and more pronounced long-term orientation and uncertainty avoidance compared to those from Russia. Such value orientations may result in a desire to follow a certain algorithm to achieve their purposes. In contrast, higher scores in systemic reflection among university students from Russia may demonstrate more flexibility in making decisions to meet the demands of the modern changing world. This is also in line with their short-term orientation and lower scores in uncertainty avoidance.

4 Conclusions

Thus, the study conducted enabled us to reveal some statistically significant differences between young people from Russia and Kazakhstan in their value orientations at the individual level and that of culture, as well as in their parameters of the psychological system of activity. Such a complex cross-cultural analysis involving young people from the post-Soviet countries has been undertaken for the first time. The paper presents only some results of the first stage of the three-year research project and this is one of the limitations. Further analysis of the data collected will be focused on identifying within-group relationships between the parameters in question to shed more light on the interplay between young people's value orientations and their parameters of the psychological system of activity. The relationship between their value orientations (at the individual level and that of culture) and personal characteristics is of particular importance for creating psychological support programs in higher educational settings to enhance university youth's personal readiness for activity.

The paper has been prepared within the research project (18-013-00781) supported by the Russian Foundation for Basic Research.

References

[1] A. Pavlenko, Russian Linguistics **32**(1), 59–80 (2008)
[2] A. Smolentseva, J. Huisman, I. Froumin, *25 Years of Transformations of Higher Education Systems in Post-Soviet Countries* (Palgrave Macmillan, Cham, 2018)
[3] A.A. Auzan, J. New Econ. Assoc. **2**(26), 243–248 (2015)
[4] L.S. Vygotsky, *The Vygotsky Reader* (Blackwell, Oxford, 1934/1994)
[5] P.A. Sorokin, *Society, Culture and Personality: Their Structure and Dynamics* (Harper, New York, 1947)
[6] S. Bogomaz, N. Kozlova, I. Atamanova, Procedia – Social and Behavioral Sciences **214**, 552–558 (2015)

[7] G. Hofstede, G.J. Hofstede, M. Minkov, *Cultures and Organizations: Software of The Mind* (McGraw-Hill, New York, 2010)

[8] I.V. Atamanova, O.N. Starichenko, S.A. Bogomaz, Tomsk State University Journal **367**, 128–135 (2013)

[9] S.A. Bogomaz, *Proceedings of V All-Russian scientific-practical conference with foreign researchers' participation "A man and the world: Psychology of conflict, uncertainty and risk of innovation"* (ERGO, Izhevsk, 2014)

[10] G. Hofstede, M. Minkov, *Values survey module 2013. Manual* (Geert Hofstede BV, 2013) `http://geerthofstede.com/wp-content/uploads/2016/07/Manual-VSM-2013.pdf`

[11] R. Inglehart, C. Welzel, *Modernization, Cultural Change, and Democracy: The Human Development Sequence* (Cambridge University Press, New York, 2005)

[12] O.S. Deyneka, R.K. Khabibulin, Modern Problems of Science and Education **1–1**, 1525 (2015)

[13] C.E. Osgood, G. Suci, P. Tannenbaum, *The Measurement of Meaning* (University of Illinois Press, Urbana, 1957)

[14] D.A. Leont'ev, *Test Smyslozhiznennykh Orientatsiy (SZhO) [The Noetic Orientations Test (NOT)]* (Smysl, Moscow, 1992)

[15] E. Yu. Mandrikova, *Oprosnik Samoorganisatsii Deyatel'nosti [The Self-organization of Activity Questionnaire]* (Smysl, Moscow, 2007)

[16] E. Diener, R. Emmons, R. Larsen, S. Griffin, Journal of Personality Assessment **49**, 71–75 (1985)

[17] E.N. Osin, D.A. Leontiev, *Materialy III Vserossiyskogo Sotsiologicheskogo Kongressa [Proceedings of III All-Russian Sociology Congress]* (Sociology Institute of RAN, Moscow, 2008)

[18] K.M. Sheldon, J.C. Hilpert, Motivation and Emotion **36**(4), 439–451 (2012)

[19] D.A. Leontiev, E.N. Osin, Psychology. Journal of the Higher School of Economics **11**(4), 110–135 (2014)

[20] D.N. Jackson, *Personality Research Form Manual* (3rd ed.) (Research Psychologists Press, Port Huron, 1984)

[21] I.M. Kondakov, Zhurnal Prikladnoi Psikhologii [J. Appl. Psychology] **4**, 99–113 (1998)

Comparison of adolescent's values: Riga and Vidzeme region

Sandra Mihailova[1], *Kristine* Martinsone[2], and *Arta* Pipcane[3]

[1] Rīga Stradiņš University, Department of Sociology and Psychology, Riga, Latvia
[2] Rīga Stradiņš University, Department of Health Psychology and Paedagogy, Riga, Latvia
[3] Riga State German Grammar School, Riga, Latvia

Abstract. The determination of individual values is the first step in the process of the personal development. Thus, the main purpose of this research is to investigate the difference between value hierarchies of Riga students and rural students and their subjective opinions about factors affecting them and compare the results with the values of previous generations. It is a pilot study with a comparative cross-sectional design. The study was carried out in several schools of Latvia in 2018: 2 schools of Riga and 2 rural schools of Vidzeme region. Participants were 11th and 12th grade students: 50 from Riga, 50 from rural schools. The research of values was based on the M. Rokeach value ranking test. It is found that the individual hierarchies of values in one group context differ more than hierarchies of adolescent groups of various years and different backgrounds. There are no significant differences related to the terminal and instrumental values between adolescents of Riga and rural areas: love, cheerfulness and education are more important for Riga students, but rural students prefer responsibility, obedience, capability. Some values have not been influenced by time or socio-economic and political situation, but some values have increased or decreased their significance with years.

Key words: terminal values, instrumental values, high school students, comparison of generations, i-generation, adolescents.

1 Introduction

Researchers have been interested in the exploration of adolescent values at all times. Several researchers have turned to research individuals values (in Latvia – Pauls Dāle, Pauls Jurevičs, Jūlijs Aleksandrs Students, Augusts Milts, Skaidrīte Lasmane, Andris Rubenis, Ārija Karpova, Ilze Koroļeva, Ritma Rungule, Viesturs Reņģe, Silva Seņkāne, Spodra Austruma, Airisa Šteinberga, Ineta Tunne, Ivars Austers u.c.).

The topicality of this research was determined by the socio-economic and political situation in modern Latvia and the changes that have impacted Latvian society in the past 30 years. The regaining of Latvia's independence and the building of a new country, joining European Union, world's integration and globalization processes, the rapid increase of digitalization impact on several social processes as well as each human's world views, stereotypes, values and statuses, and possibilities.

During the recent acceleration of development at significant areas of life, there are radical changes at human relationships, world perception, goals and ideals of life. The revaluation of values as a logical result of human dialectics is controversial and at present it takes place throughout society. New conditions set new requirements for intellectual, mentally rich, open-minded people who are able to adapt to different circumstances, change occupations, communicate with different types of people. The determination of individual values is the first step in the process of the personal development. To achieve it, one needs to understand its interests, opinions and priorities and evaluate their significance and hierarchy.

As digitalization becomes part of all areas of life, the widening of the view of the world gives adolescents significantly more opportunities. The way of how these opportunities will be fulfilled is dependent on each youngster's values and choices. With the change of circumstances comes the change of values; time is needed to understand which values fit the new situation, which will have to change and which stay the same.

Furthermore, adolescents have to adapt to new values, arranging points of view and sometimes even own life. Even though there are strict core values at the age of adolescence that have formed based on previous experiences or family values, they can be changed in the course of strong social impact. Adolescents are the social group that adapts more easily on one hand, but on the other – are more easily subjected to media news and social impacts.

The beginning of the highest level of self-confidence development is at the age of 16–19. Though the hierarchy of values is not stable, the main values of young people have been already formed, based on their experience and the values of their parents.

As showed the research of values in 2008, a whole of values is made different by place of residence; there is a significant difference in the values between the young people of city and rural regions [1]. Based on this research, there was a question whether there still is a difference, since social media has an impact on both city and rural areas, and if so, then which values differ.

The high school students of today were born at the turn of 20th and 21st centuries and started their studies when Latvia was already a member of EU. Consequently, they began to attend school when the values of EU had become popular.

These young people have grown up along with the development of Latvia, included in EU unlike the previous high school students. Therefore, it is interesting to explore the possible differences in adolescent's values of previous years and nowadays (among the high school students of 1998 and 2010). After some time, they will develop their own approach and develop the country according to their values. Therefore, it is important to study the values just of high school students to correct them if it is necessary.

Nowadays, there are a lot of discussions around the new "I-generation", which also includes current high school students. Are the values of this new generation drastically different from those of previous generations? This is the question to answer in this research. The objective was – to find out the difference between value hierarchies of Riga students and rural students and their subjective opinions about factors affecting them, and compare the results with the values of previous generations. Research issues**:** What is the hierarchy of values among adolescents? What is the difference between the value scales of Riga and rural areas? What is the difference between the values of adolescents in 2018 and the ones of previous years?

2 Values as a basic concept of the study

As it was mentioned in the introduction of this article, the values have been studied quite a lot in various fields of science, especially in the second half of the 20th century [2–7]. However, the sciences appreciate different values – the definitions of values differ in psychology, philosophy and sociology, so there are no common definitions and approaches. It has always been a topical issue about the way how other people influence on the value system of the person. A unified vision of values was formed only after the work of American social psychologist Milton Rokeach was published [8, 9] that strongly influenced further research of values. Nowadays Rokeach methodology is used as a tool for value research in several fields of science – psychology, sociology and marketing as well. Several authors developed their tools of value research based on his methodology [10, 11].

This research also is based on the value theory of Rokeach. Rokeach defines a value as long-term confidence of an individual, that a specific mode or state of existence is more acceptable than the other modes or states for this individual and the society as well [9].

According Rokeach, the value system consists of two main groups – terminal values and instrumental values. He defines terminal values as the individual's confidence in life and the desired end state of existence that the individual tries to achieve during his life. Whereas he defines the instrumental values as the principles/opinions (such as honesty and courage) that are used to achieve the desired life goals, i.e. terminal values [9]. This approach was supported and also criticized widely by several scientists [12–18]. Richard Gorsuch was critical about the breakdown of values claiming that any value can be supposed as instrumental value if the terminal value is not absolute [12]. Shalom H. Schwartz considered that the breakdown made by Rokeach is not significant, because it makes the value system too narrow and limited [14]. Unlike Schwartz's theory that included 57 values which were divided in 10 universal core value groups, Rokeach used only 36 values; and high school students filled the Rokeach's questionnaire more gladly than the Schwartz's as the author of the article has observed. Therefore, in author's opinion the Rokeach's value questionnaire is more appropriate and more often used at the research of adolescent's values.

Generally, the value system is influenced by many factors: family, society, media and authorities, various political and socio-economic conditions [1], so it is difficult to estimate which impact is more or less significant. Young people are particularly sensitive to the impact of various external factors and social pressures, because they try to find the balance between the internal and external values, they try to set their future goals and make plans [19]. Young people are the most studied group, moreover the comparative study of different groups of young people have become even more popular in 21st century [20–22]. Therefore, this research compares adolescents of different times and looks for the answer to the question: which values are influenced and which remain unchanged in the value hierarchy of adolescents in the context of time and their background.

3 Material and methods

It is a pilot study with a comparative cross-sectional design. The research of values was based on the Rokeach value ranking test. 100 adolescents at the age of 16–18 were selected as respondents for the quantitative survey (10th to 12th grade students of 2018) from two different backgrounds, 50 from each: students from Latvian schools of Riga (hereinafter called group "Riga") and students from rural schools (hereinafter called group "Rural"). The study was carried out in several schools of Latvia in 2018: 2 schools of Riga and 2 rural schools of Vidzeme region. Participants were 11th and 12th grade students, 25 from each school, 50 of them were from Riga and 50 from rural schools.

These data were compared with the selected data about 74 high school students that were collected in 2010, and the data about 110 high school students that were collected in 1998. Accordingly, there was obtained the information about the impact of various economic, political and social factors on young people values.

All groups of respondents included boys and girls, but their proportion was not stated, because it was not planned to study differences by gender. The statistical analysis of data was carried out using SPSS program and the values were compared using the Mann–Whitney U test and Kruskal–Wallis Z test.

Table 1. Comparison of terminal values of adolescent groups in 2018 according Mann–Whitney test.

Type of value	***Mdn (IQR)***		Mann–Whitney test	
Terminal values	**2018-Riga1**	**2018-Rural**	z	p
Active life	7(5–11)	7(4–12)	−1.363	0.173
Wisdom of life	8(4–11)	7(4–12)	−0.097	0.923
Physical and mental health	4(2–8)	4(2–6)	−0.177	0.860
Interesting job	11(8–14)	11(7–14)	−0.062	0.950
Beauty of nature and art	16(13–17)	16(13–18)	−1.090	0.267
Love	3(2–5)	5(2–10)	−2.043	0.016
Comfortable life	11(4–14)	9(5–13)	−0.377	0.707
Good and true friends	6(4–8)	7(4–11)	−1.193	0.233
Public appreciation and respect	15(12–16)	14(9–16)	−1.265	0.206
Cognition	12(7–15)	10(6–13)	−1.652	0.099
Productive life	9(5–12)	10(7–13)	−1.244	0.213
Development	8(6–12)	8(4–11)	−0.066	0.948
Life rich in enjoyment and fun	15(8–18)	15(10–16)	−0.108	0.914
Freedom	8(5–10)	9(5–11)	−0.668	0.504
Happy family life	2(1–6)	3(1–8)	−1.040	0.298
Happiness of other people	15(13–17)	15(11–17)	−0.392	0.695
Creativity	15(11–17)	13(9–16)	−1.280	0.200
Self-confidence	9(6–14)	10(5–13)	−0.539	0.590

The Latvian version of Rokeach Value Survey was used to perform the analysis of quantitative data (Rokeach Value Survey [9], it was adapted in Latvian by Karpova in 1994 [23]) that consisted of two parts: terminal and instrumental values. Each part included 18 values that students had to arrange according to their subjective importance and estimate (in percentage) how much this value has already been realized. Finally, each value got its position in the hierarchy. Rokeach Value Survey was used also in the researches of 1998 and 2010, so the results are comparable.

In the beginning the terminal and instrumental values were compared between both respondent groups of 2018. The statistical non-parametric Mann–Whitney test was used to find which terminal or instrumental values are statistically significant (Tables 1 and 2).

4 Results and discussion

4.1 Comparison of terminal and instrumental values of adolescent groups in 2018

Mann–Whitney test (Table 1) showed that the only statistically significant difference between adolescents of Riga and Rural groups is according the value "love" ($z = -2.043$, $p = 0.016$). It means that "love" (as mental and physical intimacy with the beloved person) is more important value for adolescents of Riga than for rural adolescents. All other values are equally important for all adolescents in 2018, regardless of where they live.

Three most important terminal values for both Riga and rural students from high schools are "happy family life", "physical and mental health" and "love" (Riga students preferred "love" to "health", but rural students had the opposite opinion). The less important values appeared to be "happiness of other people" and "beauty of nature and art".

Table 2. Comparison of instrumental values between Riga and Rural students in 2018 according Mann–Whitney test.

Type of value	*Mdn (IQR)*		Mann–Whitney test	
Instrumental values	**2018–Rīga**	**2018–lauki**	*z*	*p*
accuracy	11(6–14)	11(6–13)	−0.080	0.937
upbringing	6(4–11)	5(2–9)	−1.960	0.050
high claims	14(10–17)	15(12–17)	−1.206	0.228
joy of life and sense of humour	5(2–8)	8(5–12)	−2.691	0.007
conscientiousness	12(7–15)	8(6–12)	−2.334	0.020
independence	8(3–13)	9(5–13	−0.954	0.340
intolerance towards their own and other's deficiencies	17(12–18)	17(15–18)	−0.038	0.969
education	7(4–9)	10(5–13)	−2.518	0.012
reliability	9(4–12)	4(3–9)	−2.461	0.014
rationalism	10(5–12)	10(5–16)	−1.434	0.152
self–control	8(5–12)	9(3–14)	−0.262	0.793
courage	13(8–15)	12(6–14)	−0.641	0.522
strong will	7(3–13)	8(4–12)	−0.159	0.874
patience	6(8–12)	9(5–11)	−0.263	0.793
breadth of views	10(5–14)	11(7–15)	−0.594	0.552
honesty	6(1–12)	6(2–11)	−0.007	0.994
efficiency at work	11(8–14)	10(7–13)	−0.657	0.511
responsiveness, ingenuity	10(5–15)	8(5–11)	−1.368	0.171

Mann–Whitney test showed (Table 2) that the statistically significant differences between adolescents of Riga and Rural groups are only according following values: "upbringing" ($z = -1.960$, $p = 0.050$), "joy of life and sense of humour" ($z = -2.691$, $p = 0.007$), "conscientiousness" ($z = -2.334$, $p = 0.020$), "education" ($z = -2.518$, $p = 0.012$), "reliability" ($z = -2.461$, $p = 0.014$). It means that "joy of life" and "education" are more important values for adolescents of Riga than it is for rural adolescents. In turn rural adolescents appreciate "upbringing", "conscientiousness" and "reliability" higher than adolescents of Riga. All other values are equally important for all adolescents in 2018, regardless of where they live.

5 Comparison of terminal values of 11th and 12th grade students by years

As the previous research of values was carried out with 11th and 12th grade students from Riga and rural regions, also **comparison of terminal values by years** regarding 2018 was represented by joint group "2018" that includes adolescents from group "Riga" and group "Rural". The Kruskal–Wallis rank test (Table 3) was performed to check if there are statistically significant differences between terminal values of Latvian adolescents of Riga in 1998, 2010 and 2018.

Kruskal–Wallis test shows that the statistically significant differences between all groups are regarding only following values: "Wisdom of life", "Interesting job", "Comfortable life", "Public appreciation and respect" (from surrounding people, school mates), "Life rich in

Table 3. Comparison of terminal values by years according Kruskal–Wallis test.

Type of value	*Mdn (IQR)*			Kruskal–Wallis test	
Terminal values	**1998**	**2010**	**2018**	χ^2	*p*
Active life	8(5–11)	8(5–12)	7(5–11)	1.135	0.567
Wisdom of life	10(5–13)	9(4–13)	8(4–11)	8.043	0.018
Physical and mental health	5(2–10)	5(2–9)	4(2–8)	1.508	0.471
Interesting job	8(5–11)	9(5–12)	11(8–14)	16.378	0.000
Beauty of nature and art	16(14–17)	14(10–17)	16(13–17)	8.908	0.012
Love	4(2–7)	4(2–8)	3(2–5)	0.953	0.621
Comfortable life	6(3–11)	7(4–13)	11(4–14)	7.837	0.020
Good and true friends	4(3–7)	7(4–12)	6(4–8)	17.421	0.000
Public appreciation and respect	9(6–13)	12(8–15)	15(12–16)	33.827	0.000
Cognition	10(6–12)	11(7–14)	12(7–15)	3.743	0.154
Productive life	11(6–14)	9(6–13)	9(5–12)	1.802	0.406
Development	8(5–11)	8(5–12)	8(6–12)	0.343	0.843
Life rich in enjoyment	16(12–18)	11(7–16)	15(8–18)	16.417	0.000
Freedom	9(5–12)	9(5–14)	8(5–10)	0.869	0.648
Happy family life	6(3–13)	9(3–14)	2(1–6)	40.459	0.000
Happiness of other people	15(13–17)	15(11–17)	15(13–17)	0.744	0.689
Creativity	15(11–17)	14(8–17)	15(11–17)	3.939	0.149
Self–confidence	9(5–13)	10(6–14)	9(6–14)	1.048	0.586

enjoyment and fun" (life without duties), "Happy family life", "Beauty of nature and art" (enjoying it), "Good and true friends".

Comparing groups of respondents in pairs (Mann–Whitney test) it was found that there are statistically significant differences between the results of 2018 and 2010 regarding following values: "Interesting job" ($z = -2.356$, $p = 0.018$), "Beauty of nature and art" ($z = -2.411$, $p = 0.016$), "Public appreciation and respect" ($z = -2.729$, $p = 0.006$), "Life rich in enjoyment and fun" ($z = -2.349$, $p = 0.019$) and "Happy family life" ($z = -5.183$, $p = 0.000$).

It means that such values as "Interesting job", "Beauty of nature and art", "Public appreciation and respect" and "Life rich in enjoyment and fun" have become less important for adolescents in 2018 than it was for adolescents in 2010, though the importance of the value "Happy family life" has increased.

There are following statistically significant differences between the indicators of the value importance in 2018 and 1998: "Wisdom of life" ($z = -2.906$, $p = 0.004$), "Interesting job" ($z = -4.062$, $p = 0.000$), "Comfortable life" ($z = -2.768$, $p = 0.006$), "Good and true friends" ($z = -2.857$, $p = 0.004$), "Public appreciation and respect" ($z = -5.728$, $p = 0.000$) and "Happy family life" ($z = -5.698$, $p = 0.000$).

It proves that such values as "Wisdom of life" and "Happy family life" are more important for adolescents of 2018 than for adolescents of 1998. Whereas the values "Interesting job", "Comfortable life", "Good and true friends" and "Public appreciation and respect" seem less important than they were in 1998.

It means that comparing with the values of adolescents of 1998 and 2010 the only value, that has become more important for adolescents of 2018, is "Happy family life", but the values "Interesting job" and "Public appreciation and respect" have lost their position.

During studies of value changes of 1998, it was stated in 2015 [24] that such values of adolescents of 1998 as "Family" and "Job" have become more important in 2015, but the value "Friends" has lost its importance. It means that during 20 years the value "Family" has become more important for the new generation of high school students and for the dynamics of values of one generation as well.

Comparing the obtained data with the data of the 2015 study [25], when the data of 1998 were compared with the data of 2005, 2010 and 2015, it is evident that the value "Wisdom of life" in 1998 was of less importance than it was in 2015 and 2018, but it was as important as in 2005 and 2010. So, the attitude of young people has changed after 2010.

On the other hand, the dynamics in relation to the value "Interesting job" is reversed – it has become less important for adolescents after 2010. The banking crisis in 1995, and then the economic crisis in 2008 made many people feel threatened with losing their jobs – so the value "Interesting job" was important for young people in 1998 and 2010 as well. When the situation returned to normal, the importance of the value "Interesting job" decreased in 2018. As to the values "Good and true friends", "Public appreciation and respect" and "Productive life" – they have become less important since 2005.

The value "Life rich in enjoyment and fun" was on less importance in 1998 than it was in 2015, but it was the same as in 2018. Thus, it can be assumed that the importance of this value increased from 1998 to 2010 and after that it decreased gradually, in 2018 reaching the level of 1998. But the value "Happy family life" has the opposite dynamics – it loses its significance in the period of 2005–2010, and after that regains it gradually becoming much more important for adolescents in 2018 than in 1998. As for "Freedom", it becomes more important for young people in 2015, but in 2018 it returns to the level of previous years.

There are statistically significant differences between the groups of 1998 and 2010 regarding following values: "Beauty of nature and art" ($z = -2.851$, $p = 0.004$), "Good and true friends" ($z = -3.947$, $p = 0.000$), "Public appreciation and respect" ($z = -2.786$, $p = 0.005$), "Life rich in enjoyment and fun" ($z = -3.996$ $p = 0.000$). In addition the value "Public appreciation and respect" has become less important in 2010 and gradually loses its position in 2018 (it takes 6th–13th place of the individual hierarchies in 1998, 8th–15th place in 2010 and 12th–16th place in 2018). The other values do not show such dynamics. It is possible that this value has lost its significance because of the development of social network that may have created certain "immunity" to public appreciation.

On the other hand, the values "Active life", "Physical and mental health", "Cognition" (possibility to expand horizons, education, and cultural level), "Productive life" (maximum use of abilities, strengths and capabilities, "Development" (working with yourself, constant physical and mental development), "Freedom" (independence in thinking and judgment), "Happiness of other people" (welfare of other people, the whole nation), "Creativity" (creative work opportunity) and "Self-confidence" (internal harmony, the removal of internal contradictions and doubts) are equally significant for all young people in 1998, 2010 and 2018.

But the values "Health", "Productive life", "Development", "Freedom", "Other people happiness" and "Creativity" are equally important for all adolescents in 1998, 2010 and 2018, regardless of where they live – in Riga or in the countryside. It means the significance of these values is not affected by time and socio-economic-political situation. When analysing individual value hierarchy, it can be noticed that the value "Health" takes 1st to 10th place, "Productive life" – 5th to 14th place, "Development" – 4th to 12th place, "Freedom" – 4th to 14th place, "Happiness of other people" – 11th to 17th place and "Creativity" – 8th to17th place. It means that in terms of value importance, individual differences are greater than the differences between different groups of young people of different years.

Table 4. Comparison of instrumental values by years according Kruskal–Wallis criterion.

Type of value	*Mdn (IQR)*			**Kruskal–Wallis test**	
Instrumental values	**1998**	**2010**	**2018**	χ^2	***P***
accuracy	9.5 (6–14)	10.5 (8–15)	11(6–13)	1.451	0.484
upbringing	9 (4–12)	9 (5.8–14.3)	5(2–10)	13.617	0.001
high claims	16 (10.8–17)	15 (10–17)	15(11–17)	0.226	0.893
joy of life and sense of humour	7 (3–10.3)	5.5 (2–12)	6(3–11)	0.203	0.904
conscientiousness	13 (9–16)	13.5 (9.8–16)	10(6–14)	11.006	0.004
independence	7.5 (3–13)	6 (3–10.3)	8(4–13)	1.579	0.454
intolerance towards their own and other's deficiencies	17 (14–18)	17 (12.8–18)	17(12–18)	4.415	0.110
education	6 (3–9)	7 (3–11)	8(4–12)	10.744	0.005
reliability	6 (3–10.8)	8 (5–11.3)	7(3–12)	2.322	0.313
rationalism	9 (5–13)	9 (5–13)	10(5–14)	0.975	0.614
self-control	8 (4–13)	9 (5–13.3)	9(4–13)	1.407	0.495
courage	10 (6–14)	10 (4–13)	13(8–14)	5.617	0.060
strong will	5 (3–9)	5 (3–11)	8(4–13)	8.826	0.012
patience	10 (5–14)	10 (6–13)	8(5–12)	2.536	0.281
breadth of views	9.5 (6–13)	9 (4–13)	10(7–14)	2.218	0.330
honesty	6.5 (2–11)	7 (3–10.3)	6(2–11)	0.323	0.851
efficiency at work	13 (9–15)	10 (5.8–13)	11(8–14)	10.882	0.004
responsiveness, ingenuity	11.5 (8–14)	9 (7–14)	9(5–14)	8.000	0.018

6 Comparison of instrumental values of 11th and 12th grade students by years

The Kruskal–Wallis rank test (Table 4) was performed to check if there are statistically significant differences between instrumental values of Latvian adolescents of Riga in 1998, 2010 and 2018.

In Table 4 it was shown that there are statistically significant differences in all groups only regarding following values: "Upbringing", "Conscientiousness", "Education", "Strong will", "Efficiency at work" and "Responsiveness".

Comparing groups of respondents in pairs (using Mann–Whitney criterion) it was found that there are statistically significant differences between the results of 2018 and 2010 only regarding one value – "Upbringing" ($z = -2.170$, $p = 0.030$). It means that the value "Upbringing" has become more important for adolescents of 2018 than for adolescents of 2010, and there is no value that is less important than in 2010.

There are statistically significant differences between the indicators of value importance in 2018 and 1998 regarding following values: "Upbringing" ($z = -3.702$, $p = 0.000$), "Conscientiousness" ($z = -3.350$, $p = 0.001$), "Education" ($z = -3.004$, $p = 0.003$), "Courage" ($z = -2.418$, $p = 0.016$), "Strong will" ($z = -2.689$, $p = 0.004$), "Efficiency at work" ($z = -2.539$, $p = 0.011$) and "Responsiveness" ($z = -2.563$, $p = 0.010$).

It means that the values "Upbringing", "Conscientiousness", "Efficiency at work" and "Responsiveness" are supposed more important for adolescents of 2018 than for adolescents of 1998. On the other hand, the values "Education", "Courage", "Strong will" seem to be less important than in 1998.

Table 5. Main terminal values and their positions in the value hierarchy in different groups of respondents.

Group	1st place	2nd place	3rd place	4th place	5th place
1998	Love	Friends	Health	Happy Family	Happy family
2010	Love	Health	Friends	Comfortable life	Development
2018 (total)	Happy family	Health	Love	Friends	Active life
2018-Rīga1	Happy family	Health	Love	Friends	Active life
2018- Rural	Happy family	Health	Love	Active life/ Wisdom of life	Development

It follows from the above that the only value that has become more important for young people in 2018 compared to both young people in 2010 and 1998 is "Upbringing".

Both groups point out "reliable friends" as the most accomplished value. Adolescents of both groups note that their values are most influenced by family and school; while in the third-place rural youth mention opinion leaders, but youth from Riga – social media.

The opinions of adolescents and experts about the values important to youth differ, but they rank the value-influencing agents similarly. There is diversity among expert opinions about the values and their hierarchy that are topical to the youth that could be explained as manifestation of generation gap. Experts highlight the value "material well-being" as the most important for adolescents and emphasize the impact of social media on its development.

6.1 Changes in the hierarchy of terminal and instrumental values over the years

If we look at the main terminal values of all groups (Table 5), it is evident that the leading five values in 1998 are as follows: "Love", "Friends", "Health", "Comfortable life" and "Happy family life".

This tendency was identified also during the studies in 1998, 2005 and 2010. These values are also mentioned as leading in the author's research performed among high school students [25–30], among university students [31]and also at transnational studies [32]. These data also coincide with the research carried out with university students by Professor of psychology Ārija Karpova [23]. Similar results were obtained by Spodra Austruma in her promotion work about young people values in consumer society, where it was stated that dominant values are "Family", "Friends", "Health", "Education", "Love", "Career" and "Freedom" [33].

According to the data of this research "Happy family" is replaced by "Development" in 2010; but "Comfortable life" is replaced by "Active life" in 2018. In addition, "Happy family life" rivals "Friends" and "Comfortable life" in the hierarchy of adolescents of 2018. Apart from this, "Wisdom of life" for rural adolescents and "Self-confidence" for Russian speaking adolescents have achieved unusually high ranking in the hierarchy. These results may be explained by the background differences. To survive in the rural area people rely more on knowledge accumulated in their lives and from their ancestors; so this can be also called "Wisdom of life", but there are no statistically significant differences.

At the request of the Ministry of Education and Science, Excolo Latvia Ltd performed a study on the values of young people in Latvia in 2012–2013 and found following significant values of young people: "Domesticity", "Stable job", "Interesting job", "Happy life with the partner", "Well paid occupation", "Good education" and "Ability to provide a safe future for their children". On the other hand, such values as "Interest about politics", "Doing something for the public good" and "Learning about other cultures" were appreciated much less [34].

Table 6. Main instrumental values and their positions in value hierarchy of different respondent groups.

Group	1st place	2nd place	3rd place	4th place	5th place
1998	Imagination	Intellect	Responsibility	Honesty	Cheerfulness
2010	Imagination	Cheerfulness Independence	Honesty	Intellect	responsibility
2018 (total)	Obedience	Honesty	Cheerfulness	Responsibility	Imagination
2018-Rīga1	Cheerfulness	Honesty	Obedience	Intellect	Independence
2018- Rural	Obedience	Responsibility	Honesty	Strong will	Helpfulness

The values related to work and money do not rank high in the hierarchy, but "Work for others" ("Welfare of others" in this research) is one of the less important values in all respondent groups during 20 years.

This study shows only trends of changes in values of high school students during the last 20 years both in cross-sectional and longitudinal terms. The degree of impact of media, economic and political situation, digitization and globalization processes on the dynamics of these changes in values have to be studied in the future research. It would be worth exploring the media of young people of the related years, the changes in main messages in them and performing the qualitative study of young people values.

7 Conclusions

Individual hierarchies of values in one group context differ more than hierarchies of adolescent groups of various years and different backgrounds. There are no significant differences related to the terminal and instrumental values between adolescents of Riga and rural areas. Therefore, it can be concluded that there is no significant differences whether to study the values of adolescents in Riga or in rural areas.

Some values ("Health", "Productive life", "Development", "Freedom", "Happiness of other people" and "Creativity") have not been influenced by time or socio-economic and political situation, but there are also values with changing significance that increases ("Happy family life") or decreases ("Interesting job", "Public appreciation and respect") with years: Values "Active life", "Cognition" and "Self-confidence" are equally significant for all Latvian speaking adolescents in 1998, 2010 and 2018; "Happy family life" is the only value that has become more important for adolescents of 2018 comparing with adolescents of 2010 and 1998. On the other hand, adolescents of 2018 estimate such values as "Interesting job" and "Public appreciation and respect" as less important; Value "Public appreciation and respect" has become less important in 2010 and even much less important in 2018, comparing with 1998. Adolescents of 2018 estimate "Wisdom of life" and "Happy family life" higher than adolescents of 1998, but they underestimate such values as "Interesting job", "Comfortable life", "Good and true friends" and "Public appreciation and respect"; Values "Interesting job", "Beauty of nature and art", "Public appreciation and respect" and "Life rich in enjoyment and fun" have become less important for adolescents of 2018, but "Happy family life" has become much more important comparing with 2010.

Comparing the significance of the values of Riga and rural students we found out that love, cheerfulness and education are more important for Riga students, but rural students prefer responsibility, obedience and capability. The greatest difference in instrumental values is between 1998, 2010 and 2018, not between 2010 and 2018. "Wisdom" and "Strong will" were more important in 1998 compared with 2010 and 2018. "Obedience" is more important in 2018, compared with 2010 and 1998. "Happiness of other people" is the only value whose importance has not changed in the period of 20 years. Values "Wisdom of life", "Interesting job", "Comfortable life", "Public appreciation and respect" and "Life rich in enjoyment and fun" are equally important for all adolescents of 2018, regardless of where they live, but their importance was different for adolescents of previous years.

"Love", "Friends" and "Health" took first places in the hierarchy of terminal values, by 2010. But the value "Happy family life" competes with "True friends" and "Health" in the hierarchy of adolescents of 2018 and gets the first place among the selection of Latvian students. Young people aged 17–19 prioritize classical or terminal values (family, love, health), however there are slight differences between Vidzeme region and Riga city.

References

[1] I. Koroieva, Jauniešu vērtības un nākotnes plāni. *Latvijas jaunatnes portrets: integrācija sabiedrībā un marginalizācijas riski*. Rīga: LU Akadēmiskais apgāds, 189–210, 1pp (2009)

[2] C.K.M. Kluckhohn, Values and value orientations in the theory of action. In T. Parsons, E. Shils (Ed.), *Towarda General Theory of Action* (Cambridge, MA, Harvard University Press, 1951), pp. 388–433

[3] D. Katz, E. Stotland, A preliminary statement to a theory of attitude structure and change. In S. Koch (Ed.), *Psychology: A Study of a Science* (New York, McGraw-Hill, 1959), pp. 423–475

[4] G.W. Allport, P.E. Vernon, G. Lindzey, *Astudy of Values* (Boston, Houghton, Mifflin Co. 1960)

[5] M.B. Smith, Personal values in the study of lives. In R. W. White (Ed.). *The Study of Lives* (Atherton Press, New York, 1963), pp. 324–347

[6] H. Fallding, A proposal for the empirical study of values Am. Sociological Rev. **30**, 223–233 (1965)

[7] R.M. Williams, The concept of values. In E. Sills (Ed.) *International Encyclopedia of the Social Sciences* (New York, Macmillan, 1968), pp. 283–287

[8] M. Rokeach, *Beliefs, Attitudes and Values* (San Francisko, Jossey-Bass, 1968)

[9] M. Rokeach, *The Nature of Human Values* (Free Press, New York, 1973)

[10] S.H. Schwartz, Universals in the content and structure of values: Theoretical advances and empirical tests in 20 countries. In M.P. Zanna (Ed.) *Advances in Experimental Sochology*. Vol. 25 (San Diego, Academic Press, 1992), pp. 1–65

[11] Е.Б. Фанталова, Об одном методическом подходе к исследованию мотивации и внутренних конфликтов. *Психологический журнал*. Т.13., н1, с. 107–117(1992)

[12] R.L. Gorsuch, R.L. Rokeach's approach to value systems and social compassion. Rev. Religious Res. **11**, 139–143 (1970)

[13] R. Lynn, Review of "The nature of human values" by M. Rokeach. Br. J. Psychol. **65**, 453 (1974)

[14] S. Schwartz, Review: A Survey of Guiding Principles. Reviewed work(s): The Nature of Human Values by Milton Rokeach. In Sci. New Ser. **186**(4162), 436–437 (1974)

[15] D.M. Keats, J.A. Keats, Review of "The nature of human values" by M. Rokeach. Aust. J. Psychol. **26**, 164–165 (1974)

[16] T.M. Kitwood, A.G. Smithers, Measurement of human values: An appraital of the work of Milton Rokeach, Educational Res. **17**, 175–179 (1975)

[17] R.A. Jones, J. Sensening, R.D. Ashmore, Systems of values and their multidimensional representations, Multivariate Behav. Res. **13**, 255–270 (1978)

[18] V.A. Braithwaite, H.G. Law, Structure of Human Values: Testing the Adequacy of the Rokeach Value Survey. J. Personality and Soc. Psychol. **49**(1), 250–263 (1985)
[19] G.J. Craig, *Human Development* (New Jersey, Prentice Hall, 1996)
[20] S. Karvonena, R. Young, P. West, O. Rahkonen, Value. orientations among late modern youth – a cross-cultural study. J. Youth Stud. **15**(1), 33–52 (2012)
[21] N. Gorlova, L. Romanyuk, L. Vanbrabant, R. van de Schoot,. Eur. J. Dev. Psychol. **6**, 744–750 (2012)
[22] A. Doring, S. Schwartz, J. Cieciuch, P. Groenen, V. Glatzel, J. Harasimczuk, N. Janowicz, M. Nyagolova, R. Scheefer, M. Allritz, Cross-cultural evidence of value structures and priorities in childhood, Br. J. Psychol. **106**(4), 675–699 (2015)
[23] A. Karpova, *Personība un Individuālais Stils* (Rīga, LU, 1994)
[24] M. Sokola-Nazarenko, K. Martinsone, S. Mihailova, J. Levina, E. Karsa, The Dynamics of Value System in 1998 and 2015: Longitudinal Research in Latvia. *6th International Interdisciplinary Scientific Conference "Society. Health. Welfare"*. Available at: `https://www.shs-conferences.org/articles/shsconf/pdf/2018/12/shsconf_shw2016_01008.pdf` (2018)
[25] J. Levina, K. Martinsone, S. Mihailova, I. Gintere, Proc. Int. Sci. Conf. **3**, 577–587 (2015)
[26] S. Jirgena, *Jauniešu pašizjūta un adiktīvas uzvedības izvēle*. Promocijas darbs (Latvijas Universitāte. Rīga, 1999) 78 lpp.
[27] S. Jirgena, K. Mārtinsone, Vērtību hierarhija: jauniešu salīdzinošais vērtējums 1998. un 2005. gadā. *Sabiedrība un kultūra. Rakstu krājums IX.* A. Medveckis (Sast.) (Liepāja, LiePa, 2007, 2007) 112–119 lpp.
[28] С. Йиргена, И.Я. Михайлов, Динамика ценностных ориентаций в структуре личностных характеристик молодёжи Латвии. *Межкультурная коммуникация: концепты и модели поведения.* – Астрахань: Издательский дом "Астраханский университет". с. 191–195 (2007)
[29] С. Михайлова, Сотрудничество как ценность: сравнительная оценка латвийской молодёжи в 1998 и 2009 году. *Partnerstwo i wspolpraca a kryzys gospodarczo-spoleczny w Europie srodkowej i wschodniej.* – Lublin: Wydawnictwo. т.1., с. 607–615 (2010)
[30] S. Mihailova, Динамика ценностей латвийской молодежи во время кризиса. *Starptautiskās starpdisciplinārās zinātniskās konferences "Krīze un tās pārvarēšanas iespējas" tēžu krājums.* Rīga: RPIVA, RSU, 36–37 lpp. (2011)
[31] S. Mihailova, O. Rasnačs, Vērtību hierarhijas izpēte: Rīgas Stradiòa universitātes 1.kursa dažādu programmu salīdzinošais vērtējums 2010. gadā. *RSU 2011.gada zinātniskā konference. Tēzes.* (Rīga, Rīgas Stradiòa universitāte, 2011), 409 lpp.
[32] S. Mihailova, V. Perepjolkina, Krievvalodīgo skolēnu vērtības: Baltijas valstu salīdzinājums. *RSU 2018.gada zinātniskā konference.Tēzes.* Rīga: Rīgas Stradiòa universitāte, 295 lpp. (2018)
[33] S. Austruma, Rēzekne: Rēzeknes Augstskola, *Jauniešu vērtības patērētāj sabiedrībā Latvijā: Promocijas darbs* (2012)
[34] Excolo Latvia *Jauniešu iespēju, attieksmju un vērtību pētījums.* Available at: `http://www.izm.gov.lv/images/statistika/petijumi/23.pdf` (2013)

Significant factors of violence against women in urban and rural community

Marketa Rusnakova[1] and *L'udmila* Huskova[2]

[1] Catholic University in Ružomberok, Ružomberok, Slovakia
[2] Police of Slovak Republic, Martin, Slovakia

Abstract. Introduction. The research is focused on the issue of domestic violence against women. It deals with the definition of violence, concepts and models of individual areas by summarising the research findings and the possibilities of methodological tools for measurement of significant factors of violence against women. **Aim.** The primary goal of the research is to define significant factors which determine violence against women. **Material and Methods.** Methods of research – survey, document analysis and database statistical analysis. **Results.** Primary results of quantitative research reveal that women of all ages and educational groups are at risk. Mostly women aged 26–45, living in rural areas with middle school education without maturity exam (final exam). Violence in families involves men of all age groups with emphasis on increased incidence in specific categories: mostly men aged 36–45, living in rural areas with secondary (high school or comprehensive school) education. Another field of interest represents analysis of negative family situations, impact of violence against woman on her psychological conditions, and reasons why victims carry on marriage and/or partnership. **Conclusions**. It can be concluded that the most frequent negative phenomena during aggressor's and/or victim's childhood were physical violence of father against mother, alcohol – possibly father's alcoholism, poverty, physical punishment of children, and patriarchal model of the family. Two thirds of victims continue marriage/partnership in order to maintain the family, to keep the father for children, and for financial reasons. There is a frequent social criminality in poor biological families of aggressors, i.e., a commission of crime because of lack of finances.

Key words: significant factors, violence against women, mistreatment, victim, aggressor.

1 An overview of research activities focused on violence committed against women in the Slovak Republic

In the past research works conducted in the field of violence committed against women were solved on the agenda with the variable intensity. In our view, the present topic is currently being addressed more intensively by both governmental and non-governmental institutions. The history of domestic violence research or research of violence committed against women in the Slovak Republic (SR) is much shorter than history in western democracies, where the similar researches have been conducted since the late sixties of the 20th century. Initially, just partial explorations were realized in the Slovak Republic, which were only tangentially related with the issues, or there was some unrepresentative probe. We have chosen from the research activities in SR in nineties of the 20th century: The exploration of Democratic Union of Women of Slovakia "An analysis of intolerance towards women" (1994), The survey of the

International Centre for Family Studies (ICFS) "Family 1995" (1995), The Exploration of ICFS "Women about themselves" (1997), The state of violence committed against women in SR, The exploration Pro Families and the Centres of Counselling and Psychological Services "Beaten women in Slovakia" (1997), The Exploration of The International Organization for Migration (IOM) "The trafficking of women in SR" (2000) [1].

A good example of this is regular empirical sociological research of the Institute for Public Affairs in the years 1995, 2000, 2002 and the exploration of the group of women non-governmental organisations (MVO) in the year 2001 within the campaign "The fifth woman". The first press monitoring was realized in 1999 within the project Ester Mirror which was supported by Canada Fund in SR. The monitoring was prepared by interest association of women Fenestra with its registered office in Košice [1].

In the years 2001–2002, the International Centre for Family Studies widely tried to process information about the domestic violence cases that were in records of the Centres of Counselling and Psychological Services in SR financially supported by Friedrich Ebert Foundation, representation in SR. The research was based on clinical sample of advisory centres' clients (both men and women) who were aware of being violence victims [2]. During the period of Czechoslovak Socialist Republic and the period of federal government as well as the first 12 years of independent Slovak Republic, the issue about violence committed against women within the country was not specially researched and monitored from the legal perspective.

The first systematic research under the patronage of General Prosecutors Office of the Slovak Republic has been "*The Evaluation of using criminal law means against the domestic violence towards women and children*" ("The Evaluation") published on 30 June 2002 (General Prosecutors Office of the Slovak Republic, online). The Evaluation has proceeded from the statistical data of the all regional prosecutor's offices in the district of General Prosecutors Office of SR in the years 1999–2001. For example, the prosecution started for 1.614 acts of domestic violence against women in the year 2001, which is a rise of 5.6% than in 2001 and 14.3% more than in 1999. Apart from the quantitative research, the results have been subsequently evaluated with the proposal of consequences for more effective elimination of this widespread form of violence [2].

Slovakia was one of the last countries where the representative research was carried out, although it has not been done from the state budget resources. The first representative research was conducted by women employees of the Centre for Work and Family Studies in 2002. The research was comprised of two main parts. The first part of the research under the title of *"Domestic violence – experience and attitudes of inhabitants of SR"* focused on occurrence of various forms of domestic violence in Slovakia, measuring the attitudes and tolerance of Slovak society to domestic violence as well as the level of willingness to intervene in similar situations [2]. The second part of the research under the title of *"The violence committed against women – experience and attitudes of women in SR"* focused on finding out the occurrence, frequency, extent and essence of the violence committed against women in SR [2]. After 2002, on the initiative of the National plan of action 2005–2008 the monitoring of the organizations which provide help for victims of domestic violence was realized (2006) and five years later the second *"Representative research of the occurrence and women experience of violence committed against women (VAW) in Slovakia"* [3]. The Department of Gender Equality was the contract owner securing finances too. It was conducted by the Institute for Labour and Family Research (ILFR). These are some of the main findings of the research:

- ***There is every fifth woman in Slovakia:*** every fifth adult woman, who currently has a partner, has experienced violent behaviour from his side – to be more precise 21.2% of adult women in Slovakia aged 18–64 years, who had a partner at the time of research.

had experienced an act of violence from the partner (it was pronounced violence in 6% of the cases, it means often repeated physical and sexual violence and almost constantly present mental, social or economical violence).

- ***The incidence of the violent behaviour against women committed by former partners is even higher:*** of adult women who had a former partner, 27.9% had experienced an act of violence, of whom 12.2% had experienced pronounced violence.
- ***There is no type or form of violence against women in intimate relations that would be completely unknown to Slovak women:*** all known forms of violence have occurred – from the physical and sexual violence, through mental and social up to economical violence (physical violence has been experienced from the present partner by 15% of women and from the previous partner by almost 40% of women, 9% of women have had an experience with the sexual violence from the present partner and 25% of women have experienced the sexual violence from the previous partner).
- ***Women in Slovakia are rarely exposed to only one form of "partner's" violence:*** physical violence is very often preceded or accompanied by mental or social violence, sexual violence is very often connected with various forms of mental or social violence, economical violence is closely linked to mental violence etc.
- ***The relationships with the present violence against women are not random relationships:*** it is usually a bond of marriage (violent relationship –76% of women who are in their first marriage, clearly defined violent relationship almost 80% of women in their first marriage) and a long-standing relationship (on average more than 20 years´standing).
- ***Children are often witnesses to violence:*** almost 70% of women who have experienced violence from their present partner live together with their children or a child.
- ***The violence in intimate relationships is experienced by women of all age groups and with increased occurrence in a particular group:*** the violence occurs among women of all age groups but it is more frequent among middle-aged women (45–60 years); the violence affects women with all levels of education, however, the biggest group is comprised by women with elementary education and with apprenticeship; economically active women, as well as economically inactive women live in the violent relationships, from the group of economically active women there are mainly manual women workers and skilled women workers, from economically inactive women the highest percentage of women are retired women; according to the size of residence there are mainly women living in the smallest municipalities up to 2000 inhabitants.
- ***All groups of men participate in the violent relationships, with an emphasis on higher occurrence in particular categories:*** all age groups of men commit violence in intimate relationships, but violence is more frequent among older men (45 years and more); violent men could be found in every educational category, however there is a higher rate of committing violence among men with elementary education and with an apprentice school education without school leaving examination; there are all groups of economically active men, however blue-collar workers dominate as well as men who are out of labour market" [3], p. 10–12.

"Monitoring of the media and advertising from the viewpoint of violence committed against women and gender equality support in 2011" dealt with chosen cases of media and marketing discourses from the point of view of women sexualization and objectification – the mechanism of women devaluation and maintaining gender stereotypes [4]. The report has brought the results of the chosen media news monitoring. A standardized methodology enables to compare media coverage and the news relevance about the violence committed against women as well as reporting of gender presentation news in the years 2009–2011 [4]. The ILFR report has offered analysis of the chosen news content about the violence

committed against women and it has revealed the mystifying and misleading signs of this issue presentation. The ILFR has realized *Monitoring of the social services for women experiencing the violence and for their children* from the viewpoint of European standards and it has shown the results of the chosen social services for abused women with the children [5]. The ILFR conducted researches within the national project *The Support of the Prevention and Elimination of Violence against Women* (ILFR, online). They were the following: Special groups of women in danger of violence – older women, Special groups of women in danger of violence – women from the rural environment, Special groups of women in danger of violence – women with an experience of being raped and the sexual violence experience [14].

The researches of violence against women provide current information about the occurrence and the character of violence against women. They have revealed new findings as well as those which confirm or deepen former ascertainment. The researches of different society areas subjects, an effort to improve legislative environment, the standardization of intervention procedures and media campaign have indicated, that violence committed against women is still present and urgent problem.

Taking the mentioned facts into consideration, it is our belief that, in the Slovak Republic it is necessary to conduct specialized research aimed at particular spheres, which are connected with violence against women factors. Our survey belonged to this category. The relevant researches are being mentioned in the chapter no. 1 of the dissertation. The results of the researches will be introduced in particular topics and the chapters about the research results and conclusions.

2 The rural environment

The violence affects a large number of women in various contexts, at any age, ethnic, religious or cultural backgrounds. There are only a few data about the extent and character of the violence that affects women in rural environment, older women or disabled women.

2.1 Women from rural environment

Regardless of numerous positive attributes associated with rural environment, the presence of violence against women is, in addition to various other forms of violence, indeed a serious problem. Due to structural changes in the economy of rural areas, there has been an increase in the share of population which is at risk of poverty or social exclusion. According to data of the Statistical Authority, it is women, as a matter of fact, who are at risk of poverty, in particular women over 65 and women who are unemployed [6]. On the basis of the selected indicators it is possible to get an idea about the social as well as economic situation of rural population. The connections among social and economic development, position of women and violence against women are becoming a topic also through policies and programmes for encouraging women in achieving full economic independence, which applies particularly to women living in rural areas. We believe that attitudes of the public to violence vary according to the region. Rural population seems to be more traditionally reasoning, more religious, more conservative and in this sense also more identifying itself with tabooing of violence against women. Arguments can be found in the representative survey of the occurrence and the experience of women with violence against women, conducted by the Institute for Labour and Family Research in 2008. A non-interfering and passive approach to addressing of violence between partners was found particularly in respondents with basic education, over the age of 60, of other nationality, living in the small municipalities up to 2000 inhabitants and the widowed [3]. In order to correctly identify rural women, it is necessary to define the meaning of countryside or rural area. Sandberg (2013) draws attention to the problem of defining rural area, as it is locally very specific geographic characteristics. In general, rural environment can be defined as an area of low population density, but even this criterion has different value in different countries [7]. Even in Slovakia there are differences in the understanding of countryside. Buchta (2012) relies on the OECD definition which speaks about double method of defining a rural area. On a local level, a rural municipality is considered to be a municipality with population density of less than 150 inhabitants per km^2 [6]. According to the Act on Municipal Establishment, a rural municipality is one with less than 5000 inhabitants. At the same time, however, the act draws attention to the fact that even in the case of number of inhabitants smaller than 5000, a municipality may be declared a city if it possesses corresponding services, transport connections as well as urban character of the built-up areas and, at the same time, it is an economic, administrative and cultural centre, a tourist centre or a spa location [12].

Our experience shows that violence against women has specific and mutually differing characteristics in the case of rural and urban environment. With regard to occurrence of violence, studies from abroad report different prevalence rates. Some of them speak of higher occurrence of violence in the cities, others, on the contrary, in the rural areas. There are also studies showing comparable occurrence of violence, however, there are factors which cause this similarity to remain hidden (e.g. Thurston, Patten, Lagendyk 2006; Peek-Asa 2011; Shuman 2008) [7–10]. Certain factors may be involved in the fact that cases of violence in the countryside are reported to a lesser extent than in the cities. The literature also refers to specific attributes of rural area which may promote violence against women as well as increase the probability of returning to the violent partner in cases of women trying to escape (e.g. Thurston, Patten, Lagendyk 2006) [8]. Another problem faced by women experiencing violence in rural areas is acceptance of violence as a part of life, or even its **tolerance** by the community. Obviously, this is not a flat-rate approach. However, due to the persisting traditional views and a higher degree of conservatism, influenced also by a different religious structure compared to the urban population, it is not rare that it is indeed rural women experiencing violence who do not find understanding [7]. **Tabooing of violence** is another problem that is more typical of rural localities and that accompanies a higher degree of conservatism and traditional understanding of gender roles. There may be a predominant belief that family problems should remain and should be dealt with in the close family circle and that it is not desirable that intimate issues be brought out and discussed outside of the family. However, one of the most significant problems of rural women experiencing violence is **geographic isolation**. Such isolation is accompanied by several adverse facts, such as, for example, poorly constructed or no network of social care facilities for women experiencing violence. The geographic unavailability of certain rural settlements also facilitates **social isolation** of women experiencing violence and easier hiding of violent behaviour [9]. Insufficient education is also connected to **poor awareness** of assistance services for women experiencing violence. According to the studies by foreign authors, a lot of women are not aware of the existence of shelters for abused women. When analysing rural environment in relation to the issue of violence against women, it is necessary to mention questions of **unemployment** or **poverty and low living standard**. Several foreign studies also prove that a poor social-economic situation is one of the factors in the case of women experiencing violence as well as in the case of offenders [7, 10].

3 Differentiations of abuse signs in the context of residence of abused women and aggressors

During years 2014–2016 we conducted a survey aimed at identification of significant factors underlying violence against women in family environment. The survey sample consisted of the total of 184 respondents, 92 victims and 92 offenders committing a crime of abuse against a close person or a person entrusted in one's care. For the purposes of the conference we chose one survey question through which we tried to identify differentiating signs of abuse in abused women and aggressors from rural and urban environment. In this context we established one survey question: Are there any statistically significant differentiations in abused women and aggressors in relation to the place of residence? Is the ratio of abused women and aggressors different based on their residence compared to population distribution in the observed region? And three hypotheses:

- H1 We assume that among victims of abuse there are significantly more women from rural areas.
- H2 We assume that among aggressors there are significantly more men from rural areas.
- H3 We assume that among victims as well as aggressors there are significantly more respondents from rural areas compared to population distribution in the observed region.

For conducting of the survey, we chose the method of **content analysis of documents**, the document files of criminal offences of abuse against a close person or a person entrusted in one's care–section 208, which had female victims. Investigation files used in our survey are from Department of Criminal Police of the District Police Headquarters in Martin (hereinafter referred to as DPH). An investigation file can be defined as "*all written documents related to the same investigated criminal case (submissions, reports, records, decisions, etc.), forming the investigation file that is opened after receipt of the first submission in the case*" [11]. With respect to the objectives and the hypotheses of the submitted work survey, the relevant documents were the following:

- records from hearings of the victims, the accused and the witnesses, the court experts, motions to press charges,
- expert opinions from department of clinical psychology and psychiatry, medical reports,
- preliminary measures, criminal orders, resolutions, verdicts and screening data.

Documents are maintained in the archives of the DPH, organized by the year of filing of a criminal complaint.

In the abused women documents we focused on the following areas of study:

- social – demographic characteristics of perpetrators of violence,
- factors influencing violence in relation to the victim and the aggressor,
- specific forms of violence, duration of violence, consequences to women's health,
- history of primary family environment of perpetrators of violence.

In order to achieve the main objective and the partial objectives, we used data obtained by analysis of the investigation files regarding abuse against a close person or a person entrusted in one's care. We presented the obtained data using descriptive statistics in the form of chart and spreadsheet presentation of data which includes numeric and percentage values. To test the significance of the differences found on the basis of information obtained from quantitative data, we used the Chi-square test (with Yates's correction at low values in the

Table 1. Percentage distribution of the survey sample based on the place of residence.

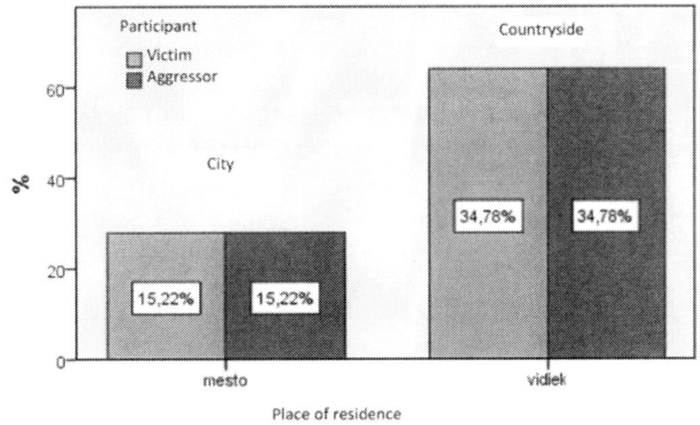

contingency table), Fisher's test, (GraphPad Instat), Shapiro-Wilk test for testing of normality of data distribution, Mann-Whitney U-test and Kruskal-Wallis test.

Table 1 shows the represented ratios of women from rural environment and from cities in our survey sample.

The hypotheses were verified by Chi-square statistical test for one choice. The test assumes that the occurrence of the observed phenomenon (residence) will be similar in each group. The results show that differences in the residence of abused women are statistically significant ($p < 0.05$).

The survey also shows that there is a statistically significant difference ($p < 0.05$) in the rate of sexual assault incidents with respect to the place of residence and the age. Victims from the cities face a higher rate of sexual assault incidents (1.21 on average) compared to victims from rural environment (0.61 on average). The average values also suggest that victims between 18 and 25 years of age face the highest rate of sexual assault incidents (1.53 on average) compared to victims between 26 to 35 years of age (0.82 years of age), between 26 and 45 years of age (0.70 on average) and victims over 46 years of age (0.08 on average). Sexual assault incidents therefore decrease proportionally with age.

Our hypotheses assumed that there are more rural respondents among victims and aggressors of abuse. This fact was confirmed, as **the majority of victims as well as aggressors live in rural areas** (64 victims and aggressors live in rural areas) ($p < 005$), despite the fact that in the towns of Martin, Turčianske Teplice and Vrútky almost 2/3 of population are adults. In comparison to population distribution in the observed region, the ratio of victims and aggressors from rural environment is significantly higher ($p < 0.05$). According to the survey by [3] women of all categories experience violence in intimate relationships, while the occurrence is higher in a specific category: with respect to the size of the residence area, these are particularly women from municipalities with population under 2,000. In the context of studying violence against women, an important attribute is locality, which is conditioned geographically and culturally. The question that remains is how much the locality, or a dichotomous relationship between the countryside and the city, represents an important attribute in identifying of abused women particularities. Studying of the relationship between countryside and gender requires detailed focus on the experience of women and men who live in this environment. In the effort to adequately capture this experience, **the affiliation to the countryside represents only one category in the entire complex of mutually overlapping categories**. Sandberg points out that the existing studies and theories, even in the intersectional context, do not look at the issue of violence against women in the context of geography and locality. He claims that "*differences that exist between urban and rural regions in relation to specific conditions and problems faced by persons experiencing violence in intimate relationships are being overlooked*" [7]. And yet the rural environment itself can often become the target of disadvantageous attitudes and believes. Cultural "representations" attribute typical values and dominant meanings to the countryside. It is usually associated with purity, harmony, safety or solidarity, simply put, the qualities which are lacked by the city or the urban environment. On the other hand, the countryside is also represented as reactionary, traditionalist and conservative (Bryant, Pini 2011). In relation to women experiencing violence, the strengthening of certain attributes might translate into these women's marginalisation, discrimination and victimisation. Localities, as geographic and cultural categories, are part of power relationships and therefore relationships which promote either exclusion or inclusion of individuals and groups, including women.

4 Summary

Socio-demographic factors as a place of living, economic inactivity, poor education, family and marital status, alcohol, social pathology in a biological family of the perpetrators as well as personality disorders can result in the formation, persistence and intensity of violence committed against women. We have formulated the research problem, the aims, survey questions and the hypotheses of the empirical part. The primary aim of the empirical part was to identify significant factors conditioning the violence against women in family environment. In order to achieve the main objective, it was used the method of content analysis of documents, the document files of criminal offences of abuse against a close person or a person entrusted in one's care - section 208, which had female victims. On the basis of the survey it could be stated that: abused women experience violent relations with repeated signs of mental, physical, social and economic violence with lower occurrence of sexual violence. All age groups of women and women with all levels of education are in danger of violence, the most often there are women aged 26–45 years who live in the countryside with elementary education or high school education without school leaving examination, married. All groups of men participate in violence in a family environment with an emphasis on higher incidence in particular categories: the most often there are men at the age of 36 to 45 years, living in rural areas, with high school education (without school leaving exam, with school leaving exam). The second researched area was the analysis of unfavourable situations in family, the impact of the violence on the woman psyche and the reasons for remaining in a bond of partnership. It could be pronounced that the most frequent negative phenomenon in both aggressors and victim's childhood was physical violence of father against mother, an alcohol problem of a father, poverty in a family, physical punishment on children and patriarchal behaviour of a father. Two thirds of victims remained in a violent relation for these reasons: an effort to preserve the family because of children and the fear of losing financial security. Social criminality, it means committing crimes as a consequence of want of money, often occurred in poor biological families of perpetrators.

It is necessary for social work as a science to search for ways of problem solutions of violence committed against women in family environment, but it requires sufficient knowledge of the problem to have at disposal. It is evident that tabooing of violence committed against women is still considerable because of the fact, that violence in a family is hidden matter. Therefore, it was necessary to aim the dissertation at problems of significant factors of violence, specifically abused women in a family environment, which were investigated in DPH in Martin. We had an approach to 92 investigation files of victims of neglecting and abuse against a close person and a person entrusted in one's care. We are of the opinion that social work can benefit from the topic of women abuse, criminal proceedings system, the importance of expert opinions, and determination of the factors from the victim´s side as well as perpetrator. Reporting and record-keeping data could be a significant source of information for social worker's needs. We can apply research showing information to identify violence more accurately, to estimate danger and to work with an abused women and aggressor. Obtained information could be useful to make practical steps to the help and prevention of the violence against women more efficient.

References

[1] Z. Bútorová, J. Filadelfiová, Násilie páchané na ženách ako problém verejnej politiky, (2005). Available at: `http://www.ivo.sk/479/sk/vyskum/nasilie-pachane-na-zenach-ako-problem-verejnej-politiky`
[2] B. Bodnárová, J. Filadelfiová, Domáce násilie na Slovensku. Priebežná správa 2/2002, (2002). Available at: `www.ceit.sk`
[3] B. Bodnárová, J. Filadelfiová, B. Holubová, Reprezentatívny výskum výskytu a skúsenosti žien s násilím páchanom na ženách. Záverečná správa VÚ č.2224 (2008). Available at: `www.ceit.sk`
[4] B. Holubová, Monitoring médií a reklamy z hadiska násilia páchaného na ženách a podpory rodovej rovnosti za rok 2011 (2011). Available at: `www.ivpr.gov.sk`
[5] B. Holubová, J. Filadelfiová, Monitoring sociálnych služieb pre ženy zažívajúce násilie a ich deti z hadiska európskych štandardov (2013). Available at: `www.genderdatabaza.wordpress.com`
[6] S. Buchta, Možnosti a perspektívy riešenia vidieckej zamestnanosti. Sociológia **39**(2), 119–135 (2007)
[7] L. Sandberg, Backward, Dumb, and Violent Hillbillies? Rural Geographies and Intersectional Studies on Intimate Partner Violence. J. Women and Soc. Work **28**(4), 350–365 (2013)
[8] W.E. Thurston, S. Patten, L.E. Lagendyk, Prevalence of Violence Against Women Reported in a Rural Health Region. Can. J. Rural Med. **11**(4), 259–267 (2006)
[9] C. Peek-Asa et al., Rural Disparity in Domestic Violence Prevalence and Access to Resources. J. Women's Health **20**(11), 1743–1749 (2011)
[10] R.D. Shuman, jr., Understanding Intimate Partner Violence Against Women in the Rural South. Violence and Victims **23**(3), 390–405 (2008)
[11] Decree no. 618/2005 Coll. of the Ministry of Interior of the Slovak Republic on compiling of files by law enforcement authorities and courts
[12] Act on Municipal Establishment no. 369/1990 Coll., section 22, paragraph 1
[13] L. Bryant, & B. Pini, *Gender and Rurality* (Routledge, New York, 2011) (9-7804-1548-8990)
[14] Inštitút pre výskum práce a rodiny (IVPR, ILFR). Podpora prevencie a eliminácie násilia na ženách. Bratislava: IVPR (2015). Available at: <`www.ivpr.gov.sk`>

Comparison between the risks in road safety behaviour in urban and rural areas in Latvia in 2016

Aija Bukova-Zideluna[1,*], *Anita* Villerusa[1], and *Iveta* Pudule[2]

[1] Rīga Stradiòš University, Department of Public Health and Epidemiology, Riga, Latvia
[2] Centre for Disease Prevention and Control, Department of Research and Health Statistics, Riga, Latvia

Abstract. The study examined the differences between respondents in urban and rural areas in respect of their self-reported attitudes and behaviour regarding taking risks in road traffic. Data of Health Behaviour among Latvian Adult Population 2016 survey was used for analysis. Results: 83.7% (N=1605) of urban respondents and 86.7% (N = 1456) of rural respondents always used seatbelts in the front seat. Only 55.3% (N = 1605) of urban respondents and 52.1% (N = 875) of rural respondents always used seatbelts in the back seat. Odds to use seatbelts in the front seat were higher for rural population (OR = 1.27; 95% CI: 1.05–1.53). Odds to use seatbelts in the back seat were higher for urban population (OR=1.14; 95% CI: 1.01–1.30). Rural residents agreed more often with suggestion that it is not necessary to fasten the seatbelt on short journeys (OR = 1.42; 95% CI: 1.15–1.75) and that it is not necessary to fasten the seatbelt travelling at speed less than 40 km/h (OR=1.22; 95% CI: 1.01–1.56). Rural respondents agreed more often than urban respondents that that driving a car under alcohol influence increases a chance of being involved in an accident (OR = 1.45; 95% CI: 1.10–1.90).

Key words: differences between urban and rural areas, road safety, use of seat-belt, alcohol, smoking.

1 Introduction

Traffic accidents were the 8th leading cause of death for people of all ages. Approximately 1.35 million people died in road traffic accidents in the year 2016. Between 20 and 50 million more people suffered non-fatal injuries, with many incurring a disability as a result of their injury [1]. According to estimates of the European Commission approximately 135 000 people sustained serious road traffic injuries on European Union roads per year. This means that for every person killed in traffic accidents, five more suffered serious injuries in European Union countries. Serious injuries were not only more common but also often more costly to society because of long-time rehabilitation and healthcare needs. It is estimated that approximately 8% of road fatalities occurred on motorways, 37% happened in urban areas and 55% on rural roads in the European Union in 2017 [2]. It has been estimated that safety belts reduce risk of fatality of front-seat occupants of passenger cars by 45 percent relative to those who choose not to wear them [3]. Failure to use seatbelts is the second biggest

* Corresponding author: `aija.bukova@rsu.lv`

cause of road deaths, after speeding and ahead of drink-driving. A study carried out for the European Commission has concluded that, in the best-case scenario, actions targeting the use of seatbelts would save 5 500 lives per year in the European Union [4].

Although progress is being made in several areas, vehicle occupant fatalities have not been falling in line with casualty reduction targets for Latvia. Latvia was the country with the second highest fatality rate among European Union countries in 2014 (around 10.6 fatalities per 100 000 population). Fatal road traffic accidents in rural areas are over-represented compared to other European countries. The use of safety seatbelts by drivers, front-seat passengers and rear-seat passengers is compulsory in Latvia. However, the rate of wearing seatbelts is lower in Latvia than in other European Union countries [5].

The prevailing risk of traffic fatalities and casualties differs between urban and rural areas. Several reasons have been suggested to explain this including road design, the proximity to emergency medical services and human factors. The main task of the study was to examine the differences between respondents in urban and rural areas in respect of their self-reported attitudes and risk-taking behaviour in road traffic.

2 Methods

The study used the data from the survey of year 2016 regarding Health Behaviour among the Latvian Adult Population. The study sample for the survey was selected using a combined sampling method – stratified random sampling and the quota method. The sample was selected from the general population of Latvian residents aged from 15 to 74 (Data of the 2011 Population and Housing Census in Latvia). The total sample size consisted of 3596 people – a representative sample. The sample was stratified by gender, age, place of residence and nationality. Data was collected by face to face standardised interviews.

The survey included self-reported questions regarding traffic safety:

- Do you use safety belts when driving in the front seat?
- Do you use safety-belts in the back seat?
- Do you agree with the following statements?
  - It is not necessary to fasten the seatbelt in short trips by car.
  - It is not necessary to use the seatbelt, when travelling at speed less than 40 km/h.
  - Driving a car under alcohol influence increases my chance of being involved in an accident.

For all respondents demographic and socioeconomic information – gender, age and education level – was collected and analysed.

The data was analysed using descriptive statistics, Pearson's chi-squared and calculations of statistical reliability using SPSS for Windows 23.0. T-test and logistic regression analyses were conducted to compare the unsafe behaviour of road users with others in terms of socio-demographic, health-related and behavioural outcomes. The 95% confidence interval (CI) was used to estimate the precision of the odds ratio (OR).

3 Results

83.7% (N = 1605) of urban respondents and 86.7% (N = 1456) of rural respondents always used seatbelts in the front seat. The use of seatbelts in the front seat was higher for rural females compared to males, higher in the age group 35–44 and those who had had more years of education. The use of seatbelts in the front seat was higher in rural population for all age groups, except for the age group of 55–64, in which the use of seatbelts was higher for

Table 1. Characteristics of respondents who always use a seat belt in the front seat.

Factor		Number (% of cases) in urban area	Number (% of cases) in rural area	*p*-value
Gender	Male	626 (81.9)	607 (84.7)	0.09
	Female	979 (84.9)	849 (86.4)	0.01*
	Total	1605 (83.7)	1456 (86.7)	0.00*
Age	15–24	220 (83.0)	245 (85.2)	0.29
	25–34	301 (86.7)	280 (89.7)	0.14
	35–44	285 (87.2)	295 (93.4)	0.00*
	45–54	287 (86.7)	233 (87.9)	0.38
	55–64	294 (82.1)	220 (81.2)	0.42
Length of education	0–9 years	123 (68.3)	148 (74.7)	0.10
	10–12 years	621 (80.1)	602 (85.4)	0.00*
	13 and more	861 (89.5)	706 (91.0)	0.17

*Differences are statistically significant.

urban respondents. The analysis of urban respondents compared to rural respondents showed statistically significant differences (Pearson's chi-squared test $\chi 2 = 20.97$; *p*-value $= 0.01$). The odds of using seatbelts in the front seat were 1.27 times higher for the rural population (OR = 1.27; 95% CI: 1.05–1.53). The characteristics of respondents who always used a seat belt in the front seat according to gender, age and level of education is shown in Table 1.

55.3% (N = 1605) of urban respondents and 52.1% (N = 875) of rural respondents always used seatbelts in the back seat. The use of seatbelts in the back seat was lower for urban males compared to females, higher in the age group 45–54 and those who had had more years of education. The use of seatbelts in the back seat was higher in urban population for all age groups, except for the age group of 24–44, where the use of seatbelts was higher for rural respondents. The analysis of urban respondents who always used seatbelts in the back seat compared to rural respondents showed statistically significant differences (Pearson's chi-squared test $\chi 2 = 15.86$; *p*-value$= 0.04$). The odds of using seatbelts in the back seat were 1.14 times higher for the urban population (OR = 1.14; 95% CI: 1.01–1.30). The characteristics of respondents who always used a seat belt in the back seat according to gender, age and level of education is shown in Table 2.

9.6% (N = 184) of urban respondents and 13.1% (N = 220) of rural respondents agreed with the idea that it was not necessary to fasten seat belt for short journeys. Rural residents – men, in the age group between 15–24 and those who had had less years of education agreed more with this statement compared to urban residents. The proportion of respondents who agree with this false statement was higher in rural population for all age groups, except in the 24–44 age group, where the proportion was higher for urban respondents.

The analysis of urban respondents in total compared to rural respondents showed statistically significant differences (Pearson's chi-squared test $\chi^2 = 14.04$; *p*-value $= 0.001$). The proportion of rural population, with the false belief in the use of seat belts for short trips, was larger (OR = 1.42; 95% CI: 1.15–1.75). The characteristics of respondents who agreed with this statement according to gender, age and level of education is shown in Table 3.

7.0% (N = 135) of urban respondents and 8.5% (N = 142) of rural respondents agreed that it was not necessary to fasten seatbelts while travelling at a speed of less than 40 km/h.

Table 2. Characteristics of respondents who always use a seatbelt in the back seat.

Factor		Number (% of cases) in urban area	Number (% of cases) in rural area	p-value
Gender	Male	381 (49.9)	321 (44.8)	0.02*
	Female	680 (59.0)	554 (57.6)	0.27
	Total	1065 (55.3)	875 (52.1)	0.28*
Age	15-24	147 (55.5)	133 (46.2)	0.03*
	25–34	200 (53.8)	172 (55.1)	0.52
	35–44	183 (56.0)	190 (60.1)	0.29
	45–54	189 (57.1)	133 (50.2)	0.04*
	55–64	194 (54.2)	133 (49.1)	0.20
Length of education	0–9 years	88 (48.9)	99 (50.0)	0.83
	10–12 years	419 (54.1)	353 (50.1)	0.13
	13 and more	554 (57.6)	423 (54.5)	0.20

*Differences are statistically significant.

Table 3. Characteristics of respondents who agree with the statement - it is not necessary to fasten seatbelts on short journeys.

Factor		Number (% of cases) in urban area	Number (% of cases) in rural area	p-value
Gender	Male	108 (14.1)	141 (19.7)	0.00*
	Female	76 (6.6)	79 (8.2)	0.15
	Total	184 (8.6)	220 (13.1)	0.00*
Age	15-24	36 (13.6)	55 (19.1)	0.04*
	25–34	45 (13.0)	32 (10.3)	0.28
	35–44	36 (11.0)	39 (12.2)	0.60
	45–54	28 (8.5)	34 (12.8)	0.08
	55–64	23 (6.4)	36 (13.3)	0.00*
Length of education	0–9 years	31 (17.2)	37 (18.7)	0.71
	10–12 years	78 (10.1)	113 (16.0)	0.00*
	13 and more	75 (7.8)	70 (9.0)	0.36

*Differences are statistically significant.

Rural residents in the age group between 45–54 and those who had had more years of education agreed more with this statement compared to urban residents. The analysis of urban respondents compared with rural respondents showed statistically significant differences (Pearson's chi-squared test $\chi^2 = 15.89$; p-value $= 0.04$). Rural residents agreed more often with the idea that it was not necessary to fasten seatbelts while travelling at a speed of less than 40 km/h (OR = 1.22; 95% CI: 1.01–1.56). The characteristics of respondents who agreed with this statement according to gender, age and level of education are shown in Table 4.

92.4% (N = 1772) of urban respondents and 94.6% (N = 1589) of rural respondents agreed that driving a car while under the influence of alcohol increased the chances of being

Table 4. Characteristics of respondents who agree with the statement – it is not necessary to use seatbelts, while travelling at a speed of less than 40 km/h.

Factor		Number (% of cases) in urban area	Number (% of cases) in rural area	*p*-value
Gender	Male	77 (10.1)	85 (11.9)	0.27
	Female	58 (5.0)	57 (5.9)	0.36
	Total	135 (7.0)	142 (8.5)	0.11
Age	15–24	28 (10.6)	35 (12.2)	0.56
	25–34	33 (9.5)	23 (7.4)	0.32
	35–44	27 (8.3)	24 (7.6)	0.76
	45–54	17 (5.1)	24 (9.1)	0.04*
	55–64	21 (5.9)	23 (8.5)	0.20
Length of education	0–9 years	23 (12.8)	22 (11.8)	0.36
	10–12 years	55 (7.1%)	74 (10.5)	0.01*
	13 and more	57 (5.9)	46 (5.9)	0.53

*Differences are statistically significant.

Table 5. Characteristics of respondents who agree with the statement – driving a car under while under the influence of alcohol increases the chances of being involved in an accident.

Factor		Number (% of cases) in urban area	Number (% of cases) in rural area	*p*-value
Gender	Male	710 (92.9)	673 (93.9)	0.47
	Female	1062 (92.1)	916 (95.2)	0.00*
	Total	1772 (92.4)	1589 (94.6)	0.00*
Age	15–24	239 (90.2)	264 (91.7)	0.54
	25–34	322 (92.9)	294 (94.2)	0.45
	35–44	303 (92.7)	304 (96.2)	0.05
	45–54	317 (95.8)	250 (94.3)	0.42
	55–64	329 (91.9)	262 (96.7)	0.01*
1264	0–9 years	158 (87.8)	181 (91.4)	0.25
	10–12 years	698 (90.1)	666 (94.5)	0.00*
	13 and more	916 (95.2)	742 (95.6)	0.69

*Differences are statistically significant.

involved in an accident. Rural residents – females, in the age group between 55–64 and those who had had 10–12 years of education – agreed more with this statement compared to urban residents. The analysis of urban respondents compared to rural respondents showed statistically significant differences (Pearson's chi-squared test $\chi^2 = 14.01$; p-value $= 0.05$). Rural residents agreed more often with the false idea that driving a car while under the influence of alcohol increased the chances of being involved in an accident (OR = 1.45; 95% CI: 1.10–1.90). The characteristics of respondents who agreed with this suggestion according to gender, age and level of education is shown in Table 5.

4 Discussion

This study has provided the analysis of the behaviour and attitudes regarding road safety displayed by urban and rural adults. Despite the existing seat belt legislation, public campaigns and regular police monitoring both of urban and rural roads, a large proportion of the population did not use seatbelts in Latvia, especially in the back seat. While 83.7% of urban respondents and 86.7% of rural respondents always used seatbelts in the front seat, only 55.3% of urban respondents and 52.1% of rural respondents always used seatbelts in the back seat. According to the study carried out for the European Commission the rates of wearing seatbelts varies from 36% to 91% in urban areas in the European Union [4]. The relatively low rates of seatbelts in Latvia have been accepted as one of the priorities in traffic safety because of the high fatality and injury rates in traffic accidents compared to other European Union countries [6].

Similar to other studies of practices regarding the use of seatbelts, demographic factors such as age, gender and education levels were associated with the lower use of seatbelts [7–9]. However, other surveys regarding the use of seatbelts emphasize that the 18 to 25 age group tend to use seatbelts less than all other age groups. This study shows those least likely to use seatbelt in the front seat are between 55 and 64 years old and 15 to 24 years old rural respondents. It is possible that respondents in the age group of 55–64 still remember cars without seatbelts and times when it was not necessary to fasten them/ Consistent to other researches [8, 10] rural males were less likely to report that they always used seat belt in the front and in the back seat compared to females. Consistent with other studies [11] there proved to be a link between levels of education and the use of seatbelts especially the responds who had had 10–12 years of education. Less use of seat belts in rural areas compared with urban areas is a factor that has contributed to the increased risk of deaths in traffic accidents in rural areas.

Apart from the demographic and socioeconomic characteristics of respondents, there were also other factors such as attitudes, beliefs, and intentions which influenced the use of seatbelts [12]. Negative attitudes and beliefs regarding the effectiveness of the use of seatbelts had a negative relationship in respect of the use of seatbelts [13]. Using seatbelts can be described as a habitual behaviour that is mainly affected by the gender, age, socioeconomic situation of a person combined with the perceived usefulness of seatbelts in a possible accident [14]. This study has shown that rural residents agreed more often with the false statement that it was not necessary to fasten seatbelts on short journeys and that it was not necessary to fasten the seatbelt while travelling at a speed of less than 40 km/h. At the same time rural residents agreed more often with the idea that driving a car under the influence of alcohol increased the chance of being involved in an accident.

5 Conclusions

Road traffic accidents, injuries and deaths are generally related to preventable risk factors. In this study the behaviour and attitudes regarding road safety were analysed based on the place of residence, demographic and socioeconomic factors. It can be concluded that there is not enough awareness of road safety in the community. The study has demonstrated that there is a greater risk of being involved in a traffic accident in a rural area than in an urban one. Consequently, more effort should be put in encouraging individuals in rural areas to use seat belts in the rear seat of the car, particularly in among males, aged between 15 to 24 and those who have studied for 10 to 12 years.

It is generally recognised that the use of seatbelts can prevent injuries and fatalities. Law enforcement and educational efforts need to be used to increase the use of seatbelts and to reduce accidents in rural areas. Stronger laws regarding the use of seatbelts and targeted enforcement may also influence the beliefs and attitudes regarding the use of seatbelts in the back seat. Combining public campaigns and enforcement with educational and development could influence behaviour among certain critical groups and further improve road safety behaviour. This research can be used to plan and design educational programmes and campaigns focused on groups at risk in order to change the road safety culture from one that is risk receptive to one that is risk protective.

References

[1] World Health Organization, Global status report on road safety, (2018). Available at: https://www.who.int/violence_injury_prevention/road_safety_status/2018/en [viewed 31.01.2019]

[2] European Commission Road safety in the European Union, Trends, statistics and main challenges (2018). Available at: https://ec.europa.eu/transport/road_safety/sites/roadsafety/files/vademecum_2018.pdf [viewed 31.01.2019]

[3] Commission of the European Communities, Proposal for a Directive of the European Parliament and of Council about facilitating cross-border enforcement in the field of road safety, Brussels, 351/2 (2008)

[4] European Commission, Proposal for a directive of the European Parliament and of the Council facilitating cross-border enforcement in the field of road safety (2008). Available at: https://ec.europa.eu/transport/road_safety/sites/roadsafety/files/enforcement/doc/full_impact_assessment_en.pdf [viewed 31.01.2019]

[5] European Road Safety Observatory, Road Safety Country Overview – Latvia, 1–16 (2016)

[6] Cabinet Order No. 180 "On Road Traffic Safety Plan for 2017-2020" (2017). Available at: https://likumi.lv/ta/en/en/id/289986 [viewed 31.01.2019]

[7] L.F. Beck, R.A. Shults, Seat belt use in states and territories with primary and secondary laws – United States, 2006, J. Saf. Res. **40**(6), 469–472 (2009)

[8] T.W. Strine, L.F. Beck, J. Bolen, C. Okoro, S. Dhingra, L. Balluz, Geographic and sociodemographic variation in self-reported seat belt use in the United States, Accident Analysis and Prevention **42**(4), 1066–1071 (2010)

[9] J. Tison, A.F. Williams, N.K. Chaudhary, Daytime and nighttime seat belt use by fatally injured passenger vehicle occupants. National Highway Traffic Safety Administration, Washington (2010)

[10] F. Laurie, J. Downs, M.R. Stevens, et al., Rural and Urban Differences in Passenger-Vehicle-Occupant Deaths and Seat Belt Use Among Adults, MMVR Surveillance Summaries **66**(17), 1–13 (2017)

[11] K. Lipovac, M. Tesic, B. Maric, et al., Self-reported and observed seat belt use – A case study: Bosnia and Herzegovina, Accident Analysis and Prevention **84**, 74–82 (2015)

[12] E.J. Chliaoutakis, C. Gnardellis, I. Drakou, C. Darviri, V. Sboukis, Modelling the factors related to the seatbelt use by the young drivers of Athens, Accident Analysis and Prevention **32**, 815–825 (2000)

[13] D.J. Begg, D.J. Langley, Seat-belt use and related behavior among young adults, J. Saf. Res. **31**, 211–220 (2000)

[14] F. Calisir, M.R. Lehto, Young drivers' decision making and safety belt use, Accident Analysis and Prevention **34**, 793–805 (2002)

Meaning of profiling of cybercriminals in the security context

Aldona Kipane

Rīga Stradiņš University, Riga, Latvia

Abstract. In this article, the author analyzes the criminological aspects of profiling of cybercriminals. The development and trends of crime are affected by both the globalization process and the development of information technology. All over the world, crime, disorder and riots are causing more and more troubles and worries for people. Nowadays, cybercrime or criminal offenses committed in the Internet environment take a stable position in the overall mass of criminal offenses. Cyberspace is easy to use to harm an individual, a group of people, or a country as a whole. The role of profiling cybercriminals is determined by the tasks of law enforcement authorities. The profile of cybercriminal combines the personal traits of cybercriminal, behavioural patterns, and demographic data based on cybercrime characteristics.

Key words: cybercrime, criminality, cybercriminal, criminal profiling, profiling cybercriminal.

1 Introduction

Vulnerability of people and threats in the virtual environment are growing, the public is concerned about security on the Internet. Cybercrime is a particularly harmful offense that manifests itself in various areas of society and has a serious impact on it in a number of forms – social disorganization, economic loss and psychological disorder. Cybercrime as a legal, practical and political problem is a direct threat to human rights, entrepreneurship, the state and the global world as a whole. Cybercriminals can disrupt or destabilize, hurt, others seek empowerment, others get data or identity, socially most dangerous persons led by political targets attack state infrastructures (electricity networks etc.).

At present, cybercrimes are the fastest – growing criminal offence compared to others. The situation is made more complex by increasing of the transnational character of crime [1]. According to experts' forecasts, cybercrime will exceed the entire drug market in 2021 and the damage is going to be six trillion all over the world. An Indian criminologist, professor Karuppannan Jaishankar reasonably points out that cybercrimes are no longer simply a hacker attack or an attack on the system, but it is an attack on people [2]. The data of Global Cyber Risk Perception Survey 2018 showed that nearly two-thirds of respondents rated cyber risk as one of the five highest risks in their organization [3]. Data from the Eurobarometer report on cyber security show that most respondents in Latvia and among European citizens are concerned about Internet offenses. The main fears of citizens are about online transactions and the use of online banking. For example, 65% of respondents in Latvia admitted that they generally feel poorly informed about the risks of cybercrime, while in countries where most Internet users feel well informed, confidence in online insecurity and inability to protect themselves is increasing [4]. The problem of cybercrime is determined by the rapid development, implementation and use of information technologies, as well as the use of these technologies for criminal purposes. For example, the rapid development of cloud computing technologies also opens up a favourable environment for cybercriminals. The criminological aspect of modern technologies is diverse.

Cybercrime is a specific, complex set of criminal offences, which, like crime phenomenon, can be assessed not only by its type, methods of committing them, harmfulness, social danger, harmful consequences, but also by the personality of a criminal. Indian advocate and Legal Scholar, Dr. Debarati Halder and Indian criminologist, professor Karuppannan Jaishankar define Cybercrimes as "offences that are committed against individuals or groups of individuals with a criminal motive to intentionally harm the reputation of the victim or cause physical or mental harm to the victim directly or indirectly, using modern telecommunication networks such as Internet (Chat rooms, emails, notice boards and groups) and mobile phones (SMS/MMS)" [5]. Thus, it can be pointed out that cybercrime is a specific, complex set of offences, and a new branch of criminology – Cyber criminology – has been developed to study it.

Since criminologists consider cybercrime as a new type of criminal activity in the cyberspace, the causes of cybercrime, the characterization of a criminal, the victimological aspects have been studied, and new theories have been raised. Cybercrime, like traditional crimes, consists of **action, action – victim/purpose – modus operandi – criminal – harm, and loss.** From a criminological point of view, cybercrime is a socially legal phenomenon, which also consists of specific elements such as an automated data processing system, cyberspace, an individual, criminal action or inaction. A cybercriminal disputes the values developed by the society. A cybercrime is being committed because of the excessive increase of the cybercriminal's need for expression in an unacceptable way for the public. In addition, it should be borne in mind that in different situations a cybercriminal has more or less possibilities to commit a crime and that there are circumstances that create a high or low level of risk. That is why cybercriminals are the biggest problem of the global web today.

Modern criminologists and law enforcement officers use criminal profiling to describe an investigative approach, which is based on analysis of behavioural and psychological patterns to predict the characteristics of a person, who probably has committed a crime. Profiling leads to the identification and interpretation of criminal behaviour or actions to identify a criminal's personality, his/her modus operandi, and possibly the motivation of a criminal offense. The purpose of profiling is not only to obtain information about character of a criminal, but also to mark a preventive measure to prevent a repeat of a similar criminal offence [6]. There are two parts of the detection of any cybercrime, both technical and legal. Both the legal and technical circumstances of the offense should be identified during the investigation of criminal offences of these categories. Depending on the circumstances of a particular offense, different traces of the offense may occur on a case-by-case basis, which helps to identify the criminal's profile. Thus, recruitment of specialists such as a criminalist, IT specialist, psychologist etc. is essential.

Professor of criminology and sociology, Ph.D. Scott A. Bonn points out that criminal profiling is a cross between law enforcement and psychology. It is still a relatively new field with few set boundaries or definitions [7]. In the last 30 years, the frequency of study of a criminal's profile has increased, the volume of literature has been growing, as well as the method of criminal profiling used in investigation is a common practice worldwide. Criminal profiling is a sub-discipline of forensic criminology. It is, therefore, a discipline within criminology, rooted in the behavioural science and forensic sciences alike [8]. Forensic criminology is currently multidisciplinary in nature [9]. Forensic criminology is a discipline characterized by an integration of material from many sub-disciplines including medicine, psychiatry, psychology; penology, anthropology, mathematics and statistics, forensic science, criminal investigation, criminology, forensic psychology, victimology, crime reconstruction, criminal event analysis, history, philosophy, practical experience, geography and more.

2 Aim

Aim is to describe the criminological aspects of the criminal profile of a cybercriminal. Profiling of a cybercriminal is a relatively new cross-disciplinary field that is in continuous development and its practical meaning is rising rapidly within the criminology.

3 Material and methods

Analysis of scientific literature and analysis of researches made before. Mainly the following scientific methods were used: analytical, comparative; historical, systemic, publications and analysis of resources.

4 Criminal profiling or profiling a criminal in the context of law enforcement authorities

Criminal profiling or profiling a criminal is a popular investigative technique where psychological achievements are also used. Criminal profiling is often reflected in detective films – by analyzing the features of a criminal offense, the investigator assumes about what a potential criminal might be. For example, a murderer is a man, perhaps a young, powerful, utterly cruel, living one, and so on. Criminal profile is a set of basic features of a person's character.

The informal process of criminal profiling has a long history. The various sources show different indications, such as the reference to the 15th century, but others that the method was already used in 1880 to make predictions about a serial killer. While some experts are discussing the effectiveness of this method, it has been used successfully in law enforcement practices for more than a century. At the end of the 20th century and in the 21st century, the method of criminal profiling was successfully applied by law enforcement authorities. Criminal profiling is typically used in crimes where the offender's identity is unknown. Nowadays there are two types of criminal profiling: the deductive and inductive approaches.

Professor of International Law and Human Rights Martin Scheinin defines profiling as "systemic linking of physical, behavioural, or psychological characteristics to specific offenses and their use as a basis for taking law enforcement decisions. Profiling in the context of law enforcement can generally be a legitimate method of investigation. Profiles can be descriptive and predictive" [10].

There is no common definition in the special literature. There are diverse interpretations of criminal profiling and different terms are used: criminal profiling; offender profiling, criminal profiling, psychological profiling, criminal investigative analysis, crime scene analysis, behavioural profiling, criminal personality profiling, socio-psychological profiling and criminological profiling, more recently, investigative psychology [6]. Generally speaking, criminal profiling involves making inferences about the physical, habitual, emotional, psychological, and even vocation characteristics of criminals [6].

The term "offender profiling" was introduced in the 1970s, linked to the activities of the FIB analysis unit, to describe their criminal investigative analysis work. Initially, criminal profiling was used for serial murders, but the boundaries of research expanded over time. Nowadays, it is linked to various criminal offenses (rape, torture, murder, terrorism, cybercrime, etc.). This includes– criminal profiling – refers to identifying and describing essential information about a suspect. Psychological profiling refers to a behavioural sketch of an individual who may or may not be a suspected offender [11].

Criminal offense has a biological, social and legal nature. Crime is a diverse, complex phenomenon and explaining many of its aspects are a real challenge [12]. It must be admitted that "criminal offense is the shadow of civilization". Similar findings are provided by David Canter, professor of psychology, who believes that profiling a criminal is a "criminal shadow". He points out that the criminal "leaves psychological traces, behavioural alarm tells that indicate personality. From the crime scene and the testimony of the witnesses, these traces are much vaguer and subtler than those examined by a biologist or physicist. They cannot be studied in a laboratory or under a microscope. They are more like shadows, which are undoubtedly related to the criminal who leave them, but they are flashing and changing, and it is not always clear where they come from [13].

For comparison, the author will refer to a number of wordings of this concept in the special literature to show the features they contain. Emeritus Professor of the University of Sheffield Peter B. Ainsworth points out that "profiling is a process where all available information about a criminal offense, the place of the crime, and the victim is used to create a (yet) unknown perpetrator's profile" [14].

Anne Davies, a specialist from the UK, writes: "Offender profiling (more technically known as Criminal Investigative Analysis) is the name given to a variety of techniques, whereby information gathered at a crime scene, including reports of an offender's behaviour, is used both to infer motivation for an offence and to produce a description of the type of person likely to be responsible" [15].

Forensic scientist and Criminal Profiler Brent Turvey, basing on behavioural evidence defines it as "the process of inferring the personality characteristics of individuals responsible for committing criminal acts [8].

Retired Lieutenant Commander of the New York City Police Department Vernon J. Geberth understands profiling as "an educated attempt to provide investigative agencies with specific information as to the type of individual who may have committed a certain crime" [16].

Latvian specialist Evija Strika writes that criminal profiling is the identification process of traits, behavioural tendencies and demographic data basing on characteristics of a criminal offence [17].

Some scientists define it as a forensic technique used by forensic investigators and law enforcement agencies to understand why criminals are committing crime, to classify criminal behaviour and to solve crimes that have already been committed [18].

Basing on the peculiarities of the research data, conclusions about the personality of a criminal are drawn and the description of a wanted person is made. This means that the main purpose of criminal profiling is to narrow the range of people, where a potential criminal has been searched for – to identify a set of information about individual peculiarities of a criminal, which are expressed as circumstances of criminal activities and traces. It is reasonable to point out that criminal profiling is a form of prediction, which helps to establish a hypothesis in order to identify a person or persons who might be liable for the crime committed. It allows dividing potential criminals into particular categories, based on particular peculiarities being observed, in order to derive other characteristics that cannot be observed [19].

Profiling is a technique or approach for solving crime. In essence, criminal profiling is primarily based on the assertion that the format in which the offender committed the offence reflects his or her behaviour and personality [18].

Summarizing it, the author acknowledges that these formulations do not differ greatly in substance. The author points out that profiling is not only a process, but also the result of analysis and research (the result of a process) – a set of features of an alleged criminal (profile). It is the method of identification of a person, who has committed a criminal offence. The criminal profile is the result of the observation and conclusion process. This is based on the analysis of evidence resulting from the criminal investigation. The author explains the profiling of a criminal **as the result of research and analysis based on the features of a criminal offense, identifying the personality traits, behavioural tendencies and demographic data of an alleged criminal.**

Criminal profiling is usually used in investigation of difficult-to-detect crimes (especially when there are indications that an offence, or a series of offences were committed by a person having a mental disorder). Criminal profiling can also be used in cases where there are several suspects; more intensive search is done for the person who corresponds to the established profile. At the general level, profiling is the classification of individuals according to their characteristics, i.e. whether they are "constant" (such as gender, age, ethnic origin, and height) or "variable" (such as habits, choice and other behavioural elements). Profiling can be a legitimate tool to apprehend suspected offenders after a criminal offense has been committed. Profiling can also be based on deliberate assumptions derived from experience and training, focusing on behaviour rather than racial, ethnic or religious characteristics. For example, police officers can work with profiles that contain instructions to search for people who repeatedly visit certain sites, that behave unpredictably or nervously, or who repeatedly make large purchases using only cash [20].

Although profiling is a popular method, there are specialists who point out that there are factors that determine its shortcomings: 1. It is wrong to think that human behaviour is constant in time and space, i.e. unchanged in different situations. 2. It is wrong to believe that specific hints and traces can always be obtained at the crime scene, enabling the psychological characteristics of the offender to be provided. However, it should be taken into account that profiling is also a means of forecasting. Based on crime scene information and the behavioural patterns or habits of the offender, the profiler tries to "predict" who the offender or offenders might be and where and how the next crime may occur [11]. In addition, it should be noted that profiling can be used to combat previously committed offenses or prevent future offenses to be committed. Two approaches can be used in the profiling process. One is prospective (related to future prospects, attempting to create a certain template/pattern), which is based on data on the characteristics of previously detained offenders. The other is retrospective approach (focused on the past, an overview of past events), which is based on the study of personality and his/her behaviour, made through the analysis of the crime scene, facts and circumstances of the criminal offense.

5 Criminological profile of a cybercriminal

Criminological research of a criminal personality is carried out in order to identify and evaluate the individual peculiarities, which lead to the commitment of a criminal offence. Criminals, including cybercriminals, have several sharpened personality traits: impulsiveness, aggressiveness (high level of aggression), difficulty predicting the consequences of their actions, rigidity, lies, selfishness, egocentricity, affluently saturated feelings, peculiar orientations and judgments, hard-to-predict behaviour, abstinence from social reality, inability to internalize moral and legal norms, hostility [21].

Personality traits play a key role in personality behaviour. A man has both inherent and typical traits. The personality of an offender is a set of negative personal traits that are specific to the type of criminals and individual criminals. A cybercriminal is not only a person with a certain status, who has rights, duties, responsibilities, but an individual as a complex system with several structures: 1) needs – interests – abilities; 2) emotions – thinking – will; 3) temperament – nature – value orientation.

The commitment of a criminal offense can be determined by a system of needs, peculiarities of temperament and nature, basic orientation of value orientation and other circumstances. The personality of a criminal, like the formation of a personality of any individual, is determined by three factors - heredity, socialization and self-actualization [22]:

1. Heredity is the succession of organism properties with genetic determination in generational shifts. Biological effects and genetic factors have been demonstrated in various studies, for example, the nature of genetics has been identified in a meta-analysis of individual traits and inheritance characteristics. Summarizing data from over 14 million twin pairs in the 50-year period in 39 countries worldwide, it is concluded that all features are heritable; they are not featuring with less impact.
2. Socialization – mastering the norms of behaviour in a particular social environment consisting of a unit of three subsystems: micro environment, macro environment and the social environment as a whole. For example, the analysis of multi-factor research on the development of criminal behaviour shows that mental health, family stability and home conditions are the decisive criteria for children aged 8–10 to experience a full adult life;
3. Self-actualization (nature, intelligence, values of life) – the interaction of personality-forming factors focuses on self-actualization, which directly forms the basis of criminogenic motivation. However, the criminogenic motivation milestone – the cause of the criminal offense – links these factors to the circumstances conducive to a particular criminal offense.

6 Criminal profiling of cybercriminal

Unlike majority of society, criminals are not able to adopt norms adequately because of diversions in the process of their socialization or they accept "the special forms" of criminal environment. It is affected by various factors: heredity, education, culture, lifestyle and socio-economic factors. A cybercriminal denies the values developed by society. In addition, it should be taken into consideration that in different situation a cybercriminal has more or less possibilities to commit a crime and that there are circumstances that present a low or high risk.

Specific common features and traits are identified in criminal profiling of cybercriminal. One has to agree that profiling is more powerful in multiple than single crime case [23]. There are traditionally distinguished three approaches of criminal profiling: 1) a criminal investigative approach; 2) a clinical practitioner approach; 3) a scientific statistical approach [24]. Each of these approaches has a discrepancy in knowledge field, based on which it is assumed to explain the criminal behaviour.

According to the criminal investigative view, it can be argued that offender profiling relies heavily on a mix of tacit and evidence-based expert knowledge. Arguably, this may make a profiler's advice more susceptible to cognitive biases and faulty decision-making [25]. Criminal profiling is reactive via the examination of behaviours exhibited at crime scenes. Offender profiling in the interring of an offender's characteristics from his or her crime scene behaviours. For example, a profiler might try to infer a criminal's age, gender or employment history from the way he or she has behaved during a crime.

Clinical profilers draw up their conclusions on the offender's traits from their clinical experience, which is obtained while working with detained offenders.

The aim of the statistical approach working with statistical data, databases is to find the relationship between the information recorded in the statistical reports and the features of the offender, using data on similar crimes and detected criminal offenses.

It is reasonable to point out that criminal profiling of cybercriminal has multidisciplinary nature. The different types of offender profiling can be broken down broadly into two types: geographical profiling[1] and the profiling of an offender's personal characteristics. The latter is what people most commonly associate with the term offender profiling [26].

The aim of geographical profiling is the reverse. Using the locations of an offender's crime as his or her starting point, the geographical profiler tries to predict the area in which the offender lives. Geographical profiling is not simply the consideration of the meaning of a dot on a map. The location has to be understood in the context of as many other aspects of the crimes as can be harnessed to the inference process [27]. Routine Activities Theory and Pattern Theory are relevant to geographical profiling. This suggests that criminals will offend in an area with which they are familiar.

One of the directions of profiling is the psychological evaluation of certain properties and traits. The profile of cybercriminal combines the description of an individual's behaviour and qualities that are created without knowing the criminal's identity. Criminal profiling involves the identification of an unknown criminal by using several techniques:

1. Analysis of the crime scene;
2. Determination of the peculiarities of criminal offence;
3. Characterization of the personality of a criminal [28].

Crime scene analysis (behaviour configuration) plays a significant research role. The analysis of the scene (also known as crime scene) is the inspection of the particular place and the objects contained therein, if it is made after receiving information about the committed criminal offense and if there is sufficient reason to believe that a crime has been committed there or continued in this place [29]. Its purpose is to find and seize traces that indicate the commission of a criminal offense, as well as to restore the mechanism of committing a criminal offense. Already in 1925, professor of Moscow State University Ivan Yakimov (Иван Николаевич Якимов) wrote that "the following stages can be distinguished in the restoration of the crime: the detection of a criminal offense, the obtaining and evaluating evidence, investigating the alleged perpetrator [30].

Nowadays, in the case of cybercrime, the investigator has to recognize a huge amount of evidence in electronic or digital form. The crime scene, contrary to the physical scene, contains computer systems or computer networks. A set of scene conditions and other investigation data can provide information about the personality, motivation, and characteristics of the offender. Taking into account the diversity of cybercrime and profiling approaches - the forensic aspect, the psychological aspect, the technical aspect, – the joint work of multidisciplinary specialists is essential. A psychologist plays an important role. The forms of cooperation may vary during the investigation. Professor Laurence Alison of the University of Liverpool has suggested a number of ways in which the expertise of a psychologist could aid the police and support the work that they do. It is important to appreciate that the ways in which psychologists can contribute extends well beyond the process of profiling offenders. Indeed, the apprehension of the offender would be assisted by enhancing police decision-making and leadership skills, improving

[1] Geographical profiling has its history in environmental criminology.

methods of interviewing witnesses and victims, developing accurate methods of recording, collating and analyzing data on pre-convictions of offenders, developing suspect prioritization systems based on empirical research and enhancing intelligence-led policing and the use of informants [31].

When predicting and profiling an individual's behaviour, his/her most significant psychological features – a system of values, an emotional state, and demographic indicators (biological parents, ethnicity etc.) are described. At the same time, the analysis of the kind of criminal activity (*modus operandi*) and action type, as well as the analysis of the crime scene are made [32]. Modus operandi reflects the nature of cybercriminal [33]. The reconstruction of modus operandi of a cybercriminal is identified from the assumption that criminal behaviour is the set of several aspects, which form a criminal's criminal experience, life experience and influential life changing events, professional skills and the level of development of intellect.

Several experts point out that this method is the most useful on the initial stage of investigation because it can help them plan and lead the process of further investigation. For example, the results of the survey carried out show that 46% of 184 cases criminal profiling specialists were invited to the initial stage of an investigation (Annon 1995; Douglas et al. 1986). The author agrees with this view and points out that by the developing modern technologies and the increasing use of them the importance of rapid and high-quality activities in the detection of a criminal offense (especially cybercrime) in the early stages of the investigation is dramatically increasing, because a timely undetected and not apprehended criminal continues his/her criminal activities. Thus, the fear of cybercrime is expanding.

The criminal profiling of an unknown cybercriminal involves three stages: the first, a law enforcement officer collects data obtained from the crime scene and transmits them to the profiler; the second, the profiler analyzes the data and the third, the profiler provides the predictions of the nature of a potential criminal. The processes that profilers use in analyzing crime scene data can be classified as either "clinical" or "statistical" in nature. Clinically oriented techniques incorporate aspects of the profilers' intuition, knowledge, experience, and training to generate predictions [34].

A profile of cybercriminal can be described by including such key elements as:

1. Personality characteristics/traits, which are specific to a particular person, and which predispose a person to commit a cybercrime. Personality traits are defined as a broad individual psychological dimension that describes the interpersonal, stable and common individual differences of the individual's behaviour, thoughts and feelings. Traits appear in individual activities in different situations and at different timescales [35]. There is a high level of legal nihilism among cybercriminals. The cybercriminal has deviations in legal consciousness; the deformation in legal consciousness is determined by the inability of a particular individual to live according to legal norms. In the author's view, it is in close connection with the enhanced internal necessity to risk in violation of the law, and with such behaviour to achieve some personal benefit or to gain material benefit, profit. The impact of the micro environment is important. For example, family factors that negatively affect personality formation, thereby increasing cybercrimes include: deficiencies in the process of bringing up children (lack of parents, lack of support and understanding), relationship of deformation in the family (neglecting children, etc.), unfavourable families (problems of addiction, financial problems) and / or social problems).
2. Criminal professionalism – it means the personality traits that contribute to the safe and effective commitment of cybercrime. It includes four mandatory features: specific

personal qualities; knowledge and skills; fearlessness, courage and self-confidence; effectiveness and viability of action; commission of a criminal offence and achieving a specific aim [36]. For example, some financially motivated cybercriminals generally have two main purposes – input data and user identity to gain access to finances through the identity they have acquired.

3. Technical knowledge related to specialized knowledge and technical skills in dealing with complex programs and devices that enable cybercrime. The Cybercrime Survey concludes that technical execution of 65% of all illegal activities is relatively simple, 13% – required medium-level technical skills, and 22% – complex technical skills. The most common cybercriminals were university students or students of other educational institutions. It is generally recognized that the level of education among cybercriminals may be higher than among other categories of criminals [37]. If a cybercriminal has the highest technical education, knowledge and skills that can be used for committing a cybercrime, then the social danger of the offense will not only increase, but increase progressively. In this case, intelligence is the main element of a cybercriminal. It must be acknowledged that a person with criminal skills based on the knowledge, skills and abilities acquired, implementing criminal behaviour, causes more harm both in everyday situations and in changed situations [36].
4. Social characteristics are demographic features, socio-economic status, socio-psychological and moral qualities. The basic elements are gender, age, nationality, socio-economic status, for example, the features of a typical fraudster are: a middle-aged man with higher education and substantial work experience in his company (almost half had six or more years of experience, almost a third – three to five years of experience).
5. Characteristics of motivation. In criminology, motivation is understood as a set of motives of action in which each of the motives determines the element of motivation and exists in both the consciousness and the subconscious. The motives are developed and formed under the influence of human emotions and feelings. The motives are internal - chosen by the person and external – driven by others. Research has shown that human behaviour is driven by a number of motives – different internal and external factors [38]. The motive is the leading and facilitating function of the activity (internal psychic encouragement), which, when creating the subject of the activity, directs the human activity. The motivation of the action is formed by a separate motive or motives. Certain types of activity can contribute to a certain motivation and different motives can contribute to a certain type of activity [22]. Juvenile offenders have their own stereotypical way of thinking – "if anyone can do it, why not me". This is one of the motives of commission of a criminal offense, the same as self-affirmation; a desire to be independent and lack of material provision in the family. The behaviour of minors has their own characteristics: insufficient experience of life, low self-criticism, superficial evaluation of circumstances, increased sensitivity, increased mobility and verbal activity, they are easy to impress, imitation, imbalance, increased sense of independence, striving for recognition by the referent group. Hackers also have the possibility of hacking for the sake of their ego, for proving a self that is different from the selves of others. Perpetrators in this category are usually frustrated in social competition elsewhere and seek an opportunity to compensate by employing their computer techniques [39].

The author points out that cybercriminals do not form homogenous group of criminals. A cybercriminal can be both a woman, a man of any age, economic status, race, religion or nationality. In addition, it should be noted that a cybercriminal has several advantages over a criminal who committed the offense in the traditional way: 1) Global accessibility; 2) Anonymity; 3) Disproportion between the perpetrator's actions and the victim's protection – the attacker chooses the time, place, type and method to offend the victim. There is no direct contact between criminals and victims, no physical use of the weapon; 4) Distance and mobility – there is no need to flee from the crime scene, low risk, but potentially great material gains or profit.

Most cybercrimes are by nature serial in that the offender habituates their behaviour and commit multiple offenses. From this, signature and modus operandi can be drawn. For example, analysis of indicators of the attack's "digital crime scene" can determine the computer hacker's intrusion activity and provide them with an insight. As such, it is an important method when it comes to classifying criminal investigations [40].

The criminal profile of a cybercriminal includes a data set of inductive and deductive profiles of a criminal. Inductive profile of a cybercriminal combines: collection of statistical data related to certain behavioural patterns; – the demographic characteristics of a criminal. Deductive profile of a cybercriminal includes the following range of data – set of the evidence obtained; evidence found on the crime scene and traces; victimological aspects; description of a criminal's personality [8].

The author supports the view that the process of profiling of cybercriminal consists of four interrelated and successive stages:

1) Victimological aspect. The current procedure within criminal profiling is the study of victimological aspects – the role and behaviour of a victim before the cybercrime, during and after it. They reveal new information; allow to study reasons and consequences in details, and to assess the risk weight of a potential victim's actions. The virtual space creates a greater chance of facing a socially dangerous person and more often than in the real world. The content entered on the Internet is eternal, providing long-term content of information, repetitious victimization of a victim, which can last for many years, often a victim not knowing about that. The researches show that a majority of the Internet users lack the experience, skills to identify, assess risk, identify and prevent harmful effects. For example, the activity of a neglected child in the cyberspace (disclosure of personal information, contact with a foreigner etc.) is of high risk. Cybercrime can harm an individual, enterprise or organization. The victimological aspects of cybercrime can be described by using Routine activity theory (RAT) and General Theory of crime and victimization (known as a general theoretical perspective).

According to the RAT guidelines, crime occurs when three elements interact in time and space: a motivated cybercriminal, a suitable victim (a suitable target) and a lack of an active, capable defender. Thus, cybercrime happens when there is a motivated cybercriminal and a suitable victim, and there is no one who might prevent a cybercrime.

In accordance with the General Theory of crime and victimization the main individual factor, which causes a criminal offence and deviance, is the low self-control of an individual. Self-control is explained as the inability by an individual to exercise personal restraint in the face of tempting immediate and easy gratification both in the short and long-term [41].The founders of this theory Michael Gottfredson and Travis Hirschiclaim that a person's self-control level is already set at an early stage, aged 8 and 10. This is a consequence of ineffective parenting and it has a variety of features that occur throughout the life cycle (bullying, bad marks, delinquency, school leaving, divorce, alcoholism, obesity, crime and unemployment). For example, the study of American specialist Fawn T. Ngo and Raymond Paternoster showed that low levels of self-control were significantly related to the likelihood of experiencing three of the five forms of cybercrime victimization – unauthorized access to one's computer, having information added, deleted or changed on one's computer without knowledge or permission, and online harassment [42].

2) Clarifying the motives of a criminal. Within the framework of criminology, several motives of cybercrime can be distinguished.

Motive of financial gain (money, financial resources): According to the data of Global Report "The State of Industrial Cyber security 2017", the average annual cumulative reported financial loss for a business affected by an ICS cyber security breach was $347,603 including the actual consequences of the incident and costs for software upgrades, staff and training [43]. According to the UK survey, 67% of hackers admitted that money is the main incentive for criminal activity. It is estimated that one cybercriminal manages to gain more than 20,000 pounds a year on average, an average of 8,600 pounds in each attack [44].

Emotional motive: Often, cybercrime is based on emotions – anger, rage, hatred, revenge, love or despair, hopelessness. The motivation spectrum for fraud is broad – greed, expression of power, vengeance, adventurousness, desire to taste the "forbidden fruit", gain satisfaction, self-affirm, gain respect, cause harm. It can also be a revenge of an unsatisfied employee by joining an employer's computer system.

A motive of self-affirmation or self-respect: It should be noted that an activity of a cybercriminal as a human being is determined by his/her cooperation with the environment. This process takes place in two basic forms: the need for self-fulfilment (self-affirmation) and the need for self-respect. The need of a cybercriminal for self-affirmation is a testimony of one's abilities, knowledge, as well as its importance and needs, such as the desire to succeed. In turn, the need for self-respect is the desire to gain recognition, appreciation and feedback from other (surrounding) people. Thus, it is an attempt to succeed, to gain appreciations, praises, which strengthens his/her self-respect.

Sexual impulses and desires: Sexual motivation or libido develops throughout the individual's life. For sexually easily excitable persons with weaker will and insufficient or improper social upbringing, the desire to satisfy one‘s libido frequently becomes prevalent, overwhelming many other tendencies and interests, and drives these people to violate the basic standards of people‘s mutual relationships, ethics and rights [45]. The results of the Cybercrime Comparative Study [37] show that: 1) The offenders, who committed crimes referring to child pornography, are between 15 and 73 years old (average age – 49 years); 2) 60% of criminals not only stored, but also distributed these materials; 3) one fifth of these criminals did not work (was retired, unemployed or received a benefit), others worked or studied; 4) 42% of criminals lived with a partner and/ or child; 5) 4% of all criminals had mental health problems; 6) all the criminals identified in the study carefully hid their activities from their relatives; 7) the registered duration of the offense – from six months to 30 years.

Political, ideological and religious motives: Hacktivism – cyber-attacks in the form of political or social protest – is the most common form of politically motivated harm. Hactivists are particularly active in the context of political events.

"Just for fun" motive is usually typical for adolescents or children [46]. The data collected from the survey "Mobile phone and Internet usage habits among children and young people" show that only 4% of parents know that their child has emotionally humiliated, hurt another child via a mobile phone, and 3% know that a child has threatened someone using a mobile phone. In turn, 12% of children admitted that they had performed above mentioned activities. Parents do not know about such activities of their children: 23% of parents do not know about emotional humiliation, 18% – about threatening by using a mobile phone [47].

3) Identification of features/properties. Inductive criminal profiles are developed by studying statistical data involving known behavioural patterns and demographic

characteristics shared by criminals. Deductive profiling uses a range of data including forensic evidence, crime scene evidence, victimology, offender characteristics etc, using such techniques seems possible in the physical world. However, in the cyber-world, their applicability might be questionable [48].

4) Digital behavioural analysis (digital evidence). Today, the field of digital forensics is rapidly advancing, a sign of encouragement to the cyber-criminal profiler. Technical ability, for the purposes of profiling, consists of a subject's expertise with digital technologies, as opposed to other technical skills (e.g., engine repair). There are two distinct subareas that are of interest in the investigative profile – general expertise and the adoption of new technologies [49]. Digital behavioural analysis is a relatively new field that applies the concepts of traditional behavioural analysis to the digital footprints of criminals. The importance of digital forensics is apparent since it is the only means of tracing the perpetrator in the absence of physical evidence. Digital evidence information of value to a criminal case that is stored or transmitted in digital form [50]. The digital evidence can indicate what category of a victim is to be searched for, where a criminal could have contacted the victims [8].

Cybercriminal behaviour is influenced by the interaction between numbers of factors. Such behaviour is a result of mutual interaction, which covers individual, social, environmental factors and conflicts between individual and society. It is influenced by various factors: heredity, education, culture, lifestyle and socio-economic conditions. Author admit that cybercrime is often caused by a person's expression of frustration and bitterness against the social structure and person's position in it.

7 Conclusions

1. Profiling of a cybercriminal is legal (criminal procedural, forensic, criminological) psychological method, which used to determine the behavioural tendencies of a criminal, personal traits and demographic peculiarities, as we as to predict the criminal's further activities.
2. Profiling of a cybercriminal shows the character of a personality of a typical cybercriminal:
   a. Technical knowledge;
   b. The degree of legal nihilism;
   c. High tolerance to risk and needs to risk;
   d. Commitment to self-assertion;
   e. A desire to manipulate or beat others;
   f. A description of motivation, which reveals the reason for committing a crime.
3. Profiling of a cybercriminal may include several criminological and criminal-law-based key elements:
   1) personal characteristics/traits that are specific to a particular person and which predispose a person to commit a cybercrime;
   2) criminal professionalism – by understanding the personality traits that contribute to the safe and effective cybercrime;
   3) social characteristics – demographic features, socio-economic status, socio-psychological and moral qualities;
   4) motivation characteristics – a system of human activity orientation that prompts people to act;
   5) the activity type of a cybercriminal (modus operandi);
   6) the way a cybercriminal has identified or contacted the victim.

References

[1] K. Jaishankar (2011) *Cyber Criminology: Exploring Internet Crimes and Criminal Behavior* (Boca Raton, FL, USA: CRC Press, Taylor and Francis Group, 2011)
[2] V. Tumalavičius, J. Ivančiks, O. Karpishchenko (2016) Issues of society security: public safety under globalization conditions in Lithuania, *Journal of Security and Sustainability Issues* **4**(9): 545–573. https://doi.org/10.9770/jssi.2016.5.4(9)
[3] *Global Cyber Risk Perception Survey 2018.* By the Numbers: Global Cyber Risk Perception Survey. Available on: https://www.marshmma.com/blog/2018-cyber-and-data-security-risk-survey-report
[4] *Special Eurobarometr 390 Cyber Security. Report. 2012.* Available on: http://ec.europa.eu/commfrontoffice/publicopinion/archives/ebs/ebs_390_en.pdf
[5] D. Halder & K. Jaishankar (2011) Cybercrime and the victimization of women: laws, rights and regulations. Information Science Reference
[6] *Criminal Profiling: International Theory, Research, and Practice* (2007), edited by Richard N. Kocsis (Human Press Inc, Totowa, New Jersey)
[7] S.A. Bonn, Criminal Profiling: The Original Mind Hunters. Available on: https://www.psychologytoday.com/intl/blog/wicked-deeds/201712/criminal-profiling-the-original-mind-hunters
[8] B.E. Turney (2012) Criminal Profiling: An Introduction to Behavior Evidence Analysis. Fourth Edition (Elsevier, Oxford)
[9] Forensic Criminology (2010) / edited by Wayne Petherick, Brent E. Turvey, Claire E. Ferguson (Elsver Academic Press)
[10] M. Scheinin, Report of the Special Rapporteur on the promotion and protection of human rights while countering terrorism', UN Doc. A/HRC/4/26, 29 January 2007. Available on: https://unispal.un.org/DPA/DPR/unispal.nsf/1ce874ab1832a53e852570bb006dfaf6/813e9af0b820e2e58525730800513beb?OpenDocument
[11] C.R. Bartol, A.M. Bartol (2011) Introduction to Forensic Psychology: Research and Application (SAGE Publications)
[12] R.J. Lilly, F.T. Cullen, R.A. Ball (2015) Criminological Theory. Context and Consequences. Sage Publication, Inc., 2015
[13] D. Canter (1994) Criminal Shadows: Inside the mind of the serial killer (Endeavour Media)
[14] P.B. Ainsworth (2001) Offender Profiling and Crime Analysis **7** (2001)
[15] A. Davies (2002) Rapists Behaviour: A three Aspect Model as a Basis for Analysis and Identification of a Serial Crime, Forensic Science International, **173**
[16] V.J. Geberth (1996) Practical Homicide Investigations: Tactics, Procedures, and Forensic Techniques, 4th edition
[17] E. Strika. Kriminālā profilēšana (2015) / Psiholoìija 3. Lietišiās jomas. Zvaigzne ABC (In Latvian)
[18] R. Saroha (2014) Profiling a Cyber Criminal. Int. J. Inf. Comput. Technol. **4**(3), 253–258
[19] J.-M. Dinant, C. Lazaro, Y. Poullet, N. Lefever and A. Rouvroy (2008) Application of Convention 108 to the Profiling Mechanism – Some ideas for the future work of the consultative committee. Available on: *http://www.coe.int/t/e/legal_aff airs/legal_co-operation/data_protection/ documents/*

[20] Uz pamattiesībām balstīta policijas darbinieku apmācība. Rokasgrāmata policijas darbinieku apmācītājiem (2018) Eiropas Savienības Publikāciju birojs (In Latvian)

[21] D.A. Andrews, J. Bonta (1998) *The Psychology of Criminal Conduct* (New York: Matthew Bender & Company, Inc.)

[22] I. Vedins (2008) *Zinātne un patiesība*. Rīga: Avots (In Latvian)

[23] K. Rossmo (2000) Geographic profiling (CRS Press LLC)

[24] D.A. Muller (2000) Criminal profiling: Real science or just wishful thinking? Homicide Studies **4**(3), 234–264

[25] D. Kahneman, P. Slovic, A. Tversky (1982) Judgment under uncertainty: Heuristics and biases (New York: Cambridge University Press)

[26] R. Bull, C. Cooke, R. Hatcher, J. Woodhams, C. Bilby, T. Grant (2006) Criminal psychology: A Beginner's Guide (Oneworld Publications, Oxford, England)

[27] D. Canter, D. Youngs (2008) *Principles of Geographical Offender Profiling* (New York: Toylor & Francis Group)

[28] R.N. Kocsis (2006) Criminal Profiling: Principles and Practice (Totowa NJ: Humana Press Inc.)

[29] Kriminālprocesa likums: Latvijas Republikas likums. Latvijas Vēstnesis Nr. 74 (3232), 11.05.2005 (In Latvian)

[30] I. Jakimov (2003) *Kriminalistika. Rukovodstvo po ugolovnoj tehnike i taktike*. Novoje izdanije perepechatonoe s izdanii 1925, Moskva: LeksEst(Якимов, И. Н. (2003) Криминалистика. Руководство по уголовной технике и тактике. Новое издание, перепечатанное с издания 1925 г. Москва: ЛексЭст) (In Rusian)

[31] L.J. Alison (2005) *The forensic psychologist's casebook: Psychological profiling and criminal investigation* (Cullompton, UK: Willan)

[32] R. Knight, J. Warren, R. Reboussin, B. Soley (1998) Predicting rapist type from crime scene variables. Criminal Justice and Behavior **25**, 46

[33] J. Lieckiewicz (2011) Cybercrime psychology – propsal of an ofender psychological profil. Problems of Forensic Sciences, Vol. **LXXXVII**: 239–252. Avalable on: `http://www.forensicscience.pl/pfs/87_Lickiewicz.pdf`

[34] B. Snook, J. Eastwodd, P. Gebdreau, C. Coggin, R.M. Cullen (2007) Taking stock of criminal profiling: A narative review and meta-analysis. Criminal Justice and Behavior **34**, 437–453

[35] S. Omārova, (2002) *Cilvēks runā ar cilvēku*. Rīga: Kamene (In Latvian)

[36] V.V. Tulegenov (2014) Kiberprestupnost kak forma virazheniia kriminalnogo professionalizma. *Kriminologiia: vchera, segodnia, zavtra*. **2**(33) (Тулегенов В. В. (2014) Киберпреступность как форма выражения криминального профессионализма. *Криминология: вчера, сегодня, завтра* № 2 (33) (In Russian)

[37] *Comprehensive Study on Cybercrime* (2013) New York: United Nations. Available on: `https://www.unodc.org/documents/organized-crime/UNODC_CCPCJ_EG.4_2013/CYBERCRIMESTUDY_210213.pdf`

[38] N. Ksheti (2010) *The Global Cybercrime Industry: Economic, Institutional and Strategic Perspectives* (New York: Springer)

[39] X. Li (2017) A Review of Motivations of Illegal Cyber Activities. Criminology & Social Integration Journal **25**(1), 110–126

[40] F. A. Zuhri, The Profile of a Cybercriminal. Digital Forensic Magazine. Available on: `http://digitalforensicsmagazine.com/blogs/wp-content/uploads/2017/05/The-Profile-of-Cybercriminal.pdf`

[41] T. Hirschi (2004). Self-control and crime. In R.F. Baumeister & K.D. Vohs (Eds.), Handbook of self-regulation: Research, theory and applications (New York: Guilford Press), pp. 537–555
[42] F.T. Ngo, R. Paternoster (2011) Cybercrime Victimization: An examination of Individual and Situational level factors. International Journal of Cyber Criminology **5**(1) (January – July), 773
[43] Global Report "*The State of Industrial Cybersecurity 2017*". Available on: `https://go.kaspersky.com/rs/802-IJN-240/images/ICS%20WHITE%20PAPER.pdf`
[44] S. Simkin, L. Ponemon, (2016) *Flipping the Economics of Attacks*. No: *Ponemon Institute*. Available on: http://www.polyscope.ch/site/assets/files/42688/06_16_53.pdf
[45] *Crime Investigation (Investigation features of the most frequently encountered crimes)* (2003) /edited by professor Anrijs Kavalieris. Riga: The Police Academy of Latvia
[46] D. Shinder (2010) Profiling and categorizing cybercriminals. Available on: `http://www.techrepublic.com/blog/it-security/profiling-and-categorizing-cybercriminals/`
[47] Aptauja "*Mobilo telefonu un interneta izmantošanas paradumi bērnu un jauniešu vidū*" (2012). Available on: `http://www.drossinternets.lv/upload/materiali/petijumi/mob-tel_interneta_izmantosana_2012.pdf` (In Latvian)
[48] H. Tennakoon, The need for a comprehensive methodology for profiling cyber-criminals. Available on: http://www.newsecuritylearning.com/index.php/archive/150-the-need-for-a-comprehensive-methodology-for-profiling-cyber-criminals
[49] C.M. Steel, Idiographic digital profiling: behavioral analysus based in digital forensics. Journal of Digital Forensics, Security and Law **9**(1), 7–18
[50] Scene of the Cybercrime: Computer Forensics Handbook (2002) 1st Edition (Syngress Publishing, Inc.)

Cybercrime and sexual exploitation of children in e-environment in the context of strengthening urban and rural security

Andrejs Vilks

Rīga Stradiņš University, Faculty of Law, Riga, Latvia

Abstract. Among various types of threats to urban and rural security, there has been an increase in cybercrime in recent years, an increase in criminal offenses in the e-environment. This also applies to the sexual exploitation of children in the e-environment. Sexual violence against children in cyberspace is linked to objectively determined factors, such as the development of Internet technologies and their wide availability. The active efforts of children to learn and use the new technologies, the attractiveness of the virtual environment in contrast to the real situation and the ambiguous and in many cases incomprehensible nature, as well as the saturation of pornographic material in the Internet, the weak control of their distribution, should be noted. Due to the increasing availability of the Internet, the expansion of e-services in cities and rural areas is relatively balanced. An Internet user may be at risk in both big cities and in a less populated rural area. Among Internet users the proportion of young people and children is relatively high, which objectively determines their increased victimization potential. Among threats to children related to the e-environment, multiple forms of child sexual abuse, including *cyberbullying, sextings, sextorings, cyberstalkings*, are increasing.

Key words: children, security, e-environment, cybercrime, sexual exploitation.

1 Introduction

Sexual abuse of children and sexual exploitation of children, including on the Internet, and child pornography as well, constitute a serious violation of fundamental rights, especially as regards the right of the child to protection and care, necessary for children wellbeing. With the use of new technologies and the Internet, child pornography containing images of sexual abuse of children, including other particularly serious forms of child sexual abuse and sexual exploitation of children, is growing and spreading. On 13 December 2011, the European Parliament adopted Directive 2011/92 / EU on combating the sexual abuse, sexual exploitation of children and child pornography and replacing Council Framework Decision 2004/68/JHA [7]. For serious criminal offenses, such as child sexual abuse and child pornography, a comprehensive approach involving the prosecution of offenders, the protection of child victims and the prevention of appropriate manifestations are essential. In taking any action to combat these offenses, the best interests of the child should be a primary consideration. The Directive provides for criminal liability for deliberate access to child pornography using information or communication technologies. In order to be held liable, a person should have both intentions to enter a site, where child pornography is available, and know that such images are found there. Sanctions should not apply to people who have inadvertently accessed child pornography sites. A deliberate offense may, in particular, be recognized by the fact that it was committed repeatedly or by means of paid services. According to Eurostat's requirements, such activities as prostitution in national accounts, which could include child pornography, should now be included [33].

2 Aim and methods

To identify the tendency in cybercrimes and sexual exploitation of children in urban and rural areas, their determining factors and prevention options. *The sexual exploitation of children in cybercrime in the context of urban and rural environments will be explored using a descriptive method that will analyze the prevalence of this phenomenon, its development trends, determinants, as well as prevention and combating opportunities. The author uses the results of statistical data and optional research in the article by doing the comparative analysis.*

3 Results

The sexual exploitation of children in the e-environment is influenced by the specific nature of the environment in which they are located. Undoubtedly, the peculiarities of urban and rural environments influence the socialization of individuals, the formation of a value system and socially important activities. The urban environment creates a qualitatively new link between people living in it. This environment has specific features such as: an advanced communication system, including high traffic; a set of industrial enterprises; a broad representation of service companies and entertainment institutions; a large number of people living permanently in urban areas; anonymity of the population that promotes the spread of deviant behaviour. In the urban environment there is a wide availability of new Internet, including informative communication technologies. In the rural environment, on the other hand, one of the greatest benefits of life is environmental safety. Lack of industrial centres and proximity to nature has a positive impact on the environment and the physical and emotional development of individuals, including children. The main disadvantage is that in some rural areas there is not a sufficiently developed system of utilities, which seriously complicates the life of the villagers. Rural areas are directly linked to agricultural production. The decline in agricultural production contributes to unemployment, which in turn leads to problems such as alcohol abuse and crime. Urban and rural environments are now increasingly linked by a unified communicative system, the Internet, and social networks.

Nowadays, the Internet has become the key component of global communications and information infrastructure, along with press, television and telecommunications. The Internet can be seen as one of the most readily available means of communication and information gathering, which is becoming more widely available in Latvia. The quality of communications provided by the Internet and the availability of versatile information resources make a significant contribution to the development of social life by raising the general level of public awareness, improving educational opportunities and methods, and ensuring the availability of free information. The Internet has become a media that effectively gives information and promotes social interaction. Any person can get any information needed. However, the use of the Internet service must also address the challenges posed by these opportunities, with a particular focus on the problems of children's rights over the Internet. In general, cyber threats, including crimes in the e-environment, are increasing in the world, and criminals' criminal assets are increasing. Cybercrime is considered to be at low risk of crime detection. Criminological studies and experts confirm this feature [13, 16]. One of the most threatened social groups due to insufficient social experience, high communicative activity, widely available informative communication technologies (ICT) are children. There are major changes in media use among minors. Minors are increasingly using the media through mobile devices, including online video games which in their turn lead to increased demand for Internet services. The phenomenon of socialization sites is becoming increasingly important both at the individual and the community level. This provides many opportunities for minors, but also poses challenges for child protection. The topicality of the issue is also emphasized in international studies [31].

The growth of child pornography is supported by a number of conditions, rapid technological development, anonymity provided by cyber space and the low cost of consumed resources. Studies have shown that around 10–20% of minors across Europe could have been sexually abused in childhood, including using modern communicative technologies [23]. Internet monitoring data show that 68% of the pornographic resources available on the Internet contain child pornography. The most common victims are children aged 7–9, children aged 10–13 – 36%, and children under 7–25%. In general, it is possible to distinguish three positions of the rights of children on the Internet: children as Internet users (www pages, chat channels, mailing lists, games); children as content of pornography and erotic web pages (visual material on websites containing pornography); children as a source of threat to the legitimate rights of other Internet users.

3.1 Trends in the prevalence of sexual violence against children

When analyzing the problem of sexual violence, including cyber-violence, it is essential to determine the operational content. The Law of 30 May 2013 "Amendments to the Law on the Protection of the Rights of the Child", which came into force on July 4, 2013, clarified the definition of violence, stating that violence is all kinds of physical or emotional cruelty, sexual violence, neglect or other behaviour, that endangers or may endanger a child's health, life, development or self-esteem [10, 11]. There is relatively little empirical research on the risks of children in the e-environment, including sexual ones. The Austrian Safer Internet Centre survey among young Europeans concluded that sexual abuse of children exists in different forms and is common in Europe and Latvia. Almost a third of the interviewed children (30%) admitted to having been exposed to *sextings* personally [27].

Although there are studies that point to risk factors that may increase the potential for sexual abuse, the age and background of victims of sexual abuse do not know the limits. Images and videos of child sexual abuse found online include both boys and girls from 0 to 18 years old. Children were under 8 years of age, and 63.40% of those children were under the age of 8 years of age. When evaluating reports for its purpose, the Canadian Child Protection Centre found that children under 12 years of age represented 78.30% of their team's pictures and videos, and 63.40% of these children were under 8 years of age [3]. In the same material, they found that 80.42% of the children were girls and 19.58% were boys.

Several images depicting the same child become a record of violence and can be circulating many times, much more than an abuse, resulting in additional adult injuries. Each year, the National Centre for Missing and Missing Children reviews 25 million images. That is 480,769 images per week. Data from the INHOPE Association's 2017 report show that almost two thirds of children (60,85) have been exposed to pornography [15]. INHOPR association data give proof that in 43 countries (including Latvia), 87,930 reports of illegal material on the Internet showing sexual abuse of children were received. These are 259,016 images or videos with illegal content. Most – (62%) illegal content was deleted within three days.

In Latvia, as evidenced by the analysis of criminal statistics, sexual abuse against children and the number of crimes committed by minors increased in 2018 [22]. State Police Chief

I. Kuzis admits that sex offenses among young people are often related to violations committed on the Internet. For example, young people can take sexual pictures that can later be used against them or come into pornographic circulation. Crimes are committed in the Internet, dealing with fraud, illegal pornography and illegal contents. We have to admit, that crimes in cyber environment are committed in a more and more refined manner, criminals use programs such as Skype or WAP and MMS on their mobile phones for camouflage purposes; they create fake profiles in websites, by means of free access online programmes they change computer configurations "fake IP", etc., fake IP addresses for use which essentially makes it harder to detect the mentioned crimes.

On the Latvian Drossinternets.lv report line, in 2018, 180 reports on child sexual abuse material were received on the Internet. It makes 31% of the total number of received reports. By 28 February, 2019, 5921reports of illegal, harmful Internet content and problem situations on the Internet were received and processed [30].

According to the Riga Christian Shelter study "On Sexual Exploitation of Children in Latvia" at least 1000 children suffer from sexual exploitation per year. It is still admitted that the number is much bigger – up to 5000. We can conclude that sexual abuse of children is a much more serious problem than depicted by official statistical data [26].

As children grow, there is a growing risk of suffering crimes against morals and inviolability. Boys are less likely to suffer from the above-mentioned criminal offenses than girls [28].

3.2 Most widespread forms of sexual violence against children

One of children sexual exploitation types is children pornography. According to Law on Pornography Restriction Section 1 children pornography is a pornographic performance in which a child is involved, or a material of pornographic character in which a child is depicted or described [29]. Any picture or a material is considered as children pornography in which:

a) A child who is involved in sexual activities, a child completely or partially without clothing in a sexual pose or in clothing of an obscene nature is depicted or described;
b) A person having the appearance of a child who is involved in the activities mentioned in the sub-Clause 1 "b" and "c", whose genitals are fully or partially depicted, or described;
c) There are realistic images with an actually non-existent child's genitals who is involved in the activities specified in sub-Clause "b" and "c" or presented in a manner specified in sub-Clause "a".

Transmission of photographic pictures of sexual character is the threat of sexual exploitation. The Criminal Law of the Republic of Latvia Section 166 provides for "punishment for downloading, acquisition, importation, production, public demonstration, advertising or other distribution of such pornographic or erotic materials as relate or portray of the sexual abuse of children" [13]. In Latvia, the use of online social networking sites with smart-phones and tablets is becoming more and more common, encouraging young people to participate in the production of pornographic material and the transmission of child pornography graphic files.

Cyberbullying is becoming more and more common among forms of sexual violence. Modern children and young people choose to fight each other using modern technology and the Internet. For example, pupils use smart-phones to photograph a classmate half-clothed in a dressing room and then publish a picture in Facebook to attract general attention. Usually the classmates actively share such pictures with the whole school to make fun of the unhappy victim. Research shows that one fifth of youngsters (21.9%) have been exposed to virtual terror – *cyberbullying* [17]. The research done in Europe among 14-17 teenagers reveals alarming conclusions – every fifth teenager in Europe has experienced bullying and mockery in the Internet. More than half (63%) of European teenagers admitted that they are contacting strangers in the Internet, but every third one has met such strangers. Virtual terror should by no means be considered a minor phenomenon. More than half of schoolchildren surveyed (53.5%) who had experienced it themselves, indicated that bullying and mockery in the Internet had very negatively affected their psyche. It is interesting that teenager *cyberbullying* victims may be both boys, and girls, though girls are attacked more, and with sexual motivation.

Cyberstalking is seen to be another form of sexual abuse, which can be defined as cybertracking by using the Internet environment, or other electronic devices [6]. It may include defamation, false accusations, and identity theft, as well as the risk of rape. Intimate partner cybercrime is the online persecution of the current or former romantic partner. Cyberstalking aims to control the victim to promote social isolation and addiction. Stalkers can be an online stranger or a person who knows the victim or the target audience. It can send repeated offensive or threatening e-mail messages to its victims, monitor or disrupt the use of their e-mails, and use the victim's account to send e-mail to others. They can also use the Internet to research and collect personal information about the victim to persecute him [20].

3.3 Factors determining sexual violence in children in the e-environment

In order to be able to successfully prevent and combat the sexual exploitation of children in the e-environment, as well as in urban and rural environments, it is essential to identify the determinants of this crime. Factors of children sexual exploitation in the e-environment can be differentiated by several, to our mind, essential criteria. First of all, depending on the scope of factors, we can divide into: 1) global factors (mega-environment); b) macro-environmental factors (mainly characteristic for the national level in macro-environment); c) micro-environmental factors which are characteristic for specific subjects whose rights are violated (children, their parents).

As to the first fact group, it is important to separate a fast Increase of Information and Communication Technologies and the Widening of their Applications. ICT have become the basis for economic growth and it is an important resource which is used in all spheres in economics and communication [5]. Representatives from the UNO International Telecommunications Union – ITU have published the data on prognoses in the information and technology fields and their consumption in the world. The prognosis is that at the end of 2018 slightly more than half or 3.9 billion of world population will use the Internet [32]. At the end of 2018 the point will be reached when more than or 51.2% of world population will use the Internet. Broad access to ICT determines that the e-environment more often is used by persons prone to sexual violence against children, developing sexual-type platforms, children pornography websites, etc., as well as minors who are active Internet users, at the same time getting acquainted with the erotic and pornographic materials on the site.

Corresponding but more pronounced tendencies are observed at macro, national level. Thus, the number of regular Internet users (at least once a week) in 2018 reached 81% [14]. More commonly the Internet users are students and schoolchildren. Each year the Internet availability in the territory of Latvia is increasing. The Internet is most often available in households in Latvian cities (83%). In the rural areas the Internet availability is increasing year by year – in 2018 it had reached 78%, this indicator, comparing to 2017 has increased by five percentage points. In households of Latgale region it was 74%. Compared to 2017, the fastest increase in the Internet access was observed in Latgale region, where it grew by five percentage points.

In micro-environment, no doubt, wide ICT and the Internet availability are observed. The practice of its use is related to the traditions, acquired and accumulated cultural values of the particular social groups, families, including children, the level of legal consciousness. It is necessary to pay attention once again to the fact that the sexual exploitation of children in the e-environment, in both the city and the rural environment, has a very high latency. Sexual abuse against children is related to the time spent by minors on computers, tablets, intensive use of mobile devices. In essence, the virtual environment largely replaces the real one. It is basically the lack of emotional support at school, unfavourable conditions in the family, inadequate child occupancy and poor control. There may be sexually explicit elements in the family, alcohol or drug abuse by parents, and passion for gambling. Poverty is an obstacle to the development of children, and so is poverty in the family. Poor and low-income families are often unable to pay for school supplies for children, out-of-school social and cultural events, leading to a delay in children's social development [21]. Sociological research also mentions the impact of poverty on violations of children's rights, including sexual violence against them [4]. The question arises of how poverty can affect sexual abuse of children in the e-environment? First of all, a poor family is not able to ensure the diversity of children's education and interest development activities. As in the city, diverse ICTs are widely used in rural areas as well. In the current circumstances, unlimited availability of large resources is not required. Secondly, inadequate material resources determine limited satisfaction of vital needs and the emergence of socially disadvantaged families, which constitute weak child control, including the use of mobile devices. Poverty has an economic, social, political and cultural aspect. In general, it threatens human rights- economic (right to work and sufficient income), social (access to health and education), political (freedom of thought, expression and association) and culture (right of residence), cultural identity and engagement in community cultural life.

Factors contributing to the increase in child sexual abuse in cities and rural areas in the e-environment can also be divided into the following groups: a) Technogenic factors: Wide availability of the Internet; Increased Internet download speed; number of users, including minors; ICT modernization; b) cultural environmental factors: accessibility of social services; level of socio-economic development; c) factors characterizing the personality of children. Opportunities to prevent child sexual abuse in the e-environment are lower in rural areas than in cities, which is associated with a lack of control resources.

3.4 Aspects of preventing and combating the sexual exploitation of children in the e-environment

In the context of preventing and combating the sexual exploitation of children in the e-environment, the following aspects need to be addressed: a) legal; (b) promoting the principle of self-regulation of the provision of Internet services and technical measures; (c) educating the public on safe Internet use and ethics [18].

As regards the legal aspects of preventing and combating the sexual exploitation of children in the e-environment, it is necessary to clarify the legal framework for electronic identification services, to introduce common legally accepted technological as well as security standards, to establish a reliable and viable monitoring system for electronic identification services as part of from a national ICT security monitoring and coordination mechanism. The European Parliament Directive 2011/92 / EU "Combating Sexual Abuse of Children, Sexual Exploitation of Children and Child Pornography and replacing Council Framework Decision 2004/68 / JHA" [7] plays an important role in preventing sexual abuse of children. The Directive defined the necessary activities of the Member States in the development of national legal acts and further development of the institutional system. In the context of criminal liability for acts related to pornographic performances, this is the case for activities which are the sexual activities of viewers. On the basis of the aforementioned Directive, the disposition of the first part of Article 166 of the Criminal Law of the Republic of Latvia was supplemented on 15 May 2014 with a new form of pornography – a pornographic performance – which in substance corresponds to pornographic performances, but differs in that pornography is not recorded in video recordings, or in any other material form, but is displayed on-site or live [12]. Thus, personal direct contact between peers with mutual consent is excluded from the definition (amendments to Criminal Law). Child pornography often contains images showing sexual abuse of children by adults. They may also be images showing children involved in sexually explicit conduct, or their genitals, if such images are captured or used primarily for sexual purposes, with or without the child's knowledge. As regards serious forms of sexual abuse and sexual exploitation of children, the national legal framework, as provided for in Directive 2011/92 / EU, should review effective, proportionate and dissuasive sanctions. This applies in particular to the various forms of sexual abuse of children and the sexual exploitation of children, facilitated by the use of information and communication technologies, such as online harassment for sexual purposes on social networking sites. The Directive requires Member States to provide for criminal sanctions in their legislation, subject to the provisions of Union law on combating sexual abuse, sexual exploitation of children and child pornography. The Directive requires that there should be criminal liability for deliberate access to child pornography using information or communication technologies. In order to be held liable, a person must have the intention to enter the site where child pornography is available and know that such images are found there. A deliberate offense may, in particular, be recognized by the fact that it was committed repeatedly or by means of paid services.

Child pornography, that is, images of sexual abuse against children, is a specific type of content that cannot be seen as an expression of opinion. In order to combat this, it is necessary to limit the circulation of child abuse material, ensuring that it is more difficult or impossible for offenders to upload such content on the public Internet. Therefore, it is essential to take proactive measures to remove such content and to apprehend those responsible for making, distributing or downloading images of child sexual abuse. We still have to admit, however, that despite such efforts, the removal of child pornography content from their source is often impossible if the original materials are not in EU countries. In some cases, the removal of materials in the country concerned is a particularly long process. It is necessary for the EU to introduce a single mechanism to block access to websites that have been identified as having child pornography or distributed through these sites within the Union. In order to remove and block child abuse content, cooperation between public authorities should be established and strengthened, in particular with a view to ensuring that correspondence with websites containing child pornography material is as complete as possible and that duplication of work is avoided. Such measures should always take into account the rights of end-users and ensure compliance with existing legal and judicial procedures as well as the provisions of the European Convention for the Protection of Human Rights and Fundamental Freedoms [8] and the Charter of Fundamental Rights of the European Union [2] unprecedented anonymity. Individuals have the opportunity to hide their true identity and personality traits such as their age. Investigative and investigative authorities on child sexual abuse, including cyber crime, and criminal prosecution should provide effective means of investigation. These tools could include special investigation activities, including interception, secret surveillance, including electronic surveillance, bank account monitoring or other forms of financial investigation, taking into account the type and severity of the offenses under investigation. Where appropriate and in accordance with applicable law, such means should also include the possibility for law enforcement authorities to use the hidden identity on the Internet.

Thomas Brewsters, a researcher at the Forbes magazine, has drawn attention to the peculiar methods of the US Federal Bureau of Investigation (FBI) to combat child pornography networks [9].

The FBI has used a completely new tactic in the fight against child pornography: taking over the suspects' online identities to enter private groups with illegal content. Agents used the identity of 23-year-old Daxton Hansen from Salt Lake City who used the KitB10 username. From April 12, 2017 to November 13, 2018, the secret FBI agent working for the Salt Lake City agency took over Hansen's online identity. Agents allowed child sexual abuse material on social network Instagram to spread about eighteen months after they had identified a pornographic distributor. This specific operation has two important aspects. First, by using this tactic, it is possible for the police to identify, document and prosecute persons who distribute pornographic material with child images and videos. Secondly, the social networking materials include specific children and young people who are hurting them repeatedly (multiple victimization). In addition, pornographic material continued to be read and downloaded by minors. It has to be admitted that conducting this type of operation should be evaluated by a legal and moral ethical approach. In a study conducted in Latvia in 2018, it was found that every second respondent has encountered violations on the Internet [1].

Among the violations in the Internet there has been found: blackmail or threat thereof; financial fraud schemes; experiencing emotional violence; interacting with pornography material without age-limiting warning. Every seventh (15%) admits that they are familiar with someone on the Internet who, using a false profile is straying on someone else. At the same time, 39% of parents have not been interested whether a child has had any negative experiences using the Internet.

Self-regulation and technical measures for the provision of Internet services are sufficiently broad and specific. The widespread availability of the electronic environment in society is inconceivable without the ability to protect the individual, especially children, from the harmful or unwanted content of the Internet and the ability to provide the necessary assistance and support in situations where such harmful web content has or is likely to occur. It is essential to stop access to harmful Internet content as quickly as possible. The Information Society Guidelines state that it is necessary to report to the State Children's Rights Protection Inspectorate on illegal and harmful content on the Internet. State Children's Rights Protection Inspectorate should provide psychologist support to children who have come across harmful information material on the Internet [24]. The Cabinet of by Ministers Order No.581 of 25 August 2009 approved the "Action Plan for the Protection of Minors from Criminal Offenses against Virtue and Intolerance 2010–2013" 1 [25]. This is also evidenced by the problems currently encountered in practice, for example, there is still no solution for dealing with persons who have sexual interest in children and want assistance, there is still insufficient involvement of the Ministry of Health in solving this problem, and there are still not enough solutions for sexual abuse in crime prevention, public, including non-governmental sector involvement in the prevention and detection of these crimes, etc.

In general, the prevention and combating of child sexual abuse in the e-environment is diverse. However, they are fragmented and included in different programs, guidelines and concepts. Political planning documents do not distinguish between the specificities of the social environment, including rural and urban areas. Preventive measures do not take into account the developments in new information technologies. It should be noted that the effectiveness and positive achievements of measures to prevent and combat sexual exploitation of children in the e-environment are underestimated.

4 Conclusions

The sexual exploitation of children in the e-environment in Latvia has not been studied in detail and comprehensively. Children's sexual cyber threats are more widespread than the real environment and have a high latency. Consequently, it is objectively difficult to determine the extent of the harm inflicted on minors in cases of sexual violence in the e-environment. Sexual abuse of children in a virtual environment distorts their socialization process, draws the value of a positive value system, and traumatizes them emotionally and psychically. The forms of sexual exploitation of children are becoming more diverse and refined. With the exception of widespread child pornography sites and links in the e-environment, new forms of juvenile sexual violence – cyberbullying, sextings, sextorings, cyberstalkings – are also spreading. We can see that, as the potential for violence increases, more and more children are subjected to sexual violence; the availability and impact of sexual information (erotic and pornographic publications) is intensifying, contributing to the development of deviant norms, ideals and values in children; a decrease in the age of sexual exploitation of children;

Factors determining child sexual abuse are social and technological. They function at global, macro and individual levels. Sexual violence against children in cyberspace is linked to objectively determined factors, such as the development of Internet technologies and their wide availability. The efforts of children to learn and use the new technologies, the attractiveness of the virtual environment, in contrast to the real situation and ambiguous and in many cases incomprehensible circumstances, should be noted. Prevalence of pornographic material in the Internet environment and poor control of their distribution are another factor. The European Parliament Directive 2011/92 / EU" Combating Sexual Abuse of Children, Sexual Exploitation of Children and Child Pornography and replacing Council Framework Decision 2004/68 / JHA' as well as the Criminal Law, which is essential for the prevention of sexual violence against children, it determines responsibility for the sexual exploitation of children. At national level, measures to prevent and combat the sexual exploitation of children in the e-environment are diverse. They are included in various programs, guidelines and concepts. However, political planning documents do not distinguish between the specificities of the social environment, including rural and urban areas. Preventive measures do not take into account the developments in new information technologies, the effectiveness and positive achievements of measures to prevent and combat sexual exploitation of children in the e-environment are underestimated. Additional detailed research is needed on the sexual exploitation of children in the e-environment in cities and rural areas, identifying their security enhancement optimization models.

References

[1] Aptauja: katrs otrais interneta lietotājs ir saskāries ar pārkāpumiem internetā. Available online at: `https://www.tvnet.lv/6516375/aptauja-katrs-otrais-interneta-lietotajs-ir-saskaries-ar-parkapumiem-interneta`

[2] Charter of Fundamental Rights of the European Union. Available online at: `https://eur-lex.europa.eu/legal-content/EN/TXT/?uri=celex:12012P/TXT`

[3] Child pornography is sexual abuse material. Available online at: `https://www.thorn.org/child-pornography-and-abuse-statistics/`

[4] Child safety online: global challenges and strategies. Available online at: `https://www.unicef-irc.org/publications/650-child-safety-online-global-challenges-and-strategies.html`

[5] Cybersecurity Strategy of the European Union: An Open, Safe and Secure Cyberspace. Available online at: https://eeas.europa.eu/archives/docs/policies/eu-cyber-security/cybsec_comm_en.pdf
[6] Davis, Joseph A. Stalking Crimes and Victim Protection. Boca Raton, Florida: CRC Press. pp. 3–19. ISBN 978-1420041743
[7] Directive 2011/92/EU of the European Parliament and of the Council of 13 December 2011 on combating the sexual abuse and sexual exploitation of children and child pornography, and replacing Council Framework Decision 2004/68/JHA. Available online at: https://eur-lex.europa.eu/legal-content/EN/TXT/?uri=celex%3A32011L0093
[8] European Convention for the Protection of Human Rights and Fundamental Freedoms. Available online at: https://www.echr.coe.int/Documents/Convention_ENG.pdf
[9] Exclusive: What Happened When the FBI Took Over the Instagram And Kik Of A Child Porn Dealer. Available online at: https://www.forbes.com/sites/thomasbrewster/2019/03/06/exclusive-the-fbi-took-over-the-online-identity-of-a-pedophile-letting-child-porn-spread-for-18-months/#6b2bee875cb1]
[10] Grozījumi Bēnutiesībuaizsardzībaslikumā Available online at https://likumi.lv/ta/id/265209-grozijumi-bernu-tiesibu-aizsardzibas-likuma
[11] Grozījumi Bērnu tiesību aizsardzības likumā", https://likumi.lv/doc.php?id=257679
[12] Grozījumi Krimināllikumā. Available online at: https://likumi.lv/ta/id/266590-grozijumi-kriminallikuma
[13] Gubene A., Barise L. Kibernoziegumu risks pārvaldes institūcijās . Available online at: http://llufb.llu.lv/Raksti/SZF-raksti-2013/SZF_RAKSTI_III_2013-19-26.pdf
[14] Informācijas un komunikācijas tehnoloìiju rādītāji mājsaimniecībās 2018. gadā. Available online at: https://www.csb.gov.lv/sites/default/files/publication/2018-12/Nr_26_Informacijas_un_komunikacijas_tehnologiju_lietosana_majsaimniecibas_2018_%2818_00%29_LV_0.pdf
[15] INHOPE Annual Report 2017. Available online at: http://www.inhope.org/tns/resources/annual-reports.aspx]
[16] Internet Organised Crime Threat Assessment 2018. Available online at: https://www.europol.europa.eu/internet-organised-crime-threat-assessment-2018
[17] Kā pasargāt bērnus no virtuālā terora. Available online at: https://www.tvnet.lv/5459102/ka-pasargat-bernus-no-virtuala-terora
[18] Koncepcijas par bērnu tiesību aizsardzību internetā. Available online at: https://likumi.lv/ta/id/68245-par-koncepciju-par-bernu-tiesibu-aizsardzibu-interneta
[19] Krimināllikums. Available online at: https://likumi.lv/doc.php?id=88966
[20] Ménard, Kim S., Pincus, Aaron L. (July 1, 2012). "Predicting Overt and Cyber Stalking Perpetration by Male and Female College Students". J. Interpersonal Violence. *Seattle, Washington:* SAGE Publications **27**(11), 2183–2207. doi: 10.1177/0886260511432144. ISSN 0886-2605. PMID 22203630
[21] Nabadzības riskam pakïauti ceturtdaïa Latvijas iedzīvotāju. Available online at: https://www.tvnet.lv/6501434/nabadzibas-riskam-paklauti-ceturtdala-latvijas-iedzivotaju

[22] Noziegumu statistika Latvijā: Pieaug nepilngadīgo noziegumi un seksuālā vardarbība pret bērniem.Availableonlineat: https://bnn.lv/vp-pieaug-nepilngadigo-noziegumi-seksuala-vardarbiba-pret-berniem-315183

[23] Par bērnu pornogrāfijas skatīšanos internetā būs bargi sodi. Available online at: https://www.tvnet.lv/5070066/par-bernu-pornografijas-skatisanos-interneta-bus-bargi-sodi

[24] Par Informācijas sabiedrības attīstības pamatnostādnēm 2014.-2020.gadam. Available online at: https://likumi.lv/doc.php?id=260931

[25] Par Rīcības plānu nepilngadīgo aizsardzībai no noziedzīgiem nodarījumiem pret tikumību un dzimumneaizskaramību 2010.-2013.gadam. Apstiprināts ar Ministru kabineta 2009.gada 25. augusta rīkojumu Nr.581. https://likumi.lv/ta/id/265209-grozijumi-bernu-tiesibu-aizsardzibas-likuma: http://likumi.lv/doc.php?id=196667

[26] Pētījums "Latvijas un citu valstu pieredze darbam ar seksuālajiem noziedzniekiem". - http://www.vpdp.gov.lv/lat/gen/58/828/ - (04.09.2007.)

[27] Pētījums: Sekstings – ierasta parādība bērnu un jauniešu dzīvē. 2011. Available online at: http://www.bti.gov.lv/lat/aktualitates/?doc=2692

[28] Plāns nepilngadīgo aizsardzībai no noziedzīgiem nodarījumiem pret tikumību un dzimumneaizskaramību 2018. – 2021.gadam. Available online at: tap.mk.gov.lv/doc/2018_08/TMPl_160818_dzimumnoz.858.doc

[29] Pornogrāfijas ierobežošanas likums. Available online at: https://likumi.lv/doc.php?id=157638

[30] Reports statistics. Drossinternets.lv (Safe Internet) Available online at: https://drossinternets.lv/lv/info/zinojumu-statistika

[31] Risk taking Online Behaviour Empowerment Through Research and Training. Available online at: http://www.childcentre.info/robert/

[32] Vairāk kā puse planētas iedzīvotāju lieto internetu. Available online at: https://www.bismart.lv/blogs/internets/vairak-ka-puse-planetas-iedzivotaju-lieto-internetu-868

[33] Vilks A. Sexual services, drugs and human trafficking, smuggling – in GDP figures: Quality of life indicators. Conference: 5th International Interdisciplinary Scientific Conference on Society, Health, Welfare Location: Riga, LATVIA Date: Nov 26–28, 2014. Available online at: https://apps.webofknowledge.com/full_record.do?product=WOS&search_mode=AuthorFinder&qid=9&SID=C1K1HXPVJr2uNpUG6Ru&page=1&doc=2

Mapping method in research of drug crime regional criminological features (Ukraine taken as an example)

Andriy Babenko[1], *Ruslan* Tarasenko[1], and *Oleksandr* Ostrohliadov[2]

[1] Odessa State University of Internal Affairs, Odessa, Ukraine
[2] Ministry of Internal Affairs of Ukraine, Kiev, Ukraine

Abstract. General national average drug crime rates on which the contemporary criminological theory and practice is based do not adequately reflect regional peculiarities in the field of illicit drug trafficking as they level high quality/quantity parameters in some areas and their lower values in others. Still, consideration of only national totals in organization of crime counteraction leads to incomplete information, neglecting its negative trends in certain areas, thus, causing aberration of the actual drug crime situation and using improper countermeasures. Under such circumstances the state law enforcement agencies do not operate preemptively against the illicit drug trafficking, do not contain in due time outbreaks of drug crime in certain areas- crime donors, and, as a result, the efficiency of steps aimed at drug addiction prevention as well as crime limitation in total becomes substantively deteriorated. Criminological mapping method enables the territorial police divisions to monitor existing criminological situation, inform the public and other law enforcement agencies on trends and locations of drug crime expansion, and reveal the most affected areas to be able to react promptly to crime pattern changes in regions. The application of this method enabled the evident demonstration of the fact that in Ukraine the Eastern and Southern areas were most affected by drug crime, wherein the affection factor in this part is twice higher than general national value.
Key words: drug crime, mapping method, regional criminological features.

1 Introduction

Drug addiction and related drug crimes today are among the most complex and most dangerous social problems of humanity, whose global character is confirmed by many negative facts. Thus, usage of drugs, production and cultivation of illegal substances and trade affect majority of countries in the world, devastatingly affecting public health, development and security of both industrialized and developing countries [1]. Drug dollars, which are flooding into the most powerful world banks, as evidenced, in particular, by the surmounting the crisis' paths of the financial world crisis of 2008–2009 [2, 3], deplete the economy of creation and development. According to expert estimates, the world drug trafficking is about $ 800 billion, which can only be compared with the impact on the world oil and gas economy [4]. Drug trafficking continues to be the most lucrative form of business for criminals worldwide. Drug trafficking flows have global dimensions linking regions and continents [5]. Criminal groups have a high level of organization, as they have good equipment, using the latest information and communication technologies [6].

Additionally, specialists in the field of combating drugs have calculated that 28 million years of healthy life (premature mortality and disability) are lost annually through drug use, and another 17 million are lost due to illnesses that arise and develop because of these substances using [7, 8]. The disaster scale is convincingly illustrated by the fact of estimated ¼ of a billion people, or about 5% of the global adult population, used drugs at least once in year (2015) [9–11].

As noted in the World Drug Report 2018, three tendencies have recently become apparent in the sphere of illicit drug trafficking: first, the so-called opiate crisis (the number of deaths caused by opiate use is 76% of all death cases, with the use of narcotic substances); secondly, the rising consumption of prescription drugs (87.0 tons of pharmaceutical opiates have been withdrawn); thirdly, record levels of heroin and cocaine production (it reached a peak and amounted approximately 1.410 tons and 87.0 tons per year respectively) [12, 13].

Thus, drug addiction [14, p. 351–352], as a negative social phenomenon, is, firstly, characterized by dynamism, since its movement takes place all over the world, quickly adapting to the new conditions associated with complications in running this business through active opposition to society, and secondly, has the ability to integrate well into the world economy (legalizing drug money, engaging in labor trafficking) and even influencing the political situation in some countries; thirdly, it is characterized by the property of certain geographical area structuring (forming the drugs production, organizing drug trafficking (their transportation) to markets, recruiting new substance consumers, which become soon drug addicts, serves as the basis for organized international criminality).

Ukraine plays a prominent role in the world drug trafficking. After all, it is, firstly, due its advantageous geographical position, in the chain of so-called Drug Silk Road; secondly, favorable climatic conditions make our country a producer of herbal drugs for the domestic and foreign markets with the simultaneous development of the production of synthetic drugs; thirdly, Ukraine is becoming louder about itself as a powerful market for these substances. Therefore, the considered issue of illicit drug trafficking is, of course, relevant for Ukraine, its dynamism, integration and structuring can be represented using the method of cartography, which allows creating maps of criminal acuteness of drug criminality in different country's regions, with the most dangerous identified zones, to point out correlations between different factors that cause crime. It can help to monitor the systematic criminal activity conduct, to carry out criminological analysis of the current situation by territorial police units, promptly reacting to changes in regions' crime rate and informing public and other state bodies about tendencies and geography of drug crimes spreading, etc.

The research aim is to reveal the possibilities and determine the effectiveness of criminological mapping method on the example of the analysis of current tendencies and regional criminological peculiarities of drug crimes in different regions of Ukraine, which seems necessary for two reasons. First, the national average indicators of drug offenses do not fully take into account the regional specificity of the latter, since these indicators offset high quantitative and qualitative parameters in some regions and low indicators in others. Secondly, taking into account only the national indicators in the organization of counteracting drug crimes leads to incompleteness of information. Under such conditions, the state law enforcement system does not work to prevent illicit drug trafficking, does not timely locate the cells of its outbreak in separate regions – "donors" of crime, as a result of which there is a significant decrease in the effectiveness of measures aimed at preventing drug addiction.

2 Methods

The methodological basis for conducting research of the possibilities of cartography in identifying of the criminological peculiarities of drug crimes in Ukraine is: 1) the official statistics of the Department of Information and Analytical Support of the Ministry of Internal

Table 1. Levels and proportion of crimes in the field of illicit drug trafficking (2011–2017).

Parameters	2011	2012	2013	2014	2015	2016	2017
Quantity of crimes	53 206	45 322	33 180	30 045	25 325	22 217	27 912
Coefficient of criminal intensities	116	99	72	70	59	52	68
Proportion (%)	10.3	10.2	5.9	5.7	4.5	3.8	5.3

Affairs of Ukraine and the Unified Register of Pre-trial Investigations and Informational and Analytical Work of the General Prosecutor's Office of Ukraine; 2) analytical materials of the UNODC; 3) demonstration of quantitative and qualitative indicators of drug crimes on geographical map; 4) tabular method that performs an additional explanatory function during the analysis of the parameters of the phenomenon under consideration. The empirical and statistical data are summarized and analyzed using the capabilities of descriptive statistics; 5) method for calculating the correlation between the phenomena of K. Pearson. Calculations and mapping are done using Adobe Photoshop CS6, Microsoft Excel 2016, Microsoft Word 2016 and QuickMap.

3 Results

The development of the method of criminological cartography [15–22] can be considered as one of the strategic directions in improving the information and analytical support of law enforcement agencies work of any state. Criminological cartography and regional criminological analysis contribute to the effective organization of preventing of illicit drug trafficking, thus envisaging rational use of forces and means of counteraction in the work of law-enforcement bodies. Such a distribution of forces and means must be carefully planned, differentiated and adapted to the local conditions of a particular territorial system.

The essence of criminological cartography lies in the hypothesis of the proposed objective and tasks of the differentiated spatial (rating, correlation, etc.) analysis of drug crimes in accordance with predefined and calculated parameters of the main indicators of this crime (its level, structure, intensity, dynamics, etc.) depending on the territory of analyzed crimes with the drawing of relevant data on thematic maps [23].

Although, according to the Ukrainian Monitoring and Drug and Drug Center of the Ministry of Health of Ukraine, as of 01.01.2017, there were 694 928 people under observation in the country with mental and behavioral disorders due to psychoactive substances using, and 115,170 people were registered for only one year, who, for the first time in their life, have been diagnosed with mental and behavioral disorders associated with the use of psychic substances [24, p. 71.] Official statistics of law enforcement agencies indicate a significant improvement in the situation in the area of counteraction to illicit drug trafficking (Table 1).

It followed from the given data that since 2013 the registration of detected drug crimes has decreased almost twice. The intensity of crimes in this category also shows similar trends. There was a noticeable decrease in the proportion of drug crimes [25–32]. This situation is the result of deterioration of the work of law enforcement agencies, which is due to: firstly, miscalculations in law enforcement reform, in particular, in the field of staff policy (dismissal of experienced workers, liquidation of territorial units for combating illicit drug trafficking, shortcomings in training staff for law enforcement agencies, etc.); secondly, the development

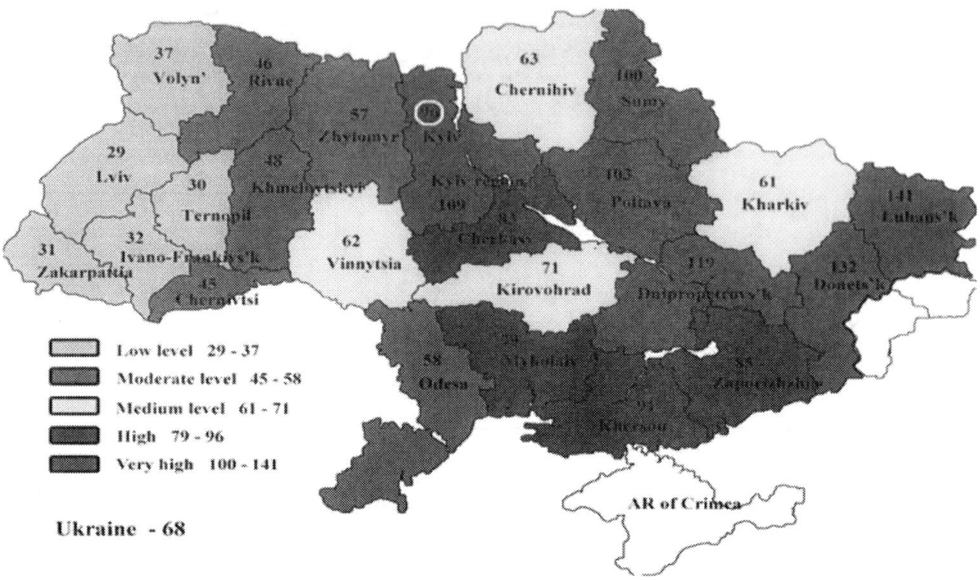

Map 1. Geography of the drug criminality intensity recorded in the regions of Ukraine (2017).

of new forms and types of illegal activities by drug offenders, excluding the direct contact of drug seller with drug buyer (for example, drug trafficking through the Internet); thirdly, using the latest methods of advertising and distribution of drugs; fourthly, with careful concealment of illegal activity and the avoidance of responsibility for the committed crimes, etc. The following were the logical consequences of this: 1) increase of drug crimes, expansion of spheres of influence, increase in the market and expansion of the offer of different drugs, involvement of new groups of population; 2) tardy reaction of law enforcement agencies to counteract drug criminals. The decrease in the efficiency of work is also explained by the lack of a scientifically based approach of law enforcement agencies to the distribution of forces and means to counter drug crimes, taking into account its territorial features; disregarding the crime specifics in the context of some settlements, etc.

With the help of the method of cartography, five groups of regions with different levels of drug criminality intensity (smear) can be identified (map 1).

The highest concentration of crimes in the area of illicit drug trafficking is seen in the eastern and southern regions of Ukraine, which make up a group of territories with very high and high levels of this type of crime (on average, from 79 to 141 crimes per 100,000 of population). Consequently, these territories can be designated as belonging to a negative crime zone, characterized by high and very high coefficients of drug criminality intensity. The middle (transitional) criminal zone includes two central, one northern and eastern regions of Ukraine (Kirovograd, Vinnitsa, Chernihiv and Kharkiv), in which there are 61–71 drug crimes per 100 thousand population. The western regions of Ukraine and one southern Odesa region are part of a relatively positive crime zone, characterized by low and moderate indicators of criminal prevalence (an average of 29–58 crimes per 100 thousand of population). Consequently, the method of cartography makes it possible to state that regional features of drug crime from the point of view of its concentration in the country's regions are very unequal, indicating significant differences in the levels of criminal prevalence of territories up to five times.

Table 2. Level and proportion of crimes in the area of illicit drug trafficking committed in some regions of Ukraine (2017).

region / parameters	Ukraine as a whole	Dniepro-petrovs'k	Kharkiv	Odesa	L'viv
Total quantity of crimes	523 911	44 343	34 986	32 576	27 526
Drug crimes quantity	29 010	3 832	1 647	1 379	732
Proportion (%)	5.5	8.6	4.7	4.2	2.7

Table 3. The level and proportion of crimes in the area of drug trafficking committed in some regions of Ukraine by type of drug crimes (2017).

Region / Kind of cases	Ukraine as a whole	Odesa region	Kharkiv region	Chernihiv region	Trans-carpathian region	Volyn region
Total	29 010	1 379	1 647	654	396	386
Art. 305 (drug smuggling)	251 (0.86%)	14 (1%)	54 (3.3%)	35 (5.35%)	19 (5%)	25 (7%)
Art. 307 (drug production, distribution and industry)	5029 (17.3%)	226 (16.4%)	338 (20.5%)	161 (24.6%)	117 (30%)	157 (40%)
Art. 309 (drug storage without industry purpose)	20 071 (69.2%)	958 (69.4%)	1 092 (66.3%)	385 (58.9%)	194 (48%)	141 (36%)
Art. 310 (illegal cultivation of drug plants)	2 017 (6.9%)	92 (6.67%)	64 (3.9%)	55 (8.5%)	51 (13%)	33 (9%)
Art. 310 (illegal cultivation of drug plants)	2 017 (6.9%)	92 (6.67%)	64 (3.9%)	55 (8.5%)	51 (13%)	33 (9%)

Tabular method is useful, for example, when analyzing such an important indicator of drug crimes' state as its structure. The regional structure of the analyzed crime differs from the corresponding national indicators. This can be clearly seen in the Table 2 (the example of separate regions of Ukraine).

The data presented in the table indicate significant differences in the proportion of drug crimes in the structure of all offenses in region. Thus, with the national average specific gravity of drug crimes in 5.5%, the share of drug crimes in the L'viv region is 2.7%; in Odesa - 4.2%; in Kharkiv –4.7%, while in Dnipropetrovs'k –8.6%. This gives grounds for conclusion that the unequal pressure of specific crimes on the regional crime situation is common, for example, in the Dnipropetrovs'k region the share of drug crimes in the structure of all crimes is three times higher than in L'viv. Regarding, in a nutshell, the domestic regional structure of drug crimes, here we also fixed significant regional features (Table 3).

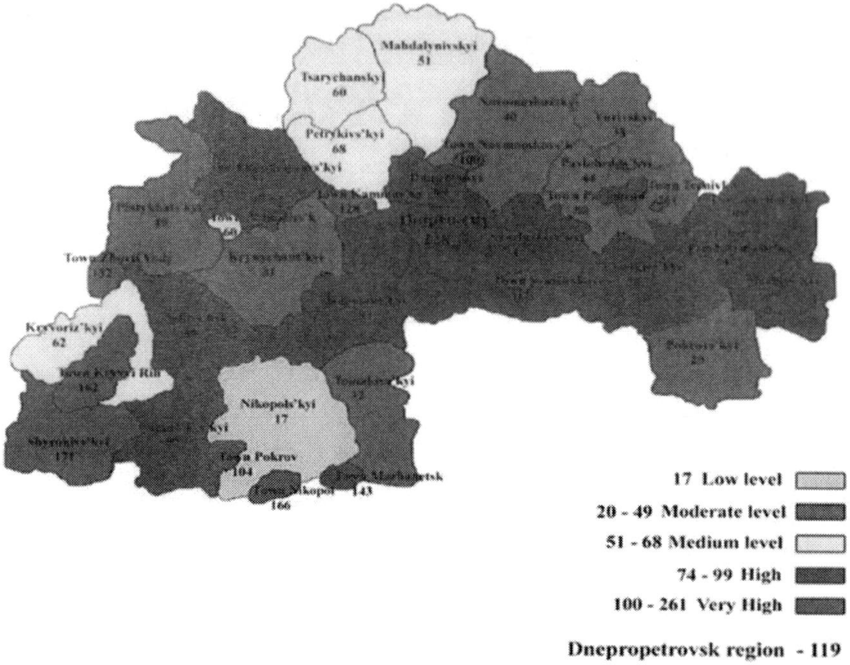

Map 2. Geography of drug criminality intensity per 100 thousand of population in towns and non-urban areas of Dnipropetrovs'k region of Ukraine (2017).

Table 4. Territorial distribution of drug criminality in towns and non-urban areas by the example of some regions of Ukraine (2017).

parameters /quantitative indicators	**Dnipro**	**Kharkiv**	**Odesa**	**L'viv**
in the region	3 833	1 647	1 379	732
in all towns of the region	3 282 (85.6%)	1 196 (72.6%)	958 (69.5%)	455 (62.2%)
in a regional town	1 283 (33.5%)	1 068 (64.8%)	734 (53.2%)	327 (44.7%)
in a small town of the region	1 999 (52.1%)	128 (7.8%)	224 (16.3%)	128 (17.5%)
in non-urban area	551 (14.4%)	451 (27.4%)	421 (30.5%)	277 (37.8%)

The analysis of regional criminological peculiarities of drug criminality in Ukraine is annexed with information on the geography of the internal intensity of the last for 100 thousand of population and this intensity rating in the context of separate several regions' towns (Maps 2, 3 and Tables 4, 5).

Thus, according to the local conditions of reproduction of crime and the attractiveness of territories for drug crimes, the focus of prevention activities, for example, for a particularly large industrial Kharkiv region, should be concentrated in regional Kharkiv city, where 64.8% of all drug crimes in the region are committed. In Odesa region, prevention of crime should be organized primarily in Odesa city, having 53.2% of drug crimes, as well as in regional agricultural districts, where 30.5% of these offenses are committed respectively.

The pointed indicators show the extremely difficult situation in the sphere of illicit drug trafficking in the Dnipropetrovsk region. Significant part of its territory is occupied by a

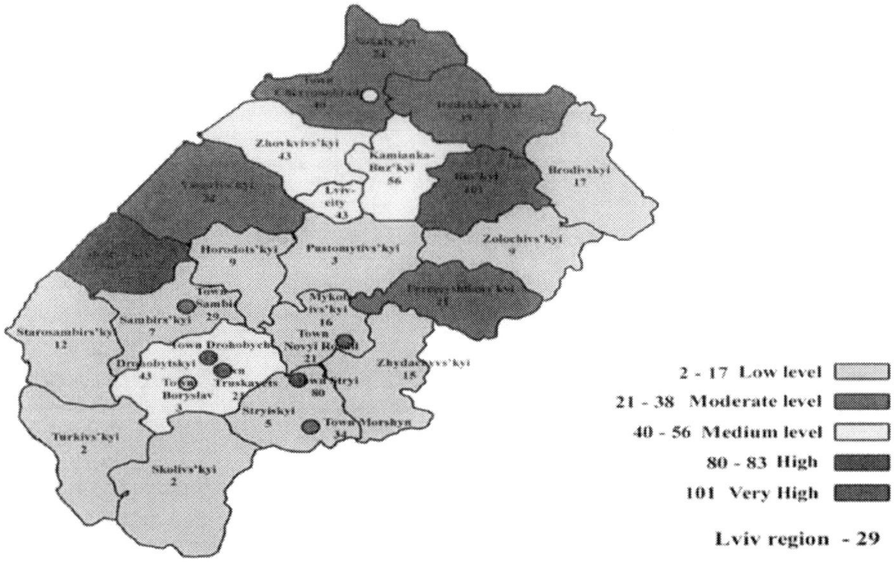

Map 3. Intensity geography of drug criminality per 100 thousand population in towns and non-urban areas of the L'viv region of Ukraine (2017).

Table 5. Rate of drug criminality intensity in towns and non-urban areas by the example of Dnipropetrovs'k and L'viv regions per 100 thousand of population (2017).

Options	City or town	Crimes' intensity ratio per 100 000 of population	City or town	Crimes' intensity ratio per 100 000 of population
Rank	**Dnipropetrovs'k region**	**119**	**L'viv region**	**29**
1	Ternivka town	261	*Busk district*	101
2	*Shyrokiv district*	171	*Mostysky district*	83
3	Nikopol town	166	Stryi town	80
4	Kryvyi Rih town	162	*Kamianka-Busky district*	56
5	Marhanetsk town	143	*Zhovkva district*	43
6	Zhovti Vody town	132	*Drohobych district*	43
7	Kamians'ke town	128	**L'viv city**	**43**
8	**Dnipro city**	**128**	Chervonohrad town	40
9	Synelnykove town	117	*Radekhiv district*	38
10	Pokrov town	104	Morshyn town	34
In Ukraine as a whole	**68**			

Table 6. Correlation coefficients of the rate of registered theft and drug crimes in Ukraine, made by K. Pearson (2009–2017).

Options	Periods								
	2009	2010	2011	2012	2013	2014	2015	2016	2017
Correlation coefficient	0,67366	0,68759	0,68135	0,66362	0,67152	0,66272	0,57227	0,37594	0,46005

negative criminological group (63% of the territory with an intensity of 74 to 261 drug crimes per 100 thousand of population), while L'viv region is dominated by relatively safe towns in terms of drug prevalence (72% of the territory with an intensity of 2 to 38 drug crimes per 100 thousand of population). In addition, in the ten most sensitive towns of Dnipropetrovs'k and L'viv regions, the town's domination tends to be a regional average, while territories with the highest parameters of drug criminality in the L'viv region (with an intensity of 80 to 101 drug crimes per 100,000 of population) are similar, with the lowest rates of the first ten towns of the Dnipropetrovs'k region (with an intensity of 104 to 117 drug crimes per 100 thousand of population). At the same time, the highest indicators of drug criminality intensity in the Dnipropetrovs'k region exceed several times similar data in the chosen region for comparing to the L'viv region [33], and in ten Dnipropetrovs'k region's towns the criminal drug offenses' level exceeds the national parameters from 2 to 4 times, while in L'viv only in the Busk district (with intensity coefficient of 101 crimes per 100 thousand of population) is 1.5 times higher than the average Ukrainian indicators (with intensity coefficient 68 crimes per 100 thousand of population).

4 Discussion

Further elaboration of regional features' methodology of crime in general and drug crime, in particular is needed. Indeed, in similar researches, the question, which appears in the geographical plan of the object for comparison, is not solved until now – whether it is the country as a whole or even several countries, or administrative territorial units of a separate country, or separate areas of one administrative – territorial unit. At the same time, we believe that the national experience of researching regional features of drug crimes through the method of cartography should be expanded with the analysis and comparison of the drug criminality situation in countries, which, for example, are connected with Ukraine by common border.

The controversial moment of the theoretical nature is the problem of regional criminology's formation as an independent area of research. If, according to regional criminology, it is nevertheless recognized as the right to be an independent direction of theoretical knowledge, then it is necessary to establish a line of conditional demarcation between the latter and the geography of crime.

It is also known that thefts and drug crimes are interrelated phenomena, since a significant proportion of thefts are committed by persons involved in drug trafficking and consumers of these substances. Their dependence is shown by the analysis of correlation dependencies, carried out by the method of calculating the correlation coefficient of K. Pearson. By 2015, there was a significant direct correlation between the intensity of theft and drug crimes, which had been established for many years. The decrease of this correlation coefficient in 2015 is likely to be the result of special units' liquidation aimed at combat drug trafficking in Ukraine the same year.

On the basis of the above mentioned, it is necessary to check the hypothesis of dependence of "violation" on correlation between thefts and drug criminality in 2015 and liquidation of special units which combat illicit drug trafficking both in Ukraine as a whole and in some regions.

5 Conclusions

The results of the study of regional criminological peculiarities of drug crimes in Ukraine with the help of the method of criminological cartography have made it possible to state that the regional peculiarities of drug crimes due to their concentration in those or other regions of the country indicate significant differences in the levels of criminal damage to territories up to five times. At the same time, the response of law enforcement agencies to drug crimes and other related offenses does not correspond to the real state of criminalization of society, because the power structures are based on insufficient knowledge of existing criminal situation. It is the result of failure to take into account the possibilities of modern technologies that allow drawing up maps of criminal damage of regions. The usage of this method allows us to step back from practice of leveling drug crimes' features in some regions of Ukraine. In this way, there is a real possibility to redistribute properly the powers of the state's law enforcement system in order to, firstly, work to advance the state of aggravation of drug situation in certain regions of country, and secondly, to work on operational location of its out-breaking cells.

References

[1] Drug prevention, treatment and care (Working towards evidence-based drug dependence treatment and care). Available: http://www.unodc.org/westandcentralafrica/en/newrosenwebsite/drug-prevention-treatment-and-care/drug-prevention-treatment-and-care.html
[2] Drug money saved banks in global crisis, claims UN advisor. The Guardian. December **13** (2009). Available: https://oldimg1.ria.ru/beznarko_help/20130626/828973392.html
[3] Деньги от наркоторговли помогли банкам во время мирового кризиса, утверждает представитель ООН. InoPressa.ru. Декабрь **14** (2009)
[4] Доходы мирового наркобизнеса. Available: https://oldimg1.ria.ru/beznarko_help/20130626/.html
[5] Drug Trafficking: a $32 billion business affecting communities globally. Available: http://www.unodc.org/southasia/frontpage/2012/August/drug-trafficking-a-business-affecting-communities-globally.html
[6] *Tendencies of drug trafficking in Russia in 2004.* Moscow: Region. Representative Office of the United Nations Office on Drugs and Crime in Russia and Belarus (2005)
[7] Executive Summary Conclusions and Policy Implications. World Drug Report 2017. Vienna: UNODC (2017)
[8] Fact Sheet on Statistics and Trends in Illicit Drugs. Available: http://www.unodc.org/wdr2017/field/WDR17_Fact_sheet.pdf
[9] World Drug Report 2015, New York: United Nations (2015)
[10] World Drug Report 2016 / United Nations Office on Drugs and Crime, 2016. Vienna: United Nations Office on Drugs and Crime (2016)

[11] Executive Summary Conclusions and Policy Implications. World Drug Report 2017. Vienna: UNODC (2017)

[12] Всемирный доклад о наркотиках 2018: опиоидный кризис, растущий уровень употребления рецептурных препаратов, рекордные уровни производства кокаина и героина. Available: `http://www.eegyn.com/pdf/WDR_2018_PressRealeaseRUS.PDF`

[13] World Drug Report 2018: opioid crisis, prescription drug abuse expands; cocaine and opium hit record highs. Available: `http://www.unodc.org/unodc/en/frontpage/2018/June/world-drug-report-2018_opioid-crisis-prescription-drug-abuse-expands-cocaine-and-opium-hit-record-highs.html`

[14] Большой юридический словарь / под. ред. А. Я. Сухарева, В. Е. Крутских. Москва: Инфра-М (2001)

[15] Бабенко А. М., Орещенко А. В. Можливості картографічних методів дослідження у інформаційно-аналітичному забезпеченні протидіы торгівлії людьми. Акт. пит. б-би з торгівлею людьми в умовах проведення антитерористичної операції (збройного конфлікту): зб. матеріалів міжнар. круглого столу. Київ: ВД «Дакор». С. 42–48 (2017)

[16] Мальцев В. В., Болотова В. О. Картографування злочинності: історія розвитку. Девіантна поведінка: соціолог., психолог. та юрид. аспекти: матеріали наук.-практ. конф. (Харків, 10 квіт. 2010 р.). Харків: ХНУВС. С. 96-99 (2010)

[17] Bianchini Å., Sicurella S. *Cases on Technologies for Teaching Criminology and Victimology: Methodologies and Practices.* Hershey, New York: Information Science Reference (2009)

[18] Camilleri G., Lazerges Ñ. Atlas de la Criminalité en France. Mappemonde. **2**, 6–9 (1992)

[19] Chainey S., Ratcliffe J. *GIS and Crime Mapping*. New York: John Wiley & Sons (2005)

[20] Cunty C., Fussy F., Perez P. Geocriminologie, quand la cartographie permet aux geographes d'investir la criminologie. CyberGeo. Jul. (2007)

[21] Hill B., Paynich R. *Fundamentals of Crime Mapping*. London: Jones and Bartlett publishers (2014)

[22] Santos R.B. *Crime Analysis With Crime Mapping.* CA, USA: Sage Publications (2008).

[23] Бабенко А.М. *Регіональна злочинність в Україні: закономірності, детермінація та запобігання*: дис. … д-ра юрид. наук. Запоріжжя (2015)

[24] Національний звіт за 2017 рік щодо наркотичної ситуації в Україні (за даними 2016 року) Київ (2017)

[25] Стан та структура злочинності в Україні (2010–2011 рр.) 2017. `http://mvs.gov.ua/mvs/control/main/ uk/publish/article/717134;14`

[26] Єдиний звит про криминальни правопорушення за сичень-грудень 2017 року. Available: `http://www.gp.gov.ua/ua/stst2011.html?dir_id=112755&libid=100820`

[27] Стан та структура злочинности в Украї ни за 2007 – по 20.11.2012 рр.: статистика МВС. Аваилабле: `http://mvs.gov.ua`

[28] Єдиний звит про криминальни правопорушення за 2013–2017 рр.: статистична информація. Available: `http://www.gp.gov.ua`

[29] *Статистичний щорічник Украйни за* 2010 *рік.* Кийв: Август Трейд (2011)

[30] *Статистичний щорічник України за* 2012 *рік*. Київ: Держаналітнформ (2013)
[31] *Статистичний щорічник України за* 2013 *рік*. Київ: Держаналітінформ (2014)
[32] Статистичний *щорічник України за* 2016 *рік*. Київ: Держаналітінформ (2017)
[33] Бабенко А.М. Сучасний стан і тенденції наркозлочинності, вбивств та самогубств у мистах і сільських районах України: кримінологічний вимір та запобігання. Наук. вісн. Днипропетр. держ. ун-ту внутр. справ: наук. журн. **1**(91). С. 16–26 (2018)

Preventative role of criminal law for traffic safety

Nataliya Gutorova[1] and *Oleksandra* Rudnyeva[2]

[1] Yaroslav Mudry National Law University, Kharkiv, Ukraine
[2] International Legal Council LEGITEAM, Kiev, Ukraine

Abstract. The purpose of the article is to raise awareness of and stimulate serious discussion about the necessity to improve the preventative role of criminal law in the field of traffic safety by adjusting criminal liability for crimes in this area, and determining the proportional amount of force that may be used by the police in order to stop the vehicle and detain the offender.
Key words: preventative role of criminal law, criminal liability, exemption of criminal liability, traffic safety, extent of justified damage.

1 Introduction

The modern world cannot exist without transport. However, the use of vehicles, in states and societies that do not pay sufficient attention to traffic safety, poses a significant threat to the lives of people. According to the Global status report on road safety 2018 of the World Health Organization (WHO), the number of road traffic deaths continues to increase reaching 1.35 million in 2016. Road traffic injury is now the leading cause of death for children and young adults aged 5–29 years, signaling a need for a shift in the current child health agenda, which has largely neglected road safety. It is the eighth leading cause of death for all age groups surpassing HIV/AIDS, tuberculosis and diarrheal diseases. Experts came to these conclusions after analyzing the situation in 175 countries [1].

1.35 million cases of death a year means that nearly 3,700 people die on the world's roads every day; every 24 seconds someone dies. Dr. Tedros Adhanom Ghebreyesus, Director-General of WHO in the Foreword to the Report underlined that tens of millions more are injured or disabled every year resulting in life-altering injuries with long-lasting effects. These losses take a huge toll on families and communities. The cost of emergency response, health care and human grief is immense [1].

European countries provide comparatively safer conditions for traffic users. For the 51 European countries included into the Report the average rate of road traffic deaths per 100,000 of population in 2016 were 9.3. The lowest rate – 2.7 – belongs to Norway and Switzerland, the highest – 18.1 – to Tajikistan. Whereas Latvia's was 9.3 – equal to the average European rate, Ukraine has a rate of 13.7. Unfortunately, along with Azerbaijan, it has the highest in Europe rate of pedestrian fatalities – 42 – and thus joined the group of countries that are the worst in terms of this indicator, namely Kyrgyzstan – 40 and Belorussia – 41.

According to the official statistics of the Prosecutor General's Office of Ukraine [2] there were 11157 crimes registered in 2016 and 10011 in 2017 stipulated in Art. 286 of the Criminal Code of Ukraine [3]. The number of victims in 2016 with at least moderate bodily injuries was 12833, including 2417 who died. In 2017 the number of victims was 12088, of which 2015 persons died.

Moreover, the statistics shows the severity of the issue as the number of traffic fatalities and injuries in Ukraine in 2015-2017 significantly exceeds the number of victims of military conflict (including civilians) in the East of the country.

Attention should be paid to the difference in traffic fatality data provided by the three state agencies in 2016. The Prosecutor General's Office of Ukraine reported about 2417, State Statistics Service of Ukraine – about 4687 and the WHO in its Global status report on road safety 2018, estimated 6089 road traffic fatalities. Such a difference in figures (more than in 2.5 times) is due to the fact that the Prosecutor General's Office of Ukraine makes relevant information about the victims of accidents in the statistical reporting immediately after the crime without counting the number of persons whose death occurred later in the hospitals. There are also some questions surrounding the reliability of the statistics. The example of Ukraine is not unique. Similar irregularities and discrepancies exist in other countries. Therefore, the development of measures to combat mortality and injury resulting from traffic accidents should be based on statistical information provided by WHO experts.

Strengthening legislation to mitigate key risk factors is recognized by the majority of governments as an important strategy to improve road safety [1]. We believe that Criminal law plays an important role in implementation of these tasks as it has the most severe tools in terms of punitive treatment of the offender, and should apply to acts that have led to a significant threat to the life or health of a person.

Criminal science mainly studies the liability for violating traffic safety rules or operating a vehicle in such a way as to cause damage to life or health of a person. Such acts are undoubtedly recognized as crimes and for their commission severe punishments are stipulated, especially in case of homicidal acts. The measures to ensure the preventive role of criminal law for traffic safety have been studied insufficiently.

Aim. The purpose of the article is to raise awareness of and stimulate serious discussion about the necessity to improve the preventative role of criminal law in the field of traffic safety by adjusting criminal liability for crimes in this area, and determining the proportional amount of force that may be used by the police in order to stop the vehicle and detain the offender.

2 Materials and methods

This study was conducted in 2018 and is based on the national regulatory acts on traffic safety, the European Convention for the Protection of Human Rights and Fundamental Freedoms, case law of European Court of Human Rights (ECHR), WHO Status Report on Road Safety 2018, criminal legislation of selected countries such as Germany, Ukraine, Poland, Latvia, the Ukrainian police and General Prosecutor's Office data on the criminal liability of those who committed crimes in the field of traffic safety and the Ukrainian Ministry of Health's statistics on traffic fatalities and injuries. Totally 34 laws and papers, 123 court judgments, 16 traffic incidents reports were analyzed.

Dialectical, comparative, analytic, synthetic and system analyses research methods were used, also for interpretation purposes.

3 Results and discussion

The effectiveness of the measures to ensure the protection of transport security by criminal law can be understood through the analysis of three relatively independent but closely related issues: 1) the establishment and application of criminal liability for offenses that pose a serious threat to traffic safety; 2) the liability for violating the traffic safety or vehicle driving rules, resulting in causing damage to the life or health of a person; 3) the proper regulation of circumstances surrounding a crime, in order to ensure the right of law enforcement officers to stop a vehicle that poses a real threat to the life or health of a human.

Drink-driving and drug-driving are at the top of the WHO reported list, among 9 types of key risk factors for the lives and health of drivers and pedestrians. It is estimated that between 5% and 35% of all road deaths are reported as alcohol-related. Driving after drinking alcohol significantly increases the risk of a crash and the severity of that crash. This conclusion is fully confirmed by the analysis of statistical data from the General Prosecutor's Office of Ukraine over the last 5 years. Yes, if among the registered crimes stipulated in Art. 286 of the Criminal Code of Ukraine for the period from 2013 to 2017, the number of crimes committed in the state of intoxication was 5.7%, among these crimes, up to 10.1% persons died.

Nevertheless a comparison of the criminal laws of Ukraine, Germany, Poland and Latvia shows that the principal difference lies in the identification and punishment for offences that did not lead to actual damage, but were still dangerous.

In Poland, drink or drug driving has been already penalized in article 178a of the Criminal Code of the Republic of Poland [4]. In par.1 of Section 262 of the Criminal Law of The Republic of Latvia it states the criminal liability for any person who operates a vehicle if he or she does not have a driver's license (the driver's license has not been obtained according to the procedures or has been taken away) and if the driver is under the influence of alcohol, or narcotic, psychotropic, toxic or other intoxicating substances [5].

Germany established a completely different determination of liability for violation of traffic safety. In Section 316 "Driving while under the influence of drink or drugs" of the Criminal Code of Germany liability is established with no conditions attached. Moreover, the list of circumstances which causes criminal liability is extensive and forms a solid preventive foundation, where even the potential for danger to traffic safety is considered a crime (Section 315c "Endangering road traffic") [6].

In Ukraine the violation of traffic safety rules and driving vehicles rules is considered to be an administrative offence, and, based on the Code of Administrative Offences of Ukraine, the following types of penalties are imposed, such as fines and, in some cases, deprivation of the right to drive vehicles. Just very recently, at the very end of 2018 the parliament of Ukraine voted to criminalize drink and drug driving, but this legislation has not been enforced yet [7].

In the meantime, beyond the statistics, incidents involving drunk driving occur daily in Ukraine. It must be assumed that the fact that drunk driving is not considered a serious crime partly determined the behavior of this person. In view of this, the criminalization of such offences is well grounded. In the case of the effective application of the criminal law, it can positively influence traffic safety.

The main question is how to achieve a relevant punishment for such an offense, with the goal of preventing traffic crashes. The Criminal Code of the Republic of Poland for this crime in Art.178a established punishment in the form of a fine, restriction of liberty or deprivation of liberty, and – in the case of a second offence –deprivation of liberty from three months to two years. The Criminal Law of the Republic of Latvia established punishment as the deprivation of liberty for a period of up to one year, or temporary deprivation of liberty, or community service, with the removal of the driver's license for a period of up to five years. Pursuant to German Criminal Code such a crime should be liable to imprisonment not exceeding one year or a fine.

Ukraine has established only a fine and the removal of a driving license, and no restrain of labour rights in the form of public work or imprisonment, so it is softer in comparison to the law of the Republic of Poland, the Republic of Latvia and the Federal Republic of Germany. To determine the right type and weight of punishment for drink and drug driving it is necessary to consider that "these policies need to be evaluated at two distinct levels. Firstly, are they effective in producing at the lowest cost the desired outcome of safer driving and less incidents? Secondly, are they defensible in terms of sentencing principles of proportionality and consistency," [8] according to John Williams.

When determining the punishment for violating the rules of traffic safety, it is not easy to find the balance between severity, which should be related to the potential or real damage to people's life and health, and the often-positive character of the offender. Alwin van Dijk and Hein Wolswijk had underlined that "Criminal traffic offences are set apart from conventional criminal offences in that they are generally not committed by hardened criminals. Many serious traffic accidents are caused by fairly ordinary people who – at least up to that moment – were perceived as well-meaning human beings. Once the criminal law has set its sights on these ordinary people, they may face penalties that are otherwise reserved for the most heinous crimes." [9]

Fines and the deprivation of the right to drive a vehicle as punishment for drink and drug driving are not, on their own, always effective deterrents. This is especially true in cases of repeated offenses committed by people who are already deprived of their driving license.

A penalty in the form of a fine is, in reality, often settled at the expense of the offender's family members, the preventive role of the criminal punishment is not actually achieved. Obviously, the possibility to be deprived of liberty, even for a short-term, or to perform community work can be more effective for alcohol and drug users. This is especially true for drivers who count on the financial support of parents or other family members.

The next traffic safety offence, which WHO experts fully justify as a key risk factor, is ***speeding***. It is extremely dangerous, most of all for pedestrians.

So, for example, a terrible traffic incident took place in downtown Kharkiv on the 18th of October 2017. A "Lexus" careered onto the sidewalk after a crash: 6 people were killed, 5 were gravely injured. Experts have identified that one of the participants of the accident, the driver of "Lexus", 18 years old Olena Z., drove into the centre of the city at the speed of 106 km/h [10].

According to the WHO Report "Even small reductions in speed affect fatal and serious crash risk; 5% reduction in average speed can reduce the number of fatalities by 30%" [1]. The relationship between the violation of speed limits on urban roads and pedestrian injuries has been calculated by WHO experts: "for pedestrians hit by car fronts, the death risk rises rapidly (4.5 times from 50 km/h to 65 km/h)." The cause-effect relationship between violation of speed limits and pedestrian fatalities is also understood by the experts. Despite that Ukraine has highest level of pedestrian fatality (42%), the level of enforcement of the national speed limit law is the lowest (3 out of 10) [1, 11].

These statistics show the necessity to penalize violation of speed limit. At the same time, criminal liability can be introduced only if all other legal measures have been exhausted [12, 13].

Such measures should be effective, and can replace criminal liability. To achieve that goal, a sufficient number of technical devices for the automatic detection of such offenses should be installed on the roads. This will create a strong barrier against corruption and allow a high level of inevitability of liability. Under such conditions, administrative punishment for speeding in the form of a fine, and deprivation of the right to drive a vehicle, will be quite sufficient.

It should also be considered that the use of criminal liability requires significant funds from taxpayers that could be more effectively used to implement an effective system to automatically apply an administrative penalty to offenders. In 1968 Nobel Laureate-economist Gary S. Becker in his work "Crime and Punishment: The Economic Approach" said that in counteracting crime, "optimal" decisions are interpreted to mean decisions that minimize the social loss in income from offenses" [14]. This theory has been confirmed and further developed in modern scientific research [15].

Turning back to Olena Z. case, the question arose – was everything done to avoid the tragedy? An eighteen and half years old adolescent girl with no fixed income, driving a luxury vehicle Lexus RX350 with already six recorded violations of the traffic rules and a fine paid by her parents, – all these factors classify her to the dangerous and/or risky group of drivers.

Also, she gained her driving experience at a time – 2015 – when the state had practically stopped controlling the speed on the roads. Devices used up to this time by the traffic police had not been certified, and new ones had not been purchased. For several years already, the introduction of automated detection of traffic violations has been prepared, but so far there is no control. That is why Olena Z. case can be used as a tragic illustration of the need for a proper traffic safety national policy, including the appropriate technical and legal measures.

The study further tackles the aspect of ***criminal laws regulating the exemption from criminal liability circumstances where it applies to the establishment of cause and scope of involuntary harm caused by law enforcement officers during the execution of traffic safety arrangements***. Analysis of this problem is important for the promotion of traffic safety, because any ambiguity in this matter may render the law enforcement totally helpless or cause grave violation of human rights.

The tragedy that stirred the entire community occurred in Kiev on 7 February 2016 around 3 a.m. A patrol car chased a BMW that flew across the city at almost 200 km/h and would not pull over. After several warnings, the police officers used firearms, killing a 17-year-old passenger. Charges of intentional homicide were brought against the officer who had made the fatal shot. Khatiya Dekanoidze, who was in charge of the National Police back then, made a statement at the trial, trying to prove that the officer in question had been legally correct to have shot his gun, but the AG representative and the court begged to disagree with it. The trial is still under way and the ruling is still pending. This accident became a milestone for the Ukrainian police after which patrolmen would not use firearms during their attempts to stop a vehicle, even when any such use appears to be absolutely warranted. Consequently, one can see obscenely expensive cars in the big cities of Ukraine racing down the night streets at around 200 km/h or witness drag races while the police appear to be remarkably unable to pull the racers over.

So, one would reasonably ask whether or not the police are allowed to use firearms so as to stop a vehicle that poses a considerable danger to the public and cannot be stopped otherwise in each particular situation.

This matter needs to be resolved through the application of the principle of proportionality and the study of exemption from criminal liability circumstances. It is about the lawful acts envisaged by criminal and other laws which outwardly resemble crime. These deeds are actually considered to be for the good of the public or socially acceptable, so they shall not entail any criminal liability [16].

Certain exemption from criminal liability circumstances are defined in applicable laws (e.g., the absolute necessity, justifiable defence, apprehension of a criminal, etc.). However, the bulk of such circumstances can be found in special laws. For example, in Ukraine pursuant to Clause 7 Part 4 Article 46 of the Law "On National Police", a policeman may in exclusive cases use firearms to have the vehicle pull over by causing damage to the same, if the driver is acting in a manner jeopardizing life and health of the people and/or the police officer.

According to criminal laws, the police officer who uses firearms to stop a vehicle is found to be exercising his/her lawful powers [17] only if and when any such use is an absolute necessity. Refusal to cause harm shall not exempt one from liability for failure to perform or improper performance of one's duties.

The grounds for the police officer to use firearms to stop a vehicle by causing damage to the latter comprise of the following factors: 1) the driver of a moving vehicle jeopardizes the life and health of the people and/or the police officer; 2) under the circumstances, the vehicle cannot be stopped in any other manner but by being damaged by firearms. Sub-clause 2 above means that previous actions failed to cause the vehicle stop, and life and health of the people are still jeopardized. The police officer may use firearms only after ordering the transgressor to stop unlawful activities and notifying the same of the intention to take the enforcement actions.

The European Convention for Human Rights and Fundamental Freedoms in the first paragraph of Article 2, specifies that everyone's right to life is protected by law, and in the second paragraph of Article 2 provides that death resulting from defending oneself or others, arresting a suspect or fugitive, or suppressing riots or insurrections, will not contravene the Article when the use of force involved is "no more than absolutely necessary".

In the ECHR decision in McCann and Others v. The United Kingdom, 27 September 27, 1995, considers that the exceptions delineated in paragraph 2 (art. 2-2) indicate that this provision (art. 2-2) extends to, but is not concerned exclusively with, intentional killing. The text of Article 2 (art. 2), read as a whole, demonstrates that paragraph 2 (art. 2-2) does not primarily define instances where it is permitted intentionally to kill an individual, but describes the situations where it is permitted to "use force" which may result, as an unintended outcome, in the deprivation of life. The use of force, however, must be no more than "absolutely necessary" for the achievement of one of the purposes set out in sub-paragraphs (a), (b) or (c) (art. 2-2-a, art. 2-2-b, art. 2-2-c) (see application no. 10044/82, Stewart v. the United Kingdom, 10 July 1984, Decisions and Reports 39, pp. 169–71) [18].

Judgment in the Case McKerr v. The United Kingdom [19], 4 May 2001, shows similar position, particularly in paragraph 200 of a Judgment in McCann and Others v. The United Kingdom provision (art. 2-2).

ECHR also approved situations where it is permitted to use force which might result in the deprivation of life in their Judgment Andronicou and Constantinou v. Cyprus [20], 09 October 1997, when the police used weapon to save the life of a woman who was kidnapped by her future husband and accidentally killed both. In certain circumstances, prevention of the loss of life, similar to the case Giuliani and Gaggio v. Italy [21] 25.08 2009, has to be properly planned and controlled by the state "taking into consideration not only the actions of the agents of the State who actually administer the force".

So, the Court accepts situations where it is permitted to use force which might result in the deprivation of life. In case where the police officer applies force based on the sincere conviction of the necessity of the response, the violation of the right to life may be justified, despite that later it can be proven as mistake.

The extent of permissible damage, subject to circumstances exempt from criminal liability, shall be determined by the extent of necessity (the need), maximum permissibility and reasonable sufficiency [16].

The necessity to cause damage exists if the police officer has no other options but to use firearms so as to stop a vehicle that jeopardizes life and health of the people.

Maximum permissible damage. Authorization to use firearms for the purpose of stopping a vehicle shall be based on the understanding that there may be human beings in the vehicle (or the driver alone) whose life and health may be harmed. This understanding is based on the assumption that a projectile from the weapons fired at the object that moves at a certain speed and along the trajectory that may be changed by the actions of the driver or by the properties of the road surface may penetrate the bodywork and hit a human body inside the vehicle. Any law, including the one considered herein, is made to the benefit of "the average man", a police officer – properly trained and skilled, whose psychic health is as sound as befits a normal human being. Without a doubt, that a highly professional sniper with skills honed in combat and with an extraordinary psychological fortitude, is capable of making exclusively clean shots without hitting a human being in the process. However, it does not mean that police academies should welcome exclusively such people. Now, therefore, maximum permissible damage caused by or resulting from the use of firearms by the police with the purpose of stopping the vehicle should be the loss of human lives inside the same vehicle.

The damage may be rated reasonably sufficient if, under specific circumstances, it served the objective of stopping the vehicle. Therefore, causing damage for any other purpose, such as apprehension of the driver or retaliation for the driver's refusal to obey the order of the police, etc., shall not be construed as damage lawfully caused.

4 Conclusions

1. Traffic safety is a responsibility of the state, which must be ensured through appropriate legal mechanisms.
2. The criminal law prevention can be achieved via: 1) the establishment and application of criminal liability for offenses that pose a serious threat to the life or health of a human; 2) clear definition of all situations, grounds and measures for the positive duty of the police to use force to prevent foreseeable loss of life which in certain circumstances might result in harm.
3. The matter of criminal liability for traffic violations or operation of a vehicle in a manner jeopardizing the life and health of a human being or causing risk of damage to other important social values shall be handled with clear understanding that: a) certain delinquent action may lead to highly dangerous consequences; b) the said action is pretty common; and c) cannot be effectively countered by other legal tools.
4. Criminal liability for drink and drug driving is essential due to its high risk to people's life and its widespread nature. For general and specific prevention purposes an appropriate basic punishment for a person who commits an offence for the first time should be a fine or restriction of liberty or a community service as its alternative. Additional obligatory punishments should be the deprivation of the driver's license for a period of up to five years. When the driver's license has been already taken away, the deprivation of liberty for a period from one to twelve months should be a basic punishment and the fine should be an additional punishment.
5. Speeding violations by drivers, although one of the key risk factors for traffic safety, do not require criminalization. It is quite possible to ensure an unavoidable administrative punishment for this act through the use of a large number of devices for automatic detection. Relevant technical solutions can significantly reduce the number of traffic offences.
6. The use of weapons by the policeman seeking to stop the vehicle by causing damage to the latter, if the driver of the same is acting in a manner that jeopardizes life and health of the people and/or the police officer, is the circumstance the circumstance that exempts from criminal liability and can be legally classified as the exercise of the official powers. The policeman may exercise the said powers in cases of extreme necessity. However, refusal to cause harm in such a manner shall not be construed to exempt the policeman from liability for failure to perform or improper performance of his official duties.
7. Lawful use of firearms for the purpose of stopping a vehicle is contingent on: 1) legal factors, such as provisions of applicable national laws (in Ukraine – Article 46 of the

Law "On National Police") that are in line with the European Convention for Human Rights and Fundamental Freedoms (Clause "a" Part 2 Article 2); and 2) a set of factors: a) danger, potential or real, to life or health of the people and/or the policeman caused by or resulting from the breach of traffic regulations by the driver of a moving vehicle; and b) inability to stop the vehicle in this specific situation in a manner other than causing damage to the said vehicle by the use of firearms, providing however that any such use is absolutely necessary at this particular place and time and under the circumstances. The ECHR does not consider the use of force as violation of the right to live, if any such power is exercised by a law enforcement officer who is reasonably convinced that his/her action is right at this particular time and place but proves to be wrong at a later time.

8. The extent of justified damage caused by the policeman under the circumstances: 1) the necessity of any such action at this time and place that is based on every legal and actual evidence of the lawful use of firearms to stop the vehicle; 2) maximum permissible extent of damage that sufficiently justifies the use of firearms, which is to take the life of individuals inside the vehicle; and 3) damage caused with the sole purpose of stopping the vehicle.

References

[1] WHO Global status report on road safety (2018). Available at: https://www.who.int/violence_injury_prevention/road_safety_status/2018/en/

[2] Prosecutor General's Office of Ukraine. Available at: https://www.gp.gov.ua/ua/stst2011.html?dir_id=113281&libid=100820&c=edit&_c=fo#

[3] The Criminal Code of Ukraine. Available at: https://sherloc.unodc.org/cld/document/ukr/2001/criminal_code_of_the_republic_of_ukraine_en.html

[4] Ustawa z dnia 6 czerwca 1997 r.Kodekskarny. Available at: http://prawo.sejm.gov.pl/isap.nsf/download.xsp/WDU19970880553/U/D19970553Lj.pdf

[5] The Criminal Law of The Republic of Latvia. Available at: https://likumi.lv/ta/en/id/88966-the-criminal-law

[6] The Criminal Code of Germany. Available at: https://www.gesetze-im-internet.de/englisch_stgb/index.html

[7] Проект Закону про внесеннязмін до деякихзаконодавчихактівУкраїнищодоспрощеннядосудовогорозслідуванняокремихкатегорійкримінальнихправопорушень. Available at: http://w1.c1.rada.gov.ua/pls/zweb2/webproc4_1?pf3511=63928

[8] Willis J. The Proper Role of Criminal Law in Road Safety. Available at: https://aic.gov.au/sites/default/files/publications/proceedings/downloads/24-willis.pdf

[9] *Criminal Liability for Serious Traffic Offences: Essays on Causing Death, Injury and Danger in Traffic,* Eleven International Publishing (2015). Available at: https://www.rug.nl/research/portal/files/28372541/Criminal_Liability_for_Serious_Traffic_Offences_final.pdf

[10] ДТП у Харкові: експертизавстановила, наякійшвидкості "летіло" авто Зайцевої. Available at: https://espreso.tv/news/2018/12/20/dtp_u_kharkovi_ekspertyza_vstanovyla_na_yakiy_shvydkosti_quotletiloquot_avto_zaycevoyi

[11] A. Bukova-Zideluna, A. Villerusa, A. Lama. An overview on pedestrians involved in traffic accidents in Latvia: Years 2010–2014. SHS Web of Conferences **40**, 01004 (2018)

[12] Н.В. Генрих *Предмет и метод уголовно-правового регулирования: история, теория и практика.* М.,Норма; ИНФА-М, 237 (2011)

[13] О. Пащенко. Неможливістьборотьби з діянням за допомогоюінших (не кримінально-правових) засобів. National Law Journal: Theory and Practice. **6**, 192–197 (2015)

[14] G.S. Becker. Crime and Punishment: An Economic Approach. Journal of Political Economy **76**(2), 207 (1968)

[15] D. Husak D. *Overcriminalization: The Limits of the Criminal Law.* Oxford, Oxford University Press (2008)

[16] Ю.В.Баулін. *Вибраніпраці.* Право, 103–321(2013)

[17] О.О. Дудоров, М.І. Хавронюк. *Кримінальне право. К,* Ваіте, 255 (2014)

[18] ECHR judgment in case McCann and Others v. The United Kingdom, application no. 18984/91

[19] ECHR judgment in case McKerr v. The United Kingdom, application no. 28883/95

[20] ECHR judgment in case Andronicou and Constantinou v. Cyprus application no. 86/1996/705/897

[21] ECHR judgment in case Giuliani and Gaggio v. Italy application no. 23458/02

Human trafficking from Ukrainian cities and villages: Current trends

Andriy Babenko[1], *Vladislava* Batyrgareieva[2], and *Alina* Kalinina[2,*]

[1] Odessa State University of Internal Affairs, Odessa, Ukraine
[2] Àcademician Stashis Scientific Research Institute for the Study of Crime Problems National Academy of Law Sciences of Ukraine, Kharkiv, Ukraine

Abstract. The official statistics of the Prosecutor General's Office of Ukraine has recorded that in 2017 the number of cases of human trafficking from Ukraine increased threefold (342 cases versus 115 in 2016). Such a sharp growth of the aforementioned crime was last observed in 2005 to 2006. Today, there are no administrative and territorial districts in Ukraine from which future victims of human traffickers would not be recruited. According to criminal proceedings, about 70% of human trafficking victims are residents of large Ukrainian cities. The aim of this research is to investigate current trends of human trafficking phenomenon from Ukrainian cities and villages and to present the ways how the victims fall into slavery. The revealed trends of the human trafficking from Ukraine allow us to affirm that whereas similarity of social and demographic, moral and physiological characteristics of urban and rural residents, the second ones have a higher level of risk to become a victim of human trafficking. The results obtained allow us to develop evidence-based recommendations for the human trafficking from Ukraine decreasing as from the hole country, as rural locals, in particular. The new feature in human trafficking is a negative trend that Ukraine is changing from the country in which human trafficking victims originate into the destination country for human trafficking.

Key words: human trafficking, slavery, exploitation, victims, residents of cities and villages.

1 Introduction

For over two decades, human trafficking has been recognized as one of the most dangerous types of international criminal business which is a flagrant violation of fundamental human rights and freedoms. According to UN experts, this type of crime ranks third in terms of profitability after drug and weapon trafficking [1, 2]. The total annual income from human trafficking is from 5 to 9 billion USD [3]. About 2.5 million people from 127 different countries are being exploited in 137 countries of the world due to their involvement in the process of human trafficking [4]. The global financial and economic crisis, which was the result of a decline in the indicators of a number of socio-economic issues around the world [5], only aggravated the human trafficking phenomenon (either for sexual exploitation, or for the purpose of attraction in forced labour). At present, almost all regions of the world are covered with this disaster. This predetermines the urgency of creating an effective system of collective opposition to this extremely negative phenomenon by efforts of the entire civilized world.

* Corresponding author: alina.kalinina@ukr.net

Fight against human trafficking is certainly part of the system of the collective opposition both from Ukraine and Ukraine itself. After all, the settlements of our country are not an exception to the analysed phenomenon, which is the product of the world wide web of human traffickers: at present, there are no such regions in Ukraine where future victims of human traffickers would not be recruited. This becomes very relevant especially in relation to the situation with the migration of labour resources from Ukraine, which, under the conditions of the extension of crisis development in the Ukrainian economy [6–8], large external debt, property stratification of society and the difference in the quality of social services in Ukrainian cities and towns, rising unemployment (especially in small Ukrainian cities and rural areas) and involuntary part-time employment, a significant gap in wage levels, on the one hand, in Ukraine and in foreign countries, and on the other hand, in Ukrainian cities and rural areas, the military conflict in the east of the country will only deepen. Ukraine is currently recognized as one of the largest donor countries in Europe where external labour migration has become an objective reality [9, p. 98; 10, p. 5]. The regulation of population's labour migration both in Ukraine and abroad has become an overcomplicated problem, as these processes are taking increasingly unlawful forms. One of these forms is human trafficking.

The goal of this research is to establish modern trafficking trends in Ukrainian cities and villages and to analyse the geography of its distribution on the basis of a comprehensive study of this phenomenon, as well as to clarify the criminological portrait of persons who have suffered from the crime provided for in Art. 149 of the Criminal Code of Ukraine, and the main reasons that contribute to the existence and further spread of this phenomenon in Ukrainian cities and villages.

2 Methods

The methodological basis of the research is, first of all, the official statistical information of the Department of Information and Analytical Support of the Ministry of Internal Affairs of Ukraine; Office of the organizational support of the National Register of Pre-trial Investigations and the Information and Analytical Work of the General Prosecutor's Office of Ukraine; sentences contained on the official website of the State Judicial Administration of Ukraine; information from the State Statistics Service of Ukraine for the period of 2014–2017; and secondly, the information of the United Nations Office on Drugs and Crime; thirdly, in 2017, a survey of ordinary Ukrainian citizens in Kharkiv and Odessa regions regarding the risk assessment to become the victim of human traffickers if these citizens decide to go abroad for earnings; fourthly, a cartographic method that gives a clear idea of ??the level of vulnerability of different regions of Ukraine by the results of the disclosure of cases of human trafficking; fifth, the tabular method, which especially allowed the disclosure of quantitative and qualitative parameters of the phenomenon under consideration.

Empirical data is processed and analysed using descriptive statistics technology, calculations and mapping using Adobe Photoshop CS6, Microsoft Excel 2016, Microsoft Word 2016 and QuickMap 2.2.

3 Results

According to the General Prosecutor's Office of Ukraine, until recently, there was certain stabilization in the dynamics of crimes in Ukraine connected with human trafficking. However, in 2017, the number of such cases registered in Ukraine has increased almost three times (Table 1). Such a sharp increase was most recently observed in 2005–2006. The situation has become so critical that Ukraine's Ministry of Internal Affairs announced 2017 as the Year of Combating Trafficking in Persons. The number of detected persons herewith is

Table 1. Level and dynamics of human trafficking or other illegal human rights (Article 149 of the Criminal Code of Ukraine).

Parameters/ Year	Amount of crime in whole	Dynamics of crimes (%)	Distribution of crimes by parts Article 141 of the Criminal Code			Amount of criminals
			p. 1	p. 2	p. 3	
2014	118	–	14	92	12	77
2015	110	−7.0	16	73	21	79
2016	115	+4.5	19	89	7	59
2017	342	+297.0	24	273	45	132

Table 2. Number of persons who have suffered from the crime, stipulated by art. 149 of the Criminal Code of Ukraine (2014–2017).

Parameters/ quantitative Indexes	2014	2015	2016	2017	Total
Parts 1-3 P. 149	98	113	86	309	597
P. 2 and P. 3 from Art. 149	89 (90.8%)	101 (89.4%)	74 (86.1%)	295 (95.5%)	559 (93.6%)

Fig. 1. Correlation of the number of registered crimes provided by Art. 149 of the Criminal Code of Ukraine, and detected criminals.

almost twice less than the number of registered law violations (Fig. 1). However, the number of victims of human trafficking is only increasing (Table 2) [11–14].

As is evident, the vast majority of crimes (87.5% on average), are committed under aggravating or particularly aggravating circumstances in circumstances (Parts 2 and 3 of Article 149 of the Criminal Code). These include cases of human trafficking: 1) under age or a minor; 2) regarding several persons; 3) repeatedly committed; 4) by prior conspiracy by a group of persons or an organized group; 5) committed by an official using the official position;

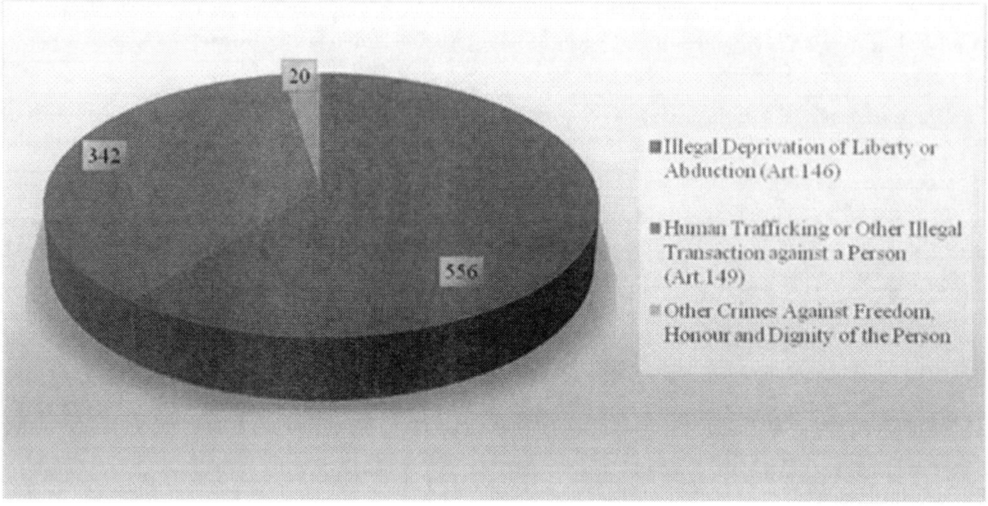

Fig. 2. Crimes Against Freedom, Honour and Dignity of the Person.

Table 3. Number of the existing population of Ukraine at the place of residence (as of January 1, 2017) [16].

Parameters/ quantitative Indexes	**Number of available population (persons)**	**Including (persons)**		**Specific gravity in total (%)**	
		Urban population	**Rural population**	**Urban population**	**Rural population**
Ukraine	42,584,542	29,482,313	13,102,229	69.23	30.77

6) committed by a person from whom the victim was materially or otherwise dependent; 7) is connected with violence, dangerous to the life or health of the victim or his relatives, or with the threat of the use of such violence, including those that caused grave consequences). 93.6% of victims of trafficking suffer from such actions.

The share of human trafficking among all crimes against the will, honour and dignity of a person envisaged by Section III of the Special Part of the Criminal Code of Ukraine constitutes 37%, taking the second step after cases of unlawful imprisonment or abduction of a person (Fig. 1).

Using the cartographic method [15], it was found that the most vulnerable regions suffering from this crime were the Kyiv, Odesa, Kharkiv and Dnipropetrovsk regions (Map 1).

The conducted criminological research has established that recruitment of about 70% of victims of human trafficking based on materials examined by the courts of criminal proceedings took place in large Ukrainian cities. Two thirds of such cases herewith fall into such cities as Kyiv, Odessa, Kharkiv and Dnipro.

This indicator proportionally reflects natural distribution of Ukrainian population in cities and rural areas.

At first sight, a similar situation seems to indicate that the inhabitants of the largest cities are more likely to become victims of human trafficking. However, the study of socio-demographic data of the victims of human trafficking showed that only 59.3% of them are inhabitants of large cities, the rest are inhabitants of settlements and small towns (40.7%). It

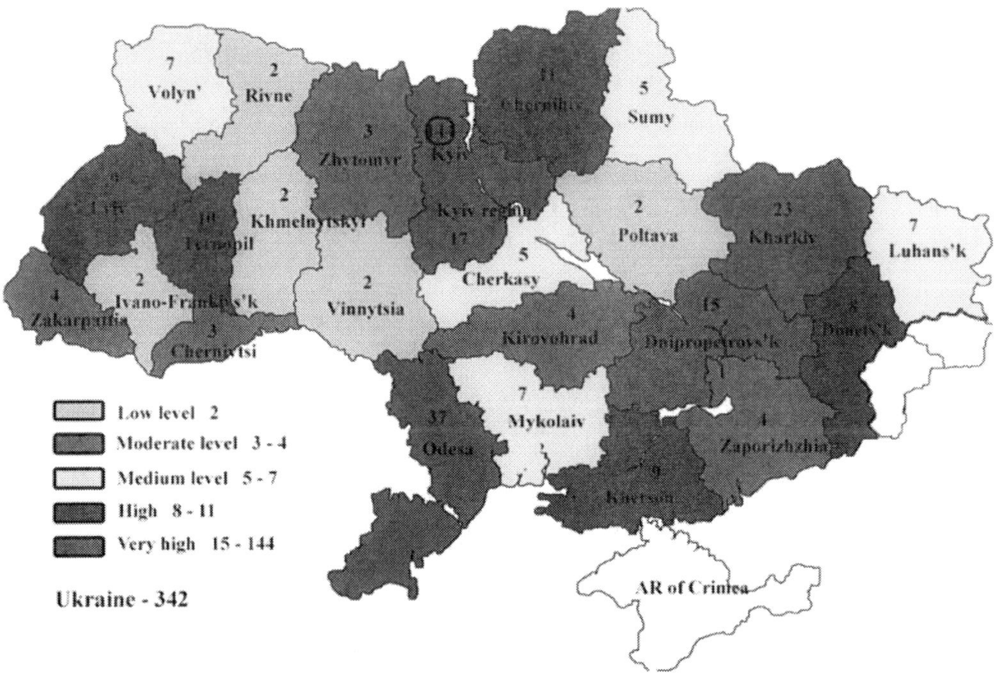

Map 1. Level of human trafficking, recorded in the regions of Ukraine in 2017.

means that the risk of becoming a potential victim from rural residents is significantly higher (1.32) than urban residents (0.86)[1], that is, approximately by half.

Pursuant thereto, the research task was to provide an explanation for the occurrence of such imbalance of risks. The study of the victims' biographies, as well as the results of the survey of ordinary citizens provided an opportunity to explain the reason for the artificial increase in cases of human trafficking from the largest cities of Ukraine. Before the victims are caught by the traffickers they try to look for work in large cities for some time, thus, becoming the inhabitants of these cities. When rural residents fail to do this or they are not happy with low wages at home, they agree on the offer of criminals to go abroad for work, sometimes deliberately looking for any, including illegal, ways to get to one or another country for earnings. Thus, the following is typical for rural residents: first, internal migration to large and especially large Ukrainian cities searching for work in a native country, further – targeted search of work abroad, and finally – exit abroad and into slavery. It is appropriate to dwell in detail on the problem of the risk of traffickers entering the trafficking network.

The researchers of general population labor migration of the and human trafficking, in particular, study these phenomena from the standpoint of different approaches, concepts, theories, among which the risk theory is assigned the significant role [see 17–26, etc.]. Risks are usually spoken when a decision was made, in case of rejection of which losses could not occur [27, p. 211]. The most dangerous among the most significant risks of labour migration abroad is the risk of being drawn into debt bondage for the unlawful acts of traffickers.

---

[1] *Note.* These coefficients are a mathematical ratio proportion of urban and rural residents who have been victims of traffickers based on criminal proceedings, and the proportion of urban and rural population in the total number of all residents of Ukraine respectively.

Table 4. Estimates of personal risk of becoming a victim of human trafficking by the inhabitants of Ukraine.

Parameters/ quantitative indicators	**Awareness of respondents of the existence of human trafficking cases**			**Probability to get in the human trafficking situation when being employed abroad according to respondents' personal assessment)**			
	heard of such cases	never heard about them	hard to answer	high	short	excluded	hard to answer
Urban Residents	136 (90.7%)	5 (3.3%)	9 (6.0%)	27 (18.0%)	52 (34.7%)	17 (11.3%)	54 (36.0%)
Rural residents	105 (70.0%)	26 (17.3%)	19 (12.7%)	13 (8.7%)	49 (32.7%)	41 (27. %)	47 (31.3%)

During the research, information was received on the awareness of ordinary citizens at working age – residents of urban and rural areas – regarding the problem of trafficking among human beings and the risk of harm from illegal actions of a trafficker if a decision was made to work abroad.

Thus, a significant part of both urban and rural residents cannot be identified with the risk of becoming victims of traffickers, which is indicative of the still insufficient awareness of citizens about the analysed problem and the criminal schemes of getting a person into slavery. Moreover, rural residents look more assertive that they will not encounter such situations if they decide to look for work abroad: their chances of avoiding troubles are estimated at two and a half times higher than urban dwellers. In this case, it is important to find out whom the risk of falling into slavery may endanger. The research found that recruitment of victims more often was carried out by acquaintances or relatives, etc. (61.1%), because the victims trust them most.

Rural and urban residents of Ukraine differ in their preferences in choosing the types of work they would agree to. Thus, urban female residents prefer work related to the service sector (tourism and show business, child care or sick care), while rural residents agree to work on farmland and as housewives. Men regardless of the region of origin prefer agriculture, construction and transport related services. The materials of criminal proceedings, however, show a somewhat different situation. Most victims were indeed invited to work as dancers in night clubs, cabaret, etc. (41.4%); 24.2% were offered to provide intimate services; 13.8% – to work as waitresses; in agriculture – 9.9%; caring for a child or patients – 3.4%; to do housework – also 3.4%; the rest – 3.9%. However, the vast majority of victims (98.2%) have been sexually exploited in one way or another, even though there are cases of such recognition by men aged 25–35 aged between the ages of 25 and 35, mostly employed in construction and agricultural production. But often researchers did not establish if the kind of "work", which the victims had been doing in Ukraine, was similar to their work abroad (in other words, if they had provided sexual services on a commercial basis). It can be said with a high degree of certainty, that such "work" had been done by 14.3% of victims, 32.2% of them had not ever done anything of the kind (this information was established in criminal files) However as for the most part of victims (more than 50%) we can say most probably, that they had prostituted in Ukraine too.

As for the analysis of the previous work activity of the victims, only 19.2% of urban and 14.2% of rural residents worked in Ukraine officially until their departure abroad. The vast majority of victims (63.3%) left Ukraine by their choice, with reliable information about the actual nature of the work they are supposed to perform, including sexual services. In other cases, travellers did not have reliable information. However, the facts of forced exportation were not stated, although such facts occurred earlier (15–20 years ago) [28, p. 4]. Therefore, nowadays the only source of danger to migrant workers from Ukraine is the mediator, to whom they apply for the purpose of employment and who undertakes the financing of the travel, especially purchasing tickets, visas processing, passports and so on. These financial expenditures of the travel are the additional motivation that does not permit an emigrant to refuse leaving even if some doubts arise as for its save. Herewith 23.8% of victims implied that abroad they had been subject to physical violence from the guilty persons: they were beaten, belaboured, abandoned without food and water. Also about 54.8% of victims said that the psychic violence was applied to them (in particular, the threat of physical violence, the threat to tell relatives and friends about what they were doing abroad, other compromising data, etc.). Both physical and psychic violence took place toward 9.7% of victims. No violence by the perpetrators, according to the victims' testimonies, was used against 11.7%. As we see, violence was used against most victims. Such behaviour with them testifies about coercion and personal restraint. Sometimes victims even had no opportunity to communicate with anyone, because during their time out of duties they were hold in a closed space. In more cases the only chance to get freedom was a police raid, after that they were deported.

The next alarming tendency, which we fixed during the victims' interview, is the fact, that when women after their returning home bring with them some money as a payment for sexual services (even if they were channel into prostitution by a fraud), they do not identify themselves as victims, despite the awful conditions in which they were abroad. Moreover, such women are now themselves trying to take out new women, having the necessary knowledge about how to better adapt to living abroad, how and where to find a "normal" job, getting good money for it, and so on. In addition, they expect that they will now be able to work without mediator, to independently seek housing, the rent for which it is easier to pay with their companions. It is interesting to note that the main reason for exposing traffickers by the victims, which we managed to establish based on the results of the study of materials of criminal proceedings, was often the banal desire of these victims to "punish" the "mediators" for not receiving the promised profit.

The risk group among women suffering from human trafficking, like 15–20 years ago, includes people aged 16–35 years. The victims (77.9%) have a low educational level, and only 25.0% have families and children. The victim's behaviour is manifested in their frivolity (52.2%), underestimation of the risks associated with working, weakening of social protection of population from the state, military events in the East Ukraine, demand for sexual services and cheap labour abroad, lack of hope for the future, internationalization of the shadow economy, formation of international criminal associations; corruption of government officials involved in the issuance of documents required for travel abroad, etc. Mass and individual psychology of consumerism, covetousness, moral degradation of a large number of potential victims and legal nihilism of the population of Ukraine play a major role in the restoration and existence of this criminal offense, in which 66.7% of crimes are committed by organized groups. Therefore, the conduct of the trafficking victim has an important determination role in the expansion of this crime. In view of the increased victimization of aggrieved persons of this crime, which in the vast majority of cases is due to the desire for high earnings, immoral way of life, frivolity and naivety of the person, special urgency is required in order to develop measures for prevention of victims of human trafficking, especially among women.

4 Discussion

A further negative trend, which arose a few years ago in Ukraine, needs to be further explored – from the country of origin of victims of trafficking, it is turning into a country of destination of living goods, as more and more citizens of other states become affected by various forms of exploitation in Ukraine: sexual, including those involved in forced prostitution and pornography, labour – in agriculture, construction, service sectors, forced begging and organ extraction [29, p. 6]. However, only a small number of foreigners are currently recognized as victims of human trafficking. Instead, such individuals tend to be expelled outside our country.

One more aspect related to this phenomenon is to be studied, namely: were there any new causes that had not been noticed before, among the reasons that led to the existence and reproduction of this phenomenon? We respond negatively to this question so far: no new causes arose except for socio-economic and personal reasons that existed before in the complex of human trafficking 30, p. 32.

Another tendency to be clarified is that, along with human trafficking, the problem of independent labour migration from Ukraine is still partly illegal. Human trafficking herewith cannot be identified as illegal labour migration, since it is a matter of somewhat different problems, although they are closely interconnected.

It is often argued in the media that Ukrainian citizens are supposedly sold for the purpose of their further use in violent conflicts. However, there is another problem – misconception of the notion of "human trafficking" of other unlawful actions against Ukrainian prisoners of war and civilians. These include cases requiring money for the release of these individuals, which is a particular problem that needs to be studied.

5 Conclusions

Human trafficking is one of the most difficult problems both in Ukraine and in most regions of the world. Its solution should be based on the combination of collective efforts of all interested countries of the world. Identification of current trends in human trafficking from Ukrainian cities and towns, reflecting regional peculiarities in particular, suggests that although socio-demographic and moral, and psychological characteristics of urban and rural residents are similar, rural residents have a higher risk of becoming victims of human trafficking than urban residents. The obtained results, firstly, allow us to familiarize the general population with the main traits of human trafficking from Ukrainian cities and villages. Secondly, they can be based on the improvement of information and analytical support of the work of state and law enforcement bodies in order to reduce the incidence of human trafficking from Ukraine in general, and from rural areas, in particular.

References

[1] Combatting Human Trafficking. Available: `http://www.austria.org/human-trafficking/`

[2] Breaking the Chain: Corruption and Human Trafficking. Available: `https://www.transparency.org/news/feature/breaking_the_chain_corruption_and_human_traffickin`

[3] Торгівля людьми. Available: `http://www.gpedia.com/uk/gpedia/Торгівля_людьми`

[4] UNODC launches Global Initiative to Fight Human Trafficking. Available: `http://www.unodc.org/unodc/en/press/releases/2007-03-26.html`

[5] *World economic situation and prospects* 2013 (2013). 1st Ed. New York: United Nations
[6] Економіка України за 2015 рік. Уряд. кур. № **24** (2016)
[7] Еконаоміка України за 2016 рік. Уряд. кур. № **28** (2017)
[8] Економіка України за 2017 рік. Уряд. кур. № **30** (2018)
[9] Васильківський Д. М., Осійчук О. А. *Зовнішня трудова міграція в Україні: сучасний стан, причини, наслідки.* Вісн. Хмел. нац. ун. № **1**. С. 97–101 (2009)
[10] *Населення України. Трудова еміграція в Україні*. Кїыв: Ин-т демографіы та соціальних досліджень ім. М.В. Птухи НАН України (2010)
[11] Јдиний звіт про кримінальні правопорушення за січень-грудень 2014 року. Available: `http://www.gp.gov.ua/ua/stst2011.html` (2015)
[12] Єдиний звіт про кримінальні правопорушення за січень-грудень 2015 року. Available: `http://www.gp.gov.ua/ua/stst2011.html` (2016)
[13] Єдиний звіт про кримінальні правопорушення за січень-грудень 2016 року. Available: `http://www.gp.gov.ua/ua/stst2011.html` (2017)
[14] Єдиний звіт про кримінальні правопорушення за січень-грудень 2017 року. Available: `http://www.gp.gov.ua/ua/stst2011.html` (2018)
[15] Бабенко А. М., Орещенко А. В. *Можливості картографічних методів дослідження у інформаційно-аналітичному забезпеченні протидіы торгівлі людьми.* Актуальні питання боротьби з торгівлею людьми в умовах проведення антитерористичної операції (збройного конфлікту): зб. матеріалів міжнар. круглого столу. Киыв: ВД. «Дакор». С. 42–48 (2017)
[16] Чисельність наявного населення України на 1 січня 2017 року. Київ: Держ. служба статистики України (2017)
[17] Яницкий О. Н. *Социальные реальности и социальные мирами. Соц. риска: ключ. идеи.* Мир Росс. Т. 12. № **1**. С. 3–35 (2003)
[18] Мозговая А. В. *Социология риска: возможности синтеза теории и эмпирическиого знания.* Риск в социальном пространстве / под ред. А. В. Мозговой. Москва: Ин-т социологии РАН (2001)
[19] Мазаник М. Н. *Мифологема «риска» в социальных науках и массовом сознании.* Философия и ценности современной культуры: материалы Междунар. научн. конф. (Минск, 10–11 октября 2013 г.). Минск: БГУ. С. 582–584 (2014)
[20] Клямкин И.М., Тимофеев Л.М. *Теневая Россия: Экономико-социологическое исследование.* Москва: Рос. гос. гуманит. ун-т (2000)
[21] Sagan S. D. *The Limits of Safety. Organizations, Accidents, and Nuclear Weapons.* Princeton: Princeton Univ. Press (1995)
[22] Beck U. *Risk Society. Toward a New Modernity.* London: SAGE (1992)
[23] Douglas M. *Risk Acceptability According to the Social Sciences.* New York: Russel Sage Foundation (1985)
[24] Douglas M. *Risk and Blame: Essays in Cultural Theory.* London: Routledge (1992)
[25] Douglas M. *Risk Acceptability According to the Social Sciences.* New York: Russel Sage Foundation (1985)
[26] Rosa E.A. Metatheoretical Foundations for Post-Normal Risk, JRR **1**(1), 15–44 (1998)
[27] Населення України. Трудова еміграція в Украыні. Київ: Ин-т демографії та соціальних досліджень ім. М.В. Птухи НАН України (2010)

[28] Батиргарејва В. С. Аналітична довідка за результатами узагальнення кримінальних справ за ст.149 КК України 2001 року (ст.124[1] КК Украыни 1960 року), розглянутих судами України за 1998–2004 рр. Харків: Архів Института вивчення проблем злочинності АПрН України (2005)
[29] Васильјва М. О., Касько В. В., Орлеан А. М., Пустова О. В. Україна як країна призначення для торгівлі людьми: матеріали для практичного використання працівниками прикордонноы служби, правоохоронних органів та суддями. Киыв: Фенікс (2012)
[30] Батиргарејва В. С. *Торгівля людьми в Україні*: *реальність та омана.* Актуальні питання боротьби з торгівлею людьми в умовах проведення антитерористичної операції (збройного конфлікту): зб. матеріалів міжнар. круглого столу. Київ: ВД. «Дакор». С. 31–37 (2017)

SHS Web of Conferences **68**, 01014 (2019) https://doi.org/10.1051/shsconf/20196801014
Int. Conf. SOCIETY. HEALTH. WELFARE. 2018

Mandatory nature of criminal proceedings in the republic of Latvia: Current issues and controversies

Sandra Kaija

Rīga Stradiņš University, Faculty of Law, Riga, Latvia

Abstract. The Criminal Procedure Law (Section 6), as one of the basic principles of the criminal procedure, provides for the mandatory nature of criminal proceedings. It determines that the official who is authorised to perform criminal proceedings has an obligation within his or her competence to initiate criminal proceedings and to lead such proceedings to the fair regulation of criminal legal relations provided for in The Criminal Law in each case where the reason and grounds for initiating criminal proceedings have become known. This research focuses on issues that relate to the legal aspects of the mandatory nature of criminal proceedings, including the competence of the officials in initiating criminal proceedings; the role of the victim; the purpose of this basic principle. This paper examines the problematic aspects of the mandatory principle in criminal procedure in Latvia. In the end, key conclusions are summarized.

Key words: criminal proceedings, mandatory, principle.

1 Introduction

The **aim** of the paper is to research legal aspects of mandatory nature of criminal proceedings, to identify problematic issues, and to propose solutions to them. **Materials and methods.** The research uses grammatical, systemic and teleological (meaning and purpose) methods of interpretation of legal norms, as well as descriptive, analytical and deduction-induction research methods. By applying these methods, legal acts and opinions of law scientists were analysed, which was followed by drawing conclusions and drafting recommendations.

2 Results and discussion

The author or this paper will determine and analyse the overall characteristics of the mandatory nature of criminal proceedings and their legal regulation in the legislation of the Republic of Latvia.

The basic principles of the criminal proceedings in the capacity of the leading norms, which application is possible during the entire course of the criminal proceedings, determine the nature and structure of the procedural order. The basic principles of the criminal proceedings form an overall system, which includes such principles of the criminal proceedings as the mandatory nature of the criminal proceedings, etc. [1].

The principles of the criminal proceedings are legal because they as general and leading procedural norms are frequently mentioned in the Constitution of the Republic of Latvia. Being connected to the objectives of the criminal proceedings, these principles indirectly follow the implementation of objectives via a special norm of the legal proceedings. The principles of the criminal proceedings are all integrated into one particular system and they determine the constraints of the conclusion of objectives of the legal proceedings and are closely linked to the Criminal Law and Court Organization guidelines. Although each principle of the criminal proceedings expresses a completely independent norm of the legal proceedings, it should be acknowledged that there are close ties and practical connection amongst them. One cannot contradict another one. Each principle is connected to another principle directly or via other guidelines [2].

One of the basic principles of the criminal proceedings is the mandatory nature of the criminal proceedings, which is in accord with Article 6 of the Criminal Procedure Law (CPL henceforth), which says that "Within the scope of his/her competences, the official who has the power of attorney to conduct the criminal proceedings has the obligation to initiate the criminal proceedings and ensure they proceed to the fair conclusion of the criminal legal relationship that is envisaged by the Criminal Law in each case when the cause and justification for the initiation of the criminal proceedings become known" [3].

Article 1 of CPL includes references that let conclude what exactly the public expects as the final outcome of the process, which is the fair conclusion of the relation set by the criminal law. Therefore, justice can be considered as both the aim and means of the criminal proceedings. It is implausible to attain the fair aim using unfair means. To ensure the justice aspect of the conclusion, fair material and procedural norms are required as well as their fair application. Article 1 of CPL determines "the basic rule" – as part of the criminal proceedings, a fair outcome must be attained, including the fair conclusion of the criminal legal relation – at the same time one must consider the restrictions on the unfair intrusion into the personal life [4].

The CPL aim is to determine the order of the criminal proceedings that will ensure the effective application of the CL norms and the fair regulation of the criminal legal relation without unfair intrusion into the a person's life, hence, it can be concluded that the most important objective of CPL is not to punish persons but to reach the fair regulation of the criminal legal relation, which is envisaged also by Article 6 of CPL "The mandatory principle of the criminal proceedings".

The inclusion of the relevant norms into the CPL section "The basic principles of the criminal proceedings" plays a significant and far-reaching role both for creation of legal norms of the criminal procedure and their legal application into real life. The legislator must realize that by announcing one position as the basic principle of the field, the legislator has not only engaged in a formal activity, but also has taken the responsibility to subordinate all other norms to this key guideline. These basic principles have been formed as a "frame", which must be completed. Specifically, the norms that are being created cannot contradict the implementation of ideas of the principles in real life under any circumstances. Therefore, for the system to be stable and sustainable, in the first place, it is important to focus on the quality of the legal regulation of the principles as well as on the fact that the meaning of the principles must be familiar to the officials working on their implementation. Although these officials must follow the specific regulation norms of the specific activity, it is also important to ensure that their work, passed decisions etc. are in accord with the basic principles. These officials must understand the special meaning and role of these norms in the overall system of criminal procedure norms. They must also comprehend the consequences when these principles are not observed. Breaching the principles of the criminal proceedings associate with serious consequences, specifically, Section 4 of Article 130 of CPL provides that the information that has been obtained by breaching the basic principles of the criminal proceedings is considered as absolutely prohibitive and cannot be used in the subsequent court hearings [5].

What follows from the above discussion is that the Latvian criminal proceedings acknowledge the so-called legality principle as oppose to the opportunity principle. To a certain extent, the expediency of the principle resonates with the proportionality principle, although occasionally these principles are not sufficiently distinguished. However, even though expediency and proportionality are the evaluative concepts and on occasion they overlap, there might also be situations when they are distinguishable because it is not always the case that something that is effective might simultaneously be proportionate, and vice versa. Similarly, it might be emphasized that the observance of the proportionality principle, when deciding on whether the criminal proceedings can be the case in a specific context or not, might be more related to the material legal norms, while the expediency principle on whether to initiate the criminal proceedings is linked to the procedural norms [6].

The mandatory nature principle of the criminal proceedings, therefore, is tightly connected to and can be analysed only in the connection to the principle of the public prosecution. The concept of "public prosecution" was already known at the KPK times, but its interpretation has changed since then. The public prosecution process can be defined as the process that has been initiated independently of the victim's wishes and public interests [7].

According to the author, the basic principle of the criminal proceedings is the mandatory principle of the criminal proceedings and it is one of the most significant basic principles, which emphasizes the duties of the officials to pass the decision on the initiation of the criminal proceedings. The reasons, grounds and order for the initiation of the criminal proceedings are determined in Section 30 of the CPL, which is why in what follows the author will further analyse the reasons and grounds of the initiation of the criminal proceedings.

The initiation of the criminal proceedings commences when the prosecution institution or the prosecuting authority receive information on that a) an offence has taken place, or b) on that there is a possibility that an offence might take place. Using such information, the official with the power of attorney to initiate the criminal proceedings who works for the prosecuting authorities, having identified the reasons and grounds as determined in CPL, decides whether indeed to initiate the criminal proceedings, while observing the CPL norms.

Having identified the grounds for the initiation of the criminal proceedings, it is possible to immediately initiate the criminal proceedings, thus allowing the prosecution to use all criminal procedure means in the context of a specific possible offence. In accord with Article 369 of CPL [8], the reasons for the initiation of the criminal proceedings might be feasible to divide into two categories:

1. The reason for the initiation of the criminal proceedings is the *information submission* to the prosecution office or the prosecuting authorities, which point to committing of a possible offence;
2. or it could also be *obtaining of such information* by the responsible authorities in pertinence to the applied procedure of the criminal proceedings.

The indicated *information can be submitted:*

1) as a submission by a person who has suffered as a result of a criminal offence;
2) by controlling and supervising institutions, in accordance with the procedures provided for in the laws and regulations governing the activities thereof;
3) by medical practitioners or institutions, as a report regarding traumas, illnesses, or cases of death the cause of which may be a criminal offence;
4) by non-governmental organisations, and authorities protecting the rights of children, as a submission regarding infringements upon the rights of minors the cause of which may be a criminal offence;

5) any natural person or legal person, as information regarding possible criminal offences from which such person has not directly suffered;
6) as a submission by any person regarding a criminal offence committed by such person.

The reason for the initiation of criminal proceedings may not be anonymous information or information whose submitter refuses to disclose the source of the information.

In turn, *information can be obtained* by the prosecution institution, prosecution authorities or court as a result of criminal procedural activities in the following cases:

1) directly determining a criminal offence at the time of the committing thereof, and discontinuing such offence;
2) directly determining clear consequences of a criminal offence;
3) performing criminal proceedings regarding another criminal offence;
4) performing other functions specified in laws: examinations, an investigative action, etc.

Upon obtaining or receiving information, there might emerge different situations.

1) In case there is a real chance that the offence has occurred or if the information is possible to check only using means and methods of the criminal proceedings (Article 370 of CPL), then the decision must be made to initiate the proceedings.
2) In case this information contains information about the violation of law, for which detection the use of means and methods of the criminal proceedings is not necessary (Section III of Article 373 of CPL), then it is forwarded for verification to the competent authority.
3) In case the official, who has received the power of attorney consistently with the procedure, determines that there is no justification for the initiation of the criminal proceedings (Section I of Article 373 of CPL), then this official either passes the decision to reject the initiation of the criminal proceedings or can pass the decision in the format of a resolution and in this case the person who has submitted the information about the possibility of an offence occurrence should be informed about this resolution, the exceptional cases being medical practioners and institutions.

Having received or obtained information, which points to the possible commitment of an offence, it is essential to pass a decision on whether this information offers the legal grounds for the initiation of the criminal proceedings. In this case two factors must be considered: 1) the criminal proceedings are initiated for the good of the public irrespective of the victim's wishes and 2) the criminal proceedings are initiated only when the victim's complaint has been received.

Justification for the category of the specific procedure type in case of a specific offence represents an especially important factor because it determines to a significant extent the outcome of such processes as the initiation of the criminal proceedings or (in many situations) the necessity to continue the proceedings, for example, in case of reconciliation.

In reference to the influence of the offence victim's on the existence of the proceedings at the time of their initiation, it is important to bear in mind the basic idea of the criminal proceedings – they are conducted in the interests of all members of the public for the offence that has been detrimental to the entire society. The acknowledgement of this fact is closely connected to the mandatory principle of the criminal proceedings, which is included into Article 6 of CPL. However, the mandatory principle of the criminal proceedings is not without any restrictions because in many cases the position of the officials who conduct the criminal proceedings is not the determining one when the criminal proceedings are initiated; instead, the decisive role belongs to the victim of the offence [9].

According to Section II of Article 96 of CPL, the one who promotes the proceedings promptly informs the person about his/her rights to be recognized as a victim in the criminal proceedings, yet Section III of the same article provides that a person can be recognized as a victim only under the condition of this person's or his/her legal representative's written consent. The person who does not wish to be recognized as a victim acquires the status of a witness, which is why CPL does not contain a mandatory requirement that a person who has suffered in an offence must be recognized as a victim in the criminal proceedings [10].

Section II of Article 370 of CPL provides that the criminal proceedings can be initiated if the information contains information on the occurrence of a possible offence and it can be verified only using the means and methods of the criminal proceedings.

This means that the institution that passes the decision to initiate the criminal proceedings in the beginning must review each received application or information in order to be able to determine if this information points to the possibility of the occurrence of the offence, and no any other features could be identifiable, for example, the features of violation of the civil rights, and that additionally it is important to determine the grounds for existence for the initiation of the criminal proceedings in accord with Article 370 of CPL.

In the initial stage of the criminal proceedings the official who must ensure the observance of the mandatory nature of the criminal proceedings should implement specific filtering functions identifying the offences which require the initiation of the criminal proceedings and those which do not. In order to ensure the effective application of the CPL norms and the fair regulation of the criminal legal relation, this official, initiating the criminal proceedings, should be able to identify features of the offence [11].

Having determined the necessity to initiate the criminal proceedings, the institution responsible for the progress of the criminal proceedings must immediately decide on the issue of competence in initiating the criminal proceedings.

It must acknowledge that Article 6 of CPL that provides an indication to the "officials who have the power of attorney to conduct the proceedings" does not have a precise definition of such an official as not all officials, conducting the criminal proceedings, have the right to initiate and complete the criminal proceedings. Section II of Article 26 of CPL provides that the power of attorney in the specific criminal proceedings is obtained by:

1) the person directing the proceedings;
2) a member of the investigative group;
3) the supervising public prosecutor;
4) an official authorised to perform criminal proceedings who executes the task of the person directing the proceedings, a member of the investigative group, or the court to conduct procedural actions;
5) an expert from an expert-examination institution;
6) an expert who does not work at an expert-examination institution, if the person directing the criminal proceedings has assigned him or her to perform an expert-examination;
7) an auditor on the assignment of the person directing the proceedings;
8) the direct supervisor of an investigator;
9) the senior public prosecutor;
10) the investigating judge;
11) the counsel for the prosecution.

However, having conducted the systematic analysis of the CPL norms, it is possible to conclude that only the investigator, the investigator's superior and the prosecutor, not all officials that are mentioned in Article 25 of CPL, have the legal rights to pass the decision on the initiation of the criminal proceedings.

For example, Section V of Article 371 of CPL states that the judge or the court forward the application, materials or information that has been obtained during the hearing, without the judgement, to the prosecution institution or in cases determined in the law to the prosecution authorities. Therefore, this paper concludes that the judge or the court does not pass the decision on the initiation of the criminal proceedings, however, during court hearing, having identified the likelihood of the occurrence of the offence, the judge or the court makes the decision to forward the case to the competent authorities for the final decision.

This is the way how the principle of separation of the procedural functions is ensured, which prohibits the court to accept the prosecution function [12].

Section I of Article 29 of CPL provides that the duties of the investigator include the obligation to review information that points to the commitment of a possible offence and to initiate the criminal proceedings as soon as the reason and grounds for doing so, which are outlined in CPL, are identified, or else to reject the initiation of the criminal proceedings.

Section I of Article 371 of CPL determines that the direct supervisor of the investigator or the investigator have the responsibility to conduct the criminal proceedings only within the scope of their competences and in accord with the reasons mentioned in Article 369. In turn, Article 31 of CPL, which describes the competence of the direct supervisor of the investigator, does not mention that fact that the duties of the direct superior of the investigator include the rights to make the decision on the initiation of the criminal proceedings. Therefore, it would be necessary to expand Section II of Article 31 of CPL by introducing a new point: "The duties of the direct superior of the investigator include passing the decision on the initiation of the criminal proceedings in connection to any reason mentioned in Article 369 of this law".

In pertinence to the competence of the prosecutor, it might be possible to distinguish a few situations:

1) Section II of Article 371 of CPL determines that the prosecutor can forward materials for the review to the prosecution institution, or
2) can initiate the criminal proceedings within the scope of own competence in connection to any reason mentioned in CPL. Having initiated the criminal proceedings, the prosecutor usually forwards the criminal case to the investigator for conducting investigation.
3) However, Section II of Article 36 of CPL provides that in the cases provided by the law the prosecutor decides on the initiation of the criminal proceedings and the prosecutor himself/herself conducts investigation. The exceptional case would be the situation when at the stage of investigation the Prosecutor General, the Criminal Legal Department of Prosecution General or the Supervising Prosecutor of the District Court can appoint the prosecutor as the promoter of the proceedings.

In this case, the prosecutor, as a promoter of the case, similarly to the investigator, organizes the direction of the criminal proceedings and at the investigation stage uses the power of the attorney given by the state by himself/herself or engaging other officials or else demands that each person complete and observe the duties outlined in the order of the criminal proceedings; furthermore, the prosecutor in the criminal proceedings ensures for all parties involved the opportunity to exercise the rights outlined in the law [13].

As it is provided by CPL, the supervision aim of the supervising prosecutor is to ensure the target-oriented investigation, not to allow unfair intrusion into the person's life or faltering, not to allow procedural violations that threaten the progress of the criminal proceedings. Importantly, the supervising tools provided by CPL to the prosecutor include giving instructions by the prosecutor, engagement in conducting of procedural activities, the rights to study the criminal proceedings materials at any time, the rights to participate in the session in which the investigating judge decides on issuing the permit to use coercive instruments un to conduct activities of special investigation, the rights to annul the decisions of the promoter of the process and of the member of the investigation group etc. [14].

Yet another aspect that should be reviewed in the context of the basic mandatory principle is the duty not only to initiate the criminal proceedings, but also "to take the case to the stage of the fair regulation of the criminal legal relation provided in the Criminal Law".

Hence, some sort of contradiction emerges between Article 6 of CPL, which in relation to the initiation principle of the criminal proceedings has the ensured mandatory principle, and Section II of Article 373 of CPL, which provides the so-called expediency principle. According to this article, an investigator, with the consent of a public prosecutor, may refuse the initiation of criminal proceedings, if a criminal offence has been committed that has the features of a criminal offence, but which has not caused such harm that would warrant the application of a criminal punishment.

CPL provides for the opportunity to reject the initiation of the criminal proceedings or stop the already initiated criminal proceedings based on the grounds that the committed offence, which has the features of the criminal offence and which has not caused such damage as to demand the criminal penalty (including the so-called "minor damage" cases). This allows the investigation institutions to "relieve" from conducting criminal proceedings in cases when the incurred damage is evidently "of no significance". In accord with CPL, this instrument, used by the promoter of the process, can be utilized with the consent of the supervising prosecutor, which points to the significant role of the prosecutor in this matter. Although this instruments is provided in the law in order not to handle criminal proceedings cases that have not caused such damage as to demand the application of the criminal penalty, in reality this norm is not sufficiently used because the unified understanding and interpretation regarding the cases which would be considered as "minor damage" cases have not yet been reached.

In accord with the assessment of the State Control of the Republic of Latvia, for legal clarity and certainty purposes it would be necessary to determine additional criteria for the application of this norm. Therefore, reaching a consensus on when not to apply the initiation of the criminal proceedings because the given cases could be considered as cases of "minor damage", could be successfully applied in real life as an instrument which provides for the opportunity to redirect freed resources to the investigation of more important cases [15].

In addition, it would be necessary to conduct the amendments of the legal regulation of the basic mandatory principle by providing for exceptional cases when the criminal proceedings might not be initiated.

Results

The mandatory principle of the criminal proceedings is one of fair components of the criminal proceedings and it must be considered by the official who has the power of the attorney (the promoter of the process), when passing the decision on initiation of the criminal proceedings.

The following findings were discovered:

1) The Criminal Procedure Law gives the authority to perform criminal proceedings to more than ten officials, however only a few of them can initiate in criminal proceedings. In addition, there are certain contradictions in the legal framework, for example, regarding the competence of the direct supervisor of an investigator;
2) initiating criminal proceedings must always remember that there are cases where criminal proceedings are initiated only if a request has been received from the person to whom harm has been inflicted;

3) contrary to the principle, there are exceptions to the law when it is possible to refuse to initiate criminal proceedings;
4) only when the person who committed the criminal offence is determined, it is possible to lead such proceedings to the fair regulation of criminal legal relations provided for in The Criminal Law.

3 Conclusions

1) It is necessary to amend the Criminal Procedure Law by resolving contradictions in the legal regulations of the direct supervisor of an investigator.
2) What could be discussed, is the purpose of the principle of mandatory nature of criminal proceedings. How to achieve a fair regulation of criminal legal relations in all cases when criminal proceedings are initiated?
3) Article 6 of the Criminal Procedure Law needs to be defined that there are exceptions to the mandatory nature of criminal proceedings.

References

[1] *Latvijas tiesību sistēma* / Aut. kol. A. Endziņš, J. Načisčonis, S. Kaija u.c. Rīga, SIA Turība Biznesa augstskola, 2017, 25 lpp
[2] Meikališa, Ā., Strada K. *Kriminālprocesuālo terminu skaidrojošā vārdnīca.* Rīga, RaKa, 2000, 123 1pp
[3] Kriminālprocesa likums: LR likums. Latvijas Vēstnesis 74 (3232), 21.04.2005, 6.pants
[4] Meikališa, Ā., Strada-Rozenberga, K. *Taisnīgums kriminālprocesā* // Jurista vārds Nr. 48 (695)
[5] Meikališa, Ā., Strada-Rozenberga, K., *Kriminālprocess raksti 2005–2010.* Rīga, Latvijas Vēstnesis, 2010, 65–66 lpp
[6] Strada-Rozenberga, K. *Publisko un privāto interešu kolīzija kriminālprocesā.* Politika un tiesības. Tiesību un juridiskās prakses ilgtspējīga attīstība.// Available: https://www.lu.lv/fileadmin/user_upload/lu_portal/apgads/PDF/Juristu-konf_PLZK_2012_.pdf
[7] Meikališa, Ā., Strada-Rozenberga K. *Pārmaiņu laiks kriminālprocesā* // Jurista vārds Nr. 11 (414)
[8] *Kriminālprocesa likums.* LV likums. Latvijas Vēstnesis:11.05.2005., 369.pants
[9] Strada-Rozenberga, K. *Cietušais un tā tiesības kriminālprocesā* // Jurista vārds Nr. 42 (547)
[10] Latvijas Republikas Augstākās tiesas Senāta Krimināllietu departamenta 2013. gada 20.marta lēmums lietā SKK –31/2013 (Krimināllieta Nr. 11840003309)// Available: http://at.gov.lv/downloadlawfile/3792
[11] Baumanis, J. *Publisko apsūdzību kriminālprocesa uzsākšanas teorētiskie praktiskās problēmas* // Administratīvā un kriminālā justīcija 4/2007 (Nr.4/41), 2007, 89. lpp
[12] Kaija, S. *Issues in Separation of Criminal Procedural Functions.* 6th International Interdisciplinary Scientific Conference SOCIETY. HEALTH. WELFARE. SHS Web Conf. Volume 40, 2018 // Available: https://www.shs-conferences.org/articles/shsconf/abs/2018/01/shsconf_shw2018_01010/shsconf_shw2018_01010.html
[13] *Rokasgrāmata prokuroriem.* Kūtra, G. zinātniskajā redakcijā, Rīga, Tiesu nama aģentūra, 2010, 9 1pp

[14] *Vai pirmstiesas izmeklēšana Valsts policijā ir efektīva?* Latvijas Republikas Valsts kontrole// Available: `http://www.lrvk.gov.lv/uploads//revizijuzinojumi/2016/2.4.1-6_2016/rz_vp_20.09.2017_bez-ip.pdf`
[15] *Vai pirmstiesas izmeklēšana Valsts policijā ir efektīva?* Latvijas Republikas Valsts kontrole// Available: `http://www.lrvk.gov.lv/uploads//revizijuzinojumi/2016/2.4.1-6_2016/rz_vp_20.09.2017_bez-ip.pdf`

Individual's level of education as a guarantee of the right to defense

Oleksandr Shevchyk* and *Andrii* Hnatiuk

Poltava Law Institute of Yaroslav Mudryi National Law University, Poltava, Ukraine

Abstract. The right to defense is enshrined in national legislation, including Art. 59 of the Constitution of Ukraine, according to which, "Everyone has the right to a professional legal assistance. In cases provided for by law, this assistance is free of charge" [1]. In addition, there is Art. 52 of the Criminal Procedural Code (CPC) of Ukraine [2]. The lack of an appropriate level of education cannot be regarded as a certain mental defect or the fact of not speaking the language, since individuals could have an appropriate oral skill in Ukrainian, but not written skills. The problem lies not only in the lack of a unified approach in the practice of law enforcement regarding the application of legal grounds for the mandatory provision of a lawyer in criminal proceedings for persons who do not have the appropriate level of education, but also in the lack of a strictly defined level of education that would be considered sufficient to ensure persons' right to protection.

Key words: level of person's education, right to defense, obligatory participation of a lawyer in criminal process.

1 Introduction

The relevant principle of the right to defense is enshrined in national legislation, including Art. 59 of the Constitution of Ukraine, according to which, "Everyone has the right to a professional legal assistance. In cases provided for by law, this assistance is provided free of charge" [1]. In addition, there is Art. 52 of the Criminal Procedural Code (CPC) of Ukraine [2]. This study is based on Ukrainian regulation acts, scientific researches using dialectical, comparative, analytic, synthetic and comprehensive methods.

2 Aim

The aim of this research is firstly – to investigate interrelation between the defendants' level of education and ensuring of their right of defense, and secondly – to propose legal adjustments in this sphere.

3 Material and methods

This study examines provisions of international documents and scientists' attitudes. Authors analyze generalized information from law database of different countries by means of scientific methods from a legal point of view. This article is based on dialectical, comparative, analytic, synthetic and comprehensive research methods.

* Corresponding author: a.s.shevchik@ukr.net

In order to achieve this goal International Covenant on Civil and Political Rights (ICCPR), Constitution of Ukraine, Law enforcement acts and legislation of other states along with practice of ECtHR have been analyzed. The international experience of Russian Federation, Kazakhstan, Uzbekistan, Belarus, Poland and other countries are investigated. During the study, methods of data extraction and content analysis were used.

The methodology of this scientific publication is based on the organic combination of philosophical approaches, general scientific and special legal research methods. In particular, the systematic method is used to carry out a systematic analysis of legislative acts of other countries. The comparative method allows the author to compare the experience of foreign countries in terms of ensuring the right to defence for persons with inappropriate level of education. The legal modelling method is used to formulate proposals in order to improve existing legislation and practice.

The analysis of scientific sources clearly shows that described in this article problems have not been studied sufficiently.

4 Results and discussions

One of the generally acknowledged rights at the international level is the right to defence from suspicion and prosecution of a crime. The International Covenant on Civil and Political Rights, ratified by Decree of the Presidium of the Supreme Soviet of the Ukrainian SSR on 19.10.73, in article 14, paragraph 3 (d), stipulates that everyone has the right to be tried in his presence, and to defend himself in person or through legal assistance of his own choosing; to be informed, if he does not have legal assistance, of this right; and to have legal assistance assigned to him, in any case where the interests of justice so require, and without payment by him in any such case if he does not have sufficient means to pay for it [3].

Provision of the accused with the right to defence under art. 129 of the Constitution of Ukraine is the basic principle of legal proceedings [1], and according to Art. 7 of the Criminal Procedure Code of Ukraine [2], ensuring the right to defence is one of the general principles of criminal proceedings.

The right to defence is enshrined in Art. 59 of the Constitution of Ukraine [1], according to which, "everyone has the right to professional legal assistance. In cases provided for by law, this assistance is granted free of charge."

This norm is reflected in other national legal acts, according to which the person is provided with the right to use the legal assistance of a lawyer, both on the basis of contractual relations and free of charge, with the involvement of advocate from the Centre of free legal aid. Such guarantee meets the requirements of international law. At the same time, it should be noted that the list of cases of compulsory appointment of a defender to the accused person is not sufficient and need some law adjustment on the basis of the following.

Thus, in the Criminal Procedure Code of Ukraine (Article 52) it is determined that a lawyer is necessarily appointed to a person by an investigator, prosecutor, judge-investigator or a court in a criminal proceeding concerning grievous crimes from the moment a person acquires the status of a suspect [2].

The following article expands the list of cases for mandatory involvement of a lawyer for defence of the following categories of persons: to persons who are suspected or accused of committing a criminal offense under the age of 18 years – from the moment of establishment the fact of a juvenile or the emergence of any doubt that the person is an adult; with regard to persons for whom enforcement of the compulsory measures of an educational nature is foreseen, – from the moment of the establishment of the fact of a minor or any doubt as to the fact that the person is of an adult age; with respect to persons who, due to mental or physical defects (dumb, deaf, blind, etc.) are not able to fully realize their rights, from the time these defects are established; with regard to persons who do not speak the language of criminal proceedings – from the moment of establishing this fact; with regard to persons for whom the use of compulsory measures of a medical nature is foreseen or the question of their application is being decided – from the moment of the establishment of the fact of mental illness or other doubts of its sanity; on the rehabilitation of the deceased person – from the moment of the right to rehabilitation of the deceased person; with regard to persons in respect of which a special pre-trial investigation or a special legal proceeding is carried out – from the moment of the adoption of the corresponding procedural decision; in case of conclusion of an agreement between the prosecutor and the suspect/accused with confession of guilt – from the moment of initiating the conclusion of such agreement [2].

In most cases, similar norms are contained in the legislation of the countries that belonged to the Union of Soviet Socialist Republics. At the same time, both in the Criminal Procedure Code of Ukraine and in the legislation of a number of other countries, cases of compulsory involvement of a lawyer as a defender of a person subject to criminal prosecution are exceptional. That is, the legislator does not provide for the possibility of an expanded interpretation of the relevant cases. In particular, the analysis of criminal procedural legislation of countries such as the Russian Federation, the Republic of Belarus, the Republic of Latvia, the Republic of Kazakhstan and the Republic of Uzbekistan also indicates that cases of compulsory involvement of a lawyer are exhaustive [4–8].

At the same time, it is considered that not all categories of persons to whom the mandatory involvement of a lawyer is provided, have the opportunity to properly exercise their right to protection. Art. 20 of the CPC of Ukraine, which reveals the content of the right to defence as a general basis for criminal proceedings, secured the right of a suspect, accused, acquitted, convicted to defence, which guarantees him with the opportunity to provide oral or written explanations on suspicion or accusation; to exercise other procedural rights provided for by the CPC [2].

The realization of procedural rights consists, among other things, in drafting by a prosecuted person of procedural documents (applications, complaints, petitions, etc.), which in turn determines the need of appropriate level of writing, the ability to read with understanding, the ability to reasonably justify the position. A person acquires the appropriate skills by education. In this case, education is associated with the formation of the key competencies necessary for each person to live. Common competences for all are the following skills: reading with understanding, ability to express their own opinion orally and in writing, critical and systematic thinking, ability to logically justify a position, etc. In turn, competence is a dynamic combination of knowledge, skills, thinking patterns, which varies depending on the level of education received [9].

It is rather clear that a person who due to certain living conditions did not acquire an adequate level of education or did not get any education at all, despite the attainment of adulthood or the absence of mental defects, is not able to properly defend himself against criminal prosecution due to the lack of appropriate skills.

The analysis of law enforcement practice shows that in case of criminal prosecution of a person, her personal data regarding education and its level are established. We believe that establishing the fact of persons' education and its level is an important fact for further ensuring proper legal procedures in criminal prosecution and further ensuring the right to defence.

The Constitution of Ukraine states that full secondary education is compulsory [1]. This provision is also enshrined in the Law of Ukraine "On Education" (Part 2, Article 12), which states that full comprehensive secondary education in Ukraine is mandatory and acquired in institutional or individual forms defined by law, mostly in educational establishments [9]. That is, the legislator has determined that a person who is living in Ukraine must have certain knowledge and competencies to be productive person in society.

It is also worth noting that the constitutional obligation to obtain a certain level of education among the countries mentioned above is enshrined not only in Ukraine, but also in the Republic of Kazakhstan, the Russian Federation, Latvia, and many other European countries, such as Poland, where at Constitution and legislative level it is compulsory to undergo training for up to 18 years [10–13]. For example, Poland has 5 levels of education that include: pre-school, primary, secondary or high school, high school and higher education. In Kazakhstan and Uzbekistan there are 7 levels of education, 11 levels of education in the Russian Federation, 6 levels of education in the Republic of Belarus, and 4 levels of education (preschool, basic (compulsory), secondary education and higher education) in Latvia.

Let us consider the purpose of education as such and why its availability, in our opinion, is a mandatory element which influences the possibility of individual to defend himself personally.

The goal of education is the comprehensive development of a person being the highest value of society, encouraging of its talents, intellectual, creative and physical abilities, formation of values and necessary for the successful self-realization of competencies, education of responsible citizens who are capable of conscious social choice and directing their activities in favour of other people, society, the enrichment on this basis of intellectual, economic, creative, cultural potential, etc. [9]. Consequently, based on the above definition, one can conclude that the education of a person is a priority of society in general, since the uneducated society is not able to realize its potential, and the development of the state in the European direction, chosen by our country, depends on an increase the number of law-conscious citizens.

The goal of comprehensive secondary education, which is mandatory in Ukraine, is the comprehensive development, upbringing and socialization of the individual capable of living in a society and civilized interaction with nature, has a desire for self-improvement and lifelong learning, is ready for a conscious life choice and self-realization, responsibility, work and civic activity. Achievement of this goal is ensured by forming the key competencies necessary for every person to succeed in life, including proficiency in national language and the ability to communicate with their native (in case of difference with the state language) and foreign languages [9].

The lack of the proper level of education can certainly indicate the lack of proper language proficiency. Judiciary in Ukraine, as well as in most countries of the world, is provided for in the state language. Accordingly, drafting of procedural documents, conducting investigatory and procedural actions take place in the state language. Language proficiency is acquired in the course of studying. Thus, according to the European Language Education Recommendations, there are six levels of language proficiency developed by the Association of Language Experts in Europe: Introductive (Breakthrough or A1), average (Waystage or A2), the Threshold, or B1), advanced (Vantage or B2), autonomous (Effective Operational Proficiency or C1) and competent (Mastery or C2). They cover the gap between the basic and fluent level of language proficiency.

A significant problem is that current legislation does not establish criteria for determining the "sufficient level of state language proficiency" needed to provide a person with the ability to protect their rights independently. There is also no single approach to the definition of a competent entity that would implement an appropriate "sufficient level of state language proficiency".

At the same time, in cases when the person against whom the prosecution is conducted does not understand the state language or the language in which the proceedings are conducted, the person concerned is guaranteed (to ensuring the right to defence) the right to an interpreter and right to have a lawyer (paragraph 18 Part 3 of Article 42 and Item 4 of Part 2 of Article 52 of the CPC of Ukraine) [2].

But the language proficiency also includes the proper level of abilities for oral and written communication.

In some cases, if there is any doubt as to the level of language proficiency, there is the possibility of appointing a linguistic expertise. Linguistic expertise is a specialized linguistic study of text materials and other linguistic objects in order to clarify issues that may have legal implications. Objects of research are texts of oral and written speech, verbal, alphanumeric, phonetic and combined signs, as well as gestures and facial expressions. However, the cases of appointment of such expertise in order to establish a sufficient level of language proficiency as the guarantee of the right to defence against criminal prosecution are essentially isolated cases. Conducting of linguistic expertise causes additional time and financial costs. Moreover, the lack of an approved methodology for conducting relevant examinations and unified approach to the minimal allowable level of language proficiency, for the person to defend him-/herself personally, is clearly a huge problem.

At the same time, individuals may even have a high level of ability to use information from various oral sources, but the relevant skills may not always ensure individual to have the appropriate level of ability to use information from written sources. A person may have no written skills at all, but may have moderate oral skills. Such cases are not uncommon among persons of Roma and other nationalities [14].

The lack of a proper level of ability to extract information from written sources is an obstacle to the realization of the right to defence, since the person concerned is deprived of the opportunity to familiarize himself with the content of the criminal procedural materials (procedural documents and written evidence).

The lack of the proper level of language proficiency also deprives an individual of drafting of procedural documents, applications, petitions, complaints, etc., which also indicates that a person is not able to use the right to defence independently.

Moreover, lack of possibility to use information from written sources and, consequently, the lack of written skills cannot be offset by the participation of the interpreter.

The ability to extract information from various oral or written sources, the ability to generalize it and make a justified statement in a coherent form is acquired while achieving education. And the fact of acquiring a corresponding level of education by a person is an adequate confirmation of language proficiency level.

In some cases, deciding on the involvement of a lawyer, according to Art. 52 CPC of Ukraine, Ukrainian national courts as the legal basis for the mandatory participation of a lawyer in criminal proceedings refer to the provisions of paragraph 4 of Part 2 of Art. 52 CPC of Ukraine, that is provide such guarantee for persons who do not speak the language of criminal proceedings. In other cases, the courts facing the fact of the absence of persons' appropriate level of education as the legal basis for the involvement of a lawyer in a criminal proceeding refer to the provisions of paragraph 3 of Part 2 of Art. 52 CPC of Ukraine, that is regulate an obligatory participation of a defense counsel in criminal proceedings for persons with mental defects [2].

The lack of an appropriate level of education cannot be regarded as a certain mental defect or the fact of not speaking the language, since individuals could have an appropriate oral skill in Ukrainian, but not written skills. In most cases of criminal proceedings, courts first establish the fact that there was no adequate level of education of the person prosecuted and decide to engage a lawyer in accordance with art. 52 CPC of Ukraine as a compulsory participant in criminal proceedings. Therefore, the fact of the revealed violation of the right to defence during the pre-trial investigation leads to the recognition of evidence gathered during the pre-trial investigation as inadmissible. Moreover, it leads to the court decision to close criminal proceedings on the basis of clause 3, part 1, Article. 284 of the CPC of Ukraine as for failure to establish sufficient evidence to prove the guilt of a person in court and to exhaust the possibilities of obtaining it [2].

We note that an important part of the right to defence in criminal proceedings is also to ensure the effectiveness of defence at all stages of criminal proceedings.

Thus, at the initial stages of the criminal prosecution of a person (the presentation of a suspicion) there must be an obligation of the state to provide effective defence to the person. At the same time, the prosecution authorities are already endowed with sufficient capabilities to promptly determine persons' level of education. Thus, even at the stage of the pre-trial investigation, in particular during certain investigative actions (interrogation), an investigator, prosecutor or another person who is empowered to conduct an interrogation is required to evaluate the level of education of prosecuted person. The corresponding obligation is conditioned, in particular, by the presence of a mandatory graph in the relevant interrogation protocol. However, determining of person's education and its level in most cases is purely formal. Also, there is no clear criterion for criminal prosecution authorities to determine minimal allowable level of education of a person to ensure its right to defence. Thus, the possibility of securing a person's right to defence is lost from the very beginning of criminal proceedings.

Determining the level of language proficiency or the presence of mental defects that could be an obstacle to the person's right to defend himself independently requires special knowledge and an expert study, and therefore – time, which could affect reasonable time frame of criminal proceedings. Determining of persons' education and its level does not require significant efforts or time expenditures for law enforcement officials. Instead, it can be a guarantee of the observance of the person's right to defence at all stages of the criminal proceedings.

Cases where a lawyer had not been appointed for the defendant or accused during a pre-trial investigation (when he must be compulsorily involved), cast serious doubt on the question of the affiliation and admissibility of the evidence on which the prosecutor's accusation was based.

The legal position of the ECtHR regarding the initial stage of securing the right to defence in criminal proceedings is set forth, in particular, in paragraph 63 of the ECtHR judgment of June 9, 2011 in the case of Lucianinov v. Ukraine, which states that for accused to exercise his right to defence he must be guaranteed with the opportunity to get effective help of the advocate from the very beginning of the proceedings (paragraph 52 of the ECtHR judgment of 27 November 2008 in the case of Salduz v. Turkey and paragraphs 90 to 91 of the ECtHR judgment of 12 June 2008 in the case of Yaremenko v. Ukraine).If necessary, the defender must be appointed officially. The fact of appointment defence counsel by the national authorities does not guarantee effective legal assistance in itself (paragraph 65 of the ECtHR judgment of 19 December 1989 in the case of Kamazinski v. Austria).

Appointment of a counsel in a court proceeding in cases where a defendant, accused during a pre-trial investigation had not been appointed, a defence counsel cannot in itself restore already violated right to defence.

At the same time, the lack of an adequate level of education, as noted earlier, is an obstacle to the realization of a full-fledged right to defence personally. We believe that in case of absence of an adequate level of education the person against whom prosecution is being pursued must be provided with the defence counsel on compulsory basis.

As was already mentioned, in the Criminal Procedure Code of Ukraine (as in a number of other countries) cases of mandatory involvement of a lawyer as a defender of a person subject to criminal prosecution are foreseen. The analysis of the criminal procedural legislation of the countries such as the Russian Federation, the Republic of Belarus, the Republic of Latvia, the Republic of Kazakhstan and the Republic of Uzbekistan demonstrates that the said criminal procedure codes, as well as the CPC of Ukraine, do not provide guarantees of protection in the form of compulsory involvement of a lawyer in criminal proceeding to a person with nonsufficient level of education. In addition, the legislator in most countries does not provide for the possibility of broad interpretation of the relevant rules, including the involvement of a lawyer in a mandatory manner. In particular, the analysis of criminal procedural legislation of countries such as the Russian Federation, the Republic of Belarus, the Republic of Kazakhstan and the Republic of Uzbekistan shows that list of cases for compulsory involvement of a lawyer are exhaustive [4–8].

Instead, according to paragraph 5 of clause 1 of Article 83 of Latvian Code of Criminal Procedure, "participation of a defence counsel in criminal proceedings is mandatory if the right to protection has an uneducated person or a person with such a low level of education that he cannot fully exercise their rights" [6]. We consider the relevant provision of the criminal procedural law to be progressive and justified.

At the same time, in our opinion, the relevant wording of the norms of Latvian criminal procedural law reveals the problem of insufficiency of clear criterion for determining minimal allowable level of education, below which involvement of the lawyer for such person is mandatory.

In our opinion, taking into account the constitutional requirements in many countries of the world, the obligation of the state to ensure the right to education of a certain level, the lack of an appropriate level of education guaranteed by the state is just that criterion for ensuring the right of a person to be protected in criminal procedure (in terms of right to defence).

In Ukraine, as already noted, it is provided at the constitutional level that full secondary education is compulsory. Therefore, we assume that absence of full secondary education is the reason for the mandatory involvement of a lawyer for person prosecuted. In cases of the lack of costs for the person concerned the said legal assistance should be provided free of charge.

5 Conclusions

The relevance of the study is justified not only by the lack of a unified approach in the practice of law enforcement regarding the application of legal grounds for the mandatory provision of a lawyer in criminal proceedings for persons who do not have the appropriate level of education (which necessitates changes to the national legislation), but also by the lack of a strictly defined level of education that would be considered sufficient to exercise persons' right to defence.

This necessitates the adjustment of national legislation by supplementing the relevant provisions of the procedural law of the following content: "obligatory participation of a lawyer is ensured in criminal proceedings against persons who do not have a level of education, which is determined by the legislation as compulsory." The relevant provision, in our opinion, will certainly be a positive factor in ensuring the observance of human rights in criminal prosecution. We believe that the relevant norm should also be introduced into the legislation of other countries with a similar to Ukrainian legal system.

References

[1] Constitution of Ukraine, 1996, available at: https://zakon.rada.gov.ua/cgibin/laws/main.cgi?nreg=254%EA%2F96%2D%E2%F0
[2] Criminal Procedure Code of Ukraine, 2012, available at: https://zakon.rada.gov.ua/laws/show/4651-17
[3] The International Covenant on Civil and Political Rights, 1973, available at: https://www.ohchr.org/en/professionalinterest/pages/ccpr.aspx
[4] Criminal procedure code of Russian Federation, 2001, available at: http://pravo.gov.ru/proxy/ips/?docbody=&nd=102073942
[5] Criminal procedure code of the Republic of Belarus, 1999, available at: http://kodeksy.by/ugolovno-processualnyy-kodeks
[6] Criminal procedure law of the Republic of Latvia, 2005, available at: https://vvc.gov.lv/image/catalog/dokumenti/Criminal_Procedure_Law.pdf
[7] Criminal procedure code of the Republic of Kazakhstan, 2014, available at: http://kgd.gov.kz/ru/content/ugolovno-processualnyy-kodeks-rk
[8] Criminal procedure code of the Republic of Uzbekistan, 1994, available at: http://www.lex.uz/docs/111463
[9] The Law of Ukraine on Education, 2017, available at: https://zakon.rada.gov.ua/laws/show/2145-19/print
[10] The Constitution of the Republic of Kazakhstan, 1995, available at: http://mfa.gov.kz/en/hague/content-view/the-constitution-of-the-republic-of-kazakhstan
[11] Constitution of the Russian Federation, 1993, available at: http://www.mid.ru/en/foreign_policy/official_documents/-/asset_publisher/CptICkB6BZ29/content/id/571508
[12] Constitution of Latvia, 1993, available at: https://www.president.lv/ru/latviiskaya-respublika/konstituciya-latvii
[13] Constitution of the Republic of Poland, 1997, available at: http://www.sejm.gov.pl/prawo/konst/angielski/kon1.htm
[14] Global Rise of Education, available at: https://ourworldindata.org/global-rise-of-education

Social and philosophical background and legal mechanism of assisted reproductive technologies regulation (by example of surrogate maternity)

Yevhen Hrekov[1], *Maryna* Hrekova[2], and *Roman* Kabalskyi[3]

[1] Poltava Law Institute of the Yaroslav Mudryi National Law University, Department of Civil, Commercial and Financial Law, Poltava, Ukraine
[2] Poltava Law Institute of the Yaroslav Mudryi National Law University, Department of Constitutional, Administrative, Environmental and Labor Law, Poltava, Ukraine
[3] Poltava Law Institute of the Yaroslav Mudryi National Law University, Department of Theoretical Law Disciplines, Poltava, Ukraine

Abstract. In many countries of the world, the practice of using assisted reproductive technologies faces a number of legal obstacles, from the introduction of restrictions to their complete prohibition. In most cases, these restrictions are due to public rejection of these methods and procedures for social, psychological and religious reasons. Indeed, these technologies add reissues of assess in galore of women in the society, its functions, problems of relationships with a surrogate mother or an egg donor, emergence and distribution of a surrogate mother's or donor's rights, as well as persons' who are beneficiaries of the results of these technologies. However, a solution of these problems cannot be achieved only by improving the legal mechanisms regulating the use of these technologies. In this situation, the solution is at the junction of the interests of the society (state) and personal interests of persons who need their use, while normative consolidation and enforcement is only a form of consolidating of the achieved balance.

Key words: assisted reproductive technologies, legal regulation mechanism, human reproductive rights, surrogate maternity.

1 Introduction

On April 26, 2012, in Strasbourg, in the framework of the spring session of the Parliamentary Assembly of the Council of Europe, the conference "Surrogate Motherhood: A Violation of Human Rights" was held. The main topic of discussion at the conference was the report "Surrogate Motherhood: A Violation of Human Rights" presented by the European Centre for Law and Justice. The document refers to the violation of the women's rights arising from their commercial exploitation, as well as the difficulty of determining parental rights since up to six people may be involved in the process of conceiving, carrying and raising a child (two donors of sex cells, a surrogate mother, her husband, a child's "legal" parents). In addition, the document emphasizes the problem of subsequent self-identification of a child born from a surrogate mother and his psychological health.

Undoubtedly, assisted reproductive technologies are influencing a change in the foundations of the social institution of family and marriage: a traditional monogamous family with heterosexual spouses is gradually losing momentum as a pattern of behaviour in relation to single-parent families (an incomplete family) and homosexual cohabitations, both male and female. Both forms of marriage and family relations are given an opportunity to have a genetically their own child through biomaterial donation and surrogate maternity. In addition, the availability of assisted reproductive technologies contributes to the expansion of asexual reproduction, which worsens the situation with gender interaction and gender identification as well.

The option of using assisted reproductive technologies for single women, men and persons with non-standard sexual orientation is also unacceptable since it deprives an unborn child of the most hypothetical possibility of having the right to a mother and a father. The current trend of the growing interest of homosexual couples to the use of the surrogate maternity technology in order to "acquire" children for themselves actualizes the issue of the most egregious violations of children's rights in such situations [1].

Another controversial issue, which currently remains unsolvable, is associated with the determination of an embryo's status. The position of perception of an embryo as a living being with all the moral and ethical consequences, rights and obligations arising from this finds both supporters and opponents.

2 Aim

The purpose of this study is to identify and analyze the main reasons for resistance to distribution of assisted reproductive technologies from a social and philosophical point of view, description of the types of legal mechanisms for securing the right to use these technologies, as well as study of law enforcement practice, primarily judicial practice, in solving problems of their application.

This study is based on international acts in the field of human rights, legislation of various countries of the world (India, the United States, Ukraine, a number of European countries), judgments, scientific works and opinions of progressive-minded people in the sphere of sociology, philosophy and law. Such methods as dialectical, comparative, analytic, synthetic and comprehensive are used in the article.

3 Surrogate maternity in religion

Reproductive technologies have created a whole layer of problems of philosophical and cultural nature which, in retrospect, are caused by a uniquely negative perception of these technologies in the religious sphere.

However, looking into the history, we can see that the Old Testament describes one of the surrogate maternity stories. Being infertile, Sarah, Abraham's wife, invited a maid to conceive and carry out Abraham's child. In Bible days, it was believed that if a wife could not give birth to an heir, she could pass on this mission to a maid who would conceive a baby with an infertile woman's husband, and after giving birth, she immediately would give the baby to the spouses. In so doing, this infertile wife would take the child as a mother after birth, putting him on her lap, and would consider the baby to be her own genetically. History is replete with examples where slaves or concubines served as surrogate mothers in different countries of the world. At that time, of course, only "traditional surrogate maternity" was used, i.e. a baby's father and a surrogate mother herself were his genetic parents since fertilization was carried out naturally [2].

At the same time, it is necessary to note that nowadays there is lack of unity in the approach to the problems associated with the use of reproductive technologies, in particular, surrogate maternity, in world-class religious denominations and some major national religions.

For instance, Judaism, which does not disapprove "in vitro" conception if sperm and oocyte are taken from a Jewish married couple. Artificial insemination by an anonymous donor can lead to unintentional incest.

Buddhism is not categorically opposed to donor sperm fertilization with the help of artificial childbirth technologies under the following conditions: 1) spouses' voluntary consent; 2) compliance with donor confidentiality; 3) absence of a remuneration arrangement since his (a donor's) participation has to be based on a desire to help his fellow human. However, even with such a liberal attitude, Buddhism indicates that raising an adopted child would be a more preferable solution to the indicated problem of childlessness.

Islam generally supports extracorporeal fertilization as it is believed that this process is similar to natural fertilization but it is allowed only under two conditions: sperm and oocyte have to belong exclusively to a husband and wife between whom a marriage agreement (nikah) has been concluded; after fertilization outside a living organism, an oocyte has to be transferred into a mother's uterus, i.e. that of a woman's whose oocyte has been fertilized.

The use of such concepts as "donor sperm", "donor oocyte", and much less "surrogate maternity" is strictly prohibited, comparable to adultery and generates a significant number of social problems. Persons involved in this in any way, if they have not complied with the two conditions above, are committing a sin and are subject to punishment [3].

The Orthodox Church approaches the assessment of various methods of insemination differentially but surrogacy is disapproved. In "The Basis of the Social Concept" of the Russian Orthodox Church, it is stated that: "Among the admissible means of medical aid may be an artificial insemination by the husband's germ cells, since it does not violate the integrity of the marital union and does not differ basically from the natural conception and takes place in the context of marital relations" [4].

The Orthodox Church strongly opposes surrogate maternity, believing that the "mother's feelings are trampled upon, and the child... may subsequently experience an identity crisis". From the Orthodox point of view, all types of extracorporeal fertilization, involving the preparation, preservation and intentional destruction of "excess" embryos, are "unnatural and morally inadmissible" [4].

However, certain changes in the degree of non-perception of the consequences of the use of reproductive technologies by religious denominations should also be noted. In this area, we can distinguish that the Church is inclined to concordance with the existing situation, namely, an increase in the number of children born using reproductive technologies, including surrogate maternity, although as it has been noted above it often used to be absolutely categorical.

For example, since 2013, the Russian Orthodox Church, whose negative attitude towards surrogate maternity has been indicated above, allows for the baptism of a child born using "surrogate maternity" at the request of persons raising him if these are either his "biological parents" or "surrogate mother", but only after they realize that, from a Christian point of view, such a reproductive technology is morally reprehensible and will do church repentance, regardless of whether they ignored the position of the Church knowingly or unknowingly [5].

In this situation, we observe a certain correlation in interpretation of the canons of the Church based on the conditions when individual followers of the Church teachings tackle the problem of impossibility of natural propagation not on the basis of their social and ethical beliefs and the religious canons but in favour of using surrogate maternity, which entails bringing them to justice. Thus, the Orthodox Church allows for "accepting Christ by faith" for a child born by a surrogate mother on the assumption that "the fact of the "surrogate birth" does not in itself constitute an obstacle to baptism of a person, since he is not responsible for his parents' behaviour" [5].

4 Moral and ethical problems of surrogate maternity in law

The predominantly negative attitude of religion towards surrogate maternity is also accompanied by a number of moral and ethical problems. Let us consider the most significant, in our opinion, of these problems through the prism of emerging relations in law and development of legal regulation of the sphere of reproductive biomedical technologies.

1. Surrogate maternity is based (moreover, sold on a commercial basis – on a remuneration basis for a surrogate mother and her intermediary) on reducing that is amoral and grossly infringes on a woman's rights, bringing down her significance and role as a mother to a significance and role of a paid live incubator in the surrogate maternity industry, essentially, an instrument of production, actual transformation of a woman into a commercially exploited "human incubator".

A number of foreign authors rightly refer to surrogate maternity as an immoral business based "on a woman womb", "uterus leasing" and "uterus hire" technology" [6;7].

In one of the editions of the World Health Organization, a surrogate mother is called a "gestational carrier" and the following definition is given: "A gestational carrier is a woman in whom a pregnancy resulted from fertilization with third-party sperm and oocytes. She carries the pregnancy with the intention or agreement that the offspring will be parented by one or both of the persons who produced the gametes" [8].

This situation is completely unacceptable as it grossly violates the human dignity of a woman and her right to be a mother, is contrary to Art. 3 and 14 of the European Convention for the Protection of Human Rights and Fundamental Freedoms dated November 04, 1950, Art. 5 and 7 of the Universal Declaration of Human Rights dated December 10, 1948, Art. 7 of the International Covenant on Civil and Political Rights dated December 19, 1966, the Convention on the Elimination of All Forms of Discrimination against Women dated December 18, 1979, a number of other international instruments on women rights.

According to the ECtHR [9], seven of the thirty-seven Contracting States of the Council Europe have permissive legislation regarding surrogacy. Some of them permit commercial surrogacy and others only permit altruistic surrogacy. Only non-commercial surrogacy is also allowed in Australia, Great Britain, Denmark, Israel, Canada, the Netherlands and certain states of America.

At the same time, such countries as Russia and especially Ukraine stand out against the rest. Ukraine has a significant advantage for married couples since in Russia, even in the absence of any genetic relationship, it is a surrogate mother who has the priority right to determine a child's fate since genetic parents' rights are recognized only after the child is abandoned by his surrogate mother. And in Ukraine, the legislation, in particular, Part 2 of Art.123 and Part 2 of Art. 139 of the Family Code of Ukraine, provides for the presumption of origin of a child from his parents – donors of genetic material. In particular, Part 2 of Art. 123 provides for that in the case of transferring a human embryo conceived by spouses (male and female) as a result of the use of assisted reproductive technologies to another woman's body, the child's parents are the spouses; and Part 2 of Art. 139 prohibits challenging maternity in such a case [10]. Based on these statutory provisions, in judicial practice of Ukraine, the position that "the law connects the fact of maternity with the origin of a child from a certain woman, therefore a child born by a surrogate mother is not her genetic descendant and therefore, she cannot be considered his mother" [11] was met with support. At the same time, the legal nature of surrogate maternity relations is considered as a service for bearing and birth of a child with the subsequent registration of customers as the child's parents [11].

As a result, a woman who has carried and given birth to a child is completely unprotected by the law in the aspect of exercising her right to maternity regarding the child, and to a great extent, it approximates the understanding of the role of such a woman as a means for producing children, which is absolutely unacceptable in a civilized society.

2. Surrogate maternity is based on the positioning a child not as a person but as an object of the law, a certain object of a transaction, essentially, as an inanimate object to whom product attributes and consumer properties of goods are ascribed.

It is not a mere coincidence that in 1992, the Court of Appeals of Michigan (USA) in the case of Doe v. Attorney General, 487 N.W.2d 484 (Mich. Ct. App. 1992), in which several potential parties to surrogacy agreements attempted to challenge Michigan law prohibiting such agreements, emphasized that the ban on surrogate maternity is aimed at ensuring several of the most important interests of the state: first, preventing a child from becoming mere commodities; secondly, the best interests of the child; and, thirdly, preventing the exploitation of women [12].

A number of foreign legal scholars assess this kind of commercial relations as distorting the nature of parental relations between a mother and a child and as unlawfully encroaching on the human dignity of women. They convincingly prove sound reasoning of the use of the notion of "child trafficking" to such kind of commercial relations [6, 7].

According to the UN's Special Rapporteur on the sale of children, child prostitution and child pornography, Maud de Boer-Buquicchio, children face becoming commodities due to spread of surrogacy in the absence of international statutory provisions to regulate it. In her report to the Human Rights Council in Geneva, Maud de Boer-Buquicchio said that "children are not goods or services that the State can guarantee or provide. They are human beings with rights." [13]... The Special Rapporteur explained that nowadays if a surrogate mother receives remuneration, the whole process corresponds to the definition of sale and purchase, as defined under international human rights law. In her opinion, this is exactly the situation currently practised in some countries. The Special Rapporteur insists that courts or other competent authorities should be involved in the regulation of the process, as private contracts generally do not provide sufficient human rights safeguards [14]. Most women employed in this business are well aware that they bear children for sale, customers and intermediaries use them as an incubator, buying a foetus as a commodity. However, they are not faced with any moral dilemmas and do not argue about the "ethicality" of their earnings [15].

However, despite the understanding of the actual nature of the relationship between customers and surrogate mothers, especially when providing such services for a valuable consideration, the legal systems of individual countries simplify the legal regime for customers of such services. The ruling of the Iowa Supreme Court can be an example, as it recently confirmed the legality of gestational surrogacy contracts (including those provided for a valuable consideration), as well as the adoption of the Uniform Parentage Act 2017 (UPA 2017) in Washington allowing compensated surrogacy services [16].

Undoubtedly, we can only welcome prescribing detailed rules in this field aimed at protecting the rights of all participants in surrogate maternity relations. However, such an approach of the states does not fully meet the interests and opinions of a significant part of society and, as a rule, transfers the responsibility for controlling the exploitative forms of using women either to lawyers who will be obligatory participants of such relations, representing interests of parties, or to the courts which will subsequently face the need to find a balance between the law and interests of society represented by participants of surrogate maternity procedures, or children born as a result of such a procedure.

In practice, it is difficult to solve the problem in case of birth of a sick child (having physical defects, pathologies, serious internal diseases, etc.) by a surrogate mother. Then customers of surrogate maternity services have the right (usually, it is a typical condition of a surrogacy service agreement) to refuse such a child (as some kind of "faulty goods" which do not comply with the original agreements), which is a gross violation of the child's rights and humiliation of his dignity. And the surrogate mother in such situation is also completely unprotected and deprived of the rights: her customers do not only leave the child but according to the model provisions of the agreement, are free to recover "compensation" from her, stating that there is only her (the surrogate mother's) guilt in the birth of the sick child. In this case, for the surrogate mother (given her lower social status in comparison with the customers of her services), it will be extremely difficult (on the verge of the impossible) to prove the absence of her guilt in this outcome.

3. Surrogate maternity is a gross violation of the child rights, first of all, to personal and family identity and related to his specific communication with his own mother because, as numerous scientific studies have shown, a child in infancy has a stable psycho-physiological connection with his mother (who has carried and given birth to the child), and such a relationship is formed already at the prenatal stage of his development (when he is in the womb of the mother – the one who is carrying him). As Sonia Allan refers, "regardless of genetic connection, the gestational mother provides many biological resources during the pregnancy" [17]. A mother is tied to a child psychologically, and a forced break of this interaction can negatively affect both a surrogate mother and a child.

Also the use of surrogate maternity, associated with exploitation of another's body, deprives a child of his own family and generally has a devastating effect on the moral foundations of the institution of the family and fundamental moral foundations of society.

As mentioned above, a significant number of nations oppose the practice of surrogate maternity directly prohibiting it, expressing special forms of responsibility, as well as fixing provisions on invalidity of contracts with persons engaged in organizing surrogate maternity services and contracts concluded directly with surrogate mothers. Ways to circumvent such prohibitions by subordinating regulation of such services under foreign law as a result of attracting foreign clinics or women from third countries, face opposition through the judicial system which uses the categories of "supra-imperative rules", "public order" or "circumvention of the law" to deny the requirements associated with the recognition of parental rights to children born as a result of the use of reproductive technologies.

However, such a ban is subject to criticism from experts who indicate that intended parents may go abroad to benefit from foreign legislation which authorize gestational surrogacy. This situation can neither be solved by a pure prohibition, nor by diplomatic discussions with countries authorizing surrogacy in order to bar access of French nationals to surrogacy as the Report of the Senate suggests, since it denies real births occurred abroad and is prejudicial to the children, who cannot be punished for the acts of their parents [18].

The French courts are known for their radical position regarding non-recognition of the parental rights of persons to a child who was born as a "result of performance of a surrogacy agreement concluded bypassing the French law, which is insignificant in connection with the violation of public order despite the fact that abroad (at the place of its conclusion) it is considered valid" [19]. The situation accompanying this ruling and other similar situations related to the non-recognition by the French courts of the fact of paternity to a child born under a surrogacy agreement gained momentum in the judgments of the European Court of Human Rights in the cases of Labassee v. France (no. 65941/11) and Mennesson v. France (application no. 65192/11) [20], as well as decisions of the Court of Cassation of France in 2015 which reversed its prior position. In accordance with the said judgments, the situation of the intended father, as long as he is genetically related with the child, is settled. However, the position of the French highest Court is still awkward towards the intended mother. Indeed, the mother is, under French law, the woman who gives birth, whatever her genetic link with the child is. Hence, the woman who gives birth to a child procreated with a donor's egg is automatically considered to be the woman's embryo as soon as it is inseminated in her uterus.

Accordingly, when a surrogate mother carries and gives birth to a child who was procreated by the intended mother's egg and intended father's gamete, she is considered to be the mother. French law continues to stick to a very old vision of maternity: the mother has to give birth. If in most births, it corresponds to the genetic reality, it is not necessary the case. The intended mother is left in a legal vacuum [21]. In this case, we are witnessing progress in changing the position of the state regarding surrogate maternity. However, so far this is carried out only through judicial practice but not via amendments of statutory provisions. A similar process can also be observed in Austria [22], and in Germany [23].

4. The business based on commercial surrogate maternity is immoral and inhumane. It can be defined as organization of children trafficking and mediation of such trafficking. In addition to the above analogy, in fact and from a moral point of view, for the purposes of discussion, this practice can be comparable with introduction of the possibility of free commercial circulation of human internal organs.

The argument given by supporters of social acceptability of commercial surrogate maternity is that remuneration for a surrogate mother only reimburses the expenses incurred during the period of carrying a child and in connection with carrying someone else's child does not stand up to criticism since there is a significant difference of surrogate maternity on a commercial basis from non-commercial surrogate maternity: a surrogate mother and, the main problem, an intermediary receive remuneration (essentially, income) by carrying out an entrepreneurial activity.

According to both the Hague Convention and the Council of Europe Convention on adoption, no one shall derive any improper financial or other gain from an activity relating to the adoption of a child. It would be contrary to the dignity of the child, because adoption would become a market. Commercial surrogacy is intrinsically contrary to this provision [24]. It must be recalled that according to Article 2a of the Optional Protocol to the Convention on the Rights of the Child on the Sale of Children, Child Prostitution and Child Pornography, "sale of children means any act or transaction whereby a child is transferred by any person or group of persons to another for remuneration or any other consideration".

According to Dr Grégor Puppinck, the Director General of the European Centre for Law and Justice, surrogate maternity is one of the modern varieties of human trafficking based on commonplace relations of purchase and sale which are disguised by the euphonic rhetoric of "helping childless people". Only poverty can stir into such a business, and the fact that women are forced to give birth to others for money fundamentally contradicts the concept of human dignity [15].

Different approaches of nations to regulation of the sphere of surrogate maternity have led to the emergence of the so-called "surrogate tourism" aimed at carrying a child by a surrogate mother in a country where surrogate maternity is allowed.

At the same time, it should be agreed that the business based on commercial surrogate maternity is immoral and inhumane. It can be defined as organization of children trafficking and mediation of such trafficking. The argument given by supporters of social acceptability of commercial surrogate maternity is that remuneration for a surrogate mother only reimburses the expenses incurred during the period of carrying a child and in connection with carrying someone else's child does not stand up to criticism since there is a significant difference of surrogate maternity on a commercial basis from non-commercial surrogate maternity: a surrogate mother and, the main problem, an intermediary receive remuneration (essentially, income) by carrying out an entrepreneurial activity. However, surrogate maternity cannot be reduced solely to the legitimacy of disposing one's body. It is a much more complicated process from the moral, ethical and psychological points of view which invariably influences the way it is enshrined in the law by either permission or prohibition.

5 Scientific capacity

Paper was developed in the field of jurisprudence, in context of legislation analysis and application, therefore, objective and scientific capacity of work can be seen in analysis of efficiency of norms and implementation thereof in further regulation. Therefore, careful scientific analysis of the subject is performed to increase scientific contribution, and conclusions obtained in result of research are consequently summarized, which authors will use as a part of issues included and reviewed in preliminary studies.

6 Conclusions

It should be stated that at the present stage of development of relations regarding the use of reproductive biotechnologies, the law as a social regulator does not effectively fulfil its functions. In fact, in the modern world, we can observe the fragmented process of penetration of regulatory provisions concerning surrogate maternity into the legal systems of different countries. On their own initiative, some "leading" centres providing surrogate maternity services (for example, India, the United States) take measures to regulate certain aspects of these services. However, such countries as Ukraine, Russia and some other "harbours of international surrogacy tourism" stand on the sidelines of these processes.

There is also no progress in settling relations in the sphere of surrogate maternity at the international, global level. As a result, when characterizing the elements of these relations, a problem arises regarding the interpretation of their content through the prism of international rules which do not take into account modern changes in public perception of such fundamental terms as a "family", "maternity", "surrogate mother" and others related to this area.

In such a situation, judicial practice is the main link between understanding the nature and scope of human rights enshrined in the provisions of international treaties and actually existing at this stage of development of public relations. Thus, through law enforcement and interpretation of rules of the law, the judicial system ensures the evolutionary development of the regulatory framework in the application of reproductive biotechnology.

In fact, we are witnessing a change in approaches to ensuring the availability (legitimization) of surrogate maternity through court judgments. And, in this case, we do not refer only to the practice of the European Court of Human Rights. Through judicial practice, national courts of different countries, including even those of France and Germany, which are the most "unfriendly" to surrogate maternity, gradually change the approaches to the invalidity of surrogacy agreements, consequences of their conclusion, recognition of the legality of methods of circumvention of the law prohibiting such agreements, permissibility of granting the status of parents to customers of surrogate maternity services through the mechanism of subsequent adoption of children, etc.

However, despite such steady progress in the development of regulation of surrogate maternity, in many countries, even despite de jure recognition of human reproductive rights, there are formal barriers to their implementation. This is a consequence of the conflict of the law and social attitudes of the society. At the same time, in some countries, there is also a reverse trend when these rights are granted an absolute status at the level of judicial practice, which results in a violation of the principle of equality of human rights, the rights of a child to personal and family identity. An initiative supported by the international community regarding the creation of a system of international legal regulation of relations connected with surrogate maternity and introduction of fundamental principles for carrying out accompanying activities can be the solution to these problems.

References

[1] I. Ponkine, N. Mikhaleva, M. Kouznetsov, *Sur les violations des droits des enfants pendant leur «adoption» par des couples homosexuels ou par des individus homosexuels.* Available: `http://strasbourg-reor.org/?topicid=1082`

[2] Н. Квіт, *Інститут сурогатного материнства: проблеми колізійного регулювання* (Альманах міжнародного права, 2014, Вип. **14**, С. 39)

[3] В. Киселев, *Религия и экстракорпоральное оплодотворение* (Биоэтический словарь, 2006, р. 99). Available: `http://dialektika-eniology.narod.ru/dictionary.0606.htm`

[4] *Основы социальной концепции Русской Православной Церкви, проблемы биоэтики.* Available: `https://mospat.ru/ru/documents/social-concepts/xii/`

[5] *О крещении младенцев, родившихся при помощи «суррогатной матери».* Available: `http://www.patriarchia.ru/db/text/3481024.html`

[6] A. Banda Vergara *Dignidad de la persona y reproducción humana asistida*(Revista de Derecho, 1998, **IX**, p. 7–42).

[7] M. Gómez de la Torre Vargas *La fecundación in vitro y la filiación*(Santiago de Chile, 1993, pp. 201, 204–205)

[8] E. Vayena, Patrick J., Rowe, P., David Griffin, *Current Practices and Controversies in Assisted Reproduction* (Report of a meeting on «Medical, Ethical and Social Aspects of Assisted Reproduction» held at WHO Headquarters in Geneva, Switzerland 17–21 September 2001), p. 20–21. Available: `http://apps.who.int/iris/bitstream/10665/42576/1/9241590300.pdf`

[9] F. Redondo, F. Valente, M. Esteves, M. Perquilhas, *Surrogacy: a clash of competing rights*, 2017, p. 7. Available: `http://www.ejtn.eu/Documents/Team%20Portugal%202%20semi%20final%20B.pdf`

[10] Н. Квіт, *Інститут сурогатного материнства: проблеми колізійного регулювання* (Альманах міжнародного права, 2014, Вип. **14**, С. 42)

[11] The Pervomaysk city court of Kharkiv region in Ukraine July, 3 2013 judgement in case No. 632/1313/13-ц. Available: `http://reyestr.court.gov.ua/Review/32185130.`

[12] Michigan Court of Appeals June 1, 1992 judgement in case Doe v. Attorney General, 487 N.W.2d 484 (Mich. Ct. App. 1992). Available: `https://www.courtlistener.com/opinion/2191740/doe-v-attorney-general/`

[13] *Children risk being "commodities" as surrogacy spreads, UN rights expert warns.* Available: `https://www.ohchr.org/en/NewsEvents/Pages/DisplayNews.aspx?NewsID=22763&LangID=E`

[14] *Эксперт ООН: с распространением суррогатного материнства дети могут превратиться в товар.* Available: `https://news.un.org/ru/story/2018/03/1325142`

[15] А. Манчук, *Украина: глобальный центр «суррогатного материнства».* Available: `http://liva.com.ua/ukraina-globalnyij-czentr-surrogatnogo-materinstva.html`

[16] *Surrogate motherhood in the news: the courtsand legislative bodies.* Available: `https://www.asrm.org/news-and-publications/news-and-research/legally-speaking/surrogacy-back-in-the-news-courts-and-legislatures/`

[17] S. Allan, *Is the Surrogate Involved in the Sale or Commodification of Children?* ("Surrogacy, Law and Human Rights", 2015, **Chapter 7**, endnote 21)

[18] H. Fulchiron, C. Bidaud-Garon,*Ne punissez pas les enfants des fautes de leurs pères. Regard prospectif sur les arrêts Labassée et Mennesson de la CEDH du 26 juin 2014* (Recueil Dalloz, 2014, p.1773)

[19] Н. Квіт, *Інститут сурогатного материнства: проблеми колізійного регулювання* (Альманах міжнародного права, 2014, Вип. **14**, С. 37)

[20] Labassée v. France, Case No 65941/11 and Menesson v. France, Case No 65192/11. Available: http://hudoc.echr.coe.int/eng#{%22itemid%22:[%22001-145389%22]

[21] C. Kleiner, *Surrogate motherhood in France: the need for a change Considerations of French Private International Law,* p. 61–62. Available: http://hdl.handle.net/10112/10980

[22] VfGH 11.10.2012 – B 99/12; VfGH 14.12.2011 – B 13/11-10, IPRax 2013, 271,275

[23] VGBerlin 5.9.2012 – I-3 Wx211/12 Rs.23 L 283.12, IPRax 2014, 81; BGH 10.12.2014 – XIIZB 463/13, FamRZ 2015, 240 mAnmTobiasHelms

[24] *The ethical case against surrogate motherhood: what we can learn from the law of other European countries*, p.16. Available: https://www.ionainstitute.ie/assets/files/Surrogacy%20final%20PDF.pdf

Problems associated with the right of access in the context of the rights of the child in Latvia

Inga Kudeikina

Rīga Stradiņš University, Faculty of Law, Riga, Latvia

Abstract. The child is a special right-holder. In the legal system, the status of the child is characterised by substantive rights that are typical only for children and by different means of exercising and safeguarding rights, i.e. the rights of a child are exercised by custodians acting on behalf of the child. It should be noted that the state, which uses the levers of public power, is also involved in the protection and safeguarding of the rights of the child in the most direct way. The right of access should be mentioned as a special right of the child. This comprises the right of a child to direct contacts with the child's parents and siblings. Communication forms an integral part of daily lives of people as social beings. Communication with parents is essential for a child undergoing the process of personality development. The right of access is an absolute right, which may be restricted only in cases specified in laws, provided that access is harmful to a child. Although the right of access is provided for by law, the existing legal framework is still deficient, which is confirmed by frequent disputes arising over the exercise of this right.

Key words: right of access, rights of the child, custodians.

1 Objectives and methodology of the thesis

The objective of the thesis is to research the legal framework dealing with the right of access in the context of the rights of the child by finding a way to make the exercise of the right of access solely in the interests of the child more effective.

The research has employed descriptive and analytical, inductive and deductive methods. Laws, the opinions of legal scholars and case-law have been analysed using these methods; conclusions and suggestions have been formulated.

It has become evident nowadays that the rights of children need to be guaranteed. However, the best and most effective ways of securing children's rights, including the right of access, still remain an open issue. The case-law shows a trend that, in certain cases, the right of access may be used by parents to settle their mutual relationships. This is evidence that the right of access does not always serve its main goal, which is to ensure the best interests of the child. The right of access is referred to in Article 181 of the Civil Law [1], whereby a child may maintain direct contact and personal relations with any of the parents, in both narrow and broad sense, and siblings, grandparents and other persons with whom the child has lived in a common household and developed a close relationship. In assessing the role of access for the development of a child, the concept of parents should be viewed more broadly, namely that a child may maintain direct contact with both biological parents and other relatives with whom the child has developed a close relationship (or an emotional bond). This is an essential aspect because it must always be established with whom of the relatives a child would prefer to maintain contact when deciding on the procedure for exercising the right of access. The notion "close relationship" is open and must be given substance in each particular case, but it must be interpreted broadly in order to safeguard the best interests of a child. In practice, there is a great deal of discussion as to whether a child should meet and communicate with a new spouse of the divorced parent and spouse's relatives. It should be concluded that this matter is not dealt with by law. It is not prohibited. The author believes that contact with members of the new family of the parent is permissible provided it is accepted by the child and does not hinder the effective communication between the child and the parent, i.e. their travels together attending cultural events and other activities should be facilitated. The law clearly implies two things. First, it is a child's right rather than a duty; second, it is a right of a child rather than of child's parents. Parents must maintain direct contact with their child, thus ensuring the right of access. It is the child who decides on how this right will be exercised. It should be stressed that a child does not possess legal capacity and, accordingly, exercises his/her rights through a guardian who, as a rule, is a parent with day-to-day care of the child. The fact that a parent and not the child brings an action regarding the right of access against the other parent creates a false impression that the dispute involves only the child's parents and not the child and a parent. The author is of the opinion that, in certain cases, this misunderstanding makes it more difficult to settle the dispute because the child's right of access may be used by parents to settle problems existing between them. Therefore, such disputes should be solved by means of mediation, with the participation of a family court that would defend the interests of a child. According to the existing legislation, a court must suggest mediation but, as regards family courts, a family court's opinion is necessary, but it is not required that their official attends court hearings. Practice shows that legislation does not always achieve its aim. Divorce, which quite often implies conflicts between parents, causes an adverse impact on the impartiality of hearings regarding the right of access. Mediation would help understand the causes of a dispute, while a family court's official would deliver an outside view and defend the child's interests effectively. Consequently, it would be necessary to amend the existing legislation and define mediation as a mandatory phase of the pre-trial dispute settlement procedure requiring the participation of a family court's official in the proceedings. This conclusion is consistent with the opinion expressed by legal scholars that a child has turned from a person to be cared for and protected into a right-holder [2]; therefore, the involvement of a child and any adult right-holder in the exercise of their rights is equally fully-fledged. It is clear that, depending on age and social maturity, children are not able to carry out procedural actions independently, and guardians of a child are persons who perform specific procedural actions required for the protection of the child's rights on behalf and in lieu of the child. This duty is a sort of "representation by legal nature" for guardians of the child [3]. It is established in the case-law that, although a child exercises the right of access independently, the child may need the other parent's assistance, with no obstacles put to exercising this right [4]. Guardians in their capacity as child's representatives must act for the child with the utmost care. It should therefore be concluded that the essence of the institution of representation consolidates the parents' duties laid down in family law.

However, the dual nature of the right of access, namely the right of a child to contact and a parent's duty to ensure such contact, should always be borne in mind. There can be no right of access, unless both these components are present simultaneously. A child must be willing to see his/her parent. For obvious reasons, legislation cannot deal with a situation when a child does not want to exercise his/her right of access. In this case, the child is considered to have waived this right voluntarily. It should be emphasised that parent's refusal to see the child cannot be treated equally. For parents, access is a duty rather than a right. Parents may not waive their duty to maintain contact with their child. The legislator, however, has not introduced any responsibility for failure to fulfil this duty. It is questionable whether any legal responsibility can be provided in the case of a duty requiring extremely personal involvement. Punishment for breach of legal duties is what makes legislation different from other types of social regulation (for example, moral standards). The author considers that legal responsibility of a parent should be introduced for refusal to exercise the right of access. The solution of this matter can be twofold. Apart from the above, competent authorities already have legal tools available for penalising parents, namely, in situations when communication is ordered by a court ruling, a parent who fails to comply with the court ruling and maintain contact with the child will be prosecuted. Article 168 of the Criminal Law lays down criminal liability for failure to comply with a court ruling regarding the right of access to the child, where a criminal offence has the following objective indicators: evasion of compliance with a court ruling, failure to comply with or delaying the execution of a court ruling in bad faith [5]. The objective aspect covers all possible forms of failure to comply with a ruling. Such criminal offences are punishable by temporary imprisonment, or community service, or a monetary fine. Punishment is adequate and serious enough. The author has analysed the Latvian case-law and established that no person has ever been prosecuted according to this article of the Criminal Law. This raises the question of whether the right of access is exercised effectively and parents do not evade their duty or law is not applied in situations when parents fail to fulfil their duty of access and, accordingly, a court ruling. Available information suggests that there are problems with the execution of court rulings. For example, 1,163 civil cases dealing with the right of access were heard in Latvia in 2018 [6]. Based on an analysis of statistical data in conjunction with publicly available information, it can be assumed that the respective provisions of the Criminal Law are not applied effectively. Considering that criminal proceedings are normally initiated on the basis of a statement lodged by a party concerned, it should be concluded that parties involved in the protection of the rights of children do not fulfil their duties properly. This mainly refers to parents providing day-to-day care of their children, whose duty would be to facilitate compliance with court rulings, including criminal prosecution of parents breaching the duty of access. Meanwhile, a family court should control parents and interfere in situations when they do not respond. From this viewpoint, cooperation between parents ensuring day-to-day care of their children and family courts should be improved.

The right of access is absolute. It means that a right-holder may not be deprived of this right; however, the right of access may be restricted or suspended temporarily under certain circumstances. This is due to the absolute nature of the right of access. Temporary suspension of access should not be regarded as a sort of punishment of parents. Its legal aim is to restrict any contacts between a child and child's parents that have become harmful to the child for a certain period of time during which the parents may eliminate violations and improve their behaviour. It should be emphasised that, in fact, the restriction of the right of access is a process, which should be divided into the following three phases: assessment of circumstances that have led to the suspension of access, assistance to parents in the elimination of such circumstances and, finally, evaluation of the achieved result, namely whether the parents have rectified the deficiencies. Circumstances are assessed by a family court and a psychologist invited by them, whose opinions are relied on by a court. The legislator has not defined the duration of such restriction or suspension. An analysis of legislation demonstrates that the right of access may be restricted or suspended on the grounds that access affects a child's interests. Circumstances should be assessed only from the standpoint of a child's best interests. Disputes over the right of access are normally referred to courts, and that is why the goal of a court is to identify relevant circumstances, i.e. whether access is detrimental to a child's interests and should therefore be restricted or suspended, and determine the duration of such restriction or suspension. Circumstances leading to the suspension or restriction of the right of access must substantially affect a child's interests and be supported by relevant evidence. Impairment of a child's interests can have both a legal and emotional or ethical nature. A child's interests can be adversely affected by communication with a parent who is striving to set the child against the other parent or encouraging the child, by either words or conduct, to act contrary to generally accepted behavioural rules (for example, not to go to school, use foul language) or commit administrative offences (for example, to beg, drink alcohol or use drugs); there is no doubt that parent violence, both emotional and physical, is harmful to the child's interests. We should agree with the opinion expressed in the case-law that violence against a child is a circumstance giving grounds for the restriction or suspension of access.

A parent's subjective interpretation or suspicions cannot be accepted as sufficient because it cannot be excluded that the parent is driven by personal negative relations with the other parent. In order to be in a position to rule on the restriction of the right of access, a court needs strong evidence that a parent has acted contrary to a child's interests. In addition to general evidence that can be normally accepted for civil proceedings, such as witness statements, pleadings of parties or material evidence, there exist two significant kinds of means of evidence for cases concerning the right of access: a child's statements and a psychologist's opinion. When necessary, reports on psychological examination can also be accepted as evidence. It is set forth in Article 13(3) of the Law of the Republic of Latvia on the Protection of Children's Rights that a child may express his/her opinion about any matters affecting the child, and the child's opinion must be given proper attention, considering the age and maturity of the child [7]. Any ruling that is contrary to the opinion of a child must be explained by a court.

When deciding on the restriction or suspension of the right of access, a court must determine the duration of such restriction or suspension. In accordance with Article 1495 of the Civil Law [8], duration can be specified as a certain period in months or years. Nevertheless, it should be acknowledged that duration defined as a certain period is not relevant for the right of access because this right can be restored as soon as circumstances that gave rise to the restriction or suspension of the right concerned no longer apply rather than upon expiry of the set duration. A psychologist, family court and the child concerned can be involved to establish these circumstances. Accordingly, the restoration of the right of access depends on whether special conditions are met. For instance, the Supreme Court of the Republic of Latvia has held that it can be difficult to enforce a judgment by which the right of access is suspended for three years provided that it can be restored in situations where, based on a psychologist's opinion, access will not be detrimental to the child's safety and health [9]. Therefore, the restoration of the right of access should be linked with a certain condition, which is the time when circumstances that gave rise to the restriction or suspension of access cease to exist. It should, however, be noted that the wording "when circumstances that gave rise to the restriction of access cease to exist", which does not refer to any specific timeframe, is too general and may unjustifiably affect the rights of both a child and a parent. It is indeed an open question who would be entitled to seek revision of the restriction imposed. How often may it be sought? While accepting the aforementioned conclusion of the court that duration based on a pre-requisite that the restoration of access will no longer be harmful to a child can hardly be defined because, to some extent, this wording is contradictory, the author still considers that special regulation is necessary. This means that legislation could define frequency of the revision of circumstances that gave rise to the restriction of access. Similar provisions have been introduced for the restriction of the capacity to act. The second paragraph of Article 364^1 of the Civil Law lays down that a court judgment regarding the restriction of the capacity to act may be reviewed at any time but not less than once every seven years from the date that the judgment has become final [1]. Certainly, a seven-year period is too long for cases dealing with the right of access because a child would become adult over this time. In this case, a period of two years would be reasonable. Therefore, a court would check whether circumstances that gave rise to the restriction of access have ceased to exist on its own initiative once every two years.

The next step is assistance to parents provided by the state. Children are interested in restoring contact with their parents, and it is the state's job to help parents improve their behaviour and correct errors they have made in communication with children. The state should formulate a clear programme of how customised assistance could be provided to parents whose right of access has been restricted or suspended by a court. It is a state's duty because the state has assumed responsibility for safeguarding the interests of children. The state is not active enough in this respect. Such a programme should incorporate mandatory state-sponsored psychological assistance, social workers' support, involvement of employment officials, namely the assistance of specialists who would be able to both identify deficiencies and provide practical advice for the rectification of erroneous behaviour, which, as a rule, takes the form of alcohol abuse, fits of rage and hatred, unemployment or lack of understanding about a child's needs. At present, parents are those who solve these matters. If parents are motivated, state's assistance can be provided, but it is only possible in situations when parents are really motivated. Meanwhile, parents who lack motivation, who do not understand why their right of access has been suspended and do not want to change their behaviour remain in the grey area. State support cannot reach these people. Cooperation and feedback among courts and municipalities are necessary for improving the existing situation. Court rulings restricting the right of access should be included in a unified register and communicated to competent municipalities, who would organise assistance for parents whose right of access has been restricted or suspended. Systemic support activities would be organised owing to these efforts. The involvement of a parent in support activities and the progress achieved would also help evaluate whether grounds for the suspension of access have ceased to exist. It is also important that decisions on the restoration of the right of access would be much more objective and require less time. Parent information would be dynamic, covering a longer period of time.

Certainly, the question remains how parents who lack motivation could be forced to take part in support activities because it can be assumed that a parent lacking self-motivation will not be willing to participate in mandatory activities voluntarily. Legislation should provide for a relevant enforcement mechanism. Interference with private rights of a child's parents is permissible for the sake of protection of the child's interests. Similar provisions have been adopted with respect to child support defaulters, whose car and boat driving licences can be suspended, except for situations when the parent is disabled or unfit for work over a long period of time, or suspension may considerably harm the child's interests (for example, a parent needs this special licence for the performance of his/her professional duties). This is provided by Article 7 of the Maintenance Guarantee Fund Law [10]. Analogous provisions could be introduced with regard to parents refusing to participate in support programmes with the aim of restoring effective communication. A child's rights to maintenance and contact are equally important; accordingly, there is no reason to consider that the adoption of a law with respect to a certain right (the right of access) in addition to that dealing with the other right would lead to excessive restriction of parents' rights.

The right of access is not identical to, but should be linked with, the right of custody. The right of access is a component of the right of custody. Custody has a broader scope than access. It should be noted, however, that the right of access can be viewed as an independent legal institution, which has a theoretical link with custody.

Deprivation of the right of custody per se does not affect the right of access. A parent having no right of custody will still have the right of access. Practice has shown, however, that actions are quite often brought for the deprivation of the right of custody and the restriction of the right of access simultaneously. This is because one of parents is confident that any contact with the other parent with whom the child does not usually reside may produce an adverse effect on the child. This concern is well-founded in certain cases but this cannot be a priori sufficient to give rise to the presumption that the parent who does not ensure day-to-day custody is not able to maintain effective communication. Deprivation of the right of custody means that a parent with whom the child does not usually reside will have no right to decide on the child's daily needs and will be released from childcare responsibilities. This, however, should not affect communication between the child and the parent and their leisure time together. In some cases, a parent may be deprived of the right of custody for objective reasons, for example, because the parentis living far away from the child, and not because the parent is unable to exercise custody or has some negative traits. Therefore, there is no reason to believe that a parent whose right of custody has been withdrawn is ipso facto unable to maintain contact or has bad traits that would prevent effective communication and affect the child's interests.

The right of custody may be withdrawn in the following two cases: based on a request of either the other parent or an institution that is responsible for the protection of a child's rights, i.e., a family court. Where the deprivation of the right of custody is sought by the other parent, the case is normally handled in a single set of proceedings and the action brought concerns access frequency and procedure. Actions brought for the withdrawal of the right of custody by family courts normally deal with both the restriction of the right of access and placing a child in a care institution because family courts are involved in serious cases, where the lowest possible point has been reached with respect to the safeguarding of a child's interests, namely in situations where the parents of the child are unable to ensure the child's rights or even abuse them, and it is therefore understandable why the suspension of the right of access is sought apart from the withdrawal of the right of custody. This situation is rather ambiguous. The author believes that the restriction of access should be used as a last resort.

As a result of being placed in an extra familial care institution subject to restricted access, the child feels estranged from his/her parents, while the parents are less motivated to restore their right of custody. Placing a child in an extra-familial care institution may lead to a situation when the right of access cannot be exercised physically. The existing legislation does not promote the exercise of the right of access in such situations because no law requires that a child be placed in an extra-familial care institution that is nearer to the parents' home. The state must promote meetings between children and their parents. Researcher A. Haružka has also noted that the European Court of Human Rights has recognised in a judgment delivered on 26 February 2002 in Case 46544/99 Kutzner v. Germany that "the fact that a child could be placed in a more beneficial environment for his or her upbringing will not on its own justify a compulsory measure of removal from the care of the biological parents; there must exist other circumstances pointing to the "necessity" for such an interference with the parents' right to enjoy a family life with their child." [11]. It is clearly indicated that a child must be brought closer to his/her parents.

Wherever the right of access is exercised, it is important to define how a child's right of access can be secured in the most effective way, and it is essential to understand the legal scope of the right of access. What falls within that scope, what specific actions are provided and what their frequency is. Practice demonstrates that access frequency and procedure may be different: at weekends, during school holidays or daily. This mainly depends on the physical possibilities of parents, such as distance, financial situation, etc. It cannot be clearly concluded that daily visits would be more useful than spending time with the other parent at weekends. It should be stressed, however, that the absence of even minimal access should be viewed negatively. The author believes that guidelines that would define the basic principles of access should be formulated. For example, the legislation dealing with minimum child support has evolved analogously, from general rules whereby parents merely had to pay child maintenance depending on their possibilities and financial situation through to the specific minimum child support payment to be made regardless of the relevant parent's possibilities and financial situation. Minimum access and minimum time to be spent with a child could be defined in a similar manner. It should be stressed that access is a two-way right, where a child has a right, while parents have both a duty and right to maintain personal relationships and direct contact with their child. Both prerequisites must be met simultaneously for the effective exercise of the right: both parents and the child must be willing to establish and maintain direct contact.

According to scientific literature, removed choices become diktat and rights are abused when interpreted too literally and forced upon people. The interpretation affecting children should be selected appropriately because children have insufficient control over their lives, which makes them susceptible to being controlled by over-interpretation [12]. Hence, it is important to establish whether a child really wishes to maintain contacts with a parent with whom the child does not live permanently and take into account the frequency and time of access desired by the child and expected activities during the time they spend together.

Legislation dealing with the right of access has been left open by the legislator. Imperatively, legislation provides for the duty and specifies persons covered by this right, such as parents, siblings, grandparents and other persons with whom the child has lived in a common household for a long time provided it is in the best interests of the child. This could be defined as a quantitative component, but the qualitative component of access, which is the frequency and quality of contacts, is just as important. Practice shows that there are hardly any disputes over the right of access within families. This is evidence that communication with grandparents, siblings and other family members is less important, namely: access to these persons is not considered individually, it is assumed when deciding on contact with a divorced parent that communication with that parent's relatives is guaranteed by default. This, however, should not be regarded as self-evident, considering the interests of a child. The right of access with respect to persons with whom the child has lived in a common household is just as important as communication with a parent; therefore, family courts should carry out both awareness-raising and control activities.

Without protecting the right of access to the fullest extent possible, the right of access is just formalism; accordingly, the safeguarding of children's rights is formal. As regards the right of access, A. Vorobjovs, professor says that any interaction between people can be called access. Means of communication include facial expressions, gestures, actions, words, results of actions [13]. It is concluded that the right of access presumes active cooperation of persons involved over a certain period of time, which includes personal presence, exchange of thoughts, talks and expression of emotions. It should be emphasised that in today's world remote access can also be implemented in the electronic environment, by using modern technologies. This, however, is surrogate access, which is not a substitute for direct physical contact.

Remote communication cannot be regarded as due access. This conclusion is also indirectly confirmed by legislation, namely Article 2(10) of the Brussels Ibis Regulation provides for a special right to take a child to a place other than his or her habitual residence in order to ensure direct contact [14]. Parents seeking to agree on the exercise of the right of access or bring an action must specify the time and place of access. It follows from both the essence of the right of access and Article 244^2 of the Civil Procedure Law [15]. The frequency and duration of access should be defined on an individual basis. It should be borne in mind, however, that a child's opinion must be taken into consideration when deciding not only on the establishment of access but also on the procedure of access, with child's interests being paramount in this respect. The author shares the opinion of Associate Professor John Tobin that sometimes children lack the ability to exercise their rights but a child's rights are his/her interests rather than the ability to exercise them [16].

The frequency of communication, which should be regular, also in the form of contact, embodies the substantive nature of the right of access. Regularity means that access must be regular and take place in certain periods of time. This is associated with the nature of access, which is to ensure dynamic information about a child and allow participation of a divorced parent throughout all stages of child's development, while the child would enjoy parent's attention based on the needs that are relevant for a child of a certain age.

It also means the right of a child to see persons with whom the child has lived in a common household because any of these persons (siblings, grandparents) are able to produce a different, but no less significant, impact on the development of the child. It should be concluded that the right of access ensures an educational effect, facilitating various qualities of a child, such as empathy, warm-heartedness and understanding of the family as a value.

2 Results

The right of access is a complicated legal institution. The right of access cannot be viewed only from the standpoint of law as it comprises psychological and emotional aspects because of which enforcement is not possible. Both children and their parents have inalienable rights to build and maintain direct contact with each other. Practice shows that it is mostly parents who do not want or are unable to ensure contact with children with whom they do not usually reside. It is also acknowledged that it is a parent's duty to ensure communication when the parent is not willing to do so. Meanwhile, a child is free to choose whether contact with a parent will be maintained, it is a child's right. This is because the rights of children have priority over those of parents as special attitude is guaranteed for children, considering that they are special right-holders.

As a result of the research, the author has arrived at the following:

1) the right of access should be exercised solely in the interests of the child;
2) the right of access is an absolute right, which should be used regardless of whether parents have custody of the child;
3) the right of access is not part of the right of custody;
4) there is a lack of legal instruments that would prevent the use of the right of access contrary to its goal;
5) the state has no tool available that would ensure systemic parent support aimed at removing obstacles that led to the restriction or suspension of access.

3 Conclusions

The right of access exercised by parents neglecting the best interests of their child, when the child is used merely as a tool to settle their mutual relationships, is a major concern.

As regards existing disputes, it is more likely that a decision on the exercise of the right of access might not fully meet the best interests of the child. The improvement of the existing legal framework must be focused more on laws that would eliminate the sources of disputes.

It should be borne in mind that the right of access is an independent legal institution governing legal relationships between parents and children in the building and maintenance of their direct contact. The state has a duty to ensure the respect of this right in the event of non-compliance for objective or subjective reasons. A child as a special right-holder has priority status in these legal relationships. Measures associated with the right of access can be qualified as ensuring access, the restriction or temporary suspension of access, the removal of obstacles that led to the restriction or suspension of access, the restoration of access. Officials involved in these stages should respect the interests of children, and their decisions should be in line with these interests. It should be borne in mind that, in fact, parents do not have their own interests in legal relationships as their interests arise from and are subordinate to the interests of children. Therefore, a policy pursued by the state in this respect should support parents, meanwhile safeguarding the interests of children in order to respect their rights.

4 Proposals

The legal framework dealing with access will definitely evolve. The first issue would be that parents should be encouraged to fulfil their duty to ensure access. Both motivational and enforcement mechanisms should be used, which would allow for the selection of the most appropriate tool depending on specific circumstances based on a parent's behaviour and the extent of the parent's motivation and involvement. In some situations, a parent who is striving to improve behaviour and restore contact with a child but is unable to do so due to external circumstances needs only motivational support. If a parent maliciously evades his/her responsibilities, enforcement or punishment is necessary.

The state has various social support tools available, such as extra benefits, development of leisure facilities, etc. It is essential to introduce a requirement that, if necessary, a child be placed in an extra-familial care institution that is nearer to the parents' home. This would facilitate communication between parents and their children, and it would be easier to implement this requirement.

The second issue to be considered by the legislator would be the introduction of adverse consequences for parents avoiding the fulfilment of their duty to provide access. Considering that administrative penalties would not be directly applicable in this case, other solutions must be found. A child as a right-holder requiring special protection has a right to state's interference with private relationships, namely parent-child relationships; therefore, proper legislation is permissible.

But, before introducing any punishment mechanisms, the state should formulate and implement a support policy, whereby the assistance of psychologists, social workers and employment officials would be provided for parents. Parents may be held responsible for their non-compliance with support activities only in the presence of due support.

The interruption of the right of access should be as short as possible. An emotional bond can be lost, and, as a result, the longer the interruption period, the harder the restoration of contact. Therefore, activities of a preventive nature should rather be facilitated. As regards the divorce of parents with minor children, a family court should be involved during the divorce process. The role of a family court would be to promote mediation, also if parents have voluntarily agreed on their divorce and all related matters, such as child maintenance, right of access and custody. Divorcing without conflict is not a guarantee that there will be no dispute over the right of access. Mediation is a means of identifying the cause of a dispute and preventing any controversies at their source.

The legal framework governing access has both the potential and need for development. It should be admitted that the existing legislation is rather declaratory. The state has defined that the right of access is an inalienable right of children, while the subsequent implementation of this right is left freely to the discretion of parties concerned. The above means that it is parents who must ensure contact. Given that the protection of children's rights is a state priority, greater involvement of the state would be necessary for the solution of these matters. The author believes that the involvement of the state might take the form of legislation that would support parents and help them to overcome problems that led to the restriction or suspension of access or, as regards parents who lack motivation and do not want to cooperate, introduce adverse legal consequences. Within the frame of this process, a family court should be an authority representing a child and safeguarding the child's interests apart from a natural guardian. Only complex solutions involving parents, state and municipal officials and, certainly, children will bring the best result for the assurance of the right of access.

References

[1] Civil Law. Part one. Family Law. Law of the Republic of Latvia. Valdības Vēstnesis **41** (1937)
[2] M. Sardenberg. *Comments on the globalization of child law, from Globalization of Child Law (The Role of the Hague Conventions)*, 7 (1999)
[3] I. Kudeikina. Institucionālā problemātika bērna tiesību aizsardzības kontekstā. *RSU Zinātniskā konference* (2017). Available at: `https://www.rsu.lv/sites/default/files/imce/Zinātnes%20departaments/2017/XIII%20sekcija/institucionala_problematika_bernu_tiesibu_aizsardziba.pdf` (visited on 8.01.2019)
[4] *Administratīvo tiesu praks elietās par bāriņtiesu lēmumiem. Tiesu prakses vispārinājums*. 22 (2014)
[5] Criminal Law. Law of the Republic of Latvia. Latvijas Vēstnesis **199/200** (1999)
[6] Court statistics. Available at: `https://manas.tiesas.lv/eTiesasMvc/lv/nolemumi` (visited on 11.02.2019)
[7] Law on the Protection of Children's Rights. Law of the Republic of Latvia. Latvijas Vēstnesis **199/200** (1998)
[8] Civil Law. Part four. Obligations Law. Law of the Republic of Latvia. Valdības Vēstnesis **46** (1937)
[9] Judgment in Case SKC-[B]/2018 delivered by the Supreme Court of the Republic of Latvia in 2018. Available at: `https://www.google.com/url?sa=t&rct=j&q=&esrc=s&source=web&cd=2&ved=2ahUKEwjO5NmMvevfAhWDqIsKHVn6AIcQFjABegQIBRAC&url=http%3A%2F%2Fwww.at.gov.lv%2Fdownloadlawfile%2F5461&usg=AOvVaw0R14bXXgPxtIJM7K66Hm10`
[10] Maintenance Guarantee Fund Law. Law of the Republic of Latvia.LatvijasVēstnesis **249**(5821) (2016)
[11] A. Haružka. PētījumsEiropasCilvēktiesībutiesasspriedumubērnutiesībujomāanalīze. 59 (2013). Available at: `http://www.tiesibsargs.lv/files/content/ECT_prakse_bernu_tiesibu_joma_2013.pdf`
[12] B. Milne. *Rights of the Child. Children's Well-Being: Indicators and Research*. 145 (2015)
[13] A. Vorobjovs. *Sociālā psiholoģija*, 23 (2002)
[14] Council Regulation (EC) No 2201/2003 of 27 November 2003 concerning jurisdiction and the recognition and enforcement of judgments in matrimonial matters and the matters of parental responsibility, repealing Regulation (EC) **1347/2000** (2003)
[15] Civil Procedure Law. Law of the Republic of Latvia. Latvijas Vēstnesis **326/330** (1998)
[16] J. Tobin. Justifying Children's Rights. Cited from *The Future of Children's Rights*, 259 (2014)

The problems of access to justice in rural areas (on the example of Ukraine)

Andrii Lapkin*

Yaroslav Mudry National Law University, Kharkiv, Ukraine

Abstract. In this scientific article, problems of access to justice in rural areas in Ukraine are considered. It is proved that for the people living in rural areas it is necessary to consider access to justice as possibility of free application to justice system for protection of their rights that is without any obstacles or difficulties. However in need of application to a court the rural dwellers face a lot of problems, such as: territorial, economic and organizational. Territorial problems relate to territorial distance of judicial authorities from rural areas. Economic problems relate to high level of court expenses and low financial well-being of rural dwellers. Organizational problems relate to absence of the necessary infrastructure in rural areas facilitating access to court (lack of transport infrastructure, absence of high-speed Internet and possibility to use technologies of "electronic justice"). The conclusion is that for the solution of this problem it is possible to offer the following steps: general development of infrastructure in rural areas; formation of judicial system on the basis of uniform territorial distribution; reduction of court expenses for dwellers of rural areas proportionally to the level of their incomes; development of institution of lay justice operating in rural areas.

Key words: access to justice, human rights, rural areas.

1 Introduction

The human right on access to justice is the main guarantee of the right to the judicial protection fixed in the International Covenant on Civil and Political Rights (1966) and the Convention for the Protection of Human Rights and Fundamental Freedoms (1950). Though the right to access to justice is not mentioned in the text of these acts directly, it is considered as a necessary component of the right to fair judicial proceedings. Originally in decisions of the European Court on Human Rights this right was allocated proceeding from the need to guarantee human rights not only during participation in already begun trial, but also concerning a reference to the court for protection of the rights. In further practice of the European Court on Human Rights the issues of refusal in access to justice or making of excessive financial or procedural difficulties to its realization were considered. These are cases "Golder v. the United Kingdom" [1], "Airey v. Ireland" [2] and others.

Especially relevant ensuring the right to access to justice is for the people living in rural areas. Ukraine is characterized by high specific weight of rural population in the general structure of the population of the country. At the beginning of 2019 in Ukraine over 30% of the population of the country, that is more than 14 million people, lived in rural areas [3]. However processes of urbanization and sharp falling of a standard of living in rural areas

* Corresponding author: an.lapkin@gmail.com

in the last decades lead to reduction of number of rural dwellers. Especially it concerns a young and able-bodied part of rural population. At the same time the level of violation of the rights of inhabitants of rural areas is rather high and level of their legal literacy – very low. In rural areas, often there are no authorities to whom people could address for protection of the rights. There are also no mechanisms of the alternative solution of the conflicts, available in the cities. As a result of this, violations of the rights of inhabitants of rural areas often have no legal consequences. Thus, the considerable part of the population of Ukraine, living in rural areas, is especially vulnerable concerning the issues of legal protection, therefore its specifics demands creation of additional guarantees of access to justice.

2 Aim

The purpose of this scientific research is allocation and classification of problems which rural population faces at access to justice. It is also necessary to offer the ways of overcoming these problems for improvement of availability of justice to inhabitants of the rural areas.

3 Material and methods

The scientific research is based on theoretical works of scientists who have considered problems of availability of justice, for example: M.A. Gurvich, I.E. Marochkin, E.N. Ovcharenko, N.Yu. Sakara, N.V. Sibileva, Yu.I. Stetsovsky and others. The empirical material consisted of statistical information of authorities of Ukraine, materials of court practice and own experience of the author's life in the rural area. In the course of the research the dialectic method and also a comparative method, methods of analysis and synthesis, a method of law modelling and others were used.

4 Results and discussion

Access to justice is defined as the ability of people to seek and obtain a remedy through formal or informal institutions of justice for grievances in compliance with human rights standards. There is no access to justice where citizens (especially marginalized groups) fear the system, see it as alien, and do not access it; where the justice system is financially inaccessible; where individuals have no lawyers; where they do not have information or knowledge of rights; or where there is a weak justice system [4]. In the Declaration of the high-level meeting of the General Assembly on the rule of law at the national and international levels adopted by the General Assembly in Resolution on 24 September 2012 it was underlined: "We emphasize the right of equal access to justice for all, including members of vulnerable groups, and the importance of awareness-raising concerning legal rights, and in this regard we commit to taking all necessary steps to provide fair, transparent, effective, non-discriminatory and accountable services that promote access to justice for all, including legal aid" [5, point 14].

It is possible to define access to justice as free possibility to use the right to judicial protection, lack of any obstacles at a reference to the court. In scientific legal literature access to justice is considered and as access to court, and as access to the judge, and as access to judicial authority as a whole. In our opinion, it is necessary to apply the widest meaning to the characteristic of this concept. Proceeding from it, for the people living in rural areas, it is necessary to consider access to justice as possibility of free appeal to justice system for protection of their rights. In a broad sense access to justice includes as possibility of address directly in a court, and to bar and prosecutor's office, bodies of execution of judgments.

At the heart of availability of justice the possibility of appeal to a court without any obstacles or difficulties lies.

Thus, guarantees of availability of justice provide, first of all, elimination or minimization of obstacles for appeal to a court. There is a direct dependence: the fewer obstacles lie in the way to a court (or the less they are expressed), the more available justice is. However in the need of appeal to a court the people living in rural areas face a number of problems. These problems are in many aspects different from the problems arising in the sphere of access of urban population to justice. On the basis of their analysis it is possible to allocate such problems of access to a court of rural population: the territorial; the economic; the organizational.

4.1 Territorial problems

Territorial problems consist in a territorial distance of judicial authorities from rural areas. So, today courts of the first instance function in Ukraine on the level of the cities and districts.

Existence of the local court in each district centre is convenient for a rural population. Regardless of the fact that the distance from some villages to the district centres can make 30–50 kilometres, in most cases within the district continuous transport connection is adjusted. Generally it is carried out by buses and also, in the presence of the railroad – by means of trains or electric trains (that represents the most budgetary option). A district centre in which a court is situated is also the centre of municipal authorities, a solicitor, etc., thus the villager can receive legal and administrative service, having arrived there.

However judicial reform provides creation of district courts which will act on the level of several districts. The law of Ukraine "On the Judicial System and the Status of Judges" of June 02, 2016 provides that instead of district and city courts country courts will operate on the local level [6]. In Ukraine, within the reforms, instead of 663 local general courts it is planned to create about 270 new country courts. The process of reorganization of local general courts in the country began in 2018 and should be completed in 2019. Thus, the number of local courts will be reduced by 2.5 times though the number of judges, who work in them, is planned to remain the same. One court will cover several areas. This would entail territorial distance of the court from the population of rural area, since it would be much more difficult for the population of other areas covered by its jurisdiction to reach it. It will cause territorial distancing of a court from the population of rural area. To the population of other areas it will be much more difficult to reach the court than of the areas where there is the court.

A similar reform was carried out concerning prosecutor's office. As a result authority of local prosecutor's offices extends to on the average 5–7 districts in rural areas. At the same time the distance between these districts and the main office of local prosecutor's office may be more than 100 kilometres. Between many districts falling under territorial jurisdiction of local prosecutor's office, there is no direct transport service. It means that to reach the prosecutor's office, the inhabitant of rural areas has to travel to the regional centre (for 100–200 kilometres), and from there – to the locality where the prosecutor's office is situated [7]. At the same time it is necessary to consider that population appeal to the prosecutor's office considerably less than in the court. That is the territorial remoteness of court from a rural population is more critical than remoteness of prosecutor's office.

Application of this principle to the local courts will lead to the fact that rural population is far away from judicial authorities. It roughly breaks requirements of territorial availability of justice. In this regard it is necessary to consider recommendations according to which within a district centre of an oblast premises of courts should be located in an administrative centre. It has to be near other premises of authorities, local governments, main enterprises, institutions and organizations. Considering absence of public transport in many cities of district significance this situation acquires special relevance [8].

Considering the reasons for such reform, its creators first of all refer to the need of economy. It is considered that the provision of smaller number of larger courts is economically more beneficial for the state than the maintenance of bigger number of smaller ones. However, if we analyze the experience of the reform of the prosecutor's office in Ukraine, we can conclude that no savings were achieved. Difficulties in placement of local prosecutor's offices led to the placement of their departments in geographically remote locations. Earlier, district prosecutor's offices functioned there, which were liquidated as a result of the reform. However, the presence of departments requires premises, staff and support staff that had operated there before. Consequently, the reform of the prosecutor's office was in many ways fictitious and did not justify hopes of saving.

It can be predicted that the reform of the local courts will be completed in this way. The need for access to justice of local population will entail creation of court departments in remote locations. They reanimate the local courts that existed there. This may solve the problem of territorial remoteness, but will sacrifice economic expediency. Otherwise, the savings from reduction of local courts, which the state will achieve, will actually be transferred to the budget of rural residents, because they will have to spend much more money to get to a remote court to participate in court proceedings.

A perspective option for solving the problem of territorial remoteness of courts from rural population is creation of magistrate's courts. The experience of their existence took place in Ukraine in the period after the Judicial Reform of 1864. According to Ukrainian researchers, the institute of lay justice makes it easier for people to access justice and relieve local courts. It is also able to speed up consideration of cases by courts and increase public confidence in the judiciary [9]. At the same time, experts point out such problems of lay justice as search for ways to finance it; determination of a network of magistrate courts; their staffing; jurisdiction of the magistrate courts; procedure for appealing their decisions, etc. [10].

These provisions can be applied to rural population. Magistrate's courts, as alternative ways of resolving conflicts, are effective and cheap means of increasing accessibility of justice in rural areas. In the conditions of territorial remoteness of official courts, lay courts are able to quickly resolve minor civil and criminal cases on the ground. Therefore, the issue of creating of magistrate's court in rural areas is perspective for Ukraine.

4.2 Economic problems

The economic aspect of access to justice includes the reasonableness of court costs for participants in court proceedings, as well as adequate funding for the judiciary. According to scientists, the economic components of accessibility of justice are: 1) the reasonableness of court costs and creation of a procedural mechanism for postponing, instalment or exemption from court costs; 2) provision of qualified legal assistance on preferential conditions to indigent persons; reasonableness of expenses for representative services; 3) adequate financing of judicial activities, ensuring the possibility of full and independent justice [11]. In the economic aspect, there are a number of limitations in access to justice for rural dwellers.

Economic problems consist in high level of court costs and low financial well-being of rural inhabitants. In comparison with cities, the average level of incomes in rural areas is much lower. So, the average salary in agriculture in Ukraine following the results of 2018 was 7129 UAH (about 222 Euros) that is below the average salary in the country (8711 UAH or about 272 Euros), as well as that which is fixed in other types of economic activity: in industry (9458 UAH or about 295 Euros), transport sphere (9787 UAH or about 306 Euros), financial activity (15986 UAH or about 500 Euros) and others [12]. Also in rural areas unemployment rate is much higher. So, according to the data at the end of 2018, the number of unemployed urban residents was 183.2 thousand, and the number of rural residents – 158.5 thousand [13] (despite the fact that rural residents, as mentioned earlier, make up one third of the population of Ukraine). Thus, the incomes of rural inhabitants are much lower than the urban ones. In addition, unemployment is much higher in rural areas, which aggravates the economic problems of rural population.

However the level of court costs is identical to all and does not consider a difference in incomes of people who are appealing to a court. According to Part 1 of the Art. 4 of the Law of Ukraine "On Court Fee", court fee is set as a percentage of the price of a claim or in a fixed amount. The fixed amount of court fee is set according to the size of the subsistence minimum for able-bodied persons, established by law on January 1 of the calendar year in which the corresponding statement or complaint was filed [14]. The minimum court fee for filing a claim in a court for individuals is 0.4 of the size of the subsistence minimum for able-bodied persons. Since January 1, 2019, the minimum of this amount is 1921 UAH (60 Euros) and the minimum court fee is 768.4 UAH (24 Euros). This amount is palpable for rural residents and limits their ability to go to court to protect their rights.

Moreover, villagers initially incur higher expenses connected with the need of trip to the city in which a court is located. Due to the frequent postponement of the proceedings in a court and the length of the process, the villager has to visit the court a lot of times. In the conditions of high prices for transport, trips to the court are burdensome. At the same time, the court does not warn the participants of the trial about transfer of the case in advance; therefore, they can learn about it only by appearing directly to a court. The legislation does not provide any compensation for the cost of travel to the court in cases where the process was postponed due to the fault of a court or for other reasons that do not depend on honest participants in the process, who appeared in the court.

Also expensive for residents of rural areas are the services of lawyers. Prices for them in Ukraine are about the same in cities and in rural areas. However, due to lower incomes of rural residents, cost of services of lawyers is often inaccessible to them. They also spend more for payment of the help of lawyers which put transportation costs in the fee. In many rural areas there are no resident lawyers. Therefore, there is a need to involve lawyers from major cities, and, accordingly, to pay for their arrival in a court.

In Ukraine, the centres for the provision of free secondary legal aid, which are being created in each district, are designed to solve this problem. The state hires lawyers to work in such centres and pays for their services from the state budget. These centres operate on the basis of the Law of Ukraine "On Free Civil Legal Aid" from June 2, 2011 [14]. In accordance with this law, the right to get a lawyer at the expense of the state have among different categories of citizens as well low-income citizens. However such centres do not provide with free legal aid of everyone today. Besides, the quality of the services provided by the centres is low as lawyers who work there, are not interested in high quality of legal services.

Thus, justice is excessively expensive for the majority of rural dwellers. In this regard, rural residents are trying to avoid participation in trials because of their high cost. Thus, they voluntarily deprive themselves of judicial protection. To overcome this negative trend, it is necessary to make justice cheaper for rural residents in order to take into account their low incomes.

4.3 Organizational problems

Organizational problems cover a wide range of issues related to creation of appropriate conditions for going to the court and obtaining judicial protection.

Organizational problems in rural areas consist in absence of necessary infrastructure facilitating access to a court. First of all, that is lack of transport infrastructure whereupon people experience difficulties to reach the court. We have already dealt with this problem above in the aspect of the territorial remoteness of courts from population. Transportation costs are expensive for rural residents and are constantly rising. Only a small part of rural population in Ukraine has a personal vehicle. Therefore, villagers are forced to use public transport to get to a court. At the same time, in many Ukrainian villages there are no normal roads and there are interruptions in transport, in this aspect a trip to a court becomes a long and difficult. In winter, the roads to remote villages are not cleared of snow and it is almost impossible to get there. The public transport schedule does not fully coincide with the court's work schedule (for example, there is no traffic until 9:00 am or after 5:00 pm, whereas at that time cases are scheduled in a court).

The solution of the problem of territorial remoteness of courts could be using of modern information technologies in the field of justice. The possibility of remote participation in consideration of cases by courts is provided for by the procedural legislation of Ukraine. However, it is used only in exceptional cases. For example, in criminal proceedings court proceedings may be conducted through video conference with transmission from another premise, including such as is located beyond the bounds of the court premises, where – 1) it is impossible for a participant of criminal proceedings to participate directly in the court proceedings for reason of health or for other valid reasons; 2) it is necessary to ensure the persons' security; 3) a minor or underage person is to be interrogated as a witness or victim; 4) such measures are necessary to ensure speedy court proceedings; 5) there exist other grounds recognized sufficient by the court. At the same time, a bailiff or a secretary of court session of the relevant court is obliged to hand such person a memorandum of his procedural rights, check his identity documents, and be with him until the end of the court session (Art. 336 of the Criminal Procedure Code of Ukraine) [16].

In a civil proceeding, participation in a court hearing via videoconference is simpler. In accordance with the Article 212 of the Civil Procedure Code of Ukraine [17], participants in a case may participate in a court hearing in the videoconference mode in the presence at court of the corresponding technical capability. For this purpose it is necessary to submit the application no later than five days before judicial session. The participants in the case take part in the court session in the videoconference mode with use of own technical means and the digital signature.

It is obvious that use of the mode of a videoconference would reduce costs of villagers of personal participation in lawsuits. As experts note, use during judicial sessions of systems of a videoconferencing will in most cases avoid delaying the court hearing of a case due to the fact that a particular participant of the process cannot appear at the court session. The efficiency of systems of a videoconferencing in the course of hearing of cases in a court is proved by experience of their application in foreign countries. Also their use will allow to citizens to save considerable sum of money necessary for trips to courts which are located in other settlements [18].

However, in a lot of villages of Ukraine there is no high-speed Internet in this connection their population is deprived of possibility to use technologies of "electronic justice". Besides, most of villagers (as well as in general residents of Ukraine) has no digital signature that does not allow them to use remote participation in lawsuit. The practice of using the videoconference mode by the courts in Ukraine is not widespread, and judges treat it with mistrust. In addition, in order to initiate the issue of participation in a court session via videoconference, a person must submit a relevant application to the court in advance, for what it needs to be personally in court. Therefore, the practice of remote participation in court sessions in Ukraine requires improvement. Its positive impact on access to justice for rural residents requires, on the one hand, the development of information technologies in rural areas, and on the other, the development of electronic justice, which level in Ukraine lags far behind other countries.

Organizational problems of access to justice are also shown by the fact that for an inhabitant of rural areas it is much more difficult to find the lawyer to receive a legal aid. Villagers have deficiency of information on lawyers who work in their area as do not use the Internet. A lot of lawyers, proceeding from low solvency of rural population and transport remoteness, do not consider rural dwellers as clients. The lawyers who work in the rural zone often have low professional level due to the lack of the competition. Personal contacts with lawyers also are inconvenient. For a meeting with a lawyer villagers are forced to overcome the same, and sometimes a bigger distance, than for a visit to a court. In the conditions of deficiency of personal contact it is difficult to lawyer to estimate a legal position of the client correctly. In turn, it is difficult to a client to determine the level of professional competence of the lawyer which they address. Therefore the villager experiences much more organizational difficulties in access to justice, than city inhabitant.

It is necessary to consider that organizational problems concern not only inhabitants of the rural zone, but also judges who work there. Today in many courts in Ukraine there is an acute issue of understaffing. So, in Ukraine at the end of 2016 there were more than 200 courts (about a third of the total) in which more than a half of judges did not work. The reasons of such situation are different: expiration of the term of office of a judge, untimely swearing in of a judge, achievement by a judge of 65 years and others. Basically, these problems concern the courts located in rural areas, since the work in them is considered less prestigious for judges. Courts operating in rural areas often lack premises suitable for implementation of justice. There are problems with heating of the premises of courts and sewage. In the rural areas housing is not under construction; therefore, the appointed judges cannot receive the room for accommodation there. The lack of jobs, kindergartens and schools creates problems for family members of judges working in rural areas. Thus, the majority of judges do not want to work in courts located in rural areas. They consider the work in such courts as an intermediate option and as soon as possible they try to transfer to the courts located in cities. On the other hand, judges with the lowest positions in the ranking get to work in courts located in rural areas, which lowers professionalism level at implementation of justice in the rural areas.

5 Conclusions

Generally, it should be noted that the people living in rural areas, face a number of additional obstacles at a reference to a court, than city dwellers therefore their access to justice is actually limited. For the solution of this problem it is possible to offer the following steps:

1) General development of infrastructure in rural areas. The more economically the village is developed – the more available justice becomes for his inhabitants. Increase of welfare of inhabitants of rural areas will give the chance to them to protect their rights more effectively. Modern roads, reliable public transport, the high-speed Internet are necessary for access to justice in rural areas.
2) Creation of judicial system on the basis of uniform territorial distribution. Defining the centres of judicial districts it is necessary to consider distance for the most remote villages which, in our opinion, should not exceed 30 kilometres. Also in courts there should be enough judges to consider cases quickly and effectively.
3) Introduction of modern information technologies. It is necessary to provide possibility of remote participation in trial for inhabitants of remote settlements. The submission

of documents in the court in an electronic form is expedient. The reliable system of informing on carrying out a court session and about its cancellation or postpone is also necessary.

4) Reduction of court costs for inhabitants of rural areas. It could be reached by calculation of the level of judicial taxes in proportion to the level of incomes of participants of process. Expansion of the system of free legal aid and its introduction in rural areas also is required. The inhabitants of rural areas being participants of the trial should obtain compensation of the transportation costs connected with a call in court.
5) Provision of villagers with the qualified legal aid. In cases of insolvency of a client such help has to be provided at the expense of the state. It can be reached by development of the centres of free secondary legal aid which are created in Ukraine. Lawyers should be interested in quality of the legal services provided by them. For this purpose it is expedient to put the size of their fee into dependence on results of consideration of a case.
6) Development of the institute of lay justice in rural areas. Possibility of solution of simple disputes by lay judges directly on the place of contract of their parties will approach justice to rural areas. Also it will simplify the trial and will reduce its terms. Another positive effect is that it will be possible to unload judicial system and to lower load of professional judges.

References

[1] European Court of Human Rights judgment in case *Golder v. the United Kingdom,* 4451/70, 21.02.1975 URL: `http://hudoc.echr.coe.int/app/conversion/pdf/?library=ECHR&id=001-57496&filename=001-57496.pdf`

[2] European Court of Human Rights judgment in case *Airey v. Ireland*, 6289/73, 09.10. 1979 URL: `https://ww3.lawschool.cornell.edu/AvonResources/Airey.PDF`

[3] *Населення України на* 2018 *рік*. URL: `https://ukrainatoday.com.ua/naselennya-ukra%D1%97ni-na-2018-rik/`

[4] *Necessary Condition: Access to Justice*. URL: `https://www.usip.org/guiding-principles-stabilization-and-reconstruction-the-web-version/rule-law/access-justice`

[5] *Declaration of the high-level meeting of the General Assembly on the rule of law at the national and international levels*: Resolution adopted by the General Assembly on 24.09.2012. URL: `https://undocs.org/en/A/RES/67/1`

[6] *Про судоустрій і статус суддів*: Закон України від 02.06.2016. URL: `http://zakon3.rada.gov.ua/laws/show/1402-19`

[7] Лапкін А.В. *Закон України «Про прокуратуру». Науково-практичний коментар*. Харків: Право, 26 (2015)

[8] Овчаренко О.М. *Доступністі правосуддя та гарантіы його реалізаціы*: Монографія. Харків: Право, 86 (2008)

[9] Цибуляк-Кустевич А.С. Мирова юстиція: дисертація ... канд. юрид. наук: 12.00 .10. Харків, 72 (2016)

[10] Дем'аненко. В. *Організаційно-правові основи діяльності місцевих загальних судів в Україні*: дисертація ... канд. юрид. наук: 12.00.10. Харків, 140 (2017)

[11] Сидоренко В.М. *Принцип доступности правосудия и проблемы его реализации в гражданском и арбитражном процессе*: диссертация ... канд. юрид. наук: 12.00.15. Екатеринбург, 8 (2002)

[12] *Середньомісячна заробітна плата за видами економічної діяльності за період з початку року у 2018 році*. URL: http://www.ukrstat.gov.ua/
[13] *Кількість зареєстрованих безробітних за статтю та місцем проживання у 2018 році*. URL: http://www.ukrstat.gov.ua/
[14] *Про судовий збір*: Закон України від 08.07.2011. URL: https://zakon.rada.gov.ua/laws/show/3674-17
[15] *Про безоплатну правову допомогу*: Закон України від 02.06.2011. URL: https://zakon.rada.gov.ua/laws/show/3460-17
[16] *Кримінальний процесуальний кодекс України* від 13.04.2012. URL: http://zakon2.rada.gov.ua/laws/main/t21
[17] *Цивільний процесуальний кодекс України* від 18.03.2004. URL: https://zakon.rada.gov.ua/laws/show/1618-15
[18] *Відеоконференція – крок у майбутнє*. URL: https://zib.com.ua/ua/print/49182-videokonferenciya__krok_u_maybutne.html

The electronic petition as a way to influence the power available to every citizen: The role and prospects in Ukraine

Dmytro Luchenko

Yaroslav Mudryi National Law University, Kharkiv, Ukraine

Abstract. The article substantiates the thesis that electronic petitions are potentially capable of acting as an effective means of challenging illegal decisions, actions, and inactivity of state authorities. In this case, the institution of the electronic petition is a mechanism that creates comfortable conditions for expressing the will of citizens, regardless of their place of residence. The article analyzes the status of the legal regulation of the institute of the electronic petition in Ukraine and the practice of its functioning. The shortcomings of the legislation on electronic petitions that negatively influence the effectiveness of this institute are systematized. The study proposes a number of measures to improve the effectiveness of electronic petitions, namely: the implementation of educational activities among the population (especially rural areas) regarding the appointment and use of the institution of electronic petition; establishing of an imperative status for petitions that overcome the minimum of votes in their support; preventing the loss of votes by the same type of petition with identical appeals; blocking irrelevant petitions; improving the interface of relevant electronic resources on which petitions are published.

Key words: electronic petition, effectiveness, progressive institution, legislative regulation.

1 Introduction

Development of ICT (Information and Communication Technologies) has created the preconditions for providing a comfortable way of expressing the will of citizens, regardless of their place of residence. People living in villages have the same opportunities to directly influence the authorities as the residents of the capital through such an important institution of e-democracy as the electronic petition. Its value and potential in ensuring access of citizens to the management of state affairs is difficult to overestimate. All the necessary for this is access to the Internet, authorization and the desire to participate in solving the urgent problems of the state. Moreover, the institute of electronic petitions provides an opportunity to influence the authorities at any moment, and not only during the elections. These factors actualize the need for further research in the issue of the institute of electronic petitions in order to increase its effectiveness.

Despite the fact that E-petition is an extremely progressive institution in modern democracies, the practice of their functioning in Ukraine is not effective. On the one hand, Ukraine ranks 32nd in the Top 50 performers in e-participation in 2016, according to the United Nations E-Government Survey 2016 [1], but many researchers point to the threats and weaknesses functioning e-petitions in Ukraine. Volodymyr Reshota, for example, points out that the current regulation of electronic petitions can become a tool for manipulating and popularizing populist ideas that have nothing to do with the rule of law [2]. Anastasia Kostyantynivska also notes that in order to become truly effective, E-petitions should be regulated more transparently and openly, and citizens should exercise effective control over the petition process [3].

Thus, the aim of the study is to identify the weaknesses of the institute of E-petitions in Ukraine as well as to find ways of improving the legislative regulation of E-democracy in Ukraine. The specific task of the article is to search for specific proposals to improve the legislative and functional regulation of the filing and petition process. The aim of the proposed innovations should be to simplify the access of the population to petitions, as well as to impose additional obligations on officials on the consideration and response to petitions.

Accordingly, for the purposes of the article, quantitative and qualitative methods of scientific research will be used. A quantitative method will allow us to determine the statistical indicators of the use of E-petitions in Ukraine and, most importantly, to find out the effectiveness of this institution and its impact on meeting the needs of civil society. The qualitative method will allow finding solutions to the shortcomings of legislative regulation of E-petitions in Ukraine and, accordingly, improve this institution.

2 Problems and prospects of electronic petitions in Ukraine

In Ukraine, the procedure for filing and reviewing electronic petitions has been regulated at the legislative level by only one article (23-1) of the Law of Ukraine "On Citizens' Appeal" [4]. Part 1 of this article stipulates that citizens may apply to the President of Ukraine, the Verkhovna Rada of Ukraine, the Cabinet of Ministers of Ukraine, the local self-government body with electronic petitions through the official website of the body to which it is addressed or the website of the public association that carries out collection of signatures in support of electronic petition. The specified norm predetermines the peculiarity of the electronic petition as a form of citizens' appeal, which is submitted exclusively to the subjects of public authority, since the rest of the appeals provided by the Law may also be submitted, for example, to enterprises, institutions, and organizations irrespective of the forms of ownership.

Electronic petition as an element of E-democracy promotes: (1) strengthening participation, initiative and citizen engagement at the national, regional and local levels of public life; (2) improving transparency of the democratic decision-making process and accountability of democratic institutions; (3) improving the sensitivity/reciprocal reaction of authorities to citizens' appeals; (4) promotes public debate and attention of citizens to the decision-making process. Therefore, it can be stated that electronic petitions as one of the elements of e-democracy are means to improve access of citizens to the authorities and their communication with officials and elected politicians.

It should also be noted that the institute of electronic petitions has such quality as prevention of sharp social tensions that may transform into armed conflicts. The experience of Ukraine shows that if the authorities listened to the voice of the people (especially those who live on the periphery, in small towns and villages), and people had an effective mechanism to convey their own will to the authorities, there would be no Maidans, and no revolutions, during which people from all over the country came to Kyiv to express their dissatisfaction with the state policy.

Even Francis Fukuyama in his famous book, The End of History and the Last Man, pointed out that the model of a state with a liberal and democratic system is the most perfect form of social organization. The scientist explained this by the fact that only liberal democracies set maximum opportunities for controlling the state authorities by population. In liberal democracies, people can convey their views to state power and thus they feel their significance role and recognition. Totalitarian countries do not have such mechanisms that would ensure the recognition of ordinary citizens, so they all sooner or later fall [5]. Electronic petitions can be an effective mechanism to ensure that people's desires are brought to government.

Whereas, electronic petitions, on the one hand, allow the authorities to quickly receive information on issues that are socially acute and resolve them without transformation into critical tensions, and on the other hand, provide citizens with a comfortable, economical but effective way of expressing their own will. Electronic petitions allow citizens to feel that they are the source of power, to realize how they are involved in the formation of the state policy and the responsibility for it. However, this is possible only under two conditions: (1) the authorities want to listen to the voice of their citizens; (2) the state has established at the legislative level an effective and transparent mechanism for the implementation of electronic petitions and for its own responsibility.

The detailed analysis of legal acts that establish the procedure for the implementation of the citizens' right to electronic petition in Ukraine, as well as the whole array of electronic petitions and their responses, which were made public on the official sites of the President of Ukraine, the Verkhovna Rada of Ukraine and the Cabinet of Ministers of Ukraine during their operation, indicate the fact that the institute of electronic petitions in Ukraine, although implemented in July 2015, does not fulfil the task assigned.

A number of shortcomings in the legislative regulation of the institute of electronic petitions turned it into fiction, distorting the very idea of this element of E-democracy. Today, electronic petition in Ukraine is simply the usual appeal of the citizen to the subject of power, which has become public, but does not cause any obligations on the part of the addressee, except for the duty to respond promptly (we emphasize – to provide an answer, not to decide asked question).

Let us consider the weaknesses in the legislative regulation of the institute of electronic petitions in Ukraine. First of all, it should be noted that the legislation does not distinguish between electronic petitions, which are complaints, does not set any particulars for them in relation to registration and consideration. This makes it impossible to emphasize the authorities' attention to socially acute problems. Such level of regulation is insufficient for the creation of an effective tool for e-democracy. Moreover, the indication in Part 17 of Art. 231 of the Law of Ukraine "On Citizens' Appeal" about the possibility of elaboration by the subjects of consideration of electronic petitions their own procedural acts, which will regulate the procedure of work with petitions, creates the risks of possible abuses.

The object of our statistical analysis is the official website of the President of Ukraine. This resource not only works with electronic petitions longer than others but also is the most popular among citizens, including complaints of decisions, actions or inactivity of public authorities. August 29, 2018, was three years since the first electronic petition was made public on the official site of the President of Ukraine. As of August 29, 2018, citizens of Ukraine registered 30,467 petitions on through this Internet resource. The biggest share (16,716) was registered in the first four months (an average of 4179 petitions per month), indicating an extraordinary public interest in this progressive achievement of world democracy.

Subsequently, the activity of citizens in the use of electronic petitions has declined significantly. For example, in 2016, only 7759 petitions were registered (that is, an average of 646.5 petitions per month). And in the first eight months of 2018, 2852 petitions were registered (an average of 356.5 petitions per month). Thus, in 2018, the number of citizens who used the procedure for addressing the President of Ukraine through the electronic petition has decreased by almost 12 times compared to 2015. We cannot say that such a decrease was the result of the reduction of socially acute issues, which are traditionally the subjects of petitions. Unfortunately, the tendency of reduction of this form of appeal to public authorities is determined by the low level of effectiveness of the electronic petition's mechanism in Ukraine.

The analysis of the data, contained on the site of the President of Ukraine, gives the following figures, which reflect the real level of effectiveness of electronic petitions. Thus, as of the beginning of 2019, with over 30 thousand petitions filed with the President of Ukraine, only 46 petitions gathered 25 thousand votes in their support for a fixed term. 43 of them received a response from the President. Thus, the low percentage of petitions that overcame the 25,000 barriers (only 1 petition from nearly 700) has led to a decrease in the interest of Ukrainian citizens in the electronic petition as a way of solving socially acute issues.

In addition, a meaningful analysis of the responses to the petition to the President of Ukraine indicates that the vast majority of the applicants received formal answers of power and not real answers. Thus, among the 43 petitions that received the response of the President of Ukraine:

- in 3 cases it was refused to satisfy the petition;
- 4 electronic petitions are satisfied;
- 32 were redirected to other entities (the Cabinet of Ministers of Ukraine, the Constitutional Commission, etc.), but neither deadlines nor the obligation to report on the results of the consideration were mentioned;
- the answer to 4 petitions does not fully answers the question.

Thus, on average, only 1 electronic petition for 10 thousand registered on the site of the president of Ukraine reaches its result. The majority of petitions considered by the President are not rejected, but redirected for consideration by other subjects of power. We believe that such an answer is inappropriate. It is worthwhile to determine at the legislative level that one of the following responses to an electronic petition may be considered appropriate:

1. When the petition is substantially satisfied and the appropriate measures are taken, in particular, the complaint is considered or directed to the bodies that play the role of the institutes of the indirect appeal. In the case of the abolition of a decision of a particular collegial body, including the Parliament, it should be about the obligation of this body to urgently consider abolishing the relevant decision or taking measures to eliminate its consequences (for example, if it means the abolition of an act already executed).
2. When the petition is not satisfied, it is exhaustively explained why it is not satisfied. The clarification should be such as to maximize the conviction of the applicant and those who supported the petition that their satisfaction is impossible or inappropriate in the specific circumstances.
3. When, on the basis of the results of the consideration of the petition, an order is drawn up to another entity competent to resolve the issue, but information on the timing and results of its execution, the measures taken and the results obtained are posted on the site. This should provide the opportunity for anyone interested in this petition and the outcome of its consideration, to obtain comprehensive information directly in response to a petition. The answer cannot be exhausted by information that the question is redirected to someone, charged with understanding, studying opportunities, etc., which gives the impression of the incompleteness of the case.

In any case, the state authority considering the electronic petition should try to solve it as much as possible, as well as substantiate its response in order to exclude any ambiguity.

To be objective, it should be noted that the low level of support for electronic petitions is due to two other factors. The first is that not all citizens of the country, especially those living in rural areas, have the technical capacity to register their own or to support an already registered electronic petition. A significant number of people (mostly older) do not have any devices. According to Newzoo, in May 2017 only 23.5% of Ukrainians have modern smartphones. The rest have either a mobile phone or nothing at all [6]. In addition, the vast majority of Ukrainian citizens are not aware of the institute of electronic petitions, do not understand its significance, and do not have the skills to use it. So, the author of this article in 2017 visited the leading law schools and faculties in the country with lectures devoted to the problems of the electronic petitions. A survey conducted during this event showed that about half of the audience (including law students, practicing lawyers and teachers) were not familiar with this institution and did not understand its meaning. Of course, this indicator is much lower among ordinary citizens, and this is a deterrent to the disclosure of the capacity of the institute of electronic petitions.

The second factor in the low level of support for electronic petitions is due to the quality of the most registered petitions. This concerns both the nature of resolved issues, which often do not cause any interest from the overwhelming majority of citizens, as well as the quality of their text. Among the disadvantages of describing the most frequently encountered e-petition text, one should separately highlight the following as the most significant: inconsistency; incomprehensibility of the wording; lack of clear proposals or requirements; the presence of numerous grammatical errors.

We emphasize that the irresponsible attitude of citizens towards the formulation of an electronic petition, the submission of it without a need, for the sake of entertainment or own PR, severely damages this kind of appeal to public authorities. On the one hand, such a frivolous approach discredits both the applicants themselves and their legal institution, and on the other, no matter how senseless and pointless electronic petition, its representation is guaranteed to launch a certain mechanism that requires the use of state resources (means and time). In this regard, it would be advisable to establish the duty of citizens to refrain from abusing their right to electronic petition, and possibly to provide for the right of the state to recover from the court costs related to the administration of a pointless petition. However, it is necessary to do so in such a way that in no way does it restrict the right of citizens to submit substantiated electronic petitions (especially those containing a complaint against a decision, action or inaction of public authorities).

The next reason that affects the effectiveness of electronic petition as an appeal to the authorities and, accordingly, its popularity, is the non-binding nature of the petition gaining the required number of votes in support. The legislation established that an electronic petition, which received support for more than 25 thousand people in three months, is considered in a special procedure. However, if the petition does not receive the above support, it will also be considered, but in the manner prescribed by law for ordinary citizens' appeals (Part 10 of Article 231 of the Law of Ukraine "On Citizens' Appeal"). The question arises: what is the meaning of an electronic petition if it does not give the applicant and the citizens who have supported it any advantage in the form of an effective mechanism of influence on power? After all, the obligation of the addressee of the electronic petition to consider it according to special order is not the obligation to make every possible effort to solve the problem. *It is obvious that under the current order, the procedure for reviewing the electronic petition does not provide any advantages, except for the possibility to address the President of Ukraine and other authorities in a public way.*

This problem is critically important for the further functioning of the electronic petition mechanism. The international community considers this way of appeal as the most progressive. The mechanism of the electronic petition is capable of ensuring the direct impact of citizens on the authorities. Also, electronic petition is an instrument for the effective implementation of the rule of law, which, along with other institutions of a democratic society, makes it possible not only to appeal to higher authorities or local self-government bodies, but also to attract other people's attention to a certain problem, get their support and force the authorities to act. As a result, state authorities will act without being guided by their own discretion, but in the direction desired by society.

There is no doubt that the level of effectiveness of the institute of electronic petitions is an indicator of the maturity of democracy in the country. Ukraine still has much to work on improving the regulation of electronic petitions, since it is clear that the existing mechanism needs to be rethought and reformatted. The leverages of power should be created. Such leverages will force the authorities to pay attention to problems that really concern society, including in those situations where citizens seek to cancel an unlawful normative or individual legal act. As an option, electronic petitions should become imperative if they overcome the minimum level of signatures, that is, they should become a draft law, as it is envisaged in a number of countries. Such a procedure exists, for example, in Finland, where petitions, which have collected 50 thousand signatures, automatically become a draft law and the parliament is obliged to consider them as a priority. In addition, the regulations of the Latvian legislature stipulate that any petition, which has received support from 10,000 citizens, must necessarily be included in the agenda of the parliament. These are vivid examples that demonstrate the state's readiness to take into account the opinions of its citizens. Such a mechanism ensures equal opportunities for the inhabitants of small towns and villages to take an active part in public administration, to draw the attention of the authorities to the urgent problems of the regions and to demand the resolution of such crisis situations.

We disagree with the norm of Ukrainian legislation (Part 10 of Article 23-1 of the Law of Ukraine "About Citizens' Appeal"), which provides that a petition that has not gained 25 thousand votes is considered as a simple appeal by citizens. This is a misleading practice since a petition that has received at least one voice in support of it is already an appeal of several individuals. We are convinced that the response to such an electronic petition should be directed not only to the applicant but also to the person who gave his vote in support of the petition. Another option is to publish petition responses through the site where the petition was posted. Such a requirement should be fixed at the level of the law.

Moreover, a complaint about a decision, action or inaction of a public authority submitted through an electronic petition must obviously be considered in a separate procedure, unlike other types of appeals. This means that in this case the petition does not concern the proposal to improve the legal regulation, for example, certain social relations, but the need to restore the violated rights, to protect the person (a group of persons) from illegal misconduct. Therefore, the timing of consideration of such complaints should be reduced, and the results of the review should be subject to appeal in court.

The next reason for the low efficiency of electronic petitions in Ukraine is the lack of a mechanism for protecting against the loss of votes. The current procedure allows registering a petition even if a similar petition has already been registered before. On the other hand, voices for similar petitions are not united. For example, the requirement to reduce customs duties when importing into Ukraine used cars within three months (from June to August 2017) was put forward in the 17 registered on the site of the President of Ukraine electronic petitions. Another 7 petitions for the specified period have similar content. In addition, on the site of the Cabinet of Ministers of Ukraine, there were also 2 petitions registered with similar appeals at the specified time. That is, 26 electronic petitions on one issue simultaneously gathered votes for their support. As a result, none of the 26 petitions received the required number of votes, although the very number of them (petitions) should signal the authorities of the high degree of social acuteness of the issue. The result of the inaction of the authorities in resolving this issue was the numerous rallies and blockages of roads that took place in Ukraine in 2018.

The unfriendly interface of sites where the electronic petition is made public is also evidence of the fact that the state has no intention of providing the effectiveness of the institute of electronic petitions. In Ukraine, these sites not only do not allow the convenient handling of petitions, promoting them through social networks, but also provide extremely limited tools for analyzing and systematizing already registered petitions.

In our opinion, recommendations for improving the toolkits of sites where electronic petitions are made public may look like this:

1. A website with petitions should block the possibility of losing votes if such statements are identical and identical. There should be a ban on registration of a petition, similar to the one already registered, where the collection of signatures began. The person registering his own electronic petition, similar to the content of a petition that is already registered and gaining votes, the service must offer to support a petition already registered or to register its own, but after the expiration of the time remaining for collecting votes for an already registered petition.
2. Currently, it is possible to sort petitions by the topics (for example, agrarian policy, land relations, transport, communications, education, etc.), by chronology, by the number of votes. Also, everyone can elect petitions with responses from authorities, petitions currently under consideration and petitions that are in the process of collecting votes. In our opinion, the following should be additionally provided:
   a) search of electronic petitions by the date of registration;
   b) search of electronic petitions for the keywords not only in the title but also in the text of the petition;
   d) grouping of petitions according to the type of appeal (complaint, statement, proposal);
   e) message users using mobile communication (SMS) or e-mail about the registration of a petition that is similar to the one that they previously supported or registered.
3. The site should have information (hints) on what issues related to the mandate of the body to which the electronic petition is addressed, as well as how to correctly write a petition (the wording of the informative and clear title, the writing of the content and the definition of the nature of the appeal (suggestions or requests)).

The electronic service should block the ability to register the following petitions:

- petitions that go beyond the mandate of the authority to which the petition is sent (in this case, the user should be offered to go to the official website of the authority authorized to solve the issues raised in the electronic petition and provide a corresponding reference to the law and the corresponding electronic resource);
- petitions that do not contain any requests, proposals or complaints;
- petitions concerning the introduction, cancellation, and reduction of taxes and other mandatory payments, the number of social benefits and wages.

3 Conclusions

Thus, we are proposing to introduce a number of innovations aimed at improving the legislative regulation and functioning of the electronic petition institute, which is designed to provide easy and comfortable access for citizens to the management of public affairs, regardless of their place of residence. In addition, our proposals will enable the authorities to promptly receive information on the urgent problems of the regions.

We propose the introduction of various mechanisms aimed at increasing the impact of electronic petitions on state authorities, which will be implemented through granting petitions the status of imperative. Also, it seems necessary to implement a number of measures aimed at improving the tools of the sites on which electronic petitions are published.

Taking into account the above-mentioned, proposals will enable the creation of a truly effective electronic petition as an element of e-democracy, which is intended to initiate citizens (especially those living outside the capital) to participate in the formation of state policy, to ensure, in a civilized way, that they effectively and effectively bring to the consciousness of the government their actual expression of will and defend their rights.

References

[1] "Top 50 Performers in E-participation in 2016." United Nations E-Government Survey 2016 United Nations E-Government Survey, 2017, 197–217. doi: 10.18356/8e967554-en

[2] Reshota, V. "Electronic Petition as a New Tool for Appeal to Public Administration Bodies." Scientific Bulletin of the International Humanitarian University, 2015, 91–94. Available: `http://www.vestnik-pravo.mgu.od.ua/archive/juspradenc15/part_1/26.pdf`

[3] Konstantynivska, A.K. "Electronic Petition as an Effective Tool for Political Participation of Citizens." Epistemological Studies in Philosophy, Social and Political Sciences **1**(1–2), 88–95 (2017). doi: 10.15421/341809

[4] About Citizens' Appeal: Law of Ukraine dated 02.10.1996 No. 393/96-VR. Information from the Verkhovna Rada of Ukraine. 1996. No. 47. Art. 256 (with the following changes and replies)

[5] Fukuyama, F. The End of History and the Last Man. London: Penguin, 2012

[6] Hrytsyna, Mykhaylo. Poll: Top 50 Countries That Use Smartphones Most. Ukraine Is at Its Lowest Point in Europe. May 2017, Available: `pingvin.pro/gadgets/news-gadgets/opytuvannya-top-50-krayin-yaki-najbilshe-korystuyutsya-smartfonamy-ukrayina-na-najnyzhchomu-mistsi-v-yevropi.html`

SHS Web of Conferences **68**, 01020 (2019) https://doi.org/10.1051/shsconf/20196801020
Int. Conf. SOCIETY. HEALTH. WELFARE. 2018

Regulatory framework of strike and its problem in Latvia

Kitija Bite

Rīga Stradiņš University, Faculty of Law, Riga, Latvia

Abstract. International legal provisions provide for human rights and freedoms, and the freedom of expression and the right to work belong to these. Considering that during any employment relationship disputes can arise between the involved parties, international legal provisions state that a strike as the final means for the settlement of a dispute can be used. Paragraph 108 of the Satversme (the Constitution of Latvia) provides that in Latvia, employed people have the right to strike. Systematically, the provisions of the Constitution are being regulated by the Labour Dispute Law and the Strike Law. It might seem that in Latvia, any employed person has been entitled to the right to strike as provided by the Satversme. However, the strike of general practitioners in summer 2017 highlighted the problem of executing strikes. Firstly, at the time being, the right to strike can only be associated with one form of employment, i.e., employment relationship. As only a part of general practitioners is employed on the basis of an employment agreement, the strike regulatory framework that is in force in Latvia can be used only by a part of general practitioners employed under an employment agreement in order to protect their collective interests. Secondly, the Labour Dispute Law provides for that a strike as the final means can be used exclusively for the protection of collective interests (within the framework of concluding a collective agreement), but not within the framework of a contract governed by public law. The strike by general practitioners showed that Latvia has complied only partially with international legal provisions because a strike can only be used by people employed under employment agreements and only in disagreements regarding a Collective agreement. In order to resolve this problem so that any employed person is entitled to the right to strike in the future, it is necessary to amend the Labour Dispute Law by expanding the range of labour dispute subjects. The aim: to analyse international and Latvia's regulatory framework for the right of employed people to strike and recommend necessary amendments to laws to solve detected problems. Materials used: international legal provisions and Latvian legal acts, publications and literature. Methods used in this article: descriptive, analysis, synthesis, dogmatic, induction and deduction as well as legal interpretation methods – grammatical, systemic, historical and teleological.

Key words: labour dispute, strike, employment, strike of general practitioners, employment agreement, collective agreement.

1 Right to strike for employed people

The natural privilege by an individual is not only limited as to undertake any action for himself or herself and to be on their own, but also it is an ability to adjust their activities to their peers and combine their efforts with the efforts of their peers and to work together. Thus, the importance of the freedom of association with other peers is highlighted by the fact that the freedom of thought, opinion, and expression would be of low significance if it does not grant any opportunity to express thoughts, ideas, beliefs and views to their peers who have unified due to similar grounds and aims [1, 4].

The right to express his or her views and to protect legal interests in a democratic society belong to human rights, i.e., they belong to civil and political rights. In Latvia, these rights are included in several legal provisions of the Constitution: the right to freedom of expression, which includes the right to freely receive, keep and distribute information and to express his or her views [2, Paragraph 100]; the right to form and join associations, political parties and other public organisations [2, Paragraph 102]; the State shall protect the freedom of previously announced peaceful meetings, street processions, and pickets [2, Paragraph 103], and employed persons have the right to a collective agreement, and the right to strike [2, Paragraph 108]. However, neither of these rights is absolute – all these may be subject to restrictions in circumstances provided for by law in order to protect the rights of other people, the democratic structure of the State, and public safety, welfare and morals [2, Paragraph 116].

This right to express views included in the Constitution may be conditionally divided into individual rights (receiving information and expressing views) and collective rights (the right of association, the right to express his or her views in an organised manner during meetings, street processions and pickets, and the right to strike). Likewise, the right associated with expressing his or her views in the field of employment and beyond it must be distinguished. These rights are systematically arranged in the Constitution: Paragraph 100 as the right to the freedom of expression in its broadest meaning; Paragraph 102 as the right to assembly and the right to association; Paragraph 103 as the right to express his or her views in public. Separately, there is the right of employed people to express their views in public, which is the right to strike.

The State protects the right of employed people to strike, which, in legal understanding, is an action associated with the freedom of assembly and an action included in the scope of the execution of the employment rights. A strike is a means of resolving a collective dispute that manifests itself in such a way that employees or a group of employees of a branch of an undertaking voluntarily, completely or in part, discontinue work in order to attain the fulfilment of demands [3, Paragraph 3, Section 1]. It can be concluded that criteria required for a strike are the following: a collective interest dispute, voluntary action, work discontinued completely or partly and there are any demands, the fulfilment of which people on strike wish to attain.

The human rights determined in the Constitution are being regulated by separate laws on the freedom of assembly and the freedom of association, and these are being regulated by the special laws and regulations. The Law on Meetings, Processions, and Pickets is one of them [5], and this is the primary legal act, by which the State guarantees and protects the freedom of peaceful assembly. The freedom of assembly can manifest itself as meetings, processions and pickets. Other state-recognised ways of assembly include public entertainment, festivity, sports and other events, and a special regulation is set therefore – the Law on Safety of Public Entertainment and Festivity Events [6]. These legal provisions are associated with the freedom of assembly. In Latvia, the freedom of association is being executed according to state-specified types – political parties, trade unions, associations of religious organisations and foundations. Any of these types of the freedom of assembly is being regulated by specific legal provision.

The Law on Meetings, Processions, and Pickets allows society expressing their views about various socially important issues of political, economic and social character, whereas the Strike Law provides for the right to active action, which is a strike, associated with the protection of collective rights of employed people.

There is a conclusion resulting from the two legal provisions that meetings, processions and pickets are manifestations of the freedom of assembly associated with the freedom of expression and it determines that society is entitled to the right to demonstrate their position towards political, economic, and social issues. Whereas the Strike Law regulates the right of employees to discontinue work in order to attain the fulfilment of demands submitted to the employer. In Latvia, these two types of association and assembly are separated because each of them has their own restrictions and demands to be fulfilled so that the State could protect the freedom of assembly and the freedom of association guaranteed in the Constitution.

The Constitutional Court has acknowledged that the right to strike has been acknowledged as one of the fundamental rights of an individual and values of a democratic society. The right to strike is the constitutional right of employed people, and the State is obliged to ensure the implementation of this right [7, Clause 7.1]. The international legal provisions determine the right to strike. The International Covenant on Economic, Social and Cultural Rights provides for that member states undertake to promote the right to strike provided that it is exercised in conformity with the laws of the particular country [8, Claused, Article 8]. The European Social Charter provides for that states acknowledge the right of workers and employers to collective action in cases of conflicts of interest, including the right to strike, subject to obligations that might arise out of collective agreements workers previously entered into. [9, Paragraph 4, Article 6].

It unambiguously arises from these legal provisions that a member state in line with national legislation is obliged to set the right for employed people to a collective action, which is a strike.

Since 1998, when the Strike Law came into force, employees representing some sectors went on strike in Latvia, i.e., educators (in 1999), medical officers (in 2002), there was a warning strike across the railway sector (in 2001) as well as a strike of general practitioners (in 2017) [10].

Traditions of going on strike have neither been cultivated nor developed in Latvia compared with other countries [11], therefore legal framework is being applied rarely, and no problems concerning the interpretation of legal framework and utilisation in practice have been detected. However, a strike announced by Latvian general practitioners in 2017 clearly highlighted legal problems, which the author will explore in the further part of this article.

2 Regulatory framework of strike

The Labour Law provides for settling disputes in employment relationship [12, Chapter 8 "Deciding of Disputes"]; however, this regulation refers to individual employment agreements and does not provide for any right to strike. The Labour Dispute Law contains a reference describing a strike as a final means for the settlement of a dispute. The Strike Law provides for that a strike is a protest of a voluntary manner so that people would attain the fulfilment of their demands, discontinuing their work completely or partially [3, Clause 3, Sect. 1]. Systematically, this regulatory framework in general provides the substantive understanding of the term 'strike' – employed people may discontinue their work and go on strike in order to attain the fulfilment of their demands if none of the other types of resolving a labour dispute has led to a favourable result.

Similarly, law determines strict restrictions for the range of subjects who may announce a strike. And, these exclusively are trade unions. A trade union has been granted exclusive rights by the State: trade unions representing and protecting the labour, economic, social and professional interests of employed people, have the right to undertake collective negotiations, receive information and consult with employers, employer organisations and associations

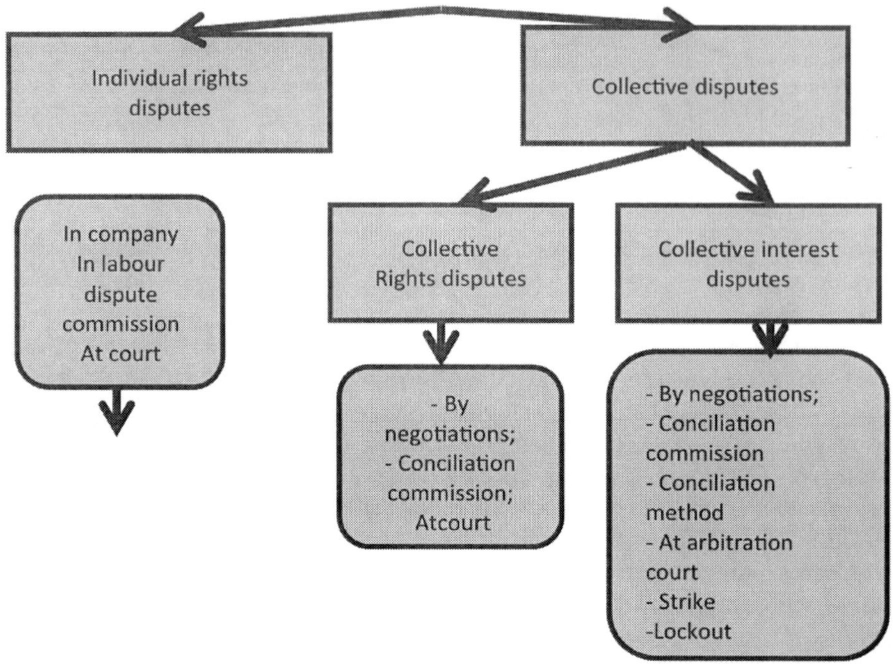

Fig. 1. Settlement of labour disputes (created by author).

thereof, to conclude collective agreement (general agreements), to announce strikes as well as to execute other rights determined in legal acts [13, Part I, Section 12].

It follows from this regulatory framework that trade unions represent labour, economic, social and professional interests of their members, and they have right to initiate bargaining and undertake negotiations with employers as well as to conclude a collective agreement. If bargaining brings no results, the State has determined the right to announce a strike as the final means for the settlement of disputes. The Law on Trade Unions is systematically adjusted to Paragraph 108 of the Constitution that determines that employed people have the right to a collective agreement and a strike. "The Labour Dispute Law and the Strike Law regulate the process of announcing a strike. The aim of the Strike Law is to correlate the right of employees to utilise a strike as a means for attaining the compliance of their interests and the fulfilment thereof with the interests of the employer. Employees are economically dependent on the employer, and there is an objective inequality between them, i.e., then employee is a weaker party in employment relations with the employer. Thus, the right to strike is an essential right of employees and their representatives in order to mitigate the inequality in powers and promote the observation of the employees' rights and well-grounded interests. A strike itself is not a fundamental instrument for attaining the observation of employees' interests during negotiations with the employers; it is a means of making the employer sit at a table and negotiate in cases when the efforts of the employees in achieving a compromise do not lead to any results." [14, 77].

The Labour Dispute Law divides all labour disputes into two big groups – individual disputes regarding rights and collective disputes and, in addition to it, collective disputes are divided into disputes regarding rights and disputes regarding interests (see Fig. 1).

It is a very important issue to understand who are entitled to the right to strike. i.e., who are subjects in labour disputes. The Labour Dispute Law names these subjects, and these are an employee, an employer or the representatives of either party. An employee is represented either by authorised representatives (trade unions) or representatives (representatives elected during the meetings of employees), but employers may be represented by an organisation of employers or associations of such organisations. The Law provides for another subject of labour disputes – the authority of the sector representatives [4, Part I, Section 2]. For example, the Ministry of Health represents medical officers, and the Ministry of Education and Science represents educators. Only the subjects specified in Section 2 of the Labour Dispute Law are entitled to use the right determined in this Law and these subjects have to execute liabilities, which have been determined.

From the Labour Dispute Law it results that labour disputes are feasible only within the framework of employment relationship [4, Section 2], but employment relationship is exclusively established by an employment agreement as a type of liabilities [12, Section 28]. Thus, the first legal problem is highlighted – in Latvia, the settlement of labour disputes has not been regulated, including the settlement with a strike, if an employment agreement between both the parties has not been concluded. And, this condition contradicts with the human rights guaranteed by the Constitution – everybody is entitled to the right to strike. There is no legal ground for restricting these human rights and determine by a law that only those employed people who have concluded employment agreements may go on strike.

Part I of Section 21 of the Labour Dispute Law and Part I of the Strike Law provide for that a strike is a collective dispute regarding interests. Thus, it has to be analysed what collective interests are, on which such a dispute may arise between the involved parties. Section 13 of the Labour Dispute Law determines subjects that may be involved in a collective dispute, and these are employees or a group of employees, or representatives of employees, whereas the other party is represented by an employer, a group of employers (2–4 employers), organisations of employers, associations of such organisations or an administrative authority of the relevant sector. The subject of the dispute is a procedure of collective bargaining, during which new working conditions or employment provisions are being set. It results from these two regulations that a collective interest dispute is a dispute arising from the involved parties determined by law about a procedure of concluding a collective agreement. And, this concerns issues discussed by the parties during bargaining.

Also Paragraph 108 of the Constitution shows that the right to strike has to be related to a collective agreement – employed people have the right to a collective agreement and the right to strike as well. The lawmaker has not separated these two rights in separate sentences or separate paragraphs; due to this, these legal provisions have to be interpreted systemically. Also Section 9 of the Labour Dispute Law defines the subject of the collective rights dispute: disputes that arise in the process of concluding, altering, terminating or fulfilling an employment agreement as well as in applying or interpreting provisions of regulatory enactments, provisions of a collective labour contract or working procedure regulations. This shows that a collective rights dispute concerns the fulfilment of effective collective agreement, interpretation and application, whereas the collective interest dispute concerns bargaining by the parties that are associated with concluding a collective agreement.

To summarise the above mentioned, collective interests are issues that the representatives of the employees wish to include into the provisions of the collective agreement (without considering that negotiations focus on altering an effective collective agreement or creating a new collective agreement) that improve the status of employees in addition to the provisions of the Labour Law or individual employment agreements.

Thus, in Latvia, the right to strike is legally regulated only for resolving those disputes concerning bargaining for concluding collective agreements that result from employment relationship. In other words, a collective interest dispute may arise only in companies in which an employment agreement is the basis for the liabilities of employed people and employers.

3 Practical problems of the right to strike in case of general practitioners' strike

The above mentioned two legal problems occurred practically in 2017 when general practitioners announced a strike. General practitioners have had negotiations with the Ministry of Health for several years about the increase in funding for providing the practice of general practitioners so that general practitioners might ensure the state-guaranteed accessibility of healthcare services for patients [12, 15].

Negotiations with the Ministry of Health were led by the Latvian Family Physicians Association (hereinafter – Association), which according to its legal status, is an association [16]. Despite the fact that the Association is a professional association representing general practitioners, this type of association cannot be considered a trade union, which is entitled to carry on negotiations about collective interests and conclude a collective agreement. As mentioned above, in Latvia, the types of association are legally distinguished – associations and foundations, trade unions, religious organisations and political parties. Each of these types of association is being regulated by a separate law, and they are registered with the Enterprise Register of the Republic of Latvia in separate registers.

On 26 May 2017, a decision to announce a strike of general practitioners was taken during the meeting of the Association members because negotiations held with the Ministry of Health and the government have not brought any results and there are no indications that this issue concerning the increase of the sate-provided funding for practicing general practitioners will be solved in the future in order to provide healthcare services for patients.

The strike actively took place from 3 June 2017 to 18 June 2017 and went on until the end of the month, in the so-called go – slow format. In 2018, a new Healthcare Funding Law [17] and the Regulations of the Cabinet of Ministers [18] were adopted. However, these resolve the fulfilment of the demands defined in the strike of the general practitioners only partially.

The strike of the general practitioners highlighted legal problems, and they have been neither solved, nor any discussions on resolving them have been initiated. The mentioned problems are as follows: regulation for strike concerning the types of employment that are beyond employment relationship and the representation of parties in negotiations covering binding issues regarding a collective agreement.

In order to understand employment issue, it has to be indicated that in Latvia, several types of employment are valid. There is a state service, in which the administrative act is the basis for the liabilities of the parties (for example, for police officers), according to which an officer is recruited in service and appointed to a post. Law is the other type of liabilities for a state service (for example, deputies, the officers of the Corruption Prevention and Combating Bureau of Latvia). Civil liability is being established by a contract (for example, an employment agreement, company contract, transport contract etc.). Similarly, a person may be employed as a sole trader according to a procedure determined in the Commercial Law [19].

Concerning the employment of general practitioners, there are several types thereof. Firstly, general practitioners can provide services individually in the general practitioners' workplace (as sole traders) or, secondly, conclude an employment agreement with a healthcare institution (a commercial company, for example, an outpatient healthcare institution), that will provide healthcare services for patients. In the first case, general practitioners receive remuneration from patients, but in the second case they receive remuneration from their employer, which is a healthcare institution that provides healthcare services.

Irrespective of the form of employment, a workplace of a general practitioner or a commercial company has to conclude a contract governed by public law with the National

Health Service [20, Clause 4.1] each year in order to provide a state-funded healthcare service. If a general practitioner does not conclude a contract with the State, a patient pays in full for services provided by a general practitioner.

The funding for general practitioners on the date of the strike of general practitioners was determined by Regulations No. 1529 by the Cabinet of Ministers "Funding Procedure for HealthCare" [21], which discontinued into force as of 1 January 2018. The funding determined by these Regulations was the basic demand of the general practitioners during that strike. A conclusion results from the above mentioned that only those general practitioners who have concluded an employment agreement with a healthcare institution are under employment relationship, and disputes arising from these general practitioners can be resolved according to a procedure specified in the Labour Dispute Law and the Strike Law. General practitioners did not undertake bargaining across the sector or concluding a general agreement, but they attained to achieve the increase in funding for healthcare.

The Association, not being a trade union (the representation of employees) could neither undertake negotiations with the Ministry of Health or the government within the sector, nor announce a strike under the existing regulatory framework. The Law on Trade Unions [Section 12 of the Law on Trade Unions] provides for that only trade unions are entitled to these activities. It has to be said that the Association has never indicated that it wished to conclude a collective agreement across the related sector, and instead, negotiations about bigger funding for healthcare were undertaken. Certainly, the Association as a professional authority for the representation of the members can resolve issues detected across the sector by negotiations, which is a direct proof of the freedom of association in a democratic society.

So that general practitioners could achieve amending the regulatory framework and changing the situation, they had to establish a trade union that would represent them in negotiations with the Ministry of Health about concluding a collective agreement across the sector; and in case of negative negotiations, if these parties might not come to any agreement about the provisions of the collective agreement and the balance of interests, a trade union exclusively is entitled to announce a strike. The second option for general practitioners was to announce a procession or a picket according to the procedure determined by the Law on Meetings, Processions, and Pickets and turn the attention of the whole society with the help of such a form of protest, to problems, which have grown bigger across the whole healthcare sector.

This situation clearly showed that resolving labour disputes in Latvia is neither well-considered nor it does cover all types of employment: the Labour Dispute Law and the Strike Law prescribe a settlement of problems regarding the parties involved in labour disputes and the object of disputes. As neither parties (not all the general practitioners are employees), nor the object (a collective agreement) complies with the Labour Dispute Law, the Latvian Family Physicians Association did not have the right to announce a strike. The Associations and Foundations Law does not give any empowerment to associations and foundations to negotiate about strike or to announce a strike.

4 Conclusion

The above mentioned situation, which involved general practitioners, highlighted the deficiency of the regulatory framework in Latvia so that it ensured the Constitutional and state-guaranteed rights of employed people to strike. The right to strike is given exclusively for the group of employed people which have concluded an employment agreement, and that does not comply with the principle of equality included in the Constitution. A strike can be announced only in cases when a trade union and employers cannot come to an agreement on collective interest disputes, thus this law allows trade unions utilising the final means, which is a strike.

In order to resolve this deficiency and so that society would be provided with the rights guaranteed by the Constitution, Section 2 of the Labour Dispute Law has to be amended, in which the concept 'labour dispute' has been defined. It is offered to define the 'labour dispute' as follows: a labour dispute is any dispute that arises from employment relationship or associated with employment relationship between an employed, employed people (a group of employed people) or representatives of employed people and an employer, employers (a group of employers), an organisation of employers or an association of such organisations, or the administrative authority of the sector.

Likewise, it is necessary to amend Section 9 of the Labour Dispute Law, by expressing it as follows: a collective rights dispute is a dispute between employed people (a group of employed people) or representatives of employed people and an employer, employers (a group of employers), an organisation of employers or an association of such organisations, or the administrative authority of the sector that arise in concluding, altering, terminating or fulfilling a collective agreement, as well as applying or interpreting provisions of laws and provisions of a collective agreement or working procedure regulations.

Expanding the subjects of employment relationship from employment relationship to legal relationship of being employed, the right to strike would be expanded to people that have other civil agreement or empowerment, a type of public service liability (law or an administrative act) as the basis of the liabilities, or would grant the right to strike to a person who is a self-employed.

Currently, it is not necessary to expand the range of objects that can announce a strike. These must be the authorised representatives (trade union) of employed people. Also, the right to strike in cases when the issue of collective interests is being resolved is sufficient in order to ensure the rights to the freedom of association and the right to strike for employed people as determined by the Constitution, but at the same time, a strike is not a means that can be utilised in all cases when the interests of the parties do not coincide.

References

[1] I. Spūle, A.Tiļļa, *Biedrošanās brīvība*, 80 / I.Spūle, A.Tiļļa, *Freedom of Association*, 80
[2] *The Constitution of the Republic of Latvija* (Satversme), Law of LV, LV **43** (1922)
[3] *Strike Law*, Law of LV, LV **130/131** (1998)
[4] *Labour Dispute Law*, Law of LV, LV **149** (2002)
[5] *Law On Meetings, Processions, and Pickets*, Law of LV, LV **31/32** (1997)
[6] *Law On Safety of Public Entertainment and Festivity Events*, Law of LV, LV **104** (2005)
[7] Decision by the Constitutional Court of 16 May 2007 on File No. 2006-42-01 "*About the Compliance of Third Part of Paragraph 24 of the Strike Law with Paragraph 108 of the Constitution of the Republic of Latvia*", LV **81** (2007)
[8] *International Covenant on Economic, Social and Cultural Rights* (1976). On Internet: `https://www.ohchr.org/en/professionalinterest/pages/cescr.aspx` (last seen 13.12.2018.)
[9] *European Social Charter*, LV **183** (2001)
[10] A. Katlaps, *Strike Movement in Latvia*. On internet: `http://providus.lv/article/streiku-kustiba-latvija` (last seen 13.12.2018)
[11] *Strike*. On internet: `https://www.lsm.lv/temas/streiks/` (last seen 13.12.2018)
[12] *Labour Law*, Law of LV, LV **105** (2001)
[13] Law on Trade Unions: LV Law. Latvijas Vēstnesis No. 60, 25.03.2014
[14] N. Mickeviča, K. Rācenājs, *Law on Trade Unions with Comments*, 160

[15] S. Olsena, *First Strike of Latvian General Practitioners: Practically, Legally, Illegitimately*, JV **41** 10–18 (2017)
[16] *Associations and Foundations Law,* Law of LV, LV **161** (2003)
[17] *Veselības aprūpes finansēšanas likums*, Law of LV, LV **259** (2017) / *Law on Healthcare Funding*, Law of LV, LV **259** (2017)
[18] *Procedure for Organisation and Remuneration for Healthcare Services*, Regulations of the Cabinet of Ministers No. 555, LV **176** (2018)
[19] *The Commercial Law*, Law of LV, LV **158/160** (2000)
[20] *Statute of the National Health Service,* Regulations of the Cabinet of Ministers No. 850, LV **178** (2011)
[21] *Procedure for Funding Healthcare,* Regulations of the Cabinet of Ministers No. 1529. Discontinued in force as of 1 January 2018

The right to private communication using telecommunication means: National and international legal aspects of protection

Olexandr Panasiuk*, *Larysa* Grynko, and *Anna* Prokhazka

Poltava Law Institute of Yaroslav Mudryi National Law University, Kharkiv, Ukraine

Abstract. Today's challenges dictate the need to strengthen the national and international legal mechanisms for the protection of personal data and the right to private communication. However, considered rights are not absolute. Legitimate restriction of guaranteed rights is possible, since these means of communication are a powerful tool in the investigation and disclosure of hard/very hard crimes, including transnational ones, especially considering the terrorist threats to Ukraine and other European countries. The possibility of restricting human rights, arising from the guarantees enshrined in the European Convention on Human Rights and consistently enshrined in the ECHR, demands from the state the least compulsory guarantee while interfering with the rights of individuals – to act "in accordance with the law". Law protection of personal data and right to privacy are researched in the context of peculiarities of conducting investigative (search), secret investigative (search) and other procedural actions in criminal proceedings, which concern access to some telecommunication means (e.g., smartphones). Taking into account different functional purposes of technical means of telecommunication, access and collecting of evidence contained therein, should be carried out on a case-to-case basis, in a different procedural form, considering specifics of telecommunication technologies in each particular case.

Key words: privacy, the secret of communication, telecommunication means, criminal proceedings, covert investigative (search) actions, due process, international law, smartphone.

1 Introduction

The protection of personal data in information and telecommunication networks/systems (telecommunication means), including using the Internet, today is one of the main tasks of the states, private institutions and the international community. Legal protection of personal data and privacy rights includes the following: constitutional law, international law, administrative law, and criminal and criminal procedural law.

Consequently, these rights on the national level are guaranteed by constitutions or special legislation. Also, today's challenges dictate the need to strengthen the international legal mechanisms for the protection of personal data and the right to private communication. However, considered rights are not absolute. Legitimate restriction of guaranteed rights is possible, since these means of communication are a powerful tool in the investigation and

* Corresponding author: alexpanasiuk.iur@gmail.com

disclosure of hard/very hard crimes, including transnational ones, especially considering the terrorist threats to Ukraine and other European countries.

General Data Protection Regulation (GDPR) (EU) 2016/679 (it. 4 of Preamble) states that "the processing of personal data should be designed to serve mankind. The right to the protection of personal data is not an absolute right; it must be considered in relation to its function in society and be balanced against other fundamental rights, in accordance with the principle of proportionality". At the same time GDPR (it. 19 of Preamble) should provide for the possibility for Member States under specific conditions to restrict by law certain obligations and rights, when such a restriction constitutes a necessary and proportionate measure in a democratic society to safeguard specific important interests, including public security and the prevention, investigation, detection or prosecution of criminal offences or the execution of criminal penalties, including the safeguarding against and the prevention of threats to public security. This is relevant, for example, in the framework of anti-money laundering or the activities of forensic laboratories [1].

Aim. The main objective of the research is to identify issues related with the legal regulation of the protection of personal data and the right to private communication, and to put forward suggestions for their solution, which are set out in the results of our study. Law protection of personal data and right to privacy are researched in the context of specifics of conducting investigative (search), covert investigative (search) and other procedural actions in criminal proceedings, which concern access to some telecommunication means.

Materials and methods. The national (Ukrainian and Latvian) and international (in EU area) legislatives, case-law of ECHR, judgements of Ukrainian courts in criminal cases and relevant legal literature are analyzed in the paper. In this research, a complex of general and special scientific methods of legal science (dialectical, comparative legal, analytical, descriptive, systemic-structural, generalizations etc.) is used.

2 Results and discussion

The stated idea of possible wrongful use, in particular of the Internet environment, for improper purposes, is consistently traced in Recommendation CM/Rec (2018) 2 of the Committee of Ministers to member States on the roles and responsibilities of internet intermediaries (Adopted by the Committee of Ministers on 7 March 2018at the 1309th meeting of the Ministers' Deputies). As specified in the Recommendation, inter alia, "the internet has facilitated an increase in privacy-related risks and infringements and has spurred the spread of certain forms of harassment, hatred and incitement to violence, in particular on the basis of gender, race and religion, which remain underreported and are rarely remedied or prosecuted". According to the Committee of Ministers of the Council of Europe, owing to the abuse, serious problems were encountered in connection with maintaining public order, national security, crime prevention, activities of law enforcement bodies, as well as the protection of other persons, including protection of intellectual property rights [2].

For example, Art. 8 Convention for the Protection of Human Rights and Fundamental Freedoms (Rome, 1950), along with the fact that it enshrines the right to respect for the right to private and family life, to own home and correspondence, also provides for the possibility of restricting it to clearly defined cases, such as "in accordance with the law and is necessary in a democratic society in the interests of national security, public safety or the economic well-being of the country, for the prevention of disorder or crime, for the protection of health or morals, or for the protection of the rights and freedoms of others". It follows from the practice of the European Court of Human Rights that interference by public authorities is possible not only when it is carried out "in accordance with the law", but also when it has a "legitimate purpose" and is "proportional" [3].

European Court of Human Rights (hereinafter, ECHR or European Court) also asserted that legitimate restriction of guaranteed rights "are tolerable under the Convention only in so far as strictly necessary for safeguarding the democratic institutions. Noting, however, that democratic societies nowadays find themselves threatened by highly sophisticated forms of espionage and by terrorism, with the result that the State must be able, in order to counter such threats effectively, to undertake the secret surveillance of subversive elements operating within its jurisdiction, the Court considered that the existence of some legislation granting powers of secret surveillance over the mail, post and telecommunications was, under exceptional conditions, necessary in a democratic society in the interests of national security and/or for the prevention of disorder or crime" [4].

All of the foregoing, taking into account the possibility, in exceptional cases, of limiting the right to privacy and private communication (as a part of the right to respect for private and family life), determines our further scientific research and analysis of legislative regulation of the presented set of problems in the criminal procedural context since nowadays, the means of criminal procedural influence and combating crime, including transnational one, are perhaps the only effective way to counteract these challenges of national security and public safety.

As Dita Plepa rightly specified in her paper, "security has been and is one of the most fundamental issues defining relations between the state and the citizen. Development of a contemporary democratic and judicial state is linked to the search for balance between the protection of constitutional values and respect for human rights" [5].

Consequently, as we see, the possibility of restricting human rights arising from the guarantees enshrined in the European Convention on Human Rights and consistently enshrined in the ECHR demands from the state the least compulsory guarantee while interfering in the rights of individuals – to act "in accordance with the law". Nowadays it is an indisputable thesis since the proper legal procedure enshrined in the law is an unconditional guarantee of enforcement of human rights and freedoms. Therefore, in our study, we will try to show that a serious problem, inter alia, for a number of European states, including Ukraine, is the imperfection of the legislative regulation of the grounds, conditions and procedure for legal interference of the state with these persons' rights, taking into account the factor of the rapid development of information and telecommunication technologies.

In this context, for example, the decision of the ECHR in the case *Liberty and Others v. the United Kingdom*, July 1, 2008, may be indicative. The applicants, a British and two Irish civil liberties' organisations, alleged that between 1990 and 1997 their telephone, facsimile, e-mail and data communications, including legally privileged and confidential information, were intercepted by an Electronic Test Facility operated by the British Ministry of Defence. They had lodged complaints with the Interception of Communications Tribunal, the Director of Public Prosecutions and the Investigatory Powers Tribunal to challenge the lawfulness of the alleged interception of their communications, but to no avail. The Court held that there had been a violation of Article 8 of the Convention. It did not consider that the domestic law at the relevant time indicated with sufficient clarity, so as to provide adequate protection against abuse of power, the scope or manner of exercise of the very wide discretion conferred on the authorities to intercept and examine external communications. In particular, it did not, as required by the Court's case-law, set out in a form accessible to the public any indication of the procedure to be followed for selecting for examination, sharing, storing and destroying intercepted material. The interference with the applicants' rights under Article 8 was not, therefore, "in accordance with the law" [4].

Consequently, as we see, the desire of the state to counteract serious challenges regarding unlawful and extremely dangerous criminal offenses against public security can, on the one hand, lead to abuses of the competent public authorities while investigating and detecting crimes and, accordingly, unlawful restriction or violation of human rights and freedoms, on the other hand.

And therefore, the law and the due legal procedure enshrined in it is a guarantee against the specified abuses. That is why we consider it expedient to analyze legal regulation of the procedural instruments (means) which enable to obtain evidence constituting personal data of a person, as well as data concerning the exchange and use of information that can be attributed to private communication. The legal analysis of our study will be based on the Criminal Procedure Code of Ukraine (hereinafter, the CPC of Ukraine) and, for example, the Criminal Procedure Law of Latvia (hereinafter, the CPL of Latvia) [6]. A number of procedural features of the criminal proceedings of these countries are very similar, which makes it possible to draw comparative parallels between the criminal procedures of these two countries.

Thus, the most effective and widespread practice of collecting evidence is, in particular, the following ways of collecting (obtaining) evidence: conducting investigative actions (both in Ukraine and in Latvia), covert investigative (search) actions (hereinafter, CISAs, in Ukraine; in Latvia, they are referred to as "specific investigative actions"), as well as special means of perpetuation of criminal proceedings (for example, temporary access to things and documents as well as temporary seizure of property) (concerning Ukraine; according to the CPL of Latvia, they are specific investigative actions). Those cases where the specified procedures in the criminal proceedings are carried out given access to telecommunication equipment, personal computers, mobile terminals of communication systems, other technical devices of processing and transmitting information, computers (including smartphones, tablets, PCs, laptops, etc., including using the Internet) are of particular interest. Interestingly enough that, according to the official data of the Register of radio electronic facilities and emitters that may be used in the territory of Ukraine in public service radio frequencies, as of June 12, 2018, there were more than 20.5 thousand specific types of radio electronic devices and emitters [7]. And these are only those ones the use of which is officially authorized by the state. According to individual estimates, by mid-2018, there were about 59% of smartphone users from the entire adult population in the world [8].

There is no doubt that information obtained as a result of communication of persons involved in criminal offences may be extremely valuable for determining circumstances surrounding a crime and can be used to detect crimes and proving guilt of persons in committing them.

The systematic approach to criminal procedural legislation both in Ukraine and in Latvia provides grounds for arguing that nowadays, there is no single unified procedural mechanism (algorithm) for accessing (collecting in the broadest sense) information contained or obtainable with the help of the indicated technical means of telecommunication. Moreover, law enforcement practice has also developed different and not always correct ways to address this issue. For example, currently, in Ukraine, the practice of detecting, recording and using information contained in mobile phones or smartphones by means of conducting such an investigative action as examination of the latter is quite common. In other cases, according to the CPC of Ukraine, such CISAs as information retrieval from transport telecommunication networks (hereinafter, IRTTN) or information retrieval from electronic information systems (hereinafter, IREIS) may be conducted. According to the CPL of Latvia, these are, respectively, specific investigative actions (hereinafter, SIA) as follows: monitoring means of communications, monitoring data contained in electronic information system or monitoring content of transmitted data (Chapter 11 of the CPL of Latvia) [6].

As it is supposed, such ambiguity is caused by the uncertainty of legal regulation of this issue, as well as by the very specific, unequal technical and legal nature of the above telecommunication means (devices) and information that is stored, processed and used with the help of the latter. For example, the mobile terminal of communication systems – "smartphone" - can be used as:

1) a means of communicating in real time (online), for example, for making telephone calls;
2) a means of access to electronic information systems using various forms of data transmission, including, as a rule, via the Internet (for example, for communication in various chats, social networks, namely, Facebook, Twitter, Instagram, Telegram, Snapchat, etc.; or for use of e-mail, including using the so-called "cloud-based technologies" ("cloud computing") for storing and processing information (for example, Google Drive or the like); or using so-called "messengers" such as Viber, WhatsApp, Skype, Facebook Messenger, Telegram, etc.). As the researchers point out, these telecommunication technologies represent network convergence supporting a wide range of access methods (traditional telephony, DSL, networks WLAN, RAN, etc.); at the level of service convergence during mobile communication sessions with the help of specialized software, mobile access to data, audio and video contraction, voice and instant messaging can be carried out. The widespread use of smartphones with installed programs that combine IP-telephony and instant messengers (Skype, Viber, etc.) or only instant messengers (ICQ, Telegram, WhatsApp, etc.) by mobile telecommunication subscribers form them as elements of distributed electronic information systems (hereinafter, DIS) [9]. DIS components are distributed, therefore, on several computers. In turn, DIS is composed of file-server information systems and client-server information systems. In the latter, for example, a database and database management system are located on the server, and client-side software is located on the workstations. As specified in the literature, both local and distributed electronic information systems can be open for the public and closed, i.e. access to which is limited to their owner, proprietor or holder, to obtain more information on the features of individual electronic information systems and IREIS [10]. At the same time one very significant feature should be mentioned. As a rule, information displayed on the device screen is not physically stored on it. It is stored in electronic information systems (information (automated) systems), on servers of the relevant companies [11] providing appropriate information and telecommunication services. In other words, in this case, a mobile device serves only as a means of access (a kind of "key") to information that constitutes communication content. Otherwise, if specified information is stored in the memory of the device itself, the features of the use of the latter will be described below, in the third group. In addition, it should also be noted that some of the mentioned messengers or social networks also allow real-time communication (online). Therefore, in such cases, the smartphone should be assigned to the previous group according to its functional (functional, technical and communication) purpose;
3) a means of storing and/or processing data (a variety of text, image, audio, video and other files and digital content). For example, audio, video, photo files, SMS messages and others sent and saved by a user can be stored in the device memory. Access to this information contained in a technical device can be gained either directly by an operating system of the latter or with the help of a specially-designed software, so-called applications (commonly known as "a mobile application" or just "an application").

As it is supposed, given the different functional purpose of the said mobile terminals of communication systems, access to the information contained therein should be carried out on a case-by-case basis in different procedural order, using diverse methods of collecting evidence, taking into account the above-mentioned features in each particular case. Thus, the due legal procedure of the procedural order, ways of obtaining evidence in the context of access to it using the specified mobile terminals of communication systems (technical means of telecommunication) should be carried out as follows.

A. In the first case, when a mobile device is used as a means of communication in real time (online), information constituting private communication content can be obtained by conducting such a CISA as IRTTN (or monitoring a means of communication, Art. 218 of the CPL of Latvia) [6a] since data transfer is carried out with the help of appropriate technical capabilities of transport telecommunication networks (communication channels). It should be borne in mind that when communication takes place in real time with the help of (by means of) software of a device that transmits data through social networks or similar online services, i.e. provides communication with electronic information systems (information (automated) systems) which are located on the servers of the respective companies, access to such information in the context of "penetration" into these systems has to be gained by means of such a CISA as IREIS (or monitoring contents of broadcasted data, Art. 220 of the CPL of Latvia) [6].

B. In the second case, when a smartphone (or other technical device) serves only as a means of access to information stored in electronic information systems and only displayed on the screen of a device, but physically not stored in it, obtaining and recording (copying) such information has to be done by means of conducting such a CISA as IREIS (such information is recognized as valid and admissible evidence in the court which is confirmed in the Ukrainian judicial practice. See, among others, for example, the judgement of Shevchenkivsky District Court in the city of Chernivtsi dated May 05, 2018 [12]), or monitoring data contained in an electronic information system, Art. 219 of the CPL of Latvia [6]. Therefore, we believe that other practice in these cases, e.g. obtaining information by means of usual examination of a technical device, i.e. as an investigative action of "examination", is illegal and ineligible. And, unfortunately, Ukrainian courts often acknowledge such materials with admissible evidence. See, for example: the judgement of Vasylkiv Town and District Court of Kyiv region dated December 04, 2015 [13]; the judgement of Avtozavodsky District Court of the city of Kremenchuk of Poltava region, dated August 13, 2018 [14]; the judgement of Podilsky District Court of the city of Kyiv dated November 07, 2016 [15], etc. That is why we fully agree with those courts of Ukraine which find such evidence inadmissible. For example: the judgement of Ordzhonikidzevsky District Court of the city of Mariupol of Donetsk region dated April 12, 2017 [16]; the judgement of the Lychakivsky District Court of the city of Lviv dated February 06, 2017 [17], etc.

C. Finally, in the third case, when a device is, virtually, a technical stored data carrier, the latter is characterized by all the features that are inherent in actual evidence or documents (in this case, these are electronic documents) in criminal proceedings. Therefore, only in this case access to the data stored in a technical device (a mobile terminal of communication systems), received and saved by a user (or automatically saved in this device), if the user is familiar with its content, can be obtained in such a procedural way as any other things or documents in criminal proceedings. In other words, namely, seizure of a mobile device and access to the information content in it can be gained in connection with a search, examining a housing or other person's possessions, searching a person during a search in a housing or other property (this applies to both criminal proceedings in Ukraine and Latvia), temporary access to things and documents, etc. (according to the CPC of Ukraine). Under the CPL of Latvia, it is also possible to carry out such investigative actions as seizure, self-initiated giving up of items and documents (if possible), storing and disclosing data contained (saved) in an electronic information system, giving up documents (Chapter 10 of the CPL of Latvia) [6]. Examination of information contents contained (stored) in a device seized in such a way, copying of the relevant electronic documents have to be carried out during examination of the device, which includes involving expert assistance (in case of the need to apply expert knowledge while examining a device or information contained therein, it is also possible to involve an expert and conduct an expert examination in accordance with the procedure established by the procedural law). This approach is also confirmed in numerous judgments of Ukraine. See, for example, the judgement of Sloviansk Town and District Court of Donetsk region dated November 19, 2015 [18]; the judgement of Zolotonosha Town Court of Cherkasy region dated September 23, 2015 [19]; the judgement of Vyshhorod District Court of Kyiv region dated May 17, 2013 [20], etc.

In our paper, while characterizing the procedural order for obtaining (collecting) evidence, as an example, a smartphone is used as a mobile terminal of communication systems (a technical means of communication, telecommunication device). However, according to the same principle, taking into account the similarity of technical and legal nature, similarity of algorithms for collecting, storing, processing, transmission and use of information, functional purpose, etc., the indicated methods of procedural order of access (collection) to evidence may be applied to other types of technical means (devices) of telecommunication (personal computers, tablets, laptops, etc.).

Consequently, taking into account different functional purposes of technical means of telecommunication, access to and collecting evidence contained therein should be carried out on a case-to-case basis, in a different procedural form, considering specifics of telecommunication technologies in each particular case.

Here is a typical example that can illustrate the problems at both the national (Ukrainian) and international (transnational) levels of legal regulation.

For example, in Ukraine, criminal proceedings are conducted against a person N. engaged in criminal activities who stores certain e-documents on an e-disk ("cloud storage"). The e-documents can greatly help to investigate the case and prove his/her guilt in committing the crime as in future they can be used as evidence in court if they are obtained lawfully. In addition, the person N. is in correspondence with his/her accomplices via social networks and e-mail. Thus, his/her technical means of communication (a smartphone, PC, etc.) are in Ukraine. The server of the company providing respective telecommunication services can physically be located both in Ukraine and (as a rule) in the territory of any country.

To obtain information contained on servers in other state, it is necessary to involve mechanisms of international legal cooperation within the framework of criminal proceedings, since powers of Ukrainian state authorities do not apply to regulation of activities of foreign telecommunication companies (taking into account sovereignty of states) and are not connected with these entities' corresponding responsibilities to comply with such instructions to grant access to the information contained on the servers of these companies (under Art. 263 of the Code of Criminal Procedure of Ukraine, competent state authorities of Ukraine have such a right and respective resident companies of Ukraine are obliged to cooperate with them). Experience shows that relevant procedures for international legal cooperation can take months or even years, which undoubtedly greatly reduces efficiency and expedition of pre-trial investigation. For instance, we can refer to the procedure of international legal cooperation, enshrined in the Agreement between Ukraine and the Republic of Latvia on Legal Assistance and Legal Relations in Civil, Family, Labour and Criminal Cases (signed on behalf of Ukraine on May 23, 1995, ratified on November 22, 1995, entered into force for Ukraine on July 12, 1996) [21].

In this context, for example, the Art. 3 of the CPL of Latvia may be indicative. According to P. 3 of Art. 3 of the CPL of Latvia, "other norms of national criminal procedure may be applied only in international cooperation on the basis of a motivated request of a foreign state, if this does not contradict the basic principles of Latvian criminal procedure" [6].

Interesting that the entry into force of the General Data Protection Regulation (GDPR) (EU) 2016/679 of 25.05.2018 improved the protection of personal data of all persons within the EU and the European Economic Area, but at the same time, from the point of view of legal regulation of the problem of access to personal data by law enforcement agencies, Directive (EU) 2016/680 of the European Parliament and of the Council, of 27 April 2016 "On the protection of natural persons with regard to the processing of personal data by competent authorities for the purposes of prevention, investigation, detection or prosecution of criminal offences or the execution of criminal penalties, and the free movement of such data and repealing Council Framework Decision 2008/977/JHA" is more specialized [22], along with Directive (EU) 2016/681 of the European Parliament and of the Council, of 27 April 2016 "On the use of passenger name record (PNR) data for the prevention, detection, investigation and prosecution of terrorist offences and serious crime" [23]. However, consistently emphasizing the protection of citizens' rights to privacy, the specified Directives also determine the direction of restricting this right in connection with the need to protect public and national interests from criminal infringements.

Therefore, at the level of international legal treaty regulation, it is necessary to envisage a specific universal mechanism in criminal procedural sphere (co-operation) which, under certain conditions, would provide legal reasons and real possibilities for authoritative bodies of one state to have a power to influence companies being under the jurisdiction of another state. Appropriate unification and harmonization of national legislation should take place on certain principles of the common framework of conventional international legal regulation.

And the EU Directives alone cannot solve the problem, given that there are non-EU countries. Such countries can easily become so-called "information offshore" providing opportunities to evade the itemized conventional IT-regulation.

Among other issues, despite the fact that The Regulation and Directives were passed and enforced, it is seriously worrying that there are still ongoing discussions in the EU countries on finding legitimate reasons for preventing the transmission of information spread over the Internet, e.g., via Skype and Viber beyond the scope of criminal proceedings (as is already the case in the United States) [24], which poses a threat to human rights.

3 Conclusions

On the basis of the research, the shortcomings of the legal regulation of the protection of personal data and the right to private communication at the national level, as well as inconsistency with international legal requirements and recommendations were revealed. This requires improvement of the legislative consolidation (in accordance with the requirements of the current level of development of telecommunication facilities), the bases, conditions and procedure for legal intervention of the state in the sphere of private life and communication; also bringing national legislation in line with international requirements and practices of international judicial institutions.

Taking into account different functional purposes of technical means of telecommunication, access to and collecting of evidence contained therein should be carried out on a case-to-case basis, in a different procedural form, considering specifics of telecommunication technologies in each particular case.

At the level of international legal treaty regulation, it is necessary to envisage a specific universal mechanism in criminal procedural sphere (co-operation) which, under certain conditions, would provide legal reasons and real possibilities for authoritative bodies of one state to have a power to influence companies being under the jurisdiction of another state. Appropriate unification and harmonization of national legislation should take place on certain principles of the common framework conventional international legal regulation.

The specified mechanism should be universal, efficient, operative, guaranteeing protection of human rights and freedoms against unlawful violation, as well as compensation for damage caused by unlawful restrictions (violations) of these rights.

References

[1] Available at: https://eur-lex.europa.eu/legal-content/EN/TXT/?qid=1549487787863 &uri=CELEX:32016R0679 (accessed on February 5, 2019)
[2] Recommendation CM/Rec (2018) 2 of the Committee of Ministers to member States on the roles and responsibilities of internet intermediaries (Adopted by the Committee of Ministers on 7 March 2018 at the 1309th meeting of the Ministers' Deputies). Available at: https://search.coe.int/cm/Pages/result_details.aspx?ObjectID=0900001680790e14 (accessed on February 5, 2019)
[3] See in particular item 133 of Judgement in Case of Benedik v. Slovenia, April 24, 2018. Available: at https://www.echr.com.ua/translation/sprava-benedik-proti-slovenii-povnij-tekst-rishennya/ (accessed on September 30, 2018)
[4] Case of *Klass and Others v. Germany*, September 6, 1978. The data is taken from: https://www.echr.coe.int/Documents/FS_Mass_surveillance_ENG.pdf (accessed on September 30, 2018)
[5] Dita Plepa. Līdzsvara meklējumi starp pamattiesību ievērošanu un valsts drošības aizsardzību Satversmes tiesas praksē. Socrates: RSU elektroniskais juridisko zinātnisko rakstu žurnāls. № 2 (5). P. 56–73. P. 70. (2016). Available at: https://www.rsu.lv/sites/default/files/imce/Dokumenti/izdevumi/socrates_5_2016.pdf (accessed on February 5, 2019)
[6] The Criminal Procedure Law of Latvia. Available at: https://likumi.lv/ta/id/107820 (accessed on February 5, 2019)
[7] The data is taken from the official web-portal of the National Commission for the State Regulation of Communications and Informatization. Available at: https://nkrzi.gov.ua/index.php?r=site/index&pg=59&id=4182&language=uk (accessed on September 30, 2018)
[8] See the research conducted by the analytical company Pew Research Center. Available at: http://www.pewglobal.org/2018/06/19/2-smartphone-ownership-on-the-rise-in-emerging-economies/pg_2018-06-19_global-tech_2-00/ (accessed on September 30, 2018)
[9] Шевчишен А. Можливості збирання доказів і розшуку при здійсненні окремих негласних слідчих (розшукових) дій у кримінальних провадженнях про корупційні злочини у сфері службової та професійної діяльності, пов'язаної з наданням публічних послуг. Право України. № С. 177–185. С. 180 (2016)
[10] Негласні слідчі (розшукові) дії та використання результатів оперативно-розшукової діяльності у кримінальному провадженні: Навчально-практичний посібник / Кудінов С. С., Шехавцов Р. М., Дроздов О. М., Гриненко С. О. Харків: Оберіг.344 с. С. 43 (2013)
[11] An exception to this is, for example, Viber and some others where information on the content of messages, sent files, etc. is physically contained (stored) in the memory of a technical device itself since, as the company claims, information content of messages is deleted from the servers of the company as soon as a message is sent to an end user. See more about Viber's privacy and security policy. Available at: https://www.viber.com/security (accessed on September 30, 2018)

[12] Available at: http://reyestr.court.gov.ua/Review/74052875 (accessed on September 30, 2018)

[13] Available at: http://reyestr.court.gov.ua/Review/54439563 (accessed on September 30, 2018)

[14] Available at: http://reyestr.court.gov.ua/Review/75827069 (accessed on September 30, 2018)

[15] Available at: http://reyestr.court.gov.ua/Review/62532485 (accessed on September 30, 2018)

[16] Available at: http://reyestr.court.gov.ua/Review/65990469 (accessed on September 30, 2018)

[17] Available at: http://reyestr.court.gov.ua/Review/64527701 (accessed on September 30, 2018)

[18] Available at: http://reyestr.court.gov.ua/Review/53596315 (accessed on September 30, 2018)

[19] Available at: http://reyestr.court.gov.ua/Review/51808318 (accessed on September 30, 2018)

[20] Available at: http://reyestr.court.gov.ua/Review/31379448 (accessed on September 30, 2018)

[21] Available at: https://zakon.rada.gov.ua/laws/show/428_627 (accessed on February 5, 2019)

[22] On the protection of natural persons with regard to the processing of personal data by competent authorities for the purposes of prevention, investigation, detection or prosecution of criminal offences or the execution of criminal penalties, and the free movement of such data and repealing Council Framework Decision 2008/977/JHA: Directive (EU) 2016/680 of the European Parliament and of the Council, of 27 April 2016. Available at: https:///eur-lex.europa.eu/legalcontent/en/TXT/%3Furi%3 DCELEX%253A32016L0680&prev=search (accessed on February 5, 2019)

[23] On the use of passenger name record (PNR) data for the prevention, detection, investigation and prosecution of terrorist offences and serious crime: Directive (EU) 2016/681 of the European Parliament and of the Council, of 27 April 2016. Available at: https://consilium.europa.eu/en/press/press-releases/2016/04/21-council-adopts-eu-pnr-directive/&prev=search (accessed on February 5, 2019)

[24] Золотар О. Інформаційна безпека людини: теорія і практика / Ольга Золотар: монографія. Київ: Тов. Видавничий дім «АртЕк». 446 с. С. 331–339 (2018)

SHS Web of Conferences **68**, 01022 (2019) https://doi.org/10.1051/shsconf/20196801022
Int. Conf. SOCIETY. HEALTH. WELFARE. 2018

Conditions for the legal commencement of an armed attack

Vita Upeniece

Rīga Stradiņš University, Riga, Latvia

Abstract. The Charter of the United Nations provides that: "Nothing in the present Charter shall impair the inherent right of individual or collective self-defence if an armed attack occurs…". In certain cases, it is difficult to find a clear answer to the question – what action is recognised as an armed attack according to the Article 51 of the UN Charter. The aim of the research is to analyze this problematic question, as well as the issue – when an armed attack begins, and as a result of the analyses of these issues, to describe the conditions for the legal use of force for self-defence.

Key words: armed attack, anticipatory self-defence.

1 Introduction

According to the Article 51 of the Charter of the United Nations the force may be used for self-defence against an armed attack. Some scholars recognize as lawful the use of force for self-defence in response to an armed attack which has not actually started. The others do not exclude the possibility of repelling an armed attack that has been launched, but not completed.

Therefore, the aim of the research is to find out what action is recognized as an armed attack, which is mentioned in the Article 51 of the UN Charter, and when such attack is considered to be started. Analysing the above-mentioned issues, by using historical, analytical, comparative and systematic methods, the author describes and analyses the main problems the international society is facing nowadays and explores the conditions for the legal use of force for self-defence.

2 Discussion

There is a point of view that the concept of an "armed attack" used in the Article 51 of the Charter has a narrower meaning than the notion of the "use or threat of force" embodied in the Article 2(4). Therefore, not all violations of the Article 2(4) can be recognized as an armed attack [1]. It is also alleged that an armed attack must reach a certain degree of gravity in order to be recognized as an "armed attack" within the meaning of the Article 51 of the Charter:

– The International Court of Justice emphasized the necessity to distinguish the gravest forms of the use of force (those constituting an armed attack) from the less grave forms and mentioned that an action would be classifies as an armed attack, if it achieved a certain scale and effect. The Court did not classify the mere frontier incidents, carried out by regular armed forces, as an armed attack, if it had not achieved a certain "scale and effect" [2]. At the same time the Court did not exclude "that the mining of a

single military vessel might be sufficient to bring into play the "inherent right of self-defence" [3].
- Some scholars base their opinion on the above-mentioned Court's decisions, and state that an armed attack constitutes the force which is used in a large scale and with essential effect [4]. However, a clear list of conditions, according to which the scale and effect of such use of force should be recognized as essential, is not provided.

Another group of scholars do not base on the assessment of the gravity of the offender's use of force in each certain case and on the definition of a hypothetical threshold of the gravity of its attack. Their view is based on the belief that the use of force by the offender triggers the right to self-defence, while the gravity of the offender's action affects the strength and means of defence (the principle of proportionality). So C.Greenwood notes that the International Court of Justice was probably concerned about the possible "opening the door" to excessive use of force in minor incidents and did the worst – it widened the gap between the use of force prohibited in the Article 2(4) of the Charter and the use of force which triggers the right to self-defence according to the Article 51. The Court highlighted that in cases where the use of force did not exceed "mere frontier incident", even if they constitute an infringement of the fourth paragraph of the Article 2 of the Charter, the victim state does not have the right to use self-defence. C. Greenwood rightly believes that there is no clear textual reason in the Charter of the United Nations to exclude such attacks, which are armed attacks but are below the threshold of the courts specified violence intensity. Moreover, the right to self-defence includes the requirement to respect the principles of necessity and proportionality, which are already prohibiting the excessive self-defence response [5].

Y. Dinstein admits that the Court's conclusion about the "mere frontier incidents" is a reason for thinking that in the cases where the state soldier fired over the border of another state hitting a tree or a cow is not considered as an armed attack. However, he emphasizes that it would be a mistake to automatically recognize all border incidents as "mere frontier incidents", because some of them may be insignificant, but others may be very serious. Many border incidents include a large military engagement and an attempt to separate them from other types of armed assaults would not be correct. Furthermore, the concept of "armed attack" means that armed attack is not legal, then any armed attack, even a small border incident, is illegal. Thus, there is no reason to remove armed attacks of a small scale from the spectrum of armed attacks. Moreover, the Article 51 of the Charter contains no restriction due to large, direct or significant attacks [6].

M.N. Shaw mentions that the Court had adopted a minimum approach to the meaning of the notion of the armed attack, which however requires some attention, because an attack can take place in different dimensions, considering current political and psychological conditions. The incident which in one context may seem relatively insignificant, may gain considerable significance in the other, creating the need to respond with the use of force in self-defence [7].

There is also a point of view that the right of states to defend themselves allows preventing armed attacks that would otherwise have occurred and to deter future armed attacks. Therefore, the Court's opinion needs some comments, because, otherwise it may be considered as a suggestion to restrict the right to self-defence, which will diminish rather than strengthen the international peace and security. The requirement that an attack reaches a certain level of gravity prior to the use of force in self-defence can make the use of force more frequent, because it can encourage the aggressor to engage in small-scale military attacks with the hope that these actions will not become the subject to the response for self-defence. Moreover, if a state is required to wait until the attacks reach a high level of gravity before using the force in response, the response will probably be much greater, making it more difficult to resolve the dispute which has turned into a full-scale military conflict [8].

There is no common point of view either between the states or in the legal doctrine about the time at which the right to self-defence arises [9]. Y. Dinstein points out that the sowing death, rather than the actual fire, is launching an armed attack. Instead of judging who has fired the first, he suggested to find out who has undertaken seemingly irreversible actions while crossing the Legal Rubicon. By analyzing three hypothetical Perl Harbor events scenarios, assuming, that the United States were clearly aware of Japanese plans, he concludes that as soon as the airplanes were launched, there would be no doubt that the United States as a target country, would have the right to consider the Japanese armed attack like started and to stop it. The similar situation would be with the Japanese navy interception. If Americans, perhaps by breaking Japanese naval codes, had got convincing evidences about the aim of the navy mission and determined the location of the Japanese fleet, they should not give up the possibility to intercept it. On the other hand, if the Americans tried to destroy the Japanese fleet before it began fulfilling its mission, that is, at a time when it participated in the military trainings for the mission, by playing the war, it would be preventive (illegal) use of force before the attack that has not yet occurred. According to Y.Dinstein trainings of warfare and previous preparations do not break the red line of an armed attack [10].

There is still no consensus about the use of force for self-defence when the radar of a missile is locked on a victim state plane. Some countries consider as the beginning of the attack the moment when the missile radar is locked on and is ready to fire. For example, the shooting of an Iraqi missile complex by the United States in 1998, because its radar was locked on the allied British planes which patrolled the no-fly zone. The United States pointed out that the Iraqi radar illuminated British planes and such action is an intention to shoot. Iraqi officials, in turn, indicated that no radar was open [11]. Thus, there was a dispute over whether the radar was actually locked on (as well as whether the planes were entitled to fly above Iraq), but the idea that the armed attack starts with the radar locked on was obviously accepted by Iraq and other countries [12].

There is an opinion that this issue relates to the problem of an imminence of the attack, though there may be circumstances where such situation just before the opening of fire should already be seen as part of an attack. In reality an armed attack can begin with an insignificant military movement (for example in an empty uninhabited zone) with the aim to deceive the future victim state before the main movement of the forces. Such victim state, correctly interpreting this initial movement, faces the question about the use of force for self-defence. Considering the rapid development of technologies, the question may be answered that this situation, or the situation when the radar is locked on an aircraft, can be considered as an imminent attack or even as the beginning of an attack [13]. In any case the state that sees itself as a victim, will be obliged to prove the fact of the beginning of an attack or the imminence of an attack.

As the International Court of Justice mentions in the case *Islamic Republic of Iran v. United States of America (2003)* the obligation to prove the existence of an attack rests on the country which has exercised the right of self – defence. [14] In the case *United Kingdom v. Albania* (1949) the Court emphasizes that the evidence may appear from the findings of the facts if they do not leave the place for reasonable doubt [15].

In the case *Nicaragua v. United States of America* (1986) the Court stated that the evidences should be adequate and direct. Examining whether the rebels acted on behalf of leadership of the United States, the Court indicated that despite subsidies and the other support provided by the United States there were no clear evidences that the United States had taken such level of control in all areas that would justify rebels' action on behalf of the United States [16]. Therefore the state's responsibility for rebels' actions can only arise in the case of the existence of evidences which the Court will find clear. So, the Court did not provide a clear solution to the cases where the third countries aid rebels by supplying them with weapons or by providing them with the other forms of assistance, and where the actions of these rebels endanger the territorial integrity of the victim state. Such an approach left the unresolved question – is a country that is the victim of other state supporting grouping attacks entitled to respond with a military force, if this support does not reach the threshold set by the Court [17]?

This issue was not also resolved in the case *Bosnia and Herzegovina v. Serbia and Montenegro* (2007). In this case the Court referred to the case *Nicaragua v. United States of America* (1986), when considering the issue of state control over the activities of non – governmental armed groups and pointed out that it must be established that a person or entity is considered as a State organ, even if it has not been granted this status under national law. It must be also established that it acted in accordance with the instructions of the state or under the "effective control" of the state. In addition, the Court considers essential to establish that this "effective control" was carried out or that the state instructions were given for each individual operation resulting in an alleged infringement [18].

At the same time the problem of obtaining evidences does not prevent the Court from trailing the case. In the case *Cameroon v. Nigeria* (1998) Nigeria responded to the initial allegations of unlawful use of force by claiming that the Court must decline to exercise its jurisdiction in the light of the ongoing nature of the conflict and the possible difficulties connected with it. Nigeria stated that the complexity of the situation was reflected in the fact that Cameroon's claims were vague and unfounded. These difficulties in obtaining evidences would make it impossible to guarantee a fair trial. The Court stated that the applicant should be responsible, if it substantiated the claim with inadequate facts and justifications. Therefore, the Court rejected Nigerian objections [19].

There is a growing consensus amongst both states and scholars that pre-emptive action for self – defence against an imminent attack may be lawful in strictly limited instances of the principles of imminence, necessity and proportionality which derive from Carolina case [20]. The point of view of the legal positivists that the apparently started armed attack may not be actually finished is partly justified. The scholars commonly base this view on the optimistic expectation that the potential aggressor can always withdraw the order to attack given to its armed forces or that launched missiles are guided and their course can be changed. However, it is difficult to believe that all states in all cases will be ready to wait until the attack actually takes place and causes damages. Doubts about the alleged attack can be justified by possible technical errors. In this case, after assessing all evidences in their context (namely, the incident is one; the guilty state has immediately provided the incident explanatory information or has expressed regret about what had happened; the suffered country has no information about the intention of the guilty state to attack) the suffered country's response with the use of force most likely would be seen as aggression.

There are legal positivists who have stepped aside from the narrow literal reading of the Article 51. On the one side they emphasize that the purpose of the United Nations is to maintain international peace and security and that they do not recognize the right of anticipatory self-defence. But on the other side, they do not exclude the possibility of extended interpretation of the Article 51, depending on the actual aspects of each individual case, if the state offers the United Nations convincing evidences of the impending attack, which it considers to be a justification for the use of armed force and proves that the pre-emptive attack is proportional to the threat and limited by the elimination of these threats [21].

The use of principle of imminence in the case of preventive self – defence is evaluated critically, because this principle is related to the question of the proximity of the anticipated attack to the moment of the use of self – defence. The further in future the attack is foreseen, the more opportunities are to prevent it by other means, without the use of military force. In any case, the potential victim state should obtain convincing evidences that none of the peaceful means of the attack prevention has succeeded and could not succeed in the future. The evidences of the planned attack will also be more convincing if they are close to the date of an attack. On the other hand, if the moment of the future attack is so close that it is no longer avoidable, then the self–defence from preventive becomes pre-emptive.

3 Conclusions

The need for achievement of a "certain threshold" may serve as a deterrent mechanism for the use of force. However, it is difficult to define precisely this threshold. Among other issues, it is important to consider the geographic features of the victim state. As M. Shaw has pointed out, Russia and China can absorb the initial attack and are able to regroup and attack in response even if they lose hundreds of miles of their territory. Other countries, which are not so well-geographically endowed, will not be able to do that [22]. The point of view that if this threshold is clearly defined, the offender will deliberately and purposefully use force that borders but does not violate it and, in this way, will disturb the peaceful existence of the victim state for a long time, is also correct. In such situations the cumulative approach provided by Y. Dinstein may be used. According to it a group of attacks is considered as one. However, such approach could provoke the rebukes about the failing to comply with the principle of proportionality [23]. In practice, this approach was used by Israel. But it was not accepted during the discussions in the Security Council [24]. It is also important to take into consideration the aim of the certain military action of the offender (the victim state should find an answer to the question – does the offender use the force against the territorial integrity or political independence of the victim state or is it just an unplanned incident, which is connected with technical mistakes and does not contain the intention to attack?). Thus, the right to self-defence should not primary depend on the severity of the force used by the offender. It should be based on the principles of necessity, imminence and proportionality. At the same time the severity of the armed action of the offender state should affect the nature and extent of self-defence activities applied within the framework of the principle of proportionality.

The diverging understanding of the criteria and conditions for anticipatory (preventive and pre-emptive) self – defence poses a greater threat to international peace and security then possible "opening the door" to pre-emptive self–defence in the regulatory framework. The hope that a bright case of anticipatory self– defence will be brought to the International Court of Justice and that it will not only assess the factual and legal circumstances of the case, but will set out a comprehensive set of conditions and criteria for the legal use of anticipatory self–defence may not be fulfilled. The present reality confirms the need for sufficiently clear legal framework for the use of force for self – defence before the victim state has actually suffered from an armed attack.

Thus, the use of force for pre-emptive self-defence could be based on the following conditions:

1) the existence of convincing evidences, proving soon completion of the preparations for the devastating attack or proving the attack itself at the moment of its actual commencement;

2) it is not possible to prevent the attack by other means and methods:
   - the victim state has used the peaceful methods of settlement of conflict available to it and these actions have not changed the offender's intention to attack;
   - the use of the other methods of peaceful settlement of the conflict is an obvious waste of time or there is so little time until the beginning of an attack that its use only for the methods of peaceful attack prevention is inappropriate, namely, there is a reasonable belief that these methods will not change the intent of the offender to attack. However, the victim state does not have to fully abandon the search for and the use of peaceful methods of resolving the conflict after the use of force for self–defence.

These conditions for the use of pre-emptive self–defence correspond to the criterion of imminence and necessity and include a requirement to make efforts to settle a dispute peacefully before the use of force.

In addition to the above-mentioned conditions the principle of proportionality should be obeyed, namely, the response should be at the same time proportional to:

- the means and the scale of an imminent attack;
- the purpose of defending the victim state.

References

[1] A. Randelzhofer, *The Charter of the United Nations*, a commentary, Edited by Bruno Simmma, Second edition, volume I, Oxford University Press, 796 (2002)

[2] International Court of Justice, Case concerning military and paramilitary activities in and against Nicaragua (Nicaragua v. United States of America) Merits, §191, 195, 228 (1986). Available at: `http://www.icj-cij.org/files/case-related/70/070-19860627-JUD-01-00-EN.pdf`

[3] International Court of Justice, Case concerning oil platforms (Islamic Republic of Iran v. United States of America), §72 (2003). Available at: `https://www.icj-cij.org/files/case-related/90/090-20031106-JUD-01-00-EN.pdf`

[4] A. Randelzhofer, *The Charter of the United Nations*, a commentary, edited by Bruno Simmma, second edition, volume I, Oxford University Press, 796 (2002)

[5] C. Greenwood, Self-Defence, Max Planck Encyclopedia of Public International Law, Oxford Public International Law (2011). Available at: `http://opil.ouplaw.com/view/10.1093/law:epil/9780199231690/law-9780199231690-e401?prd=EPIL`

[6] Y. Dinstein, *War, Aggression and Self – Defence*, New York: Cambridge University Press, Fourth edition, 195 (2005)

[7] E. Wilmshurst, *Principles of International Law on the use of force by states in self – defence*. ILP WP 05/01, Independent Royal Institute of international Affairs: Chatham House,18 (2005). Available at: `https://www.chathamhouse.org/publications/papers/view/108106`

[8] W.H. Taft IV., *Self-Defense and the Oil Platforms Decision*, Yale J. Internat. Law, **29**(2), article 3,295, 300 (2004). Available at: `http://digitalcommons.law.yale.edu/cgi/viewcontent.cgi?article=1232&context=yjil`

[9] A. Randelzhofer, *The Charter of the United Nations*, a commentary, edited by Bruno Simmma, Third edition, volume I, Oxford/New York: Oxford University Press, 1421 (2012)

[10] Y. Dinstein, *War, Aggression and Self – Defence*, New York: Cambridge University Press, fourth edition, 190–191 (2005)

[11] S. Schaffer, *U.S. jet fires on Iraqi site*. The Associated Press (1998). Available from: https://products.kitsapsun.com/archive/1998/07-01/0075_u_s__jet_fires_on_iraqi_site.html

[12] C. Gray, *International Law and the Use of Force*, New York: Oxford University Press, 96 (2000)

[13] E. Wilmshurst, *Principles of International Law on the use of force by states in self – defence*. ILP WP 05/01, Independent Royal Institute of international Affairs: Chatham House, 18 (2005). Available at: https://www.chathamhouse.org/publications/papers/view/108106; Shaw M.N., International Law, Fifth edition. United Kingdom: Cambridge University Press, 2004, p. 1027

[14] International Court of Justice, Case concerning oil platforms (Islamic Republic of Iran v. United States of America), §64, 72 (2003). Available at: https://www.icj-cij.org/files/case-related/90/090-20031106-JUD-01-00-EN.pdf

[15] International Court of Justice, Corfu channel case (United Kingdom v. Albania), §48 (1949). Available at: http://www.worldcourts.com/icj/eng/decisions/1949.04.09_corfu 1.htm

[16] International Court of Justice, Case concerning military and paramilitary activities in and against Nicaragua (Nicaragua v. United States of America), §109, 111 (1986). Available at: https://www.icj-cij.org/files/case-related/70/070-19860627-JUD-01-00-EN.pdf

[17] M.N. Shaw, *International Law*, fifth edition, United Kingdom: Cambridge University Press, 1026–1027 (2004)

[18] International Court of Justice, Case concerning application of the Convention on the prevention of the crime of genocide (Bosnia and Herzegovina v. Serbia and Montenegro), §398-401 (2007). Available at: http://www.icj-cij.org/files/case-related/91/091-20070226-JUD-01-00-EN.pdf

[19] International Court of Justice, Case Concerning the Land and Maritime Boundary between Cameroon and Nigeria (Cameroon v. Nigeria), §101–102 (1998). Available at: http://www.icj-cij.org/files/case-related/94/094-19980611-JUD-01-00-EN.pdf

[20] J.A. Green, *The rationetemporis elements of self–defence*, Journal on the Use of Force and International law **2**, 104–105 (2015). Available at: https://www.tandfonline.com/doi/abs/10.1080/20531702.2015.1043097

[21] A. Cassese, *International Law*, Oxford University Press, 357–362 (2005); S.S. Alexandrov, *Self-Defense Against the Use of Force in International Law*, The Hague/London/Boston: Kluwer Law International, 100–101 (1996); R. Värk, The Legal Framework of the Use of Armed Force Revisited, Baltic Security and Defence Review, **15**(1), 73 (2013)

[22] E. Wilmshurst, *Principles of International Law on the use of force by states in self – defence*. ILP WP 05/01, Independent Royal Institute of international Affairs: Chatham House, 48 (2005). Available at: https://www.chathamhouse.org/publications/papers/view/108106

[23] Y. Dinstein, *War, Aggression and Self – Defence*, New York: Cambridge University Press, Fourth edition, 230–231 (2005)

[24] C.J. Tams, The Use of Force against Terrorists, Eur. J. Internat. Law, **20**, 370 (2009) Available at: http://www.ejil.org/pdfs/20/2/1793.pdf

SHS Web of Conferences **68**, 01023 (2019) https://doi.org/10.1051/shsconf/20196801023
Int. Conf. SOCIETY. HEALTH. WELFARE. 2018

Legal status of the honorary consular officers: Theory and practice in Latvia

Janis Grasis

Rīga Stradiņš University, Faculty of Law, Riga, Latvia

Abstract. Honorary consuls are not professional diplomats, it means, they do not make a living as diplomats. They usually do live and work as well as pay taxes in the country where they operate while they do represent their native country on a voluntary/not-salaried/ basis until their appointments are revoked. Honorary consuls have limited authorization to act and conduct on behalf of their native country; they have the honour to serve their respected country and they are usually selected/appointed by their merits. Legal regime relating to honorary consular officers and consular posts headed by such officers is regulated by Vienna Convention on Consular Relations from 1963. According to the Article 59 of the mentioned convention the receiving State shall take such steps as may be necessary to protect the consular premises of a consular post headed by an honorary consular officer against any intrusion or damage and to prevent any disturbance of the peace of the consular post or impairment of its dignity. Similarly, the receiving State is under a duty to accord to an honorary consular officer such protection as may be required by reason of his official position. If criminal proceedings are instituted against an honorary consular officer, he must appear before the competent authorities. Comparative and analytical research methods are used for this paper. The author of this article proposes to introduce a special number sign for cars of the honorary consuls in Latvia, thereby showing special respect for foreign honorary consuls.

Key words: honorary consuls, Vienna Convention, protection of consular premises.

1 Material and methods

The aim of the article is to analyze legal norms from the Vienna convention and its practical application in the Republic of Latvia, especially concerning functional immunity of honorary consular officer. The research object of the article is international consular law. The tasks of the research are the following: 1) to make general characteristics of the status of honorary consuls; 2) to analyse court practise in Latvia relating to honorary consuls; 3) to work out suggestions to overcome shortcomings in the practise. Comparative and analytical research methods are used for this paper.

2 Results

The status of honorary consuls is rarely discussed in the books on international law. For example, M.Shaw looks at consular privileges and immunities, but honorary consuls are not mentioned at all [1]. A similar situation can also be found in compilation of articles of Oxford University [2]. The status of honorary consuls is analyzed in some articles, for example, in a very good article of K. Stringer [3], as well as by the very recent article by Janis Zelmenis [4].

The honorary consul does not receive any remuneration from his sending country and is usually an entrepreneur or public servant who has close links with the country to be represented. It is a serious advantage for a country that chooses honorary consuls to represent its citizens. Thus, very often in countries where it would not be economically beneficial to maintain diplomatic missions or career consuls, honorary consuls are appointed [3].

According to the data available on the home page of the Ministry of Foreign Affairs of the Republic of Latvia, 25 honorary consulates are operating in Latvia [5]. During collection of materials for this article, the author sent e-mails to all honorary consuls in Latvia to the all e-mail addresses indicated on our Ministry's home page. The 3 addresses did not exist; the fact which makes doubts on whether all the honorary consuls listed actually perform their functions. Latvia is also represented by 189 honorary consuls and 13 general consul general abroad (data from August, 2017) [5]. The honorary consul or general consul of the Republic of Latvia shall be a person to whom the Republic of Latvia entrusts the performance of state representative and consular functions in accordance with the 1963 Vienna Convention [6], international customs, consular conventions concluded by the Republic of Latvia and other international treaties in force in the Republic of Latvia, the Diplomatic and Consular Service Law and other regulatory enactments. The honorary consul may be both a citizen of the Republic of Latvia and a foreign state; the honorary consul must have an excellent reputation. The honorary consul is not an official and is not an employee of the Ministry of Foreign Affairs of the Republic of Latvia.

In most European countries, honorary consuls in the particular country in question create professional associations. In Latvia, since January 2015, there has been an association – the Honorary Consular Corps in Latvia, which brings together more than ten of the 25 honorary consuls accredited in Latvia [7].

In accordance with subparagraph 2 of the Article 3 of the Law on Consular Assistance and Consular Services [8], consular functions shall be performed by civil servants and employees of the diplomatic and consular service authorised thereto, as well as by honorary consuls of Latvia authorised thereto. Career consuls shall be paid officials of the State, and honorary consuls shall be appointed by the Minister of Foreign Affairs persons worthy of such office who are not remunerated from State resources. Consular missions led by honorary consulates shall be called honorary consulates. As a remuneration for expenses related to the maintenance of the honorary consulate, the Ministry of Foreign Affairs may leave at the disposal of the honorary consul either a certain part of the consular revenues or all of them. The honorary consul shall perform separate consular functions on the basis of a special mandate from the Minister for Foreign Affairs (State Secretary of the Ministry of Foreign Affairs). This mandate can be very broad-based, similar to that of a career consul, or very narrow-based, symbolic. This is determined by the specific interests of the country and by the agreement of the honorary consul himself; customs law and bilateral consular agreements between states can also play an important role. Honorary consuls generally do not have permission to issue visas or passports; as a rule, the honorary consul cannot perform functions where the use of public authority is required, for example, to carry registers or perform notary functions [9]. As a general rule, honorary consuls have two main tasks:

1) development of economic, scientific and cultural relations between countries (in particular, promoting commerce in both directions),
2) protection of the interests of their sending State and their citizens.

However, the functions of the honorary consul may be extended or reduced as necessary [3]. For example, in accordance with the Article 17 of the Law on Consular Assistance and Consular Services the consular officers abroad and honorary consuls shall perform individual notary functions in accordance with the Notary Law [10]. Notary deeds drawn up by consular officers and honorary consuls shall constitute the same public credibility and have the same legal consequences as they would if such deeds were drawn up by a sworn notary in Latvia [8].

Without the pursuit of classical consular functions, the institute of honorary consul, in good cooperation with entrepreneurs, has created, creates and will continue to create success stories in business. According to the chairman of the board of the Honorary Consular Corps in Latvia and the honorary consul of Malta in Latvia Janis Zelmenis, hotel "Riga" turned into a five-star hotel thanks to the honorary consul: "Nothing happens in business if you don't know a person. Latvia has a distinguished honorary consul in Armenia. Any businessman who has decided to start in Armenia shall be advised to meet him. This consul has excellent contacts and excellent information on both Latvia and Armenia. Question: what does the honorary consul in Armenia has common with a €50 million investment in Latvia? Answer: Our honorary consul told one wealthy Armenian he had to buy the shabby hotel "Riga" for €25 million. The wealthy Armenian trusted to the opinion of honorary consul, bought the hotel, took credit from one of Latvia's commercial banks, carried out the hotel renovation, attracted it to the Kempinsky chain, and now the hotel works as an excellent five-star hotel" [7].

In general, if we compare with diplomatic immunity, the total immunity of honorary consuls and consulates is less, but there is no difference of the immunity performing consular functions. According to the article 59 of the 1963 Vienna Convention on Consular Relations, the receiving State shall take such steps as may be necessary to protect the consular premises of a consular post headed by an honorary consular officer against any intrusion or damage and to prevent any disturbance of the peace of the consular post or impairment of its dignity. The term "Consular premises" is explained in the subsection j) of the first part of the article 1 of the 1963 Vienna Convention on Consular Relations, and means "the buildings or parts of buildings and the land ancillary thereto, irrespective of ownership, used exclusively for the purposes of the consular post"; besides "The consular premises shall not be used in any manner incompatible with the exercise of consular functions (Part 2 of the article 55 of the 1963 Vienna Convention on Consular Relations).

There are not too many cases in Latvian case-law regarding honorary consuls, but individual cases are interesting enough. One of these is an administrative case for a domestic dispute between husbands at the Indonesian honorary consulate, where the administrative case was launched against the Indonesian honorary consul in Latvia/pers. B/, which was largely dominated by the question of the immunity of consular premises [11].

According to the information provided by the honorary consul, his spouse/pers. D/provoked /pers. B/to the dispute on June 10, 2014 at the consular premises/address/-95 of the Indonesian honorary consulate, in which/pers. D/had also declared. After/pers. D/sneered on /pers. B/, he took away the computer which his wife had tried to retrieve. About the circumstances of the dispute/pers. B/explained that/pers. D/ was not very stable after alcohol use; /pers. B/ presumed that she had slipped and slipped into the big marriage bed. No violence was done by /pers. B/, so there is no clear reason for minor bodily injury because the bed has no sharp edges or places to get bruises. They may have been obtained under other circumstances. Also to the police he said that never have pushed/pers. D/. In addition, after the incident, they both slept at night in one bed and there was no fear of doing it for /pers. D/. After this incident there has also been another case in which /pers. D/ tried to provoke a dispute to person /pers. B/, but it failed due to the peaceful action. By /pers. B/. Meanwhile, the testimony of the injured woman differed over the events of the evening of June 10, 2014 at Indonesia's honorary consulate. She told her husband that she wanted to divorce and divide the property because she suspected her husband in adultery since 2008.

She has obtained this information from the computer; /pers. B/ took his wife's computer and threatened to throw it out of the window. Then he put the computer on the bed and pushed her into the bed as she passed, as a result. /pers. D/ breathed to the tallest end of the bed. The head of /pers. D/ was against the mattress, then/pers. B/gripped it and started choking, then /pers. B/ somewhere gone, according to her opinion, after links to tie her up, as he had done it before. /Pers. D/ had enough time to call her mother, and when he returned to the room, there was a call from her mother, after which /pers. B/ wife had not been more physically affected. The most important aspect of this violent dispute, in which the Indonesian honorary consul and his spouse participated, is whether the premises of honorary consul, for which information can be found on the home page of the Ministry of Foreign Affairs, and in which a country's consulate is officially registered, enjoy consular immunity. This was specifically indicated by defender of /pers. B/ Oscar Galander, stressing that the premises in which a honorary consulate is officially registered, enjoy consular immunity. Criminal Justice Board of Riga Regional Court acknowledged that/pers. B/ as the Indonesian honorary consul, as well as the premises in which he performs the role of honorary consul, do not enjoy consular immunity in respect of the administrative violation to be examined. The Criminal Justice Board took into account a definition from the subsection j) of the first part of the article 1 of the 1963 Vienna Convention on Consular Relations that "consular premises" means the buildings or parts of buildings and the land ancillary thereto, irrespective of ownership, used exclusively for the purposes of the consular post; as well as second part of the article 55, which determines that "The consular premises shall not be used in any manner incompatible with the exercise of consular functions. It follows from the files of the case that the apartment by address/address A/which/pers. B/has designated as consular premises, also is declared place of residence of his wife/pers. D/, which is also their actual place of co-residence. In his explanatory notes of 8 July 2014 /pers. B/ stated that during the conflict the examination works were corrected, and we could presume that the premises were not used solely for the purposes of the consular authority. Article 59 of the 1963 Vienna Convention on Consular Relations provides, hat the receiving State shall take such steps as may be necessary to protect the consular premises of a consular post headed by an honorary consular officer against any intrusion or damage and to prevent any disturbance of the peace of the consular post or impairment of its dignity. In the present case, it was found that the actual actions from/pers. D/ in connection with /pers. B/shall not be recognised as intrusion or damage to the consular premises of the consular post, and that action by /pers. D/ has not resulted in a disturbance of peace or a disturbance of the peace to the consular authority within the meaning of the above article of 1963 Vienna Convention, but in a violent dispute case between /pers. B/ and /pers. D/ in their residence and as a result of this dispute wife has suffered minor bodily harm from /pers. B/. It was concluded that the provisions of Article 59 of the 1963 Vienna Convention on the Protection of Consular Relations are not applicable in this case and therefore also indication of /pers. B/ on the obligation of the state of Latvia to take protective measures of the consular authority from peace disturbances are not justified. The author agrees in general with the court's arguments in this case.

The receiving State is under a duty to accord to an honorary consular officer such protection as may be required by reason of his official position (Article 64 of the 1963 Vienna Convention on Consular Relations). However, it should be taken into account that privileges and immunities provided in the 1963 Vienna Convention shall not be accorded to members of the family of an honorary consular officer or of a consular employee employed at a consular post headed by an honorary consular officer(third part of the Article 58 of the 1963 Vienna Convention).Article 61 of the 1963 Vienna Convention, by analogy with career consuls, prescribes that the consular archives and documents of a consular post headed by an honorary consular officer shall be inviolable at all times and wherever they may be, provided that they are kept separate from other papers and documents and, in particular, from the private correspondence of the head of a consular post and of any person working with him, and from the materials, books or documents relating to their profession or trade.

If criminal proceedings are instituted against an honorary consular officer, he must appear before the competent authorities. Nevertheless, the proceedings shall be conducted with the respect due to him by reason of his official position and, except when he is under arrest or detention, in a manner which will hamper the exercise of consular functions as little as possible. When it has become necessary to detain an honorary consular officer, the proceedings against him shall be instituted with the minimum of delay. Also, Article 119 of the Criminal Procedure Law of the Republic of Latvia, which guarantees consular immunity in conformity with international treaties, applies to honorary consuls. The right of a person to consular immunity shall be attested by a certificate issued by the Ministry of Foreign Affairs, in which the privileges and immunities of the relevant person are indicated in accordance with international agreements entered into by the Republic of Latvia. If there is a need to detain an honorary consular officer, proceedings against him should be initiated as soon as possible (Article 63 of the 1963 Vienna Convention). Article 63 of the 1963 Vienna Convention does not provide for the immunity of the honorary consul against administrative infringements, but it should be noted that the honorary consul also enjoys functional immunity in the performance of his duties. In practice it is not so easy to draw the border when a person performs the duties of honorary consul and when he does not enjoy functional immunity. This is well characterised by the following example from Latvian case-law.

On 21 November 2015, R. Gabalis, senior inspector of the local municipal police of Riga, compiled an administrative infringement report – notification No 511249554 (hereinafter – initial decision), about the fact, that on 21 November 2015 at 11.25 vehicle Mercedes Benz, with/number/parked outside the building in Riga, Meldru Street 1a, in operation area of traffic sign no. 532 and secondary traffic sign no. 833 without time indication, thus failing to comply with the requirements of the article 4., 307.17 of the Road Traffic Regulations [12] (hereinafter – RTR), thus committing an administrative violation for which responsibility is provided in the part 7 of the article 149.10 of the Latvian Administrative Violations Code [13] (hereinafter - LAVC). Administrative fine of Euro 30,- was imposed by initial decision of an administrative infringement report – notification. Decision was appealed to a higher institution by honorary consul – /pers. A/, later in the Vidzeme suburb court of Riga City, which left the court decision of the first instance unchanged. After the judgment of the Vidzeme suburb court of Riga City /pers. A/ (hereinafter referred to as "the Applicant") lodged an appeal in which the Applicant appealed the judgment in full. The Applicant pointed out that at the time the vehicle was parked there was a municipal police vehicle in front of it, where police officers were present and they saw him parking the vehicle. According to his view, the activities of police officers, i.e. immediately drawing up an administrative infringement report – notification without waiting for exceeding 1 hour, are deemed to be a dishonest or even abusive exercise of powers. It is believed that the time mark of parking was formal because police officers saw the moment when /pers. A/ left the vehicle after parking. The Applicant considered that there was no administrative violation in the particular case, since he was not considered as the honorary consul to be the subject of it. The court shall be deemed not to have provided its assessment and justification for when consular functions are or are not performed, but merely referred to a letter from the State Protocol which cannot provide the court with an analysis of binding legal norms. Moreover, it was noted that Article 5 of the 1963 Vienna Convention lists activities (for example, ascertaining by all lawful means conditions and developments in the commercial, economic, cultural and scientific life of the receiving State, reporting thereon to the Government of the sending State and giving information to persons interested) which are considered to be consular functions and cannot be carried out without representative means, and their acquisition is therefore considered to be an integral part of the exercise of consular functions. He has also had no intention of committing an administrative violation, since the vehicle has been parked for 20 minutes and in fact no administrative violation has been committed. The Applicant deems to be absurd the situation when police and court try to divide the time when consular functions are carried out and when not; an opinion of dr. iur. Inese Kalnina was also submitted on the immunity of honorary consuls: under the circumstances, his prosecution for administrative violation should not be accepted, as the honorary consul at the time of the events identified in the particular case was granted consular immunity. Representative of local municipal police of Riga Girts Aleksans explained that he did not agree with the Applicant's claim that he had fulfilled the duties of honorary consul of Malta since the Ministry of Foreign Affairs of the Republic of Latvia has indicated that doing shopping does not constitute an activity to be performed within the consular functions. The Court of Appeal held that, pursuant to the first part of Article 43 of the 1963 Vienna Convention, consular officers and consular employees shall not be amenable to the jurisdiction of the judicial or administrative authorities of the receiving State in respect of acts performed in the exercise of consular functions. Article 5 of the 1963 Vienna Convention provides that consular functions are to promote the development of commercial, economic, cultural and scientific relations between the sending State and the host State and to otherwise promote friendly relations between them in accordance with the provisions of this Convention. It is therefore concluded that consular officials have immunity only at the time when they perform their consular functions directly. The acquisition of means of representation does not constitute a consular function but serves as a means to enable these functions to be performed at a later stage [14]. However, the author of this article does not want to agree with the opinion of the local municipal police of Riga, the Ministry of Foreign Affairs of the Republic of Latvia and the decision of the Riga Regional Court. In this particular case the actual circumstances of the case had to be taken into account: the honorary consul had not parked the car in an unauthorised place, police officers of the local municipal police of Riga saw the honorary consul parking the car, and in fact punished not for being misplaced of the car, but for not having a time indication in the window. As mentioned above, even in criminal cases, the honorary consul must be treated with respect due to his official status; in this particular case, there is an administrative violation. Unfortunately, contrary to diplomats, the vehicle registration rules do not provide for special registration signs for consuls and honorary consuls in Latvia. Informally, honorary consuls are given number signs marks with the letters "CH", but the background of the sign is the same as the normal signs, which makes it more difficult to distinguish between such numbers in practice. In many foreign states, special number signs are also given to honorary consuls to make it easier for national law enforcement authorities to identify honorary consuls in their country. For example, the following photograph contains the special car sign of the Latvian honorary consul in the state of Karnataka in India (Fig. 1).

The author of the article considers that special number signs for honorary consul cars should also be introduced in Latvia. As for honorary consuls, it could be left to 'CH', and additionally letters indicating the country what they represent, and, like diplomats, at least on a red background.

Among other examples, when Latvia infringes the provisions of the 1963 Vienna Convention, the situation with the Mexican honorary consul in Latvia should be noted. He had to have hard fight to meet in prison with a detained citizen of his country. This means that not all employees of the Latvian Internal Affairs ministry have a clear view of what honorary consuls are dealing with in Latvia [7].

Fig. 1. Special car sign of the Latvian honorary consul in the state of Karnataka in India.

Fig. 2. The author attending meeting of the Association of Honorary Consular Corps in Latvia on October 10 2018.

Honorary consuls sometimes want to find out how many citizens of the country they represent are in Latvia. The Office of Citizenship and Migration Affairs stated to the honour consul of the Philippines that: "We will not provide you with any data, we have the General Data Protection Regulation [15]". Honorary consul made a remark: "I am not a by-passer, but an honorary consul appointed by the State of the Philippines and accredited in Latvia"... Then the following answer follows: "Okay, we can provide this information, but you have to pay Euro 15 for each person ..."This creates a confusion for the honorary consul, as he does not know whether 300 or 3000 Filipinos are staying in Latvia and how much he will pay [7]. In any case, the honorary consul of a country in Latvia has the right to know the total number of nationals of the country to be represented officially in Latvia, and this is certainly not in contradiction with the General Data Protection Regulation [15].

3 Conclusions

It should be acknowledged that the legal status of honorary consuls has not been studied enough in Latvia and in the world. The article shows that there are some practical problems in practical application of the 1963 Vienna Convention on Consular Relations in the Republic of Latvia, especially concerning functional immunity of honorary consular officer. The author of this article proposes to introduce a special number sign for cars of the honorary consuls in Latvia, thereby showing special respect for foreign honorary consuls. In certain cases such step would also help to resolve disagreements between the Latvian authorities and honorary consuls.

I wish to say a warm word and thanks to the Association of Honorary Consular Corps in Latvia and, especially to the honorary consul of Malta in Latvia Janis Zelmenis for supplying practical materials during writing of this article (Fig. 2).

References

[1] M.N. Shaw, *International Law* (Fifth edition, Cambridge University Press, 2006), pp. 688–690
[2] *International Law*, Second edition. Edited by Malcolm. D. Evans (Oxford University Press, 2006), pp. 404–405
[3] K.D. Stringer, *Honorary Consuls in an Era of Globalization, Trade, and Investment*, Consular Affairs and Diplomacy, Edited by J. Melissen, A.M. Fernández (Leiden: Brill, 2011), p. 63–96
[4] J. Zelmenis, *Institute of Honorary Consuls*, Moscow J. Internat. Law, No. 2, p. 91–101 (2018)
[5] Ārvalstu pārstāvniecības Latvijā, Electronical resource: `https://www.mfa.gov.lv/vestniecibas-un-parstavniecibas?type=foreign` (Overviewed:19.12.2018)
[6] Vienna Convention on Consular Relations (1963): International agreement. Latvijas Vēstnesis, 2003, July 16
[7] R. Rozenbergs, *Goda konsulu institūts tiecas ciešāk sadarboties ar uzņēmējiem.* (Neatkarīgā rīta avīze, 2018.gada 28.jūnijā)
[8] Law on Consular Assistance and Consular Services (2018): Law of the Republic of Latvija. Latvijas Vēstnesis, 2018, May 15
[9] Dela M, *Legal status of the honorary consul.* Wroclaw Review of Law, Administration & Economics. Der Gryuter open. **4**(1), 71 (2014)
[10] Notariate Law: Law of the Republic of Latvia. Latvijas Vēstnesis, 1993, July 9
[11] Administratīvās rajona tiesas 2013.gada 16. decembra spriedums lietā Nr. A420422412. Electronical resource: `https://manas.tiesas.lv/eTiesasMvc/lv/nolemumi` (Overviewed: 26.11.2018)
[12] Road Traffic Regulations (2015): Regulations of the Cabinet of Ministers of the Republic of Latvia. Latvijas Vēstnesis, 2015, June 27
[13] Latvian Administrative Violations Code (1984): Law of the Republic of Latvia. Latvijas Padomju Sociālistiskās Republikas Augstākās Padomes un Valdības Ziņotājs, 1984. December 12

[14] Rīgas apgabaltiesas Krimināllietu tiesas 2016.gada 13. decembra spriedums lietā Nr. 130030416. Electronical resource: `https://manas.tiesas.lv/eTiesasMvc/lv/nolemumi` (Overwieved: 31.01.2019)

[15] Regulation (EU) 2016/679 of the European Parliament and of the Council of 27 April 2016 on the protection of natural persons with regard to the processing of personal data and on the free movement of such data, and repealing Directive 95/46/EC (General Data Protection Regulation): EU regulation. Official Journal of the European Union, 2016, April 5

Legal and practical aspects of establishing servitude of right of way by court judgment within the framework of urban and rural development and changing global conditions

Jolanta Dinsberga* and *Valdis* Savickis

Rīga Stradiņš University, Faculty of Law, Riga, Latvia

Abstract. As a result of urban and rural development new land areas are being acquired. However, many proprietors face challenges in terms of inability to access their immovable properties due to the reason that they are not adjacent to public roads. One of the ways how a person can acquire the right to use his or her property is to request a court to determine servitude of right of way to access the property. The goal of this study is as follows: to identify the problems of the described topic, to draw conclusions and to offer solutions to solve the identified problems by studying the legal framework of establishing servitude of right of way and the specific features of the application thereof. Methodology applied in the study: 1) general scientific – descriptive, analytical, modelling, dogmatic, deductive, and inductive; 2) interpretation of legal norms – grammatical, systematic, and teleological. Issues covered in the study: jurisdiction of the disputes over servitude of right of way; time limits for dispute resolution by courts; and impact of the initiation of insolvency proceedings on the progress of cases related to establishing servitude of right of way by a court. The following conclusions were drawn by the authors at the end of the study: global changes generate the necessity to facilitate urban and rural development in Latvia, inter alia, to tackle the issues of property accessibility.

Key words: servitude of right of way, land reform, public road, property rights.

1 Introduction

In 2010, according to the assignment given by the Ministry of Regional Development and Local Government of Latvia, the group of experts headed by R. Ķīlis developed the Sustainable Development Strategy of Latvia until 2030 (Latvija 2030) (hereinafter – Latvija 2030). After approval thereof at the *Saeima* on 10 June 2010 [1], the elaborated Strategy became the primary national planning instrument having statutory effect. It is stated in Paragraph 270 of Latvija2030 that "Latvia is a relatively small country with limited economic and natural resources, therefore national competitiveness and development in the most direct way are linked with the ability to follow up and to adapt rapidly to global changes by using its privileges in the most efficient way" [2]. In the opinion of authors, global changes force the country to move forward and serve as a constant driving force for continuous development of urban and rural areas of Latvia. According to Paragraphs 331 and 332 of

* Corresponding author: dinsbija@gmail.com

the aforementioned Strategy, "The interaction between urban and rural areas gives greater opportunities to the population and diversifies access to high-quality environment regardless of the place of residence [..]" [3]. Evidently, one can agree that urban and rural development cannot be attained separately from each other. The mutual interaction of urban and rural areas is one of the key development factors. Therefore, also Latvija2030 authors emphasise that by using specific advantages of each territory and by addressing the challenges jointly, various interaction directions of both urban and rural areas will increase where the opportunities for mobility would be one of such directions, inter alia, ensuring access of the population of rural areas to development centres, as well as generating opportunities for the sale of agricultural and forestry products as specified in Paragraphs 333 and 334. It must be noted that the aforementioned also correlates with Paragraphs 278–307 of Latvija2030 which define the improvement of accessibility, for example, the improvement of external and internal accessibility, development of paths and solutions in planning transport infrastructure, and planning of the road network, etc. It is forecast that over 80% of Europeans will be living in urban areas in 2030 alongside with the ageing of the population and increase in immigration flows. It is therefore essential to consider health and welfare matters in such areas as urban planning, construction, land use, transport, [..] a.o. [4] not only in the national planning documents, but also in urban management.

According to authors, it is impossible to achieve the strategic goals defined in the highest-level national planning documents and to perform the tasks subordinated to the achievement of these goals if even seemingly insignificant problems are not prevented in certain areas of law. The authors, therefore, in this publication offer a brief analytical view of the problems in Latvia in particular in relation to accessibility issues (ensuring access to immovable properties) that must be addressed within the scope of implementing Latvija2030. Only then Latvia will be able to integrate and participate fully in the already mentioned global environment.

Accordingly, as a result of rural development [5] and urban development [6], new territories of land are constantly being acquired, industrial, agricultural and forestry sectors experience development, intensive trade of immovable property takes place in Latvia [7].

However, in the context of the aforementioned processes immovable property owners are facing problems related to inability to access their property due to the reason that they are not adjacent to public roads, and the only way to access the property is to use the plot of land included in the adjacent immovable property. In order to obtain an overview of the nature of problems and the external manifestation thereof, authors would like to name the following examples:

1) The owners of adjacent plots of land do not agree voluntarily to authorise access to their immovable properties, even if they realise that the neighbour has no other possibility to access the property;
2) Even in cases where the road leading to the immovable property is planned according to an administrative act (but has not been built yet), it is marked as an encumbrance in the relevant urban planning documents and the land border plan of the adjacent immovable property, the owner of the adjacent immovable property precludes the possibility to move around the plot of land and refuses to agree thereon voluntarily;
3) The person whose consent is required to cross the plot of land owned is absent or cannot be reached. Consequently, the consent cannot be obtained;
4) In situations where insolvency proceedings/legal protection proceedings of the owner of the adjacent immovable property – a natural or legal person – are initiated, the solution of the issue related to ensuring access rights may take some time or stop

completely as this person no longer has the right of action and all decisions related to encumbering immovable property with property rights is taken by an insolvency administrator;

5) At the time when land reform was implemented in Latvia, when restoring proprietary rights to nationalised properties to the owners or heirs thereof, as well as transferring immovable properties for privatisation, the State Land Service, land commissions and local governments were entitled to decide on ensuring access to the immovable property [8]. The aforementioned entities established numerous servitudes of right of way on the basis of an administrative act, and these servitudes are corroborated in the Land Register in the form of a remark, instead of an entry. Due to the legal loopholes of such decisions and remarks, currently the servitudes of right of way recorded in the Land Register in the form of a remark cannot be recognised as effective and the owners of adjacent immovable properties have a legal basis to prohibit entrance in their property.[1]

The abovementioned and other cases which are not mentioned in the study cause the restriction of the right to own property referred to in Article 105 of the Constitution of the Republic of Latvia [9], and frequently also significant material losses to the owners of immovable properties that are not adjacent to public roads. For instance, a farmer who is involved in grain cultivation may face losses with all the resulting consequences, considering that the farmer cannot access his or her property and to harvest the crops in due time;; an entrepreneur cannot deliver the goods manufactured in its factory to the customer if the new owner of the adjacent territory near the exit from the territory builds a fence along the border without any warning. It must be noted that this causes losses not only to immovable property owners, but also to national economy – as a result of discontinuing or terminating the economic activity, taxes payable within the scope of the economic activity will not be transferred to State budget.

In the aforementioned situations a certain mechanism to ensure access rights is stipulated in the laws and regulations of Latvia. This mechanism is governed quite comprehensively. Servitude of right of way must be established so that a person could access his or her immovable property. It is stated in Section 1130 of the Civil Law that "Servitude is such right in respect of the property of another as restricts ownership rights regarding it, with respect to utilisation, for the benefit of a certain person or a certain parcel of land" [10]. It should be noted that encumbered immovable property is referred to as *a servient immovable property*, whereas immovable property in favour of which servitude is being established – *a dominant immovable property*. By means of servitude of right of way rights may be granted: 1) to a footpath; 2) to a livestock path; and 3) to a roadway [10]. If servitude of right of way is not established by law, a contract or a will, it shall be established by a court.

In view of the aforementioned, authors set forth the following **goal of the study**: to identify the problems of the described topic, to draw conclusions and to offer solutions to solve the identified problems by studying the legal framework of establishing a servitude of right of way and the specific features of the application thereof.

---

[1] Information on a more extensive analytical study on servitudes established by administrative acts and related problems in Latvia is available at: Dinsberga, J. Bite K. Legal consequences and problems of the servitudes of right of way established by administrative acts in Latvia. 6th International Interdisciplinary Scientific Conference SOCIETY HEALTH. WELFARE. SHS Web of Conferences **40**, 01011 (2018)

Due to limited size, the study covers only a few problems identified during legal proceedings and suggests potential solutions thereof.

2 Research materials

The study is based upon the analysis of laws and regulations, case-law and conclusions drawn by specialists in the field of civil law. The following legislative acts are used in the study: Civil Law, Constitution of the Republic of Latvia, Insolvency Law, Latvija2030. Document analysis and case analysis were used as primary data of the study. In addition, the following was used as secondary data: study by Mārtiņš Auders on servitudes of right of way established by administrative acts during land reform conducted in 2014; conclusions expressed by specialists in the field of civil law from the book by E. Kalniņš (2005) and the article by E. Meļķisis included in a collection of articles (2003). The study is based upon a case analysis and unpublished documents were analysed as part of the case analysis.

3 Methods

General scientific study methods: Descriptive – by identifying the existing problems in relation to access possibilities to immovable properties, studying the legal framework for establishing servitudes of right of way, compiling information and relying on study results and providing explanations. Analytical – by studying laws and regulations, case-law, opinions of specialists in the field of civil law in order to identify problems in the context of the matters examined in the study from different aspects. Dogmatic – by conducting a comprehensive analysis of the legal framework that defines development areas of Latvia, determination of servitude by a judgment of a court, insolvency and legal protection proceedings. Modelling potential situations and resolving them basing on the newly created legal norms, offered by the authors of the study. Induction and deduction – by expressing separate judgments during the entire study.

The methods of interpretation of legal norms have been also applied in the study: Grammatical – by examining the meaning and substance of the notions included in legal norms, assessing legal norms from the grammatical aspect. Systematic – by examining the procedures for establishing servitude of right of way in the Civil Law in conjunction with other laws and regulations governing the determination of servitudes of right of way. Teleological – by examining the intention of the legislator when adopting laws and regulations related to establishing servitudes of right of way.

4 Results and discussion

4.1 Subordination of disputes over establishing servitude of right of way

Servitudes may be established: 1) by law; 2) by a judgment of a court; 3) by a contract or a will [10]. The establishing of servitude of right of way by a judgment of a court is a measure of last resort for establishing property rights. The disputes over establishing servitude of right of way are qualified as disputes under civil law. The allocation and jurisdiction of civil legal disputes are governed by Chapter 3 of the Civil Procedure Law. It is stated in Section 23 of the aforementioned Law that "All civil legal disputes shall be allocated to the court, unless otherwise provided for in law." This shall not deprive the parties of the right to apply, upon mutual agreement, to an arbitration court or to use mediation in order to settle a dispute" [11]. It is apparent from the content of the mentioned Section that civil legal disputes can be subject to resolution by a general jurisdiction court and also by other institutions. Although Section 23 of the Civil Procedure Law refers to the right of the parties "to apply, upon mutual agreement, to an arbitration court or to use mediation in order to settle a dispute", it should be noted that in Latvia claims for establishing servitude of right of way are submitted to and resolved by a general jurisdiction court. Unfortunately, authors could not find judgments on establishing servitude of right of way which would be delivered by an arbitration court. In view of this, authors are wondering why the belief that servitudes-related issues are to be resolved by a general jurisdiction court prevails in Latvia. It is likely that this mistaken belief derives from Section 5, Paragraph one, Clause 5 of the Arbitration Law which prescribes the competence of an arbitration court "(1) An arbitration court shall resolve any civil legal disputes, provided that the parties have voluntarily agreed upon and concluded an arbitration agreement, except for the following dispute: [..] 5) over the establishment, alteration or termination of property rights in regard to immovable property, if a party to the dispute is a person whose rights to acquire the immovable property for ownership, possession or use are restricted by law; [..]" [12]. While analysing the mentioned legal norm, authors came to the conclusion that arbitration courts have the right to resolve issues related to the establishment, termination or alteration of servitude of right of way. Though servitude is a property right that can be said to be linked with Section 5, Paragraph one, Clause 5 of the Law, when interpreting this Clause according to the method of grammatical interpretation, it is understandable that the focus is kept on persons without the right to obtain the immovable property in ownership, possession or use. It can be therefore concluded that this Clause is applicable to a narrow range of persons. And, if such a restriction does not exist in relation to a party to the dispute, the dispute is to be resolved also by an arbitration court. For instance, when analysing the contract concluded between persons "S" and "M" [13], authors discovered that "S" and "M" have agreed to establish servitude of right of way in the future. Clause 3 of the contract states that "If "M" purchases the immovable property under cadastre No. [..], located in [..], the parties undertake to conclude the contract on establishing a servitude of right of way, agreeing on the length, width, location of a servitude of right of way and other mutually beneficial and acceptable conditions to the parties". The parties have the right to conclude a transaction on establishing a servitude of right of way in the future in accordance with Section 1413 of the Civil Law, which prescribes the following: "[..] a transaction may also apply to future property" [14]. In addition, "S" and "M" have included an arbitration clause in the contract that is a precondition for the dispute to be resolved by an arbitration court. Consequently, if "S", who is the owner of the servient immovable property, refuses to conclude the contract on establishing servitude of right of way after the property has been purchased by "M", "M" would have the right to address to an arbitration court with the request to conclude the contract on establishing the servitude of right of way. Furthermore, Section 1587 of the Civil Law stipulates as follows: "A contract legally entered into shall impose on a contracting party a duty to perform that which was promised" [14]. Of course, according to Section 24, Paragraph one of the Arbitration Law, "The arbitration panel, upon receipt of an application for a claim, shall decide on the arbitral jurisdiction of a civil legal dispute [..]", assessing whether the dispute is within arbitral jurisdiction and there are no other exceptions for the resolution of disputes referred to in Section 5 of the Arbitration Law. Different regulatory framework and approach in resolving issues concerning the jurisdiction of the dispute over establishing servitude of right of way exist in the Russian Federation. In accordance with the Code of Administrative Judicial Procedure of the Russian Federation disputes of economic nature and other matters related to commercial activities and other economic activities are within arbitral jurisdiction [15]. According to the case-law collection of the Supreme Court of the Russian Federation, when resolving a matter on the allocation of a case on a servitude of right of way either to a general jurisdiction court or an arbitration court, the status of the subject of the dispute has the decisive moment in this regard.

4.2 Duration of dispute resolution at courts

In accordance with the principle of procedural economy enshrined in Section 28 of the law On Judicial Power "A judge shall try a matter as fast as possible" [16]. The case-law of the Supreme Court of the Republic of Latvia is summarised by authors in the table below: judgments in cases related to the establishment of a servitude of right of way delivered by the Civil Cases Department of the Senate of the Supreme Court for the purpose of determining the duration of the resolution of cases of the referred to category from the moment a judgment has been delivered by a court of first instance until the moment a judgment is delivered by the Civil Cases Department of the Senate of the Supreme Court. Authors did not select and analyse all judgments delivered by a court of first instance and appellate instance court, as it is believed that the number of judgments is sufficient to obtain sufficient information and to obtain basis to draw conclusions and to identify problems related to duration of the resolution of cases at courts.

Based on the analysis of court judgments specified in the table above, authors conclude that the resolution of disputes in relation to the establishment of servitude of right of way on average lasts for a period of 3 years from the moment a judgment has been delivered by a court of first instance until the moment a judgment is delivered by the Civil Cases Department of the Senate of the Supreme Court. The time of examination of the case by court of first instance can last from one to four years (authors' experience), and it is excluded from this period. The period when the case is re-examined by an appellate instance court is excluded, too. Out of six court judgments included in the table above, five court judgments have been revoked either in full or in part and referred for re-examination by an appellate instance court. This means that the duration of the resolution of a case is extended. The analysis resulted in authors' conclusion that cases related to establishing servitude of right of way are examined excessively long. In this period, the right to own property of the owners of dominant immovable properties stipulated in Article 105 of the Constitution of the Republic of Latvia and also Article 1 of Protocol No. 1 to the Convention for the Protection of Human Rights and Fundamental Freedoms [23] is breached, i.e., completely "paralysed" due to the reason that in the course of legal proceedings there is no access to the immovable property and it is not possible to use the property. This legal situation where a property cannot be used over several years only due to the flaws in laws and regulations that leads to inability to ensure unimpeded access to any immovable property is not acceptable to contemporary society and is not typical to this era. It is therefore necessary to introduce a legal mechanism whereby the respective deficiency can be rectified.

At authors' disposal there are documents that indicate to the first attempts of lawyers to seek a temporary solution and to ensure temporary access possibilities to the immovable property in the course of the legal proceedings related to establishing servitude of right of way. The dispute is based upon the fact that the plaintiff "A.R." has purchased immovable property that could be accessed by crossing the immovable property owned by "I.B." along the territory intended for servitude of right of way. "I.B." prohibited "A.R.\" from using the territory of servitude of right of way.

At the moment when the immovable property was purchased, "A.R." relied on the territory of a servitude of right of way being recorded in the plan of immovable property borders, layout and encumbrances and the part of the Land Register, and was convinced that only agreement on the procedures for the right of use of a servitude of right of way

Table 1. Judgments by the Civil Cases Department of the Senate of the Supreme Court in cases related to establishing servitudes of right of way [by authors].

Duration of legal proceedings calculated in years	**Date of a judgement by district (city) courts**	**Date of the judgment by the Civil Cases Department of the Senate of the Supreme Court of the Republic of Latvia Resolved as follows:**
longer than 3	Balvi District Court – 05.12.2014	04.07.2017, judgment in case No. C09030814 SKC 229/2017 [17] **To revoke** the judgment of 16.03.2015 by the Civil Matters Division of Latgale Regional Court and to refer the case for re-examination thereof by an appellate instance court.
longer than 3	City of Riga Central District Court – 08.10.2012 *(action brought on 06.09.2010)*	12.04.2016, judgment in case No. C27209810 SKC-33/2016 [18] **To revoke the appealed part** of the judgment of 08.04.2013 by the Civil Matters Division of Riga Regional Court and to refer the case for re-examination thereof by an appellate instance court.
3	Ventspils City Court – 31.03.2011 *(action brought on 16.10.2010)*	31.03.2014, judgment in case No. SKC-11/2014 [19] To uphold the judgment of 04.01.2012 by the Civil Matters Division of Kurzeme Regional Court, but to reject the cassation complaint of A.R.
almost 3	Daugavpils City Court – 19.10.2011 *(action brought on 28.05.2010)*	09.10.2013, judgment in case No. SKC-458/2013 [20] **To revoke** the judgment of 02.02.2012 by the Civil Matters Division of Latgale Regional Court **in the part** where the claim is satisfied and the counter-claim is rejected, and to refer the case in this part for re-examination thereof by an appellate instance court.
longer than 1	Preiļi District Court – 28.02.2008	12.08.2009, judgment in case No. SKC – 192 [21] **To revoke** the judgment of 27.01.2007 by the Civil Matters Division of Vidzeme Regional Court and to refer the case for re-examination thereof by an appellate instance court.
less than 1	Valmiera District Court – 19.08.2004	08.06.2005, judgment in case No. C39125103 SKC-320/2005[22] **To revoke** the judgment of 27.01.2005 by the Civil Matters Division of Vidzeme Regional Court and to refer the case for re-examination thereof by an appellate instance court.

with “I.B.” would be required. Moreover, when dividing and merging immovable properties in accordance with the laws and regulations effective in the Republic of Latvia (Cabinet Regulation No. 182 of 20 March 2007, Regulations Regarding Specification of an Immovable Property Object [24], Binding Regulation No. 60 of 27 April 2006, Graphical Display of the Spatial Plan of Ogre Municipality and Regulations Regarding the Use and Construction of the Territory of Ogre Municipality [25]), it was necessary to provide for access to the divided immovable property. The property of “A.R.” emerged as a result of actual division of the immovable property owned by the mother of “I.B.”. The territory of the respective servitude has been marked off in the land border plan of the servient immovable property and in the form of a remark in the Land Register as an encumbrance based on an administrative act regarding the division of the immovable property into actual units and determining of the address. However, the latter does not give the right to move across the respective plot of land, because as the result of dividing the plot of land the territory of servitude of right of way is planned for access, but the contract thereon has not been concluded between "I.B." and "A.R.".

When interpreting the norms of the aforementioned regulations according to the method of historical and teleological interpretation, it can be concluded that the goal set by the legislator in regard to the actual division of the immovable property was to ensure to the owners of divided immovable properties unimpeded use of the property and access thereto. Unfortunately, the legislator has not regulated all aspects of these matters. In the opinion of authors, the fact that the territory of the planned servitude of right of way is determined in an administrative act does not impose an obligation on the owner of the servient immovable property to conclude the contract on establishing servitude of right of way with the owner of the dominant immovable property. From the point of view of authors, the norms adopted by the legislator only serve as the so-called "formal reference" which has the informative role. This may be also classified as a "legislative gap", because the attempts of the legislator to regulate the particular situation do not work in practice and the respective norms do not fulfil their function.

The request to take a provisional decision in a civil case on establishing servitude of right of way was submitted to Zemgale District Court (Ogre) on 26 September 2018 [26]. In the respective request it was demanded to take a provisional decision and, until giving the final judgment, to determine access to the immovable property on foot or by means of transport [..] of the immovable property to be used [..] land border of 18 April 2008, territory of a servitude of right of way marked off in the plan of the layout and encumbrances (area of 0.16 ha, length: 300 meters, width: 4.5 meters). The request was grounded by "A.R." with a "legislative gap" in the legal framework and "A.R." also referred to Section 5, Paragraph five of the Civil Procedure Law that states: "If there is no law governing the contested relation, a court shall apply a law governing similar legal relations, but if no such law exists, a court shall act in accordance with general legal principles and meaning". However, Paragraph six of the referred to Section of the Law stipulates as follows: "In applying legal norms, a court shall take into account the case law" [27]. "A.R." also mentioned in the request that "in individual cases the Civil Procedure Law allows taking of a provisional decision until the giving of a judgment". It is stated in Section 238.[1], Paragraph one of the Civil Procedure Law that "(1) Upon a request of a party a court or judge may take a decision which temporarily, until the judgment on divorce or annulment of marriage is given, specifies the place of residence of the child, the procedures for child care, the procedures for exercising access rights, child maintenance, prohibition to taking the child out of the State, means for the provision of the previous welfare level of the spouse, procedures for use of the joint home of the spouses or instructs one of the parties to issue to the other party household and personal articles." Furthermore, also Section 244.[10], Paragraph one and Section 267.[1], Paragraph one of the Civil Procedure Law provides for taking a provisional decision upon request of a party.

Just like taking of a provisional decision, Section 238.[1], Paragraph one, Section 244.[10], Paragraph one, and Section 267.[1], Paragraph one of the Civil Procedure Law stipulates and safeguards the legitimate interest of particular persons until the final judgment, also "A.R." requires a provisional regulation to ensure the guaranteed rights to the property thereof [27].

E. Meļķisis shares the following opinion: "Considering that the court has identified a lack of regulation, there are grounds to discuss the matter on the existence of a legislative gap and elimination of this gap. As stated in the legal doctrine, there exists a legislative gap if the law, considering the interpretation thereof, does not cover positive legal framework, even though such regulation is necessary from the point of view of the legal system [28].

E. Kalniņš indicates: "The legislative gap can be identified if the law in respect of the particular group of occasions does not cover positive legal framework, even though such regulation should have been covered therein, considering the plan and objective of the respective law. The identified legislative gap can be eliminated by analogy" [29].

Having assessed the quoted conclusions expressed by prominent specialists in the field of civil law in Latvia, authors consider that as a result of extensive interpretation of the legal norms and general legal principles, the application of analogy would be an effective temporary solution to ensure access to the immovable property. However, from the perspective of judges, this is probably a complicated and bold decision in terms of providing grounds thereto. Unfortunately, this is also demonstrated by the case-law. By decision on October 1, 2018 in case No. C73405818 Zemgale District Court judge rejected the request of "A.R.". It was stated that "[..] neither the cases regarding annulment or divorce of marriage nor cases arising from the custody rights and access rights, or restriction of trusteeship are in any way similar to cases regarding establishing a servitude of right of way. Furthermore, also the mentioned norms of the Civil Procedure Law [..] do not provide for such a provisional regulation [..]" [30].

The ancillary complaint submitted by "A.R." concerning the decision by Zemgale District Court judge was also rejected and left without amendments by Zemgale Regional Court in decision of 21 November 2018 in case No. C73405818. It is stated in the decision that "The fact that the Civil Procedure Law does not provide for a provisional regulation in other categories of cases, inter alia, establishing of servitude, cannot be recognised as a legislative gap that should be eliminated in accordance with the provisions laid down in Section 5, Paragraph five of the Civil Procedure Law. The court is not entitled to establish a provisional regulation in cases where the legislator has not established such a regulation. These matters are within the competence of the legislator" [31].

Considering the grounds for refusal stated in the decisions by Zemgale District Court and Zemgale Regional Court, authors conclude that the court, however, narrowly interprets the statutory provisions when examining the possibility to apply analogy. The following theses are stated in judgment of 14 February 2006 in case No. SKA – 34 of the Civil Cases Department of the Senate of the Supreme Court of the Republic of Latvia: "Thesis: The entity granting rights prior to considering the application of analogy must examine whether in the particular case it is possible to apply analogy at all [32]. In the analysed case the court based on the belief that analogy cannot be applied and this is not within the competence of the court. However, at the same time authors want to emphasise that "ECHR has declared that the first and foremost [..] requirement stipulated in Article 1 of Protocol No. 1 to the Convention is that any interference by a public authority with the right to property should be lawful" [33]. However, ECHR has not claimed that the legislator only has exclusive competence to impose restrictions on the right to property of a person. Also, other public authorities, inter alia, courts have such an entitlement, considering that the principle of legality is respected. Also, scientific literature states that the restriction of the right to property conforms to Article 1 of Protocol No. 1 to the Convention not only in the case it is prescribed by law, but also in the case where rights are restricted based on a law [34]. Thus, considering all conditions, also the court should have the entitlement to restrict the rights of "I.B.".

Having examined the specific nature of the application of analogy, current position of the court and the diversity of the cases related to the establishment of a servitude of right of way, authors conclude that reference to the application of analogy and the general principles of law in other cases related to the establishment of a servitude of right of way, most likely, would not yield the desired results. In view of the aforementioned, the authors believe that the Civil Procedure Law requires amendments that stipulate an entitlement of a court to determine a provisional regulation in the cases related to the establishment, alteration and termination of a servitude of right of way upon request of a party – allowing to take a provisional decision until the giving of a final judgment.

It is not possible to define criteria in the law by which such decision ought to be taken. Consequently, the court would have to examine each individual request and the grounds thereof in compliance with the principles of proportionality, fairness, good faith and other principles. Furthermore, it will be necessary to examine whether the possibility to satisfy claim prima facie is more likely than the possibility of its rejection.

By simulating the situation where the owners of the dominant property bring a claim before a court wherein the court would have to examine the matter on taking a provisional decision and prima facia possibility to satisfy the claim, authors will illustrate this situation with the help of cases that frequently occur in practice. For instance, it is evident from documents and the actual situation that: 1) the dominant immovable property is surrounded by a wide and deep river, swamps or other immovable properties; 2) in order to access this property it is necessary to cross the servient immovable property as this is the only way to physically access the property; 3) already historically there is a road on the servient immovable property that leads to the dominant immovable property; 4) the road is used also by other neighbours and a servitude of right of way has already been established in favour of these immovable properties and recorded in the Land Register; 5) it is necessary to mention also such cases where there is even a remark on a servitude of right of way in the Land Register of the servient immovable property [35]; 6) the owner of the servient immovable property denies access to the road that at the same time is used by other neighbours and the owner refuses to conclude the contract on establishing a servitude of right of way. Authors are certain that, in such legal and factual circumstances, a court would have grounds to declare that the possibility to satisfy a claim prima facie is more likely than the possibility of its rejection. Moreover, there would be sufficient grounds for taking a provisional decision until the giving of a final judgment. Of course, it will be a completely different situation if a road suitable for walking and driving had not been constructed within the servient immovable property or there would be only a narrow path, or the owner of the dominant immovable property would be willing to establish a servitude of right of way at a specific location, although it could be established elsewhere where the owner of the servient immovable property is disturbed the least. By applying the modelling method, which is applied in the field of law quite rarely, authors reaffirm the aforementioned conclusion that it is not possible to define criteria in the law by which the provisional decision ought to be taken.

By outlining the described problem, authors want to launch debate on the necessity of a provisional regulation in cases related to servitudes of right of way. Consequently, potential amendments to the Civil Procedure Law are developed on a conceptual level. Authors propose to supplement the Civil Procedure Law with Section 230.[2] by including therein: 1) the definition of a provisional decision, because the respective notion has been used in the Civil Procedure Law (Sections 238.[1], 244.[10] and 249.[3]) [36]), but it has not been defined; 2) when listing the cases subject to the submission of a request to take a provisional decision (already existent and cases on the establishment, alteration and termination of a servitude of right of way); 3) also to envisage a special scheme for the submission of a request and taking of a decision in cases related to the establishment, alteration and termination of a servitude of right of way (according to the principle stipulated in Sections 238.[1], 244.[10] and 249.[3] of the Civil Procedure Law).

Next question which draws authors' attention is related to the impact of insolvency proceedings and legal protection proceedings on the progress of the case on the establishment of servitude of right of way by a court. It should be noted that due to the limited length of this article the problems related to the impact of insolvency proceedings and legal protection proceedings are described briefly as authors plan to conduct an individual study for full disclosure of the respective problem.

According to the legal norms laid down in the Insolvency Law [37], in situations where insolvency proceedings or legal protection proceedings of a natural or legal person are initiated, the debtor is deprived of the right of action in respect of the immovable property. This means that during the respective proceedings the debtor is prohibited from alienating the immovable property or encumbering it with property rights. The respective rights can be used if harmonised with the supervisory person of legal protection proceedings or by the administrator of insolvency proceedings. However, taking into account the overall activity of the mentioned persons, according to the purpose of the Insolvency Law – to promote the honouring of the obligations of a debtor in financial difficulties and, where possible, the renewal of solvency (Section 1 of the Law) [37], the encumbering of the immovable property of the debtor with a servitude of right of way in favour of another person or immovable property is not in the first position of the list of priorities. Also, the legal norms of the subjects of insolvency proceedings (natural and legal persons) that define the proceedings contain the following substance: this is an aggregate of measures of a legal nature, within the scope of which the claims of creditors are settled from the property of a debtor, in order to promote the honouring of the debtor's obligations [37]. Consequently, it can be concluded that the change of the owner of immovable property is inevitable that as an additional factor can complicate the progress of proceedings to establish servitude of right of way by a court. Thus, also creditors will have no interest in establishing servitude of right of way. For instance, one of the principles underlying the legal protection proceedings is that methods which restrict the interests of creditors more than required to restore solvency of the debtor must not be included in the plan of measures of legal protection proceedings. Encumbrances of the immovable property reduce the value thereof. Thus, cause a potential risk to creditors in terms of seeking to satisfy their claims to the fullest possible extent.

It should be noted that the minimal time frame for insolvency proceedings is set at six months, as stipulated in Section 111, Paragraph six of the Insolvency Law [37] in respect of the sale of the property of a debtor, Section 73, Paragraphs two and three of the Insolvency Law stipulates a period of six months during which creditors are allowed to submit claims, if the deadline specified in Section 73, Paragraph one of the Insolvency Law (one month) has been missed. However, the time period for implementation of legal protection proceedings can last even four years from the day of entering into effect of the court judgment regarding the implementation of the plan of measures of the legal protection. [37]. This can significantly prolong the legal proceedings in the dispute over establishing a servitude of right of way, because the debtor is not allowed to encumber the immovable property with property rights (except for cases where this is stipulated in the plan of measures of legal protection proceedings), and this is a restriction of the debtor's action during the implementation of legal protection proceedings [37].

5 Conclusions

During the study authors discovered that accessibility is one of the most important issues to be resolved within the scope of urban and rural development. This could partly be resolved by legislative changes (see Clause 4.2). Conceptual amendments regarding the introduction of a provisional regulation when taking a provisional decision in cases related to establishing (including alteration and termination of) servitude of right of way to the Civil Procedure Law proposed by authors, probably, in most cases will enable the owners of dominant immovable properties to continue their economic activity, as well as families to freely use their property without causing a particular burden or loss to the owners of servient immovable properties.

In respect of cases related to insolvency and legal protection proceedings it should be noted that the period of implementation thereof could prolong and hinder the process of establishing servitude of right of way. Unfortunately, there is no legislative act which defines the practice of the supervisory person of legal protection proceedings or the administrator of insolvency proceedings in particular in cases when servitude of right of way is to be established in respect of the debtor's immovable property.

References

[1] *Paziņojums par Latvijas ilgtspējīgas attīstības stratēģijas līdz 2030.gadam apstiprināšanu.* 10.06.2010. Latvijas Vēstnesis, Nr. **101** (4293), 29.06.2010
[2] *Latvijas ilgtspējīgas attīstības stratēģija līdz 2030.gadam (Latvija2030).* 270.punkts. Available at: https://www.pkc.gov.lv/sites/default/files/inline-files/Latvija_2030_7.pdf
[3] *Latvijas ilgtspējīgas attīstības stratēģija līdz 2030.gadam (Latvija2030).* 270.punkts. Available at: https://www.pkc.gov.lv/sites/default/files/inline-files/Latvija_2030_7.pdf
[4] *Eiropas pilsētu un regionu attīstība, lai tās kļūtu veselīgākas, drošākas un ilgtspējīgākas.* 12.10.2017. (2018). Available at: http://www.vm.gov.lv/lv/tava_veseliba/vides_veseliba1/eiropas_pilsetu_un_regionu_attistiba_lai_tas_klutu_veseligak/
[5] *Eiropas revīzijas palāta. Īpašais ziņojums Nr.16/2017.* (2017) Available at: http://publications.europa.eu/webpub/eca/special-reports/rd-programming-16-2017/lv/
[6] *Pētījums "Latvijas pilsētu sociāli ekonomiskās attīstības tendences"* (2008) Available at: http://www.vraa.gov.lv/uploads/documents/petnieciba/petijumi/Petijums_Latvijas%20pilsetu%20sociali%20ekonomiskas%20attistibas%20tendences1.pdf
[7] *Valsts zemes diensts. Latvijas Nekustamā īpašuma tirgus – statistiskie rādītāji.* Available at: http://kadastralavertiba.lv/tirgus-dati/statistika/
[8] *Likums "Par zemes lietošanu un zemes ierīcību".* 20.panta trešā daļa. Ziņotājs, Nr. **31/32**, 15.08.1991., Zaudēja spēku 30.12.2009; likums "Par zemes privatizāciju lauku apvidos" 22.panta trešā daļa. Ziņotājs, Nr. **32/33/34**, 20.08.1992
[9] *Latvijas Republikas Satversme.* 15.02.1922. Latvijas Vēstnesis, Nr. 43, 01.07.1993. Valdības Vēstnesis, Nr. **141**, 30.06.1922
[10] *Civillikums* (1937). 1130.pants. Latvijas Republikas likums. Valdības Vēstnesis, Nr. **44**, 24.02.1937
[11] *Civilprocesa likums* (1998). Latvijas Vēstnesis, Nr. **326/330**(1387/1391), 03.11.1998
[12] *Šķīrējtiesu likums* (2014). Latvijas Vēstnesis, Nr. **194**(5254), 01.10.2014
[13] 2018. gada 9.septembra Līgums Par servitūta nodibināšanu nākotnē. (not published)
[14] *Civillikums* (1937). Latvijas Republikas likums. Valdības Vēstnesis, Nr. **44**, 24.02.1937
[15] *Обзор судебной практики по делам об установлении сервитута на земельный участок* (утв. Президиумом Верховного Суда РФ 26 апреля 2017 г.) (2017). Available at: http://www.supcourt.ru/documents/all/15912/
[16] *Likums "Par tiesu varu"* (1992). Ziņotājs, Nr. **1/2**, 14.01.1993
[17] Latvijas Republikas Augstākās tiesas Senāta Civillietu departamenta 2017.gada 4.jūlija SPRIEDUMS Lieta Nr.C09030814 SKC 229/2017. Available at: http://at.gov.lv/lv/judikatura/judikaturas-nolemumu-arhivs/civillietu-departaments/hronologiska-seciba?year=2017
[18] Latvijas Republikas Augstākās tiesas Senāta Civillietu departamenta 2016. gada 12. Aprīļa SPRIEDUMS Lietā Nr. C27209810 SKC-33/2016. Available at:

http://at.gov.lv/lv/judikatura/judikaturas-nolemumu-arhivs/civillietu-departaments/hronologiska-seciba?year=2016

[19] Latvijas Republikas Augstākās tiesas Senāta Civillietu departamenta 2014. gada 31.marta SPRIEDUMS Lietā Nr. SKC-11/2014. Available at: http://at.gov.lv/lv/judikatura/judikaturas-nolemumu-arhivs/civillietu-departaments/hronologiska-seciba?year=2014

[20] Latvijas Republikas Augstākās tiesas Senāta Civillietu departamenta 2013. gada 9.oktobra SPRIEDUMS Lietā Nr. SKC-458/2013. Available at: http://at.gov.lv/lv/judikatura/judikaturas-nolemumu-arhivs/civillietu-departaments/hronologiska-seciba?year=2013, [viewed on 21.11.2018.]

[21] Latvijas Republikas Augstākās tiesas Senāta Civillietu departamenta 2009. gada 12.augusta SPRIEDUMS Lietā Nr. SKC – 192. Available at: http://at.gov.lv/lv/judikatura/judikaturas-nolemumu-arhivs/civillietu-departaments/hronologiska-seciba?year=2009, [viewed on 21.11.2018.]

[22] Latvijas Republikas Augstākās tiesas Senāta Civillietu departamenta 2005.gada 8.jūnija spriedums Lietā Nr. C39125103 SKC-320/2005. Available at: http://at.gov.lv/lv/judikatura/judikaturas-nolemumu-arhivs/civillietu-departaments/hronologiska-seciba?year=2005, [viewed on 21.11.2018.]

[23] *Cilvēka tiesību un pamatbrīvību aizsardzības konvencija.* Eiropas Padome (EP). 04.11.1950. Ratificēts Latvijā 27.06.1997. Latvijas Vēstnesis, Nr. 143/144 (858/859), 13.06.1997

[24] *Ministru kabineta 2007.gada 20.marta noteikumi Nr.182 "Noteikumi par nekustamā īpašuma objekta noteikšanu",* 16.1.punkts. Latvijas Vēstnesis, Nr. **50** (3626), 23.03.2007. Zaudēja spēku: 02.06.2011

[25] *Ogres novada domes 2006.gada 27.aprīļa saistošie noteikumi Nr.60 "Ogres novada teritorijas plānojuma grafiskā daļa un Ogres novada teritorijas izmantošanas un apbūves noteikumi"*. Pieejams Ogres novada domes arhīvā

[26] "A. R." 2018.gada 26.septembra Lūgums Lietā Nr.C73405818 (par pagaidu lēmuma pieņemšanu civillietā par braucamā ceļa servitūta nodibināšanu). (not published)

[27] "A. R." 2018.gada 26.septembra Lūgums Lietā Nr.C73405818 (par pagaidu lēmuma pieņemšanu civillietā par braucamā ceļa servitūta nodibināšanu) (not published)

[28] "A. R." 2018.gada 26.septembra Lūgums Lietā Nr.C73405818 (par pagaidu lēmuma pieņemšanu civillietā par braucamā ceļa servitūta nodibināšanu) (not published)

[29] E. Kalniņš *Tiesību tālākveidošana // Privāttiesību teorija un prakse.* Rīga: Tiesu namu agentūra, 319–320, lpp. (2005)

[30] 2018.gada 1.oktobra Zemgales rajona tiesas Lēmums Lietā Nr.C73405818 (not published)

[31] 2018.gada 21.novembra Zemgales apgabaltiesas Lēmums Lietā Nr.C73405818 (not published)

[32] Latvijas Republikas Augstākās tiesas Senāta Administratīvo lietu departamenta 2006.gada 14. februāra spriedums Lietā Nr. SKA – 34. Available at: http://at.gov.lv/lv/judikatura/judikaturas-nolemumu-arhivs/administrativo-lietu-departaments/hronologiska-seciba?year=2006

[33] Eiropas Cilvēktiesību tiesas spriedums: Iatridis v Greece, judgment of 25 March 1999, no.31107/96, ECHR 1999-II, para 58; Skrzyński v Poland, judgment of 6 September 2007, no.38672/02, Available at: http://cmiskp.echr.coe.int/tkp197/view.asp?item=1&portal=hbkm&action=html&highlight=38672/02&sessionid=17139593&skin=hudoc-en

[34] C. Grabenwarter, Europäische Menschenrechtskonvention. München, C.H. Beck, S. 365, 369, 370. (2005) Available at: http://www.satv.tiesa.gov.lv/wp-content/uploads/2016/02/2008-11-01_Spriedums.pdf
[35] M. Auders, *Pētījums par zemes reformas laikā ar administratīvo aktu nodibinātiem ceļu servitūtiem.* 7.lpp. (2014) Available at: http://petijumi.mk.gov.lv/sites/default/files/title_file/petijums_varam_2014_par%20zemes%20reformas%20laika%20ar%20administrativo%20aktu%20nodibinatiem%20celu%20servitutiem.DOCX
[36] *Civilprocesa likums* (1998). Latvijas Vēstnesis, Nr. **326/330** (1387/1391)
[37] *Maksātnespējas likums* (2010). Latvijas Vēstnesis Nr. **124** (4316), 06.08.2010, 40.panta 4.daļa, 49.pants, 63.pants

Peculiarities of differentiation of legal regulation of pharmaceutical workers' labour

Anna Sydorenko and *Olena* Melnyk

Poltava Law Institute of Yaroslav the Wise National Law University, Poltava, Ukraine

Abstract. The paper deals with peculiarities of legal regulation of pharmaceutical workers' labour. Exercising the right to work is determined and analyzed. The authors prove that differentiation originates from special legislation. They determine what globalization is in the modern society. The paper analyzes the differentiation of pharmaceutical workers' labour through the prism of such objective reasons as industry affiliation, working conditions, the nature of labour relations between an employee and an employer, and their content is determined. The study draws attention to the fact that industry affiliation is characterized by the importance of the industry for society and the degree of provision of public health institutions with highly skilled specialists, as well as the production process. The paper determines that working conditions are specific for each industry and workplace. It is proved that pharmaceutical workers have harmful and dangerous working conditions. Working conditions have to satisfy the requirements to the fulfilment of labour duties without life and health hazard. It is analyzed that the nature of labour relations between an employee and an employer are determined by the nature of labour and employment contract term. It is proved that the differentiation of pharmaceutical workers' labour is expressed through special rules, which allows them to exercise their rights and obligations and to enjoy certain guarantees.

Key words: unity, differentiation, globalization, pharmaceutical workers, industry affiliation, working conditions, nature of labour relations.

1 Introduction

One of the most important tasks of modern labour law is creating conditions for reconciling interests of participants in social and labour relations, protecting rights and legitimate interests of the subjects of labour law. The fulfilment of this task should, first of all, take place by securing proper rights and obligations in statutory acts both internationally and nationally.

Under current conditions of social development, the actual problems of exercising the right to work by a person are caused by gaps in the rules of labour legislation concerning regulation of labour of various categories of employees, inconsistency of the legal principles enshrined in general, and special and local statutory acts. Relations in the labour field are governed by a large number of statutory acts of both international and national legislation, which causes conflicts in the regulation of certain issues related to the implementation of the principle of the right to work. This, in turn, raises the problem of the relevance of the study of the unity and differentiation of legal regulation of labour, which, in our opinion, is one of the main trends in the improvement of labour legislation and lies in the peculiarities of the nature, working conditions, providing different categories of workers with guarantees, privileges and benefits.

In today's world, the issue of public health remains topical. The right to health is defined by international rules and national legislation of all countries without exception. In the realm of exercising this right, it is interesting to study the legal regulation of pharmaceutical workers' labour since they are the subjects directly assisting in exercising the right to health.

2 Aim

The paper is aimed at studying peculiarities of legal regulation of pharmaceutical workers' labour through the prism of labour differentiation in the context of medical reforms in many countries and in drawing conclusions aimed at improving the current legislation.

3 Material and methods

Scientific and theoretical studies of scientists, international statutory acts on labour are the empirical foundations of the study. The conducted research caused the need for application of general scientific and special methods of scientific knowledge.

4 Results and discussion

International rules and national legislation of most developed countries determine the right to work as a priority and one of the basic rights of every person since it is one of the fundamental rights. The right to work has been enshrined in the main international documents being the basis of legal rules in all countries. Thus, it should be emphasized that the Universal Declaration of Human Rights provides that everyone has the right to work, free choice of work and protection from unemployment [1]. The International Covenant on Economic, Social and Cultural Rights adopted in 1966 contains a similar provision which "includes the right of everyone to the opportunity to gain his living by work which he freely chooses or accepts" [2]. According to scientists, the international document referred to above forms the standards that the member states have to strive for achievement and enforcement but does not impose any legal obligations on them [3, 4].

The right to work is also defined by the International Covenant on Civil and Political Rights [5]. The Resolution of the United Nations General Assembly on "Setting International Standards in the Field of Human Rights" defines a general system of human rights, including the right to work and opportunity to provide for oneself by virtue of such work [6]. This right is also enshrined in the Convention for Protection of Human Rights and Fundamental Freedoms [7], the European Social Charter (Revised) [8], the Community Charter on Fundamental Social Rights of Workers [9], the Charter of Fundamental Rights of the European Union [10] since the basis the development of the social policy of the member states of the European Union is precisely the definition of their citizens' social and labour rights in many other international instruments, including the Conventions and Recommendations of the International Labour Organization, etc. According to the above-mentioned international instruments, exercising the right to work takes place through the principles of freedom of work, free choice of place of work and the duty of governments to ensure measures for exercising the right to work by all able-bodied people without exception.

It should be emphasized that the acts of the European Union contain a clear list of rights which should be laid down as the basis for the development of statutory acts and be respected and observed while implementing the domestic and foreign policies of the member states of the European Union [11].

It is necessary to turn attention to the fact that there are a large number of international rules regulating the issue of exercising the right to work by individuals. However, B. Hepple rightly observes that the problem is not in quantity but in quality of the international labour standards since acts adopted in recent years are not aimed at overcoming the negative effects of fulfilling one's work responsibilities and the effects of globalization [12].

We agree that the development of modern labour legislation should meet the new conditions of social development, technologies and other processes within the framework of the modern period of globalization.

The scholars understand globalization as integration and development of the economy, society under the influence of state-of-the-art technologies, economic relations in the society, as well as the development of national and international policies conducted by governments of countries, international organizations, associations of workers and employers and civil society [13].

It should be noted that the problem of improving the legal regulation of labour and closely related relations is primarily due to the need for differentiation. It is with the help of this phenomenon that general rules acquire more specific features, which approximate the real relations between their subjects while exercising the right to work in various fields.

Before considering the issue of differentiation of legal regulation of pharmaceutical workers' labour, it is necessary to answer the question of what the phenomenon of differentiation is since specifics of the legal status of this category of workers are determined by the general rules of labour law expressed in their unity, that are applied to all employees without exception, and special statutory acts, that are characterized by differentiation of legal regulation of labour. As for a legal definition of the notion of differentiation of legal regulation of labour, it is absent in statutory acts.

In French, the term "différenciation" means a division of the whole into parts, forms, degrees. In turn, differentiation should be understood as legal regulation of labour of various categories of workers which is provided for by special legislation, i.e. certain statutory and subordinate acts in which specific conditions of work in various fields of the economy are manifested and there is division of statutory provision in accordance with any labour relations supplementing the general rules of labour law. This concept contributes to exercising the inalienable right to work by individuals.

Legal regulation of pharmaceutical workers' labour is carried out on the ground of unity and differentiation. In this regard, we consider it necessary to disclose the peculiarities of pharmaceutical workers' legal status, to study the legislation on this issue and to determine possible improvements of labour legislation concerning legal regulation of these workers' labour. As pharmaceutical aid is one of the main components of the public health system, its quality depends on a large number of factors. One of them is the specialists' professional level and the procedure for regulating their work.

We understand unity as a single approach to labour regulation. For example, the main statutory act regulating medical and pharmaceutical workers' labour in the Republic of Moldova is the Labour Code of the Republic [14] which is characterized by unity. The act extends to all employees without exception. It determines the unified procedure for employment contract conclusion, grounds and procedure for dismissal, principles of the labour formation, working time, rest time, etc. The Labour Law of the Republic of Latvia and the Labour Code of the Republic of Lithuania are also examples of unity [15]. However, in addition to the general rules, pharmaceutical workers' legal status falls within the following category of differentiation since differentiation of legal regulation of labour is expressed, first of all, through special rules allowing clarifying, expanding and exercising their rights, obligations and guarantees in the workplace by the relevant categories of workers. The legislation of Estonia can be attributed to the category of differentiation of legal regulation of labour since there is no single act of labour legislation in this country. Labour relations are regulated by separate laws, regulations and procedures.

Consequently, we can say that legal regulation of pharmaceutical workers' labour is carried out precisely on the grounds of unity and differentiation, which are interrelated categories. The problem of differentiation is of particular importance in the study since it provides an opportunity to take into account the specifics of different groups of relations. Namely, peculiarities of pharmaceutical workers' legal status are to be considered through the prism of the grounds for differentiation which are divided into two groups: subjective and objective. Particular attention should be paid to the objective reasons. Among the main objective grounds for differentiation of legal regulation of pharmaceutical workers' labour, it is necessary to highlight industry affiliation, working conditions caused by labour specifics, the nature of labour relationship between an employee and an employer. To characterize the objective grounds for differentiation of legal regulation of pharmaceutical workers' labour, it is appropriate to analyze each of them.

The first reason to be observed is the industry affiliation of an enterprise, institution or organization. For this reason, the importance of the industry and a degree of providing a particular enterprise, institution or organization with highly skilled specialists are taken into account. As for the importance of the industry, public health has a significant role in society since each person has the inalienable and indispensable right to health protection.

As far as pharmaceutical workers' qualifications are concerned, the degree of such qualifications in public health is taken into account even more than in any other field and is determined by the level of professional training, practical experience, individual qualities due to the fact that pharmaceutical workers' labour has to do with human life and health. Referring to the Conclusions of the World Health Organization's Second Meeting (held in Tokyo in 1993, "The Role of the Pharmacist: Quality Pharmaceutical Services – Benefits for Governments and the Public"), the concept of pharmaceutical services were formed, according to which the patients' interests have to be in the centre of the pharmaceutical workers' activities given such their proper training.

It should be noted that under the modern conditions of social development, qualified pharmaceutical workers should understand both organizational and economic issues as well as the modern regulatory framework which is changing rapidly enough. They should know the current trends and processes of the pharmaceutical market, understand the purpose, goals, objectives of the institution they work at, comply with the requirements of pharmaceutical activity, bear the corresponding responsibility for the quality of such activities, adhere to appropriate ethical standards in their work, etc.

Training pharmaceutical workers in various countries is carried out differently. Thus, in European countries, higher education is obtained in accordance with the provisions of the Bologna Process. For example, pharmacists' diplomas issued by all European countries are recognized in such countries as the Czech Republic, Poland and Germany. The practice of Poland is interesting enough as students there are granted the right to choose English-language curricula and their curricula exclude subjects not applicable for future professionals. Their special feature is a fairly long period of practice and internship in Germany or France. In Bulgaria, there are training centres set up by pharmacies and pharmaceutical firms. Practice of the United States is different as there is not a federal law on pharmaceutical education and qualifications, so each state has its own requirements for such education. It should be noted that the International Pharmaceutical Federation with the World Health Organization and UNESCO are constantly developing and improving the Global Competency Framework for services provided by pharmaceutical workforce. One of its versions has been developed on the basis of standards, legislative acts and recommendations of pharmaceutical workers' associations. It is based on the areas as follows:

scientific and managerial knowledge, public health and patient orientation, system and applied competences. The global model of pharmaceutical workforce's competences is the basis of Continuing Professional Development, adopted by the majority of countries and the professional standards that are to be followed by such countries rest on it.

Industry affiliation is characterized by the fact that employees perform work only within a certain industry and in accordance with special regulations. Pharmaceutical workers' legal status includes their professional rights, duties, and responsibilities. For example, the right to rest is fixed in the first place in one of the main international documents, namely, Article 24 of the Universal Declaration of Human Rights, which states that "everyone has the right to rest and leisure, including reasonable limitation of working hours and periodic holidays with pay" [1]. Taking into consideration the fact that pharmaceutical workers have harmful and dangerous working conditions, the length of their holidays is prolonged in comparison with other categories of employees.

Furthermore, our attention should be paid to Directive 2003/88/EC of the European Parliament and of the Council of 4 November 2003 concerning certain aspects of the organization of working time which states that long working hours and inadequate rest periods may result in fatigue or other irregular working patterns and lead to further accidents, significant errors, cause stress for a person working in this mode and lead to health risks [16]. It should be noted that pharmaceutical workers often work overtime and this problem is drawn attention to in the Report of the Health and Medical Committee of the International Labour Organization. It refers to a wide range and diversity of the rules for overtime work for medical and pharmaceutical workers depending on the required amount of services provided to the public, as well as the traditions and culture of health care in a country or region [17]. In addition, certain limitations are imposed on pharmaceutical workers in the performance of their duties, for example, the prohibition on receiving unlawful benefits from their activities, providing inaccurate information about the availability of medicines, etc., for which they can be brought to a certain degree of responsibility.

Consequently, industry affiliation as the basis for differentiation is determined, first of all, by the nature of the production process.

Working conditions are the next reason for differentiation of legal regulation of pharmaceutical workers' labour. Each type of specific labour conditions requires labour costs and appropriate efforts. Working conditions are specific to each production, each workplace since they affect their body state, health and work capacity.

As for the issue of normative consolidation of the concept of adequate and safe working conditions, the main international rules are contained in the regulations as follows: the International Covenant on Economic, Social and Cultural Rights, the European Social Charter (Revised), the International Labour Organization Convention No. 155: Occupational Safety and Health, 1981, Occupational Safety and Health Recommendation, 1981 (No. 164) concerning Occupational Safety and Health and the Work Environment [2, 8, 18].

As noted above, pharmaceutical workers deal with harmful and dangerous factors. The factors causing damage to human life and health should be understood as harmful working conditions. Researchers of the European Agency for Safety and Health at Work conclude that as a result of special conditions of work, medical and pharmaceutical workers are exposed to occupational risks such as: biological (when performing duties imposed on pharmaceutical workers, they are in contact with a large number of people as a result of which they may become infected or have other illnesses); chemical (pharmaceutical workers contact with medications and remedies that can cause damage to their health; they are influenced by disinfectants that can cause allergies, etc.); physical (there are effects of ionizing radiation, uncomfortable climate, insufficient lighting of the workplace and other factors affecting this category of employees in the performance of their professional duties); psychological (they are characterized by emotional and psychological stress); ergonomic risks (they suffer long standing and constant walking in the workplace) [19].

It should be emphasized that working conditions have to meet the requirements for carrying out employment duties without life and health hazards, avoid fatigue and ensure high labour productivity. Researchers from the World Health Organization point out that an unsatisfactory working environment has a negative impact on providing the public health sector with human resources and quality of medical and pharmaceutical care as the work environment is a factor of great importance for employing and retaining human resources [20]. As for labour remuneration, differentiation of wages for this category of workers, depending on working conditions, should increase since it is intended to compensate for additional labour costs and create material benefits for workers having harmful working conditions.

It should also be noted that the impact criteria of the working environment, first of all, characterize special conditions of pharmaceutical workers' labour and derive from the workers' right to proper, safe and healthy working conditions. The above criteria are characterized by intra-industry differentiation of legal regulation of labour.

The next reason for differentiation of legal regulation of pharmaceutical workers' labour is the nature of labour relationship between an employee and an employer. The nature of labour relations should be understood as a kind of work as well as a period which necessitates issuing differentiated legal rules [21]. Such differentiation is characterized by a special order of arising and terminating labour relations. In the case of pharmaceutical workers, labour relations usually arise on the basis of an employment contract in accordance with the law. Consequently, their differentiation based on the nature of labour relations is more specific when concluding fixed-term employment contracts.

5 Conclusion

Legal regulation of pharmaceutical workers' labour is expressed through unity and differentiation. The latter is manifested in a significant number of statutory acts since special legislation reflects the peculiarities of work, both objective and subjective.

A fairly large number of statutory acts and international standards do not always speak for their quality. We are sure that, as far as pharmaceutical workers' labour is concerned, common standards for education, training, performance evaluation and adequate working conditions should be binding, enforceable and developed to amend national legislation of all countries.

As for working conditions, they are determined by relevant criteria of production and labour process. Without the creation of proper production and social conditions, introduction of a new system of labour remuneration, protection of medical and pharmaceutical workers, the situation within public health can only deteriorate. Since a number of private pharmacies is increasing, the turnover of personnel is quite significant.

Differentiation of legal regulation of pharmaceutical workers' labour is intended to provide additional privileges, benefits and guarantees that are not clearly regulated by law. In our opinion, it would be reasonable to systematize a significant number of statutory acts into one general document for more rational use since the development of modern labour law should take place within the framework of modern globalization.

References

[1] Universal Declaration of Human Rights. Available at: `http://www.un.org/en/universal-declaration-human-rights/`

[2] International Covenant on Economic, Social and Cultural Rights. Available at: www.ohchr.org/en/professionalinte-rest/pages/ceser.aspx
[3] A. Robertson Human Rights in the world. Manchester (1972)
[4] D. Trubec Economic, Social and Cultural Rights in the Third World: Human Rights and Human Needs Programs // Human Rights in International Law: Legal and Policy Issues: Oxford (1984)
[5] International Covenant on Civil and Political Rights. Available at: www.ochr.org/en/professionalinterest/pages/ccpr.aspx
[6] A/Res/41/120 Sitting International standards in the field of human rights The General Assembly. Available at: www.un.org/ga/res/41/a41r120.htm
[7] European Convention on Human Rights. Available at: www.en.m.wikipedia.org/wiki/European_Convention_on_Human_Rights
[8] European Social Charter (Revised). Available at: www.en.m.wikipedia.org/wiki/European_Social_Charter
[9] Community Charter of the Fundamental Social Rights of Workers. Available at: www.eurofocend.europa.eu/observatories/eurwork/industrial-relations-dictionary/community-charter-of-the-fundamental-social-rights-of-workers
[10] Charter of Fundamental Rights of the European Union. Available at: www.en.m.wikipedia.org/wiki/Charter_of_Fundamental_Rights_of_the_European_Union
[11] Communication from the Commission "On the Charter of Fundamental Rights of the European Union" COM (2000) 559 final. Brussels, (13.09.2000)
[12] B. Hepple. Labour Laws and Global Trade. Hart Publishing. Oxford and Portland, Oregon (2005)
[13] B.Gunter, R. Hoeven, The Social dimension of globalization; A review of the literature// International labour Review, vol.143, No. – 1-2, Geneva (2004)
[14] Labour Code of the Republic of Moldova. Available at: www.cis-legistation.com/document.fwsc?rgn=3836
[15] Republic of Lithuania Labour Code. Available at: www.e-seimas.lrs.lt
[16] Directive 2003/88/EC of the European Parliament and of working time. Available at: www.eur.-lex.europa.eu/smartapi/cgi/sga_doc?smartapi!cebxapi!pred!CELEXnum doc@lg =en&numdoc=3200340088&midel=guichetf
[17] International Labour Organization Standing Service, First Session, Geneva (1992)
[18] Occupational Safety and Health Convention, 1981 (No.155). Available at: www.ilo.org/dyn/normlex/en/f?p=NORMLEXPUB:12100:0::NO::p12100_instrument_id:312300
[19] European Agency of Safety and Health at Work (OSHA). Health and Safety of Healthcare Staff. Available at: www.osha.europa.eu/en/sector/ healthcare.
[20] К. Вискоу, Т. Альбрехт, де К. Прието Как создать для медицинских работников привлекательную и благоприятную рабочую среду // Мат.конф. «Инвестирование в кадровые ресурсы здравоохранения Европы завтрешнего дня: возможности для инноваций и сотрудничества» (Ла-Юльп, 9-10 сентября 2010)
[21] С.А. Иванов Советское трудовое право: вопросы теории / С.А. Иванов, Р.З. Лившиц, Ю.П. Орловский. – М.: Наука (1978)

Legislative framework of in-house procurement

Anatolijs Krivins

Daugavpils University, Daugavpils, Latvia

Abstract. The article "Legislative Framework of In-house Procurement" examines the importance of the concept of the in-house transactions in Public Procurement. The purpose of the work is to analyse the Legislative Framework from the perspective of the principle of free competition. The study of free competition is based on works and the main ideas of the following authors: Adam Smith, Jean-Baptiste Say, David Ricardo, Thomas Robert Malthus, John Stuart Mill, John Maynard Keynes, Friedrich August von Hayek and Milton Friedman. Having done the analysis of the EU directives and regulations concerning In-house Procurement (Directive 2014/24/EU; Directive 2014/25/EU; Directive 2013/34/EU; Case C-107/98, Case C-26/03, Case C-84/03, Case C-295/05, Case C-480/06, Case C-324/07, Case C-573/07, Case C-196/08, Case C182/11 and C183/11), as well as normative regulations of several countries, the author made a conclusion, that In-house Procurement contains considerable threats to the principle of free competition in Public Procurement. In-house Procurement contains considerable threats to the principle of free competition in Public Procurement. The data obtained confirm this hypothesis. The results obtained allow us to develop recommendations for the Legislative Framework of In-house Procurement. The results of the research can be used by procurement theoreticians and practitioners.

Key words: legislative framework, in-house procurement, free competition.

1 Introduction

The aim of this research is to analyse the Legislative Framework from the perspective of the principle of free competition.

The main pillar of the in-house doctrine is the Treaty on the Functioning of the European Union [1] – Article 345 (ex. Article 295 TEC) prescribes: "The Treaties shall in no way prejudice the rules in Member States governing the system of property ownership." Based on the admissibility of the application of the in-house exception, national courts of the EU member countries emphasize rather wide exception limits. For example, Section 16 of the judgment in the Case No. A420612510 rendered by the Department of Administrative Cases of the Supreme Court of the Republic of Latvia on December 29, 2014, states that in the question whether to pass the provision of services to the market as such or to ensure them on one's own, any public person possesses vast freedom of action based on politically defined goals and considerations that may depend on the level of social orientation of the policy it implements. Thus, the control over such determination and action does not abide by the criteria of rights and thus – the influence of the court [2]. Likewise, the Polish Constitutional Tribunal pointed out that the possibility of introducing in-house contracts envisaged by EU law does not prevent the national legislator from putting in place tendering procedures in order to select economic operators to perform specified public tasks [3]. The task of research is to analyze the concept of the in-house transactions in Public Procurement. The object of research is a Legislative Framework of In-house Procurement.

2 Materials and methods

The methods of the present research were: literature review, discourse analysis, logical, deductive and descriptive methods.

3 Results

3.1 In-house doctrine in the European Union rights

According to Directive 2014/24/EU [4], a public contract awarded by a contracting authority to a legal person governed by private or public law, shall fall outside the scope of this Directive where all of the established conditions are fulfilled. Equal regulation (Contracts between contracting authorities) is contained also in the Directive 2014/25/EU (see Directive's Subsection 3, Article 28) [5]. The aim of Directive 2013/34/EU is designing and delivering regulation of the highest quality whilst respecting the principles of subsidiarity and proportionality [6].

The abovementioned provisions of the Directives become of particular significance in the context of the feasibility regarding the current criteria of the in-house doctrine. While analysing the abovementioned Directives, several crucial considerations should be taken into account. First, there is considerable legal uncertainty as to how far contracts concluded between entities in the public sector should be covered by public procurement rules. Second, the relevant case-law of the Court of Justice of the European Union is interpreted differently between Member States and even between contracting authorities. In this context the relevant case-law of the Court of Justice of the European Union becomes a game-changing decision for two main reasons: 1) The sole fact that both parties to an agreement are themselves public authorities does not rule out the application of procurement rules; 2) the application of public procurement rules should not interfere with the freedom of public authorities to perform the public service tasks conferred on them by using their own resources, which includes the possibility of cooperation with other public authorities [4].

The Directive 2014/24/EU emphasises several significant nuances. On the one hand, the exemption should not extend to situations where there is direct participation by a private economic operator in the capital of the controlled legal person since, in such circumstances, the award of a public contract without a competitive procedure would provide the private economic operator with a capital participation in the controlled legal person an undue advantage over its competitors. On the other hand, where there is private capital participation in the controlling contracting authority or authorities, this does not preclude the award of public contracts to the controlled legal person, without applying the procedures provided for by this Directive as such participations do not adversely affect competition between private economic operators [4].

The second module of information under study that carries considerable significance for the understanding of the in-house doctrine is judgements by the CJEU. The Court's case-law shows that a call for tenders is not mandatory where a public authority, which is a contracting authority, exercises over the separate entity concerned control similar to that which it exercises over its own departments, provided that that entity carries out the essential part of its activity with the public authority or with other controlling local or regional authorities. A public authority has the possibility of performing the public interest tasks conferred on it by using its own resources, without being obliged to call on outside entities not forming part of its own departments, and that it may do so in cooperation with other public authorities [7, 8].

In any of these cases, it should be carefully verified whether the local authorities at issue were contriving to circumvent the rules on public procurement. [7] According to settled case-law, the derogations from the rules intended to ensure the effectiveness of the rights conferred by the Treaty in connection with public works contracts must be interpreted strictly [9, 10]. Directives on public procurement are, in general, applicable in the case where a contracting authority plans to conclude a contract for pecuniary interest with an entity which is legally distinct from it, whether or not that entity is itself a contracting authority. Likewise, that directive applies both where a contract is awarded for the purposes of fulfilling the task of meeting needs in the general interest and where it is unrelated to that task [11].

However, there is an exception to that general rule. The directives on public procurement are not applicable, even if the contracting party is an entity legally distinct from the contracting authority, where two conditions are met. First, the public authority, which is a contracting authority, must exercise over the distinct entity in question a control which is similar to that which it exercises over its own departments and, secondly, that entity must carry out the essential part of its activities with the local authority or authorities which control it [11–14].

There is another option for a public authority falling within the definition of a 'contracting authority' to avoid the application of that directive. That is a situation in which a public authority performs the tasks conferred on it in the public interest by using its own administrative, technical and other resources, without being obliged to call on outside entities not forming part of its own departments. Since in such case there can be no question of a contract for pecuniary interest concluded with an entity, legally distinct from the contracting authority, there is therefore no need to apply the Community rules in the field of public procurement [11, 15]. That means that public authorities are not obliged, when performing tasks in the public interest, to turn to the market to obtain the provision of a service. They have the option of choosing between using their own resources or of turning to the market [11].

In order to determine whether a public authority exercises over the other party to the contract a control, similar to that which it exercises over its own departments, it is necessary to take account of not only all the legislative provisions but also the relevant circumstances. That must be a power of decisive influence over both strategic objectives and significant decisions of that entity [8, 11, 13, 16, 17].

In this aspect, one should agree with the opinion that is presented in the article "In-house Procurement – How it is Implemented and Applied in Poland", namely: EU law sets some limits on in-house procurement but does not in itself guarantee uniform application of this modality, leaving a great deal of freedom to Member States. Moreover, European Union law does not in itself guarantee uniform application of in-house procurement rules, as it leaves a great deal of freedom in this respect to the Member States. Member States still have the sole power to decide whether to make an exception in their legal systems concerning in-house procurement [3].

For example, on 23 February 1996, Asemfo lodged a complaint against Tragsa for a declaration that Tragsa was abusing its dominant position in the Spanish forestry works, services and projects market. According to Asemfo, Tragsa's special status enabled it to carry out a large number of works at the direct demand of the Administration, in breach of the principles relating to public procurement and to free competition, which eliminates any competition on the Spanish market [18]. The set prejudicial question remains without answer since the Court has held the view that there is no precise information in the order for reference concerning the existence of a dominant position, its unlawful exploitation by Tragsa or the effect of such position on trade between the Member States.

In any event, it is important to recall that, according to the Court's settled case-law, a call for tenders, under the directives relating to public procurement, is not compulsory, even if the contracting party is an entity legally distinct from the contracting authority, where two conditions are met. First, the public authority which is a contracting authority must exercise over the distinct entity in question a control which is similar to that which it exercises over its own departments and, second, that entity must carry out the essential part of its activities with the local authority or authorities which control it [12, 15, 17, 19, 20]. The market for possible private candidates from elsewhere in the Community is then correspondingly restricted [21].

In cases, where the relationship between a contracting public authority and an executive service or entity is not governed by the Community rules on public procurement the general provisions of the Treaty [16, 22], and in particular the provisions on fundamental freedom of movement and competition, remain applicable. Nevertheless, this can result in substantial sub-markets being removed from the operation of primary and secondary Community law on public procurement [21]. Those risks lie in the fact that the useful effect of the Community directives on public procurement may be seriously undermined or, where those directives are not applicable, serious obstacles may arise to the movement of goods and services between States and to freedom of establishment, and competition between those executive services or publicly controlled entities and private undertakings on the relevant markets may be seriously distorted [21].

Let us review the following example – Judgement of the Court in Case C-573/07 *Sea* [23]. In this aspect, exceptionally, special circumstances may require events occurring after the date on which the contract in question was awarded to be taken into consideration. Such is the case, in particular when shares in the contracting company, previously wholly owned by the contracting authority, are transferred to a private undertaking shortly after the contract at issue has been awarded to that company by means of an artificial device designed to circumvent the relevant Community rules [20, 23].

It is apparent from the case-law that the award of a public contract to a semi-public company without a call for tenders would interfere with the objective of free and undistorted competition and the principle of equal treatment, in that such a procedure would offer a private undertaking with a capital holding in that company an advantage over its competitors [15, 20, 24].

In the case C196/08 *Acoset*, the Advocate General has expressed several crucial comments. Namely, the Advocate General considers that Unlike Commission v Austria [20], in which the Court found that there was an artificial construction comprising several distinct stages and revealed the unlawful nature of the award of a contract to a semi-public company under the guise of an in-house relationship, in the present dispute the issue of competition comes into play and displays unprecedented potential in a procedure in which a measure of economy is evident, for, in a single instrument, a company is incorporated and assigned a task, and the criteria for inclusion of the private industrial participant are adopted. What more can one ask? The difficulty thus arises of determining whether that procedural short-cut complies with the requirements of Community law /…/ An affirmative answer to the question referred for a preliminary ruling is unavoidable given the need for efficient administration, which is an essential requirement, provided that the safeguards deriving from Community law are not disregarded [25].

The answer to the question must therefore be that Directive 93/36 is applicable in the case where a contracting authority, such as a local authority, plans to conclude in writing, with an entity which is formally distinct from it and independent of it in regard to decision-making, a contract for pecuniary interest for the supply of products, whether or not that entity is itself a contracting authority [12].

On the basis of the analysis performed it can be stated that the EU member countries have too much freedom in the interpretation of the in-house exception. Furthermore, although the Directives 2014/24/EU and 2014/25/EU, as well as court judicature determine certain in-house limits, yet the most important decision – to decide in favour of one's own firm or the free market – still remains up to the customer.

The article will continue with the analysis of the risks set by the in-house doctrine, among them questions whether the customers always decently use this freedom of action and if it is really necessary to have this amount of freedom of action.

3.2 In-house doctrine within the principle of free competition

In the classical understanding "Free competition" meant absence of grants of exclusive privilege, freedom of trade and freedom of entry into occupations [26]. For example, to Adam Smith, "competition" meant competition unhampered by governmental grants of exclusive privilege - rivalry between two or more independent persons or firms, the competition of sellers, which bids prices down when supply is greater than demand [26, 27].

The concept of the free competition traditionally includes liberty to choose one's employment, free competition, free trade, beyond as well as within the frontiers of a single country, free banks, and a competitive rate of interest; and on the negative side it implies resistance to all State intervention wherever the necessity for it cannot be clearly demonstrated [28]. In the opinion of Classical writers, free competition was the sovereign natural law.…It secured cheapness for the consumer, and stimulated progress generally because of the rivalry it aroused among producers [28].

Taking into account the facts mentioned, it is logical to conclude that artificial limitation of competition (also as in-house deals) causes significant risk of rise in the price and hindering the progress. The majority of classical economists emphasize the negative manifestations of the competition restriction. For example, Ricardo – there are several pages of attack on the British colonial monopolies — grants of exclusive privilege such as the British East India Company [29]. John Stuart Mill predicts misallocation of resources in markets where monopoly power prevails [30, 31].

Jean-Baptiste Say believed that the love of domination never attains more than a factitious elevation, which is sure to make enemies of all its neighbours [32]. While studying the impact of the in-house doctrine on the principle of the free competition, it is very important to define the compulsory traits of the free competition. According to Lewis Henry Haney, perfectly equal competition is rare, and elements of differential advantage abound on all hands [33]. "Perfectly free competition" implied not only absence of price discrimination, but also a large number of buyers and sellers in each industry [34]. "Free market" implies that at each step of the way, terms of exchanges, or prices, are determined by the voluntary interactions of suppliers and demanders. This market is "free" because choices, at each step, are made freely and voluntarily. The market, then, is not simply an array; it is a highly complex, interacting latticework of exchanges [35].

According to Alfred Marshall, we do not assume that competition is perfect. Perfect competition requires a perfect knowledge of the state of the market [36]. John Bates Clark [37] emphasizes that labour and capital must be absolutely mobile [35]. The free competitive market also rewards and stimulates technological innovation. Does the restriction of competition have its positive aspects? In the philosophical view, a common charge against the free-market society is that it institutes "the law of the jungle," of "dog eats dog," that it spurns human cooperation for competition. Yet the historical development of the world has proven that the peaceful market competition of producers and suppliers is a profoundly cooperative process in which everyone benefits and where everyone's living standard flourishes [35].

Though in particular conditions monopolies and protectionism can also ensure particular benefits (Clark saw the advantages that could come from mergers and large firms: a vast corporation that is not a true monopoly may be eminently progressive [38]), yet these probable benefits do not compensate for obvious drawbacks. As it is justly indicated by Alfred Marshall "a man of sound judgment . . . will keep a watchful eye on sources of possible competition, direct and indirect" [39], albeit the difficulty does not lie in finding a producer, but in finding a consumer [32].

The in-house question affects not only the aspect of offer, but also that of demand. If services of a particular supplier are used only by the only big customer, such order will hardly be efficient. Already in the 19th century this issue was highlighted by Thomas Robert Malthus: thirty or forty proprietors, with incomes answering to between one thousand and five thousand a year, would create a much more effectual demand for the necessaries, conveniences, and luxuries of life, than a single proprietor possessing a hundred thousand a year [40]. Like in the past landowners had assured the best places in Malthus's banquet through force, and that they could benefit from it without contributing to the growth of the national Wealth [31], also today businessmen who enjoy the in-house privilege, gain advantage without any risk. Here we can quote Mill: "They grow richer, as it were in their sleep, without working, risking or economizing" [41].

While outlining the drawbacks of limiting the free competition, the following points should be mentioned. In the long run, coercion is a negative-sum game that leads to reduced production, saving, and investment; a depleted stock of capital; and reduced productivity and living standards for all, perhaps even for the coercers themselves [35].

The economic freedom (as the freedom to produce, trade and consume any goods and services acquired without the use of force, fraud or theft), that correlates with the highest standard of living, economic growth, income equality, less corruption and less political violence is based on three cornerstones: 1) Rule of law; 2) Private property rights; 3) Freedom of contract. Freedom of contract is the right to choose one's contracting parties and to trade with them on any terms and conditions one sees fit. Contracts permit individuals to create their own enforceable legal rules, adapted to their unique situations [42]. Parties decide whether contracts are profitable or fair, but once a contract is made they are obliged to fulfil its terms, even if they are going to sustain losses by doing so [43]. Government, in every society, is the only lawful system of coercion [35].

Smith believed that any interference with free competition by government is almost certain to be injurious. If profits in one line of business are unduly large, there will be a rush of other businessmen into that field until competition has lowered surpluses [44]. The nature of the market has changed vastly since the 18th century – put into today's language, Smith explicitly recognizes the usefulness of public investment for projects that cannot be undertaken by the private sector – he mentions roads and education as two examples. Smith is against government laws that shelter industry from competition. The great enemy to Adam Smith's system is not so much government per se as monopoly in any form [44].

John Maynard Keynes provided a specific rationale for government's taking a bigger role in the economy and a more general confidence in the ability of government to intervene and manage effectively. However, in1970s that evidence began to accumulate in many countries that Keynes's theories, at least as implemented by Keynes's advocates after his death, might not perpetually yield the favourable outcomes Keynes himself had predicted [45].

Friedrich von Hayek believed that the problem of knowledge defeats central control of economies: where effective competition can be created, it is a better way of guiding individual efforts than any other. It does not deny, but even emphasizes, that, in order that competition should work beneficially, a carefully thought-out legal framework is required. And it regards competition as superior not only because it is in most circumstances the most efficient method known, but even more because it is the only method by which our activities can be adjusted to each other without coercive or arbitrary intervention of authority [46]. It is necessary in the first instance that the parties in the market should be free to sell and buy at any price at which they can find a partner to the transaction and that anybody should be free to produce, sell, and buy anything that may be produced or sold at all. And it is essential that the entry into the different trades should be open to all on equal terms and that the law should not tolerate any attempts by individuals or groups to restrict this entry by open or concealed force [46].

Exceptions to the free competition should be admitted where no legal arrangements can create the main condition on which the usefulness of the system of competition and private property depends: namely, that the owner benefits from all the useful services rendered by his property and suffers for all the damages caused to others by its use, [46] where the implementation of that cooperation is governed solely by considerations relating to the public interest and that no private service provider is placed in a position of advantage vis-à-vis its competitors [4].

It is not unusual for a number of public authorities to establish, for the execution of certain public tasks such as sewage treatment, a partnership which is responsible for managing a joint executive service [21]. But the fact that we have to resort to the substitution of direct regulation by authority, where the conditions for the proper working of competition cannot be created, does not prove that we should suppress competition where it can be made function [46]. The only exception is where there is objective justification for such work, such as in the case of natural disasters or similar exceptional circumstances [21].

The role of the State in ensuring the free competition is of extreme significance: 1) to create conditions in which competition will be as effective as possible, to supplement it where it cannot be made effective, to provide the services, 2) an effective competitive system needs an intelligently designed and continuously adjusted legal framework as much as any other. [47] In Milton Friedman's view, if you have a central direction you inevitably have coercion. The only way that has ever been discovered to have a lot of people cooperate together voluntarily is through the free market [48].

The article "In-house providing and (minimum) "effective" public control: Sunset or breaking dawn for purely public (commercial) service providers? (C182 and 183/11)" has analysed statements that are present in the Judgment of the Court of Justice of the EU of 29 November 2012 in Joined Cases C182/11 and C183/11. According to this line of case law, contracting entities can award contracts directly (without a competitive tender) where they exercise over the contractor a control similar to that which they have over their own departments, and the contractor carries out the essential part of its activities with the contracting authorities to which it belongs. In those cases, it is assumed that there is no potential for competition and that the market is not affected by the decision of the contracting authority to retain the activity "in-house" [49].

Particular attention should be paid to Section 31 of the judgment: where the position of a contracting authority within a jointly owned successful tender does not provide it with the slightest possibility of participating in the control of that tender, that would, in effect, open the way to circumvention of the application of the rules of EU law regarding public contracts or service concessions, since a purely formal affiliation to such an entity or to a joint body managing it would exempt the contracting authority from the obligation to initiate a tendering procedure in accordance with the EU rules, even though it would take no part in exercising the 'similar control' over that entity [22, 50].

This risk has not been duly evaluated in the EU Directives. No doubt that the in-house doctrine limits the free competence in any case by attributing the procurement beyond open procedures. This considerably increases the risk of corruption. For instance, in Poland, what has been observed on the market is the tendency for municipal companies to use the privilege created for them in in-house procurement regulations to encroach on a market that is not related to the tasks of their owner and compete with private operators. (One of the practices enabling local authorities to bypass statutory restrictions when contracting their tasks to municipal companies is to set up subsidiaries (or sister companies) of a municipal company) [3]. Undoubtedly, if the contracting authorities want to disengage from the direct management of those services and take the back seat (for example in a board of directors), there is no reason to see why public contractors should be shielded from the competition of private contractors, since both would be offering a commercial relationship to the outsourcing contracting authority and there would be an effective risk of generating relevant distortions of competition [49, 51]. Therefore, in the lack of a sufficiently active involvement, in the absence of an actual organic link between the contracting authority and the "in-house" entity, there is no good reason to exclude the application of the EU public procurement rules [49].

One should agree with the conclusion expressed in the article, yet not only because of the justification provided in the article. Danger lies not in the aspect who governs the supplier with the preferences, but in the restriction of the competition per se. From the business point of view, there is no considerable difference in who the preference is given to – a businessman with the State/municipal capital or a private firm, since in both cases the consequences are the same – the principle of the free competition is deformed. In this aspect the primary issue is not that of control (or formal conformity), but that of expedience.

4 Conclusions

1. On the basis of the analysis performed it can be stated that the EU member countries have too much freedom in the interpretation of the in-house exception. Furthermore, although the Directives 2014/24/EU and 2014/25/EU, as well as court judicature determine certain in-house limits, yet the most important decision – to decide in favour of one's own firm or the free market – still remains up to the customer.
2. The in-house exception should be interpreted as narrowly as possible. Otherwise the regulation regarding ensuring of the free competition included in the Directives will carry declarative character only.
3. The in-house doctrine, being in philosophical contradiction with the principle of the free competition, considerably endangers both the concept of the free economy and the free market. No doubt that the in-house deals definitely limit the free competition. The question is only about the fact to what extent this limitation is justifiable and required.
4. The author of the article holds a view that the legal, economical or philosophical necessity for such an exception cannot be justified in standard fields that are widespread in the market – such as construction work, designing services and other. Therefore, EU Directives shall clearly and unmistakably state what particular types of procurement contracts (fields) can enjoy the in-house advantages. Only specific fields should be left on this list of exceptions – for instance, waste management, production of heat energy, water supply, etc.
5. Administrative convenience the customer may gain as the result of an in-house deal (savings on indirect costs in case public procurement is not realized, faster process for concluding a procurement contract, more convenient administration of the concluded contract, etc.) cannot be more important than the free competition and efficient

use of public funds. Therefore, in fields where competition exists, verification of economic efficiency of the planned procurement contract must be performed as an open procurement procedure where any supplier could participate.

6. The in-house doctrine has considerable impact not only on the free competition, but also on the EU fundamental values regarding economy without borders (since in this way the locals can ensure protectionism), it affects risks of corruption.

References

[1] Consolidated version of the Treaty on the Functioning of the European Union // Available at: `https://eur-lex.europa.eu/legal-content/EN/TXT/HTML/?uri=CELEX:12012E /TXT &from=LV` (seen on 05.02.2019)
[2] Available at: `http://www.at.gov.lv/downloadlawfile/4425` (seen on 05.02.2019)
[3] W. Hartung, K. Kuzma, In-house Procurement – How it is Implemented and Applied in Poland. European Procurement & Public Private Partnership Law Review **13**(3), 171–183 (2018), `https://doi.org/10.21552/epppl/2018/3/4` (seen on 05.02.2019)
[4] Directive 2014/24/EU of the European Parliament and of the Council of 26 February (2014)
[5] Directive 2014/25/EU of the European Parliament and of the Council of 26 February (2014)
[6] Directive 2013/34/EU of the European Parliament and of the Council of 26 June 2013
[7] Judgement of the Court in Case C-480/06 Commission v. Germany, [2009] ECR I–04747
[8] Judgement of the Court in Case C324/07 Coditel Brabant [2008] ECR I0000
[9] Judgement of the Court in Case C-57/94 Commission v Italy [1995] ECR I-1249
[10] Judgement of the Court in Case C-318/94 Commission v Germany [1996] ECR I-1949
[11] Opinion of Advocate General in Case C-480/06 Commission v. Germany
[12] Judgement of the Court in Case C107/98 Teckal [1999] ECR I8121
[13] Judgement of the Court in Case C337/05 Commission v Italy [2008] ECR I0000
[14] Judgement of the Court in Case C-371/05 Commission v. Italy [2008] ECR I–00110
[15] Judgement of the Court in Case C26/03 Stadt Halle and RPL Lochau [2005] ECR I1
[16] Judgement of the Court in Case C458/03 Parking Brixen [2005] ECR I8585
[17] Judgement of the Court in Case C340/04 Carbotermo and Consorzio Alisei [2006] ECR I4137
[18] Judgement of the Court in Case C-295/05 Asemfo [2007] ECR I-2999
[19] Judgement of the Court in Case C-84/03 Commission v Spain [2005] ECR I-139
[20] Judgement of the Court in Case C-29/04 Commission v Austria [2005] ECR I-9705
[21] Opinion of Advocate General in Case C-295/05 Asemfo
[22] Judgement of the Court in Case C231/03 Coname[2005] ECR I-7287
[23] Judgement of the Court in Case C-573/07 Sea [2009] ECR I-8127
[24] Judgement of the Court in Case C-196/08 Acoset [2009] ECRI-9913
[25] Opinion of Advocate General in Case C-196/08 Acoset
[26] Ludwig von Mises Institute, Strictly Confidential: The Private Volker Fund Memos of Murray N. Rothbard (2010). `https://mises.org/library/ strictly-confidential-private-volker-fund-memos-murray-n-rothbard`
[27] A. Smith, Wealth of Nations. Modern Library (2000)
[28] C. Gide, C. Rist, A History of Economic Doctrines from the Time of the Physiocrats to the Present Day. Heath (1930)

[29] D. Ricardo, On the Principles of Political Economy and Taxation. Piero Sraffa (Ed.) Works and Correspondence of David Ricardo, Volume I (Cambridge University Press, 1951)
[30] J.S. Mill, Principles of Political Economy (Appleton, 1901)
[31] E. Trincado Aznar, A turning point in the concept of justice: from Adam Smith to John Stuart Mill. Las Torres de Lucca. Int. J. Political Phil. **1**, 75–96 (2012)
[32] J.B. Say, A Treatise on Political Economy. Augustus M Kelly (New York, 1964)
[33] L.H. Haney, History of Economic Thought (Macmillan, 1949)
[34] W.S. Jevons, The Theory of Political Economy, 3rd ed. (Macmillan, 1888)
[35] The Library of Economics and Liberty, Free Market by Murray N. Rothbard (2009). Available at: `http://www.econlib.org/library/Enc1/FreeMarket.html` (seen on 05.02.2019)
[36] A. Marshall, Principles of Economics, 8th ed. (Macmillan, 1938)
[37] J.B. Clark, The Distribution of Wealth: A Theory of Wages, Interest and Profits (The Macmillan Company, NY, 1908)
[38] J.B. Clark, Essentials of Economic Theory: As Applied to Modern Problems of Industry and Public Policy Macmillan (1907)
[39] A. Marshall, Industry and Trade: A Study of Industrial Technique and Business Organization; and of their Influences on the Conditions of Various Classes and Nations (Macmillan, London, 1919)
[40] T.R. Malthus, On the Progress of Wealth. Reprinted from his Principles in "Studies in Economic Development". Edited by Okun and Richardson, Holt, Rinehart and Winston (1961)
[41] J.S. Mill, The Collected Works of John Stuart Mill, Volume XX – Essays on French History and Historians, ed. John M. Robson, Introduction by John C. Cairns. Toronto, University of Toronto Press, London, Routledge and Kegan Paul (1985)
[42] J.V. Orth. Contract and the Common Law. The State and Freedom of Contract (Stanford University Press, 1998)
[43] D.A. Harper. Foundations of Entrepreneurship and Economic Development. Routledge, 82–88 (1999)
[44] The Lives, Times, and Ideas of the Great Economic Thinkers by Robert L. Heilbroner, 7th ed. (1999)
[45] Available at: `https://www.pbs.org/wgbh/commandingheights/shared/minitext/ess_keynesiantheory.html` (seen on 05.02.2019)
[46] The Road to Serfdom: Text and Documents–The Definitive Edition (The Collected Works of F. A. Hayek, Vol. 2 (2007)
[47] J.C. Wood, R.D. Wood, F.A. Hayek: 2nd series (Taylor & Francis, 2004)
[48] Available at: `http://www.pbs.org/wgbh/commandingheights/press_site/people/friedman_intv.html` (seen on 05.02.2019)
[49] In-house providing and (minimum) "effective" public control: Sunset or breaking dawn for purely public (commercial) service providers? (C182 and 183/11) (2012). Available at: http://www.howtocrackanut.com/blog/2012/12/in-house-providing-and-minimum. html (seen on 05.02.2019)
[50] Judgement of the Court in Joined Cases C-182/11 and C-183/11, Econord [2012] ECR I-00000
[51] S. Graells, Public Procurement and the EU Competition Rules (Oxford, Hart Publishing, 2011)

Legal implementation of blockchain technology in pharmacy

Vitalii Pashkov[1] and *Oleksii* Soloviov[2]

[1] Poltava Law Institute, Poltava, Ukraine
[2] Shupyk National Medical Academy, Kiev, Ukraine

Abstract. Author analyzes the possibilities of using blockchain technology in pharmaceutical sphere with the purpose of detection of counterfeit and poor-quality medicinal products and discusses the legal acts concerning the use of this technology. Author examines the grounds for the legal provision of blockchain technology in pharmaceutical sector in order to protect the rights of patients. The necessity of legal regulation of the use of blockchain technology by private parties is substantiated taking into account the private legal nature of this system. The mechanism of application of this technology in pharmaceutical activity is revealed.

Key words: counterfeit medical products, fake medicine, blockchain technology.

1 Introduction

The study of legal aspects of preventing the spread of falsified medicines is highly relevant from both theoretical and practical point of view. Negative consequences, caused by the spread of falsified medicines are so obvious that there is a need for modern technologies for their detection. The use of blockchain technology in pharmacy could provide a reliable legal protection of the supply of pharmaceutical products and its tracking. Such system could provide significant simplification of audit and control of pharmaceutical products, including by the specially authorized bodies. Healthcare and the pharmacy industry specifically, could employ blockchain to streamline efficiencies, improve privacy and confidentiality, enhance patients' safety, and provide a higher level of clinical care to consumers. The main advantages of the blockchain technology are the maximum transparency, speed and accuracy of transactions. Although the investment in blockchain technology is relatively low, it is rapidly growing. For example, IBM and Microsoft are incorporating blockchain platforms in their customer support systems. Additionally, major companies are actively exploring the value of integrating blockchain technology into their operations. Examples are Walmart – vendor payment and digital shopping; Starbucks – tracing beans from various countries; American Express – collaborating with Boxed to customize rewards for cardholders; JD.com – use blockchain technology to accelerate AI development.

2 Aim

The aim of this study is to explore the potential of blockchain technology in ensuring patients' rights to quality pharmaceuticals.

3 Material and methods

This study examines provisions of international documents and scientists' attitudes. Author analyzes generalized information from scientific journals by means of scientific methods from medical and legal point of view. This article is based on dialectical, comparative, analytic, synthetic and comprehensive research methods.

In order to achieve this goal Directive 2011/62/EU of the European parliament and of the Consul of 8 June 2011 amending Directive 2001/83/EC on the Community code relating to medicinal products for human use, as regards the prevention of the entry into the legal supply chain of falsified medicinal products has been analyzed. The international aspects of pharmaceutical activity of the USA and certain European countries are investigated. During the study, methods of data extraction and content analysis were used along with scientific publications of leading specialists in the pharmaceutical law field.

The methodology of this scientific publication is based on the organic combination of philosophical approaches, and general scientific and special legal research methods. In particular, the systematic method is used to carry out a systematic analysis of European legislation on the circulation of medicines and the protection of patients' rights. The comparative method allows the author to compare the experience of foreign countries in the patients' rights legal protection area. The legal modelling method is used to formulate proposals for improvement of existing legislation and practice in the field of patients' rights protection.

The analysis of scientific sources clearly shows that the issues of pharmaceutical activity and patients' rights have been studied by recognized scholars, namely: A. Harkusha [1, 2], N. Gutorova [3], A. Olefir [4] and others. Meanwhile, the problems of legal provision of Blockchain technology use in pharmacy have not been studied sufficiently.

4 Results

The legal basis for the introduction of Blockchain in pharmacy is Directive 2011/62/EU of the European Parliament and of the Consul of 8 June 2011 amending Directive 2001/83 / EC on the Community code relating to medicinal products for human use in respect of the prevention of the entry of falsified medicinal products into the legal supply chain. This document, which came into force on January 2013, stipulates that the manufacturer can mark drugs registered in the country with the unique identifier.

The directive also provides for the introduction of a system that helps the consumer to identify legitimate websites that sell legal drugs through a recognizable logo. In the United States, Great Britain, Ukraine and most EU countries such rules are based on the legal act for quality and safety of medicines.

Substandard and falsified medical products, as defined by the World Health Organization, are major threats to public health. Through increasing globalisation, the problem has expanded in both developed and developing countries [5].

Sale of fake medicines is rising rapidly and is considered a global epidemic. It is estimated WHO that 10–30% of drug market is counterfeit, with the upper end of that range predominantly in emerging regions [6]. For example, in 2015 as part of Interpol's Operation Giboia more than 150 tons of counterfeit and illicit medicines for nearly USD 3.5 m were seized in seven African countries.

With regards to internet sales over 20.7 million counterfeit and illicit medicines, including products for blood pressure medicines, erectile dysfunction and cancer as well as nutritional supplements, were seized by Interpol's Operation Pangea in 2015 [7].

The falsification and distribution of counterfeit medicines and other medical products is one of the most serious crimes against public health. As any other fraudulent economic activity, this crime is harmful to legal producers, distributors and pharmacy establishments that lose their income and sometimes also are compelled to compete with distributors of counterfeit goods.

But the worst consequences are the damage to people's life and health, which cannot be overestimated. The range of such damage vary from distrust of doctors, resulting from the ineffectiveness of treatment, to causing death or serious harm to health due to the use of poor quality medication and/or the lack of proper therapeutic effect, false diagnostic results, etc. These effects are not always easy to establish, it is even more difficult to prove the existence of a causal link between the use of counterfeit medicines and caused harm to human life or health [8]. However, existing global pharmaceutical supply chains are complex and work with outdated platforms that do not provide effective partnerships for regulatory compliance and cannot guarantee the security of drug distribution [9]. That is why researchers say that an automatic drug tracking system should be an effective component of preventing the spread of counterfeit medicines through the individual coding of each package, as is the case in EU countries [10].

Another effective reason for the introduction of blockchain technology is the Council of Europe Convention of 28.10.2011 on the counterfeiting of medical products and similar crimes involving threats to public health. In accordance with this international legal instrument, crimes in this area are deemed deliberate fabrication, supply or proposal of supplies, including mediation, trade, including storage in the warehouse, import and export of counterfeit medical products (medicines for human and veterinary use and medical devices), active substances, parts, materials and accessories (Articles 4, 5 and 6 of the Convention).

The implementation of the provisions of this Convention provides for the imposition of sentences of imprisonment for such crimes, strengthening of penalties for persons who have committed crimes in their professional or official duties (through the abuse of trust given to them as professionals or as producers or suppliers). An important aspect of Convention is the provision of the possibility of bringing legal entities to justice for crimes established in accordance with this Convention committed in their favor by any natural person who acted in person or as a part of a legal entity and held a leading position within this legal entity.

Meanwhile, researchers say: "Digital technology has been one of the most significant drivers impacting our businesses and our lives. Blockchain has the potential to be very disruptive to our healthcare supply chain. It's important for healthcare stakeholders, including pharmacy, to really understand the potential implications and be strategically positioned to evaluate and potentially adopt the "fast-moving" commercial applications as blockchain integrates with mainstream healthcare supply chain" [11].

The US experience shows that there are risks of using the digital system – fraud and forgery. That is why the United States continues to look for the ways to protect personal data of patients when they issue prescription drugs due to the risk of their digitization [12]. Meanwhile, in May 2018, updated personal data processing rules established by the General Data Protection Regulation (EU Regulation 2016/679 of April 27, 2016 or GDPR – General Data Protection Regulation) entered into force in the EU. This regulation has direct effect in all 28 EU countries, will replace the Framework Directive on personal data protection 95/46/EC of October 24, 1995. An important nuance of GDPR is the extraterritorial principle of the new European rules for the processing of personal data. That is, GDPR applies to all companies that handle personal data of residents and EU citizens regardless of location of such company.

Blockchain enables that data sharing capability without sacrificing privacy while offering pharmacists the ability to verify the source of the medicine. Blockchain could provide pharmacists with appropriate level of ensuring of patient safety by verifying the source of supplies of medicines.

In addition, the blockchain technology allows to create new decisions to control the circulation of drugs, in particular, in terms of compliance with the temperature regime in transferring of drugs [13].

Blockchain Technology is a public registry that is based on cryptographic system algorithms. It stores data about all operations and changes that have occurred in system, is decentralized and grounded on public Internet services.

The basic principles of the Blockchain system are: 1) decentralization; 2) the absence of intermediaries; 3) public availability (publicity); 4) the consensus of the participants [14].

It should be noted that supply chain management (SCM) has been used for a long time before [15]. Supply Chain Management (SCM) or Supply Chain Management is the term that came into circulation relatively recently. This concept is voluminous and includes the definitions used earlier, such as ECR (Efficient Customer Response) and DRP (Distribution Resource Planning).

Blockchain enables that data sharing capability without sacrificing privacy while offering pharmacists the ability to verify the source of the medicine [16]. In addition, for example in the United States the federal government more and more often puts the pharmaceutical industry in charge of controlling the supply chain for illicit and non-standard drug orders [17].

To date, the patient cannot track which drug is being sold to him, both at an independent procurement and in inpatient treatment. Although the patient has the right to demand for example a quality certificate for a medicinal product, nobody prevents an unscrupulous vendor or employee of a pharmacy institution from tampering with it.

However, if all the information is stored in the blockchain from the moment of clinical research, production, further storage and delivery to the pharmacy and/or its structural subdivision, including the medical institution, it will be possible to trace the entire chain.

It is clear that this system can be modified by adding of additional functionality, for example, the rules for drugs storage and treatment of inpatients. It is possible to establish to whom drugs were transferred and, in general, to what extent they are needed. For this purpose, the packaging of medicinal product (for instance) has a barcode using which one could via special application enter blockchain. And there is second important characteristic of Blockchain – it is irremovable. The information once stored in it cannot be modified or deleted, which is a huge advantage over the standard database in which data can easily be falsified also with hiding the fact of making any changes in it.

Thus, checking the reliability of information on the effectiveness of drugs, their effectiveness, safety, including adverse reactions may become available to the patient by dint of blockchain system [9].

Huge pharmaceutical conglomerates like Pfizer and Genentech with a total market capitalization of over $ 300 billion announced the creation of a blockchain project MediLedger. It is based on the Ethereum-based Quorum platform, the purpose of creating a blockchain – to block access to the supply chain for counterfeit goods [18].

In addition, thanks to the new Ethereum system pharmaceutical manufacturers will be able to detect poor-quality medicines before they enter the supply chain [19].

A separate sphere of blockchain technology use is clinical research. Such aspects of blockchain as decentralized transaction verification, transparency of information are useful for clinical research. And that is why in January 2018 pharmaceutical companies Pfizer,

Amgen and Sanofi announced the creation of a blockchain platform for clinical trials to reduce the cost of developing and testing of new drugs [20].

According to experts, the confidentiality and transparency problems in clinical trials of medicines may hinder future drug development and increase costs. In order to overcome the confidentiality issues, it was decided to turn to such a new technology as Blockchain [21].

When conducting clinical studies with the help of Blockchain technology it is monitored: 1) the place, time, period, performers of clinical trials; 2) the order of testing and the level of patient health at different stages of research of pharmaceutical products; 3) the results of the tests.

Disclosure of clinical trial data is rapidly becoming a top priority for both large and small pharmaceutical companies in the United States [22].

The Blockchain system will allow interested patients to make their personal data available to pharmaceutical companies and research institutions. At the same time, the research results will be automatically recorded in the blockchain with no possibility of making changes or falsifying research results. In general, the clinical trials' registration taking into account unclear legal status of the experiment participant is a relevant problem [14].

Blockchain technology allows you to observe the chain of supply: 1) all information is stored in Blockchain from the time of release and till entering to the retail network, then you can add functional to this system, for example, to control the drugs in medical institutions; 2) the conditions of storage of medicinal products during their delivery (temperature regime) are also monitored; 3) it is possible to trace to whom concrete drugs were given, in what amount (for that purpose there is a bar code on a medicine package).

Thus, Blockchain technology allows: 1) to improve the clinical trials of drugs; 2) to improve the procedure of licensing of pharmaceutical products; 3) to track the quantity of pharmaceutical products from the manufacturer with the further selling thereof; 4) to track sources of origin of pharmaceutical products and the procedure for their use; 5) to monitor the registration period, expiration date, conditions of transportation and storage of pharmaceutical products; 6) to limit the activities of online pharmacies that do not have permits; 7) to minimize the possible shortage of pharmaceutical products; 8) to ensure openness to all partners.

In addition, the use of Blockchain technology in pharmacy can provide reliable legal protection for the supply of pharmaceutical products, tracking it's transferring from manufacturer to patient. The system provides for simplification of audit and control of pharmaceutical products, including by the specially authorized bodies.

However, it must be noted that the existing legal framework for the use of Blockchain technology is imperfect, although de facto used in the field of pharmaceutical activity. The use of this technology not only minimizes the risks of supplying counterfeit medicines to patients but also provides for transparency of delivery and reduced costs for logistics. As a consequence, there is a reduction in prices of final product for patients. However, application of Blockchain technology requires appropriate legal support.

Analysis of the prospects for using Blockchain technology allows us to establish that the objects of legal relationships in this technology are: 1) computer software; 2) telecommunication networks; 3) informational resources, products and services; 4) patients' health rights; 5) informational security.

Meanwhile, Blockchain's relationship in healthcare has a private-law nature (dispositive), despite the publicity of using Blockchain. As a result, the lack of imperative principles for the use of Blockchain technology in pharmacy, including in terms of patient rights ensuring can undermine the ability to use such cutting-edge and advanced technologies.

The technology of blockchain in pharmacy allows the use of so-called smart contracts between contractors. Smart contract is an electronic protocol written using a computer code. His appointment is the transfer of information and enforcement of contract terms by both parties.

Smart contracts provide security for the transaction and devoid of the risk of ambiguous interpretation of the conditions, due to the fact that they are based on cryptography. They are much more beneficial deals financially because a person does not need to pay lawyers, intermediaries etc. The terms of the agreement are implemented automatically with the minimum cost for their support, without the involvement of third parties (intermediaries) because smart contracts are based on blockchain technology.

This is a distributed registry, which is a decentralized system that exists due to the myriad of computers that are integrated into a single network. Blockchain allows users to execute transactions, transfer information and material value without banks and intermediaries. If information is stored in a blockchain, which everyone trusts, most of the process can be automated. As soon as the records enter the chain, the smart contract checks whether all conditions are fulfilled and exercise payment. If there's not enough information it asks for an additional study of the situation. The widest dissemination of smart contracts is observed in business sector in payments and payment-related actions [23].

5 Conclusions

One must admit that the existing legal basis for the use of blockchain technology is imperfect, although it is de facto used in the field of pharmaceutical activity. The use of blockchain technology not only minimizes the risks of supplying counterfeit medicines to patients, but also ensures transparency of delivery and reduction of logistics costs. Consequently, there is a reduction in price of the final product for patients. However, full engagement of blockchain technology in pharmacy requires adequate legal support.

References

[1] Vitalii Pashkov, Andrii Harkusha, Certain aspects on medical devices software law regulation. Wiadomości Lekarskie **LXIX**(6), 765–767 (2016)

[2] Vitalii Pashkov, Alla Kotvitska, Andrii Harkusha, Legal regulation of the production and trade of medical devices and medical equipment in the EU and USA: experience for Ukraine. Wiadomości Lekarskie **LXX**(3) cz. II., 614–618 (2017)

[3] Vitalii Pashkov, Nataliya Gutorova, Andrii Harkusha, Medical device software: defining key terms. Wiadomości Lekarskie **LXIX**(6), 813–817 (2016)

[4] Vitalii Pashkov, Andrii Olefir, The impact of the legal regime of intellectual property protection in the pharmaceutical market. Wiadomości Lekarskie **LXIX**(3) (cz. II), 582–587 (2016)

[5] International Pharmaceutical Federation (FIP), FIP is the global federation representing four million pharmacists and pharmaceutical scientists worldwide. URL: `https://www.fip.org/counterfeit_medicines`

[6] WHO. Substandard and falsified medical products, URL: `https://www.who.int/news-room/fact-sheets/detail/substandard-and-falsified-medical-products`

[7] International Pharmaceutical Federation (FIP), FIP is the global federation representing four million pharmacists and pharmaceutical scientists worldwide. URL: `https://www.fip.org/counterfeit_medicines`

[8] Гуторова Н. Відповідальність за фальсифікацію лікарських засобів: чи створена в Украині належна правова база? Щотижневик АПТЕКА. 2018 № 2 (1173). С. 11. URL: https://www.apteka.ua/article/485029?fbclid=IwAR3TM8hbtOou30daeR9FTRyNxNG4zO_4z4jhSLwvwMfjqZD5TL7Hcq_jByQ
[9] Blockchain Tech Could Track Pharmacy Supply Chain. Drug Topics. Nov 9, 2017/ URL: http://www.drugtopics.com/latest/blockchain-tech-could-track-pharmacy-supply-chain
[10] Гуторова Н. Фальсифікація лікарських засобів – як в Україні суворість кримінально-правової норми поєднується з фактичною безкарністю фальсифікаторів? Щотижневик АПТЕКА. 2018. № 4 (1175). С. 8–9. URL: https://www.apteka.ua/article/486071?fbclid=IwAR2EXhpqPnhxFnaZ6ujzeh8VTsN_5mz4elXJbOIamwU0G5ahNDSIKFYh-hk
[11] David J. Fong. Blockchain and Its Potential Impact to Healthcare and Pharmacy. URL: https://www.wolterskluwercdi.com/blog/blockchain-potential-impacts/
[12] How Blockchain Can Reduce Waste for Pharmaceutical Companies. Finance magnates. Tuesday, 16/01/2018. URL: https://www.financemagnates.com/thought-leadership/blockchain-can-reduce-waste-pharmaceutical-companies/
[13] ЕгороваН. ЕС вводит новые правила для фармацевтов, применяющих блокчейн. URL: https://getway.io/ru/eu-vvodit-novyye-pravila-dlya-farmatsevtov-primenyayushchikh-blokcheyn
[14] ПашковВ.М. Блокчейн Системі Охорони Здоров'я: проблеми правового забезпечення. Медичне Право 2018. №2(22). С. 32–40
[15] Supply Chain Management (SCM) или управление цепочкой поставок. Logistics.ru. URL: http://www.logistics.ru/21/7/1/i8_466.htm
[16] Blockchain Tech Could Track Pharmacy Supply Chain. Drug Topics. 2017. № 9. URL: http://www.drugtopics.com/latest/blockchain-tech-could-track-pharmacy-supply-chain
[17] Bill Siwicki. The next big thing in pharmacy supply chain: Blockchain. Healthcare IT News. 2017. № 12. URL: https://www.healthcareitnews.com/news/next-big-thing-pharmacy-supply-chain-blockchain
[18] Pfizer и Genentech создает блокчейн-проект MediLedger для борьбы с поддельными медицинскими товарами. Подробности. Опубликовано: 05 Октябрь 2017. Tutellus. URL: http://www.coinfox.ru/novosti/7780-pfizer-i-genentech-sozdayut-blokchejn-proekt-mediledger-dlya-borby-s-poddelnymi-meditsinskimi-tovarami
[19] В России будут бороться с подделкой лекарств с помощью блокчейна. Forklog. 31.01.2017. URL: https://forklog.com/v-rossii-budut-borotsya-s-poddelkoj-lekarstv-s-pomoshhyu-blokchejna/
[20] Полина Гиверц. Потенциал Применения Технологии Блокчейн В Фармации. DigitalMedInfo. URL: http://digitalmedinfo.ru/blockchainpharm
[21] Как блокчейн может улучшить фармацевтическую промышленность. Coinspot. URL: https://coinspot.io/technology/kak-blokchejn-mozhet-uluchshit-farmacevticheskuyu-promyshlennost/

[22] Mohamad Zahreddin, How Blockchain Will Revolutionize the Pharmaceutical Industry. Forbes Technology Council. 2018. №14. URL: https://www.forbes.com/sites/forbestechcouncil/2018/11/14/how-blockchain-will-revolutionize-the-pharmaceutical-industry/#468c242726e5
[23] Что Такое Смарт-контракты простым языком. Prosto coin. URL: https://prostocoin.com/blog/smart-contract

Medical practitioner's legal regulations in the practice from urban-rural development perspective

Karina Palkova*

Rīga Stradiņš University, Faculty of Law, Riga, Latvia

Abstract. The role of medical practitioners in community safety area is huge. Especially huge responsibility medical practitioners have in medical treatment process from the point of view of treatment quality and legal liability. Medical practitioners are professionals regardless of territoriality. The mission and functions they perform do not depend on practice place. The responsibility is equal in both urban and rural level. To protect community safety from the medical practitioners' perspectives they should know their own rights, as well as patient's rights in healthcare. From the particular legal regulations arise duties and responsibilities of medical practitioners, which are described in the paper. The aim of the research is to study legislation that regulates medical practitioner's rights and duties in Latvia. The task of the paper is to study the characteristics of rights of medical practitioners in Latvia and the medical practitioner's knowledge of patients' rights in Latvia. In the research, the author has used general scientific methods such as analysis, synthesis, modelling, comparative method, inductive and deductive method, methods of interpretation of legal norms – grammatical, teleological, historical, as well as empirical method – survey. The paper describes the most important legal regulations in medical practitioner's activity, as well as defines problems and provides potential solutions that are necessary to protect community's safety. The results of the paper show that the regulatory framework exists in the field of healthcare of Latvia. Also in the field of professional activity, legal regulations exist, which govern the professional activity of medical practitioners. They are included in various law sectors (civil rights, labour rights, criminal rights, etc.). Apart from that, there exist a large number of legislative regulations of national level, which govern the professional activities of medical practitioners (special laws and Cabinet Regulations). It should be noted, that the competences of medical practitioners are not consolidated, but stipulated in separate legislative provision for each group of medical practitioners (for a doctor, doctor's assistant, nurse, etc.). The medical practitioner's knowledge of patients' rights is not good enough, especially in rural areas of Latvia.

Key words: medical practitioners, medical law.

1 Introduction

The modern society is not imaginable without a provision of the healthcare service. The availability of healthcare is a ground of any legal state. Healthcare is a special and quite complicated social institute, which in its creation process has become an internal part of the modern society.

* Corresponding author: karina.palkova@inbox.lv

Currently healthcare and the persons involved therein provide to the society the execution of such tasks as society health provision, individual's right to health and life, etc. For the provision of a successful healthcare process a series of measures should be implemented at national level, including organisation, arrangement and adjustment of public relations in the field of healthcare.

As it is provided in the Conceptual Message of the Cabinet of Ministers "About the Reform of the Healthcare System" from 7 August 2017 [1], the public health indicators in Latvia significantly stay behind other member states of the EU. For example, the number of oncologic patients is 1.3 times larger (3.55%) than in average in EU (2.7%), the number of the patients who are first time recognized as disabled twice exceeds the average number in EU, while the people, who evaluate their health as good or very good in Latvia are – 46.2%, average in EU – 67% [1]. Therefore the abovementioned Conceptual Message included such important healthcare aspects as human resources in healthcare and the main planned activities for their development, provision of medical practitioners in 2016 and 2025, further development of the doctor's profession, as well as medical practitioners' salaries, and the regulatory framework, governing the industry. It should be pointed out that regarding the latter the information provided in the Conceptual Message is insufficient. Nowadays a medical practitioner is the one who must know, understand and must be able to apply legislation, which governs his/her as a specialist's work. The involvement of medical practitioners in the improvements of the healthcare process at various levels, including the socially legal, will allow increasing the efficiency of the healthcare system, as well as the mechanism for the protection of professional activity of medical practitioners.

During last few years, legislation governing the activity of medical practitioners has been actively developed. It can be explained not only with a development of the healthcare system in general, but also with rapid changes in the society's attitude regarding the particular field. The content of the legislation, governing the activity of medical practitioners, in the perfect version is pointed at the improvement of healthcare quality. However due to the hard legislative base and its inaccurate arrangement, the target of the idea might not be achieved. Therefore it is important to systematically arrange (create and systemize) the legislative base, which governs the professional activity of medical practitioners. The professional activity of a medical practitioner is directly related to the legislative provisions determining the industry. For example, the technologies of medical manipulations are related with skills and knowledge of a medical practitioner, while the requirements for their realization always have been determined and governed by legislative provisions [2]. The paper gives insights into the questionnaire results that show the level of the medical practitioner's knowledge in the area of specific legal documents about the patient's rights from urban-rural perspective.

2 Material and methods

In the paper, questionnaire results are used, as well as descriptive, analytical and deduction-induction methods. Using these methods, legal acts are analysed as well as recorded opinions of legal scientists, drawn conclusions and made suggestions. The descriptive method is used to describe characteristics of a specific phenomenon of medical practitioner's rights and responsibilities. The method is used to describe the most important of the legal regulations in practice of medical practitioners. There is also used the analytical method, which helps to determine and then evaluate the legal concept or the issue and the content of the legal principles and norms. But deduction-induction methods are used to research current problems from more general to more specific, and opposite - moving from specific observations to broader generalizations and theory. It helps to clarify the outcome of the research. In the research, is also used a systematic review, including database search, management and acquisition of relevant literature, and literature review. Applying these methods, legal acts and opinions of legal scientists are analysed and conclusions made.

3 Results

In order to clarify the range of the legislative provisions, which govern the professional activity of medical practitioners in the context of protection of rights, a study was carried out and analysis made of the existing legislative provisions of the field. During the study, legislative provisions were identified, which due to the professional activity of medical practitioners regulate the complex of public relations. The legislative provisions by their nature and content are divided into national legislative provisions and international ones (European Justice Portal, Member State Rights – Latvia). From the Articles of the Constitution of the Republic of Latvia, for example, Articles 89, 68, 98, etc., we understand that the international legislative provisions are a part of Latvian legal system [3]. Therefore the activity of medical practitioners is regulated more widely than at national level. In Article 53 of the Vienna Convention on the Law of Treaties, to which Latvia joined on 4 May 1993, it is stated that "the general imperative norm of the international rights is a norm, which is adopted and recognized in the international community of nations as a norm, from which no tolerance is allowed and which might be changed only by a successive general international rule of law, which has the same nature" [4].

As such rules of law, for example, shall be considered the rules, which are stated in the Constitution of the United Nations Organization. In other cases, additional legislation is necessary, for example, the Guidelines of the World Health Organization for Development of National Drug Policies (1988) [5]. This document contains guidelines for the national drug policy. The guidelines cover all aspects which a country needs to know in order to carry out policy effectively, such as legal issues, financial resources, development, and monitoring. As an example should also be mentioned the Convention of the European Council for the Protection of Human Rights and Fundamental Freedoms, to which Latvia joined on 27 June 1997 [6]. The legal provisions of the mentioned convention are the sources of the international rights. Here should be mentioned the Directive 2010/45/EU of the European Parliament and Council from the 7 July 2010 on standards of quality and safety of human organs intended for transplantation [7]. The Directive entered into force on 26 August 2010, and Directive 2011/24/EU of the European Parliament and of the Council from the 9 March 2011 on the application of patients' rights in cross-border healthcare [8]. Consequently the professional activity of medical practitioners is not the issue of the national level only, but it is largely affected by the international aspect, as the principles, stated in the international legislative provisions, are reflected in the legislative provisions of national level. In the study, more attention is paid exactly to national legal sources, which govern the professional activity of medical practitioners and by their nature is an important legal institute in medicine.

The most important rule of law, which governs the professional activity of medical practitioners, is the **Constitution** [3]. Article 106 of the Constitution states that "everyone has right to freely choose an occupation and workplace according to his/her abilities and qualification..." Therefore the professional activity of medical practitioners shall be supported and protected regardless the institution, public, municipal or private, where the medical practitioner works. Selection of an appropriate form of activity is exclusively in the competence of a medical practitioner. Article 92 of the Constitution states that everyone is allowed to defend his/her rights and legal interests in a fair trial. Everyone shall be considered as innocent, until his/her fault is not adjudged in accordance with law. In case of a groundless infringement of rights everyone has right to an appropriate reimbursement. Everyone has right to the lawyer's assistance [3]. The abovementioned is also related to a medical practitioner.

The basic law, which provides the terms for society treatment and a procedure for its provision, is the **Medical Treatment Law** [9]. In accordance with the Article 2 of the mentioned law, the target of the law is to regulate public relations in medical treatment, in order to provide qualified prevention and diagnostics of diseases or injuries, as well as qualified patient's treatment and rehabilitation and to determine special terms for the legal regulation of medical institution's economic activities [9]. In the **Medical Treatment Law** there are stipulated ethic issues, medical practitioner's rights and their activity in medical treatment, etc. From the abovementioned arises that the **Medical Treatment Law** regulates the basic terms related to the healthcare [9].

Also, the **Civil Law** governs the professional activity of medical practitioners. However, unlike special legal acts, the Civil Law in some ways transforms the institute of the medical practitioners' activity from the social to civil [10] thus providing also the regulation of the medical practitioners' activities in the context of rights of the Civil Law liabilities. Depending on the medical practitioners' activity form medical practitioners may establish, for example:

- Contractual relations,
- Relations with persons, who realize a business in a provision of services of an appropriate field of specialization.
- Relations, which arise from the provision of human basic rights and basic freedom, etc. [10].

At the same time in Article 1635 of the Civil Law it is stated that every infringement of rights, i.e., each individual unauthorised action, which results in a damage (also a moral damage), shall give a right to the victim to demand from an offender a satisfaction to the extent it can be blamed for this action [10]. Therefore, legal relations between a medical practitioner and a patient are regulated. Article 1635 of the Civil Law also states that as a moral damage shall be considered physical or mental suffering, which is infringement of a victim's non-material rights or non-material benefits, caused due to unauthorised action. The amount of remuneration for a moral damage shall be determined by a court according to its discretion, considering the severity and consequences of a damage. While the note provided in the Article points out that an activity here shall be understood in the broadest sense, covering not only an action, but also abstention from it, i.e., inactivity [10]. Article 1644 of the Civil Law states that the person, "who causes damage to other party without an evil intent, if he/she shall be blamed for this damage, comes to negligence."

The negligence may be gross or mild. From the above mentioned one can conclude that in the case of a medical practitioner in their professional activity is allowed negligence, and he/she might come to a civil liability. The medical practitioner's negligence in this case might show up as, for example:

1. A medical practitioner was obligated to take care of a person;
2. A medical practitioner has failed to fulfil or has fulfilled inappropriately the obligation to take care;
3. Due to failure to fulfil the obligation, could foresee a possible damage;
4. Real damage is caused.

Therefore, it can be concluded that a medical practitioner in their professional activity shall be guided by norms of the Civil Law [10].

Latvian Administrative Violations Code, which states the requirements for a medical practitioner in the field of provision of medical services, and control mechanisms of a specific field. Within the Latvian Administrative Violations Code the medical practitioner shall be put to trial, which is one of the national compulsory powers and types of responsibility, which is related to impose of administrative penalty [11]. It is a liability for a violation, which harms national or public procedure, property, human rights and stated management procedure. The Latvian Administrative Violations Code determines a series of prohibitions of legally administrative nature. For instance, in accordance with the Article 45.[3] of the Latvian Administrative Violations Code to a medical practitioner for an illegal disclosure of confidential information obtained during the treatment process shall be applied a fine up to three hundred and fifty *euro* [11]. At the same time there is stated a liability in it, of which the medical practitioner's obligation arises to operate with the medical private practice, which is registered according to the procedure stipulated by law. A medical practitioner also must be careful with person's data, as for any illegal actions with person's data, including data collection, registration, entry, storage, arrangement, modification, use, transfer, disclosure, blocking or deleting in the Article 204[7] of the Latvian Administrative Violations Code there is provided a liability. A medical practitioner in their professional activity must also observe the principle of equality, i.e., not to allow discrimination, as in Article 204.17 of the Latvian Administrative Violations Code a liability is provided also for the violation of non-discrimination [11].

Additionally should be noted that the professional activity of medical practitioners is guided also by labour rights. For example, the legal provision of the **Labour Law** regulates relations of medical practitioners as employees and medical institutions as employers. But it shall be applied exclusively on the condition that from the side of an employee or a medical practitioner a work function is realized [12].

The **Criminal Law** provides the criminal liability of medical practitioners for the offense against human life and health [13]. Part 1 of Article 138 of the Criminal Law states that for the failure to fulfil the medical practitioners' professional obligations or for their negligent fulfilment, if this negligence has caused to a victim a severe body injury or a moderate body injury, the person shall be punished with a deprivation of liberty for a time period up to one year or with a short-time deprivation of liberty, or with a forced labour or a fine. The Article 2 of this law states that for the same offense, if it has caused the victim's infection with a Human Immunodeficiency Virus or B or C hepatitis virus, or has been a reason for the victim's death, the person shall be punished with a deprivation of liberty for a time period up to five years or with a short-time deprivation of liberty, or with a forced labour or a fine [13]. The regulation of the medical practitioners' professional activity is provided also in Article 136 of the Criminal Law for a forced abortion, in Article 135 of the Criminal Law for the unauthorised abortion, in Article 139 of the Criminal Law for illegal removal of living or deceased person's tissues or organs for the use in medicine, if it is done by a medical practitioner [13].

Article 155 of the Criminal Law provides for deliberately illegal person's placing in psychiatric hospital, Article 154.1 of the Criminal Law for a human trafficking, Article 141 of the Criminal Law for a failure to give a necessary and obvious help to a person, who is in life threatening condition, if a guilty party was aware that they could give it without a severe harm to themselves and other persons, and if the failure to give a help has caused person's death or other severe consequences. The Article 140 determines the medical practitioner's liability for the violation of sanitary hygiene and epidemiological safety regulations [13].

In the practice, common are also problem issues of a legal nature of the medical practitioner's operations with person's data. The mentioned issue is topical also in connection with the fact that on 25 May 2018 in Latvia new Data Protection Regulations of the European Union enter into force [14]. The mentioned data protection terms shall be observed by all enterprises, including medical institutions.

Considering the abovementioned, special attention shall be paid to Article 145 of the Criminal Law regarding the analysis of legislative regulations, governing the professional activity of medical practitioners. In it, liability for illegal actions with natural person's data is stated, as well as for the influencing the personal data processing manager or operator or data subject, using a violence or threat, or abuse of confidence, or craftily for a purpose to perform illegal actions with a natural person's data. The extent of a medical practitioner's professional activity has been stated also in Article 315 of the Criminal Law, including the medical practitioner's obligation to inform about a gross crime or especially gross crime [13]. In Article 198 of the Criminal Law, in relation with medical practitioners, a prohibition for the acceptance of benefits is determined [13]. It should be noted that until April 2013 the Article 3 of the Criminal Law Convention on Corruption of the European Union stated that the passive bribery is a deliberate action, by which any public official of the country requests or receives, directly or indirectly, any of his/her undue advantage, which is meant for both this official and any other person [15]. In the Article it is stated that the benefit can be meant for both an official and other person. With amendments in the law "On the Criminal Law Convention on Corruption of the European Council", which are made in the law "On the Criminal Law Convention on Corruption of the European Union" (Reporter of the Parliament and Cabinet of Ministers of the Republic of Latvia, 2001, no.3) the mentioned Article was excluded.

The extent of the professional competence of medical practitioners, including regarding a patient, is determined in the Cabinet Regulations. It should be noted that the regulation, stipulated in the Cabinet regulations, has been implemented considering the opinions of social partners and profiled organizations [16]. Cabinet Regulations No.268 from 24 March 2009 "Regulations on the competence of medical practitioners and students who study first or second level professional higher education programs in medicine and the extent of their theoretical and practical knowledge" determine the competences of students and persons, registered in the Medical Person Register, as well as their extent of knowledge of healthcare [17]. Cabinet Regulations No.264 from 23 May 2017 "Regulations on the Classification of Occupations, the basic tasks fulfilling the profession and the basic qualification requirements", the main tasks of medical practitioners are determined [18].

For example, in the Article 160 of these regulations it is determined that professions and appropriate main duties of an individual group "2211 Family (general practice) doctors and doctors" are to examine a patient; to set for a patient diagnostic examinations and analyze their results; to advise a patient on health promotion; to provide a general treatment and its continuity, to provide and receive the information necessary for a patient's treatment from other healthcare specialists, etc. [18]. The Article 167 of the Regulations states that one of the main tasks in the profession of a Nurse is to educate patients [18].

Making the analysis of above stated legislative regulations, author has succeeded to summarize and identify the most significant problems regarding the professional activities of medical practitioners and legislative regulations governing them.

There are a huge number of legal documents that state and provide medical practitioner's regulations in particular field, as well as on patients' and medical practitioner's rights and duties. To find out whether the medical practitioners are informed on the legal documents mentioned above the questionnaire was developed and specific data was collected. The questionnaire offers objective means of collecting information about medical professional's knowledge in the area of patients' rights, attitude to minor patients and relevant behaviour in cities and the countryside of Latvia.

The total number of medical practitioners who participated in the questionnaire is 230. Of them 83% – women (or 190) and 17% – men (or 40) were interviewed. The main place of work of the medical practitioners who participated in the study is Riga, i.e. 62% (or 143), and outside of Riga 38% (or 87) of respondents. The results of the questionnaire show that the level of medical practitioner's knowledge of patients' rights in Riga and Regions is low and that the medical practitioners in Regions are more tended to professional growth than in Riga. Besides, the competences of medical practitioners are not consolidated, but stipulated in separate legislative provision for each group of medical practitioners (for a doctor, doctor's assistant, nurse, etc.). This does not promote the patient's understanding of the competences of medical practitioners and opposite. In the Medical Practitioner Law the competences of a medical practitioner are not stipulated.

4 Conclusions

Making the analysis of the regulations, which govern the professional activity of medical practitioners, it can be concluded that they are based on both national and international legal sources. A big role in the activity of medical practitioners is played by the standards and guidelines in the field of healthcare. Studying the content of the legal regulations, provided in the paper, the following features can be distinguished:

1. Healthcare issue (including the relations between a medical practitioner and a patient) is included in Article 111 of the Constitution. It indicates to the vitality, significance of the tasks, performed by medical practitioners in the cut of international relations;
2. The regulatory framework exists in the field of healthcare of Latvia, and also in a field of professional activity regulation of medical practitioners;
3. Legal regulations, which govern the professional activity of medical practitioners, are included in various law sectors (civil rights, labour rights, criminal rights, etc.);
4. A big number of legislative regulations of national level exist, which govern the professional activities of medical practitioners (special laws and Cabinet Regulations);
5. The competences of medical practitioners are not consolidated, but stipulated in separate legislative provision for each group of medical practitioners (for a doctor, doctor's assistant, nurse, etc.). This does not promote the patient's understanding of the competences of medical practitioners and opposite.
6. The abovementioned inconveniences not only the simplified availability of the information important for the medical practitioner, but does not promote also the patient's understanding of the competences of medical practitioners. In the Medical Practitioner Law the competences of a medical practitioner are not stipulated.
7. The existing legislative regulations are not directed to the interest of medical practitioners to improve their professional knowledge, and to increase their potential with a target to improve the quality of medical services provided to the society.

References

[1] Conceptual Message of the Cabinet of Ministers "About the Reform of the Healthcare System" (2017). [Accessed 20.09.2018], Available on the Internet: `http://www.vm.gov.lv/lv/aktualitates/par_veselibas_aprupes_sistemas_reformu/`

[2] Злобина О. Ю.,. Солодун Ю. В (2015), Правовое регулирование деятельности врача. Иркутск ИГМУ, p. 4

[3] Constitution of the Republic of Latvia [Accessed 20.03.2018], Available from Internet `https://likumi.lv/doc.php?id=57980`

[4] The Vienna Convention on the law of Treaties. [Accessed 20.10.2018], Available on the Internet: `https://likumi.lv/ta/lv/starptautiskie-ligumi/id/1`

[5] Guidelines of the World Health Organization for Developing National Drug Policies (1988). [Accessed 20.10.2018], Available on the Internet: `http://apps.who.int/medicinedocs/en/d/Js19151en/`
[6] Conventions of the European Council for the Protection of Human Rights and Fundamental Freedoms. Available on the Internet: `https://likumi.lv/ta/lv/starptautiskie-ligumi/id/649`
[7] Directive 2010/45/EU of the European Parliament and Council from the 7 July 2010 on standards of quality and safety of human organs intended for transplantation. [Accessed 20.03.2018], Available on the European Justice Portal, Member State Rights – Latvia, [Accessed 20.03.2018], Available on the Internet:10 `https://e-justice.europa.eu/content_member_state_law-6-lv-lv.do?member=1`
[8] Directive 2011/24/EU of the European Parliament and of the Council from the 9 March 2011 on the application of patients' rights in cross-border healthcare. [Accessed 20.09.2018], Available on the Internet: `https://publications.europa.eu/lv/publication-detail/-/publication/377ab800-836b-4fde-85ff-2e4aac7dee94/language-lv`
[9] Medical Treatment Law: 12.06.1997 (valid from 01.10.1997). [Accessed 28.11.2018], Available from Internet `https://likumi.lv/doc.php?id=44108`
[10] Civil Law: 28.01.1937 (valid from 01.09.1992). [Accessed 20.11.2018], Available from Internet: `https://likumi.lv/doc.php?id=225418)`
[11] Latvian Administrative Violations Code: 07.12.1984 (valid from 01.07.1985) [Accessed 20.03.2018], Available on the Internet: `https://likumi.lv/ta/id/89648-latvijas-administrativo-parkapumu-kodekss`
[12] Labour Law: 20.06.2001 (valid from 01.06.2002). [Accessed 20.10.2018], Available on the Internet: `https://likumi.lv/doc.php?id=26019`
[13] Criminal Law: 17.06.1998 (valid from 01.04.1999). [Accessed 20.11.2018], Available from Internet: `https://likumi.lv/doc.php?id=88966`
[14] Regulation (EU) 2016/679 of the European Parliament and of the Council (April 27, 2016) on the protection of individuals with regard to the processing of personal data and on the free movement of such data and repealing Directive 95/46/EC (The General Data Protection Regulation) [Accessed 20.09.2018], Available on the Internet: `https://eur-lex.europa.eu/legal-content/LV/TXT/?uri=CELEX%3A32016R0679`
[15] Criminal Law Convention on Corruption. European Treaty Series – No. 173. Strasbourg, 27.I.1999 Available from Internet: `https://www.coe.int/en/web/conventions/full-list/-/conventions/treaty/173`
[16] S. Ašnevica- Slokenberga (2015). Medical Rights. Courthouse Agency, p. 192
[17] Cabinet Regulations No.268 from the 24 March 2009 "Regulations on the competence of medical practitioners and students who study first or second level professional higher education programs in medicine and the extent of their theoretical and practical knowledge" determines the competences of students and persons, registered in the Medical Person Register, in healthcare, as well as their extent of knowledge. [Accessed 20.09.2018], Available on the Internet: `https://likumi.lv/doc.php?id=190610`
[18] Cabinet Regulations No.264 from 23 May 2017 "Regulations on the Classification of Occupations, the basic tasks fulfilling the profession and the basic qualification requirements". [Accessed 20.09.2018], Available on the Internet: `https://likumi.lv/doc.php?id=291004`

How will future health policy impact on regional disparities and healthcare in Latvia

Ieva Bikava[1] and *Ilga* Kreituse[2]

[1] Rīga Stradiņš University, Riga, Latvia
[2] Rīga Stradiņš University, Department of Political Science, Riga, Latvia

Abstract. Latvia is a country with high level of regional disparities and as researches have approved – income, education, and environment significantly affect the status of people's health.

The reforms in healthcare have been carried out under the flags of accessibility, efficiency, effectiveness, and quality since 1991. Some groups support the idea that the state should take over almost all services, the others claim that more responsibilities should be transferred to the private sector, and dispute on service provider network and financing system.

In accordance with the coalition theory and legislation of Latvia, the initiators and designers of the structural reforms are politicians, elected by the citizens. In theory, every politician represents the political party with its ideology and vision that is expressed in the Programme of the party.

Due to Parliament elections in 2018, to understand which way the country will go, and how the chosen strategy will impact on the regional disparities, analysis of the middle-term development strategy was made.

Research results showed that political parties in Latvia are not based on the ideology and the strategical development vision of the State, so the development and changes made in legislation are mainly ad-hoc decisions or the result of lobbying activities of influential groups.

Due to the lack of middle term vision as the gap between long-term development strategy and short-term decisions, unfortunately, the prognosis of future health politics leads to the conclusion that regional disparities and the problem with Health care accessibility in next decades will only rise.

Key words: health policy, regional disparities, political party ideology.

1 Introduction

Due to coming Parliamentary elections in October 2018 new political parties were established, with different visions and solutions how the state should develop, based on diverse ideological roots, but all with the same aims – to work on better policies, to raise the prosperity, to make the state become more effective and efficient.

The Health policy issues were on every party's agenda – to provide health care accessibility to inhabitants, through different ways – some by promoting more money to battle the quotas, others – to provide better salaries to medical specialists, others to work on cheaper medications.

According to medium-term policy planning the document "Public Health Guidelines 2014-2020" – "the overarching objective of public health policy is to increase the number of healthy years of inhabitants of Latvia and to prevent premature death, preserving, improving and restoring health". For this purpose, one of the sub-objectives is "to prevent inequality in healthcare by taking measures for ensuring equal health promotion and health care opportunities for inhabitants of Latvia". The other objective is "to ensure efficient

Table 1. Average monthly wages and salaries by statistical region (in euro) [4].

Region/Year	2013	2014	2015	2016	2017
Riga region	583	633	679	710	758
Vidzeme region	410	444	480	502	544
Kurzeme region	441	479	512	529	568
Zemgale region	436	476	509	539	579
Latgale region	358	388	419	440	471
Latgale: Riga	61.41%	61.30%	61.71%	61.97%	62.14%

Table 2. Unemployment rate by statistical region in age group 15–64 (%) [5].

Region/Year	2013	2014	2015	2016	2017
Riga region	10.8	8.3	7.5	7.8	7.9
Vidzeme region	13.4	13.3	11.8	9.4	9.8
Kurzeme region	10.4	13.0	10.8	10.1	9.0
Zemgale region	15.4	13.0	11.5	12.3	9.5
Latgale region	17.3	17.3	19.0	17.9	14.4

management of the health care system and rational utilisation of resources in order to improve the sustainability of operation of the health care system and equal access for all inhabitants of Latvia to good quality health care services, which are funded from the State budget resources" [1].

To improve public health means not only actions on clinical health care issues, but also "interventions with actions on equity and social determinants of health and necessary key health system inputs, such as human resources and medicines". Countless researches have indicated that there is a direct impact of social determinants on health outcomes in an individual as well as the whole society. "Health inequalities arise from the societal conditions in which people were born, grow, live, work and age referred to as social determinants of health" [2].

Income inequality during the last decade has dropped from 37.5 in 2008 to 34.5 in 2017, but is still very high, in 2017 being the third largest value in European Union, just after Bulgaria (40.2) and Lithuania (37.6) [3].

In Latvia's case, income inequality has regional features with significant difference comparing the centre – Riga with the rural areas, especially the Latgale region as represented in Table 1.

The regional difference in average monthly income comparing the centre – Riga and Latgale – the region that is on the European Union border with Russia, differs significantly in a long period, and the average salary in Latgale is only 2/3 of average income in Riga.

There is not only difference in income level, but also different possibilities to get work, which means difference in unemployment rate, as represented in Table 2.

According to data on the unemployment for five-year period, the rate of unemployment in Latgale is twice higher than in Riga. Both of these conditions – higher unemployment rate and lower salaries lead to higher poverty rate which also has dramatic regional differences as represented in Table 3.

Table 3. Share of persons under minimum income level by statistical regions (%) [6].

Region/Year	2012	2013	2014	2015	2016
Riga region	3.5	4.1	3.6	3.4	3.7
Vidzeme region	14.4	13.4	13.1	9.5	10.9
Kurzeme region	8.0	8.3	8.0	7.4	8.0
Zemgale region	10.1	10.3	12.7	7.8	9.7
Latgale region	15.1	14.5	18.6	17.1	16.1

Table 4. Reasons for unmet needs for medical care for persons 16 or over (%) [7].

Reason	Could not afford to (too expensive)		Wanted to wait and see if problem got better on its own		Too far to travel/ no means of transportation	
Year	**Urban**	**Rural**	**Urban**	**Rural**	**Urban**	**Rural**
2013	60.9	58.1	13.7	21.5	1.8	4.4
2014	53.7	56.1	15.1	27.1	1.7	3.6
2015	57.2	59.1	14.5	18.4	2.2	4.2
2016	41.7	51.2	16.4	20.9	2.9	6.1
2017	42.5	44.6	21.6	25.4	3.3	6.6

In the circumstances of twice higher unemployment rate and significantly lower average income level, the share of the poverty rate in Latgale region is five times higher than in Riga. As known from several researches, the higher poverty rate leads to worse health outcomes as one of the most influential social determinants. Those data leads to the conclusion, that health status of Latgale region inhabitants is significantly worse than in Riga.

In addition to worse health conditions, affected by social determinants, the availability of health care services is also more restricted in the rural region due to the lack of human resources – availability of medical doctors and nurses. The analyses of unmet needs in urban and rural regions by the leading causes are shown in Table 4.

Analysis of reasons for unmet needs indicates that leading obstacle is high costs for both urban and rural inhabitants. In the course of time, the main cause "too expensive" is getting lower, but other worrying trend arises – the share of people who postpone the medical care rises significantly, that can be explained also with financial issues. Too far to travel is not one of the leading obstacles, but at the same time is twice higher for rural than urban population. However all indicators show that the level of unmet needs is higher for rural than urban population.

Summarizing the above mentioned – there are significant regional disparities in health status and health care availability in Latvia, that are based on social determinants and disparities in medical service availability in regions.

2 Political parties as the agents of change

According to the political theory, the initiators of structural reforms are politicians, elected in Parliament and Government. In accordance with the Constitution of Latvia – the members of the Parliament are elected from political parties, which participate in elections. The parties that get more than 5% of votes get places in Parliament and take part in forming the Coalition. The Coalition forms the Government by nominating the candidates of the Ministers and prepares the Government Declaration – the document, that contains main aims and tasks that Government undertakes to achieve during their governance term. The Cabinet of Ministers and the Government Declaration afterwards are affirmed by the vote of Parliament, if the majority of parliament members vote for.

Thereby, the idea of changes or reforms raises in the political party that is elected in the Parliament and presented through the elected deputy. The theories on political parties, political power and formation of the coalition emphasize that every politician is motivated to be elected for the next term, so they are concerned about the voters' confidence, trust and loyalty. To gain and enhance the trust of the voters, every politician tries to fulfil the promises that were given to voters.

In multiple political party systems, which is the case of Latvia, the competition is among the political parties, which are represented by the members of parties, who are nominated as the candidates for deputy. Every political party is presented not only by the people but also by the pre-election Programme that contains the main party commitments to society.

In political theory, "the parties ensure that the citizens are permanently capable to act politically. They articulate and integrate different interests, visions and opinions. They are also the main source for the recruitment of political elites. In order to participate successfully in elections, the political parties have to be the voice of broad sectors of society" [8].

Thereby, political parties that not only participate in elections, but also are elected, should present the voice of majority of society and embody the will of the nation, they should be the persons who understand the needs and problems to be solved and have ideas and visions that are acceptable by their voters.

All political parties have some kind of core values that consolidate all the members and distinct one party from another. For example, in the USA, there are two main political divisions – Republicans and Democrats. Depending on which party get the majority or win the elections, it can be predicted what will be the main values on which the further development of the state will be based on. Will it serve more to interests of the capital and free market – if the winning party is Republicans, or will it be towards the welfare of the state – if the lead is taken by Democrats.

According to Political theory, there are several courses of ideologies – Liberalism, Conservatism, Socialism, Nationalism, as well as others more specific, like Christian Democrats, Pacifism, Ecology, etc.

In classic politics, ideology is the determining factor, based on which every party declares their main values and vision of the state development. In contemporary politics, the role of ideology has diminished and blurred, and nowadays it is not the main identity factor for the political party. Furthermore, there even are parties, which are based on opposite ideologies – they are liberal on human rights issues, but are conservative in economic development issues.

Nevertheless, the ideology is one of the most significant factors that indicate the core values of the party, their vision of society and social developments. In accordance with Coalition building theories, political parties form alliances and associations, based on a common ideological platform. The alliances of parties are established by unifying values and goals, even if they are not quite similar in their vision of the state development and in some issues can even be of the opposite point of view. However, the question is – can these disagreements be put aside to reach more important unifying goals [9]?

The main document of party's vision, mission, and tasks to achieve is the programme. "Party programme constitutes the identity of the party and offers the general orientation to the citizens and voters on the guiding principles and ideas of a party. With its programme

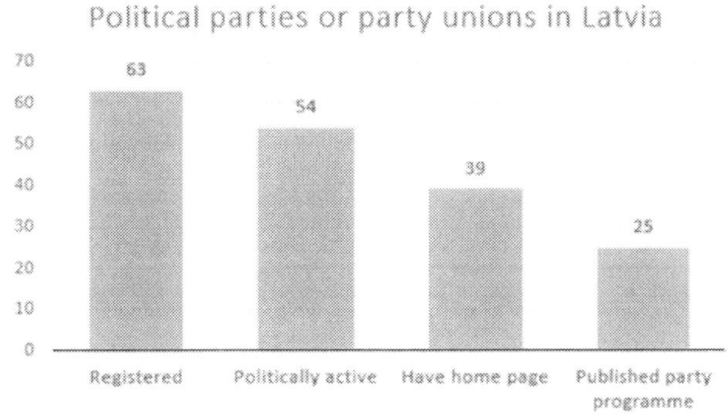

Fig. 1. Characteristics of political parties or party unions in Latvia.

the party justifies its existence and explains how it differentiates itself from others. The party programme illustrates political ambition and basic values, demands and suggestions for a party" [8].

Significant is not only political party's programme, but also the requirement to register a political party in Latvia. The Law of Political parties in Latvia states that political party is the organization that is established in order to take political activities, participate in election campaigns, nominate deputy candidates, and through elected deputies realize political party programme [10].

Considering the Constitution of Latvia, the Law of Political Parties in Latvia, as well as attainments of political theory, the research of political parties' programmes that are campaigning for the participation in the government should be carried out in order to understand the development of future policy and its main values.

3 The vision of the future in the political parties' programmes

To understand the vision of the future state according to political parties that are competing for the governance, the research of the programmes of the active political parties in Latvia was conducted.

All political parties are registered in Enterprise Register of Latvia. According to data from the Register, before the 13th Parliament elections, there were 51 political party and 10 political party unions in Latvia, that can be considered as active (party or party unity that is registered and there is no information about liquidation process) [11].

Considering the fact that not all of the registered parties or party unions are politically active, four factors were taken into account to identify whether a political party or party union can be considered as politically active: 1) political party or a union of political parties have participated in the previous, 12th Parliament, elections; 2) whether the party participated in municipal elections in 2017; 3) whether it has submitted Annual Report of the year 2017 to KNAB; 4) was it registered in 2018 [12, 13]. According to these criteria – 54 or 88.5% of parties and party unions can be considered as politically active.

Nowadays the main sources of information are web and social networks thereby assumption was made that every active political party should be on the Internet, that means to have the home page. The research indicated that only 39 or 72% of active political parties or party unions have their home page. The further research of the content of the party or party union home page led to results that only 25 or 46% of active parties or party unions have published their programme on the web page (Fig. 1).

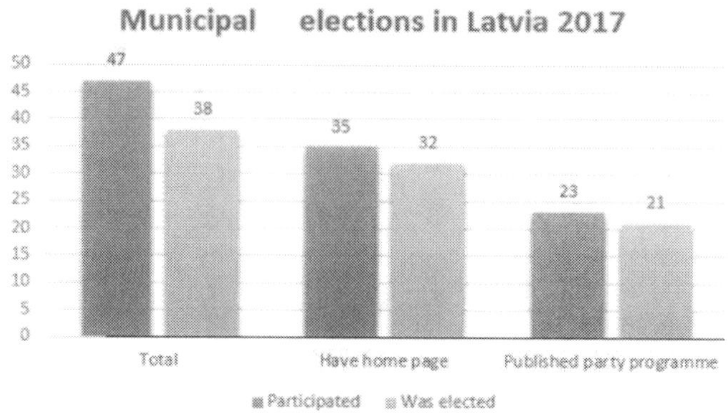

Fig. 2. Political parties or party unions in Municipal elections in Latvia, 2017.

Despite the fact that the programme of the political party is a compulsory document that has to be submitted to Enterprise registry to register the political party, the programme is not the document that can be easily found. Even of the parties that have their web page, only a few have published their programme.

These findings indicate that the party programme, the document that according to political theory – justifies the party's existence and explains how it differentiates itself from others, is not meaningful in Latvia. Even more, we can say that the development of party programme as the document that has to be submitted to Enterprise registry to register the political party, for a significant share of parties is done formally, not meaningfully.

To understand voter's expectations and the requirement for the existence of party programme, the further research was carried out analyzing political parties which participated and were elected in municipal elections in 2017.

According to the data from the Central Election Commission of Latvia, in Local elections in 2017, 47 political parties and party unions participated. Exploring their websites, we found out that 35 of participating parties or party unions had their web pages, and 23 of them had published their programme. The parties or party unions that had been elected in at least one municipality – were 38, of them 32 had a web page and 21 had published their programme on their web page (Fig. 2).

The results, comparing participated and elected parties or party unions, showed that publicly available party programme is not a significant factor for the voters, because there were elected 17 (45%) parties or party unions, which had not published their programme and 6 (16%) of them even did not have their web page.

This phenomenon can be explained by the local character of the elections. The local voters are familiar with members of local parties, and before elections, every political party organizes local events and activities to introduce and attract potential local voters.

Especially in local scale elections, direct mail, posters, and handouts are the tactics that are used to attract the voters.

To gain a deeper understanding of the value of the party programme for a potential voter, analysis of the political parties and unions that took part in the previous 12th Parliament elections was made.

According to the data from the Central Election Commission of Latvia, in the 12th Parliament election in 2014 12 political parties and party unions participated. Exploring their websites, we found out that 11 of participating parties or party unions had their web pages, and 7 of them had published their programme on the web page.

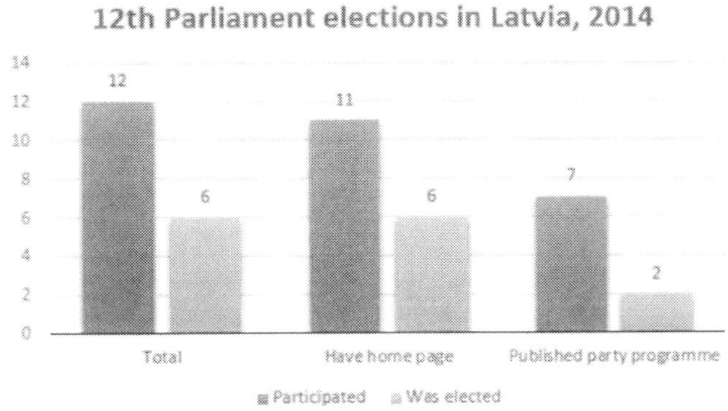

Fig. 3. Political parties or party unions in 12th Parliament elections in Latvia, 2014.

The results, comparing participated and elected parties or party unions, showed that publicly available party programme is not a significant factor for the voters in both municipal and in Parliament elections, since there were 6 parties or party unions elected, all of them had web pages, but only 2 of them had published their programme on web page.

According to results, in Latvia, the political party programme is the significant factor neither for political parties nor for the voters. With reference to political theory, programme is the document that deals with the ideology of the party, their vision, and mission. The research reveals that ideology is not the backbone of the political party in Latvia.

As there is not the possibility to predict the future of the state basing on theoretical approach – on the ideology and the programme, analysis of pre-election programmes available on the Central Election Commission of Latvia home page was made in order to find out what promises political parties offer to voters.

Since the problem issue for research was regional inequalities, the context analysis of all 16 party pre-election programmes was made to search issues on inequality or regional development.

Although inequality and regional disparities are the issues that are on the agenda for a long time, only 11 of 16 political parties or party unions have mentioned the terms "regional policy" or "regional development". Taking a closer look at the content, only 7 of 16 programmes identified the need to work on regional disparities in the Healthcare sector in their programmes.

From those 16 parties and party unions that participated in elections – seven were elected. Work on regional development was mentioned in 4 party or party union programmes and only 3 of them had on their agenda goal to achieve better health care accessibility for every inhabitant in all regions. From those 7 elected parties or party unions, only one with 13 mandates of 100 Parliament members had declared that they would assess the impact of every decision on inequality and would not support the reforms that increase the inequality.

4 Conclusions

Results of research on political party programmes indicated that political parties in Latvia are not formed on ideology bases, and since there is the lack of party programmes, parties do not have vision of strategic development of the State.

As the results of elections showed – a significant part of Citizens made their choice in elections basing on other factors, not on the programme and ideology of the political party.

So, the political party programme and ideology is a significant factor neither for political parties, nor voters.

That can lead to the rise of populism in society, but in the legislative process can lead to the situation that state development and changes in the legislature are mainly ad-hoc decisions or the result of lobbying activities.

The lack of middle-term objectives and goals is threatening the possibilities to attain long-term state development plans and can deepen the problem of regional disparities and inequality.

Evaluation of the election results of the 13th Parliament elections in Latvia in 2018 indicates that most of the elected parties are "right-wing parties" as they describe themselves – that means that they are mostly based on liberal or conservative ideology. As the research of pre-election programmes indicated – the work on regional disparities and inequality is not a priority for the majority of the Parliament.

Due to the lack of middle term vision as a gap between long-term state development strategy and short-term strategy, there is no road map of activities to achieve those long- term vision and goals.

Unfortunately, the prognosis, based on the research results, is that regional disparities and healthcare accessibility problems for rural areas will rise.

References

[1] Cabinet of Ministers, *Cabinet Order No. 589 of 14 October, 2014. Public Health Policy Guidelines 2014–2020.* Available at: `https://likumi.lv/ta/en/en/id/269591-on-the-public-health-policy-guidelines-2014-2020` [viewed 27.01.2019]

[2] World Health Organization Regional Office for Europe, *Health 2020. A European policy framework and strategy for the 21st century.* Available at: `http://www.euro.who.int/__data/assets/pdf_file/0011/199532/Health2020-Long.pdf?ua=1` [viewed 27.01.2019]

[3] Eurostat, Gini coefficient of equivalised disposable income – EU-SILC survey. Last updated 19-01-2019. Available at: `http://appsso.eurostat.ec.europa.eu/nui/submit ViewTable-Action.do` [viewed 27.01.2019]

[4] Central Statistic Bureau of Latvia, Social Processes. Wage and Salaries. Annual Data. *DSG050. Average monthly wages and salaries by statistical region (in euro).* Latest update 3/1/2018 [viewed 29.01.2019]

[5] Central Statistic Bureau of Latvia, Social Processes. Employement and Unemployement. Economic Activity. Annual data. *NBG040. Activity rate, employement rate and unemployememt rate by statistical region (%).* Latest update 2/22/2018 [viewed 29.01.2019]

[6] Central Statistic Bureau of Latvia, Social Processes. Poverty and Inequality (EU-SILC). Minimum income level. *MIG020. Share of persons under minimum income level by statistical regions of Latvia (%).* Latest update 3/2/2018 [viewed 29.01.2019]

[7] Central Statistic Bureau of Latvia, Social Processes. Health. Self-perceived Health Status (EU-SILC). *VPG072. Reasons for unmet need for medical care (except dental) for persons aged 16 or over (%).* Latest update 1/17/2019 [viewed 29.01.2019]

[8] W. Hofmeister, K. Grabow, *Political Parties. Functions and Organisation in Democratic Societies* (Konrad Adenauer Stiftung, 2011)

[9] H. Noel, *Political Parties and Ideology. Interest Groups in Context.* New Directions in Interest Group Politics (Routledge, 2014)

[10] Saeima, Politisko partiju likums. No 01.05.2017. Available at: `https://likumi.lv/ta/id/139367-politisko-partiju-likums` [viewed 30.01.2019]

[11] Uzņēmumu reģistrs, *Politisko partiju un to apvienību saraksts*. Available at: `https://www.ur.gov.lv/lv/registre/organizaciju/politiska-partija/politisko-partiju-un-to-apvienibu-saraksts/?filter=ppa&search=` [viewed 30.01.2019]

[12] Centrālā Vēlēšanu komisija, *Vēlēšanas*. Available at: `https://www.cvk.lv/lv/velesanas/saeimas-velesanas` [viewed 30.01.2019]

[13] Korupcijas novēršanas un apkarošanas birojs, *Partiju Finanšu datu bāze. Partiju deklarācijas*. Available at: `https://www.knab.gov.lv/lv/db/declaration/` [viewed 30.01.2019]

Implementation of health promoting policies through tailored interventions at health promoting schools and municipalities in Latvia

Inese Gobina[1,*], *Santa* Pildava[1], *Elina* Millere[1], *Dita* Heiberga[1], *Margarita* Apine[1], *Ilze* Straume[2], and *Aigars* Miezitis[1]

[1] Rīga Stradiņš University, Institute of Public Health, Riga, Latvia
[2] Centre of Disease Prevention and Control, Riga, Latvia

Abstract. Schools and municipalities play an important role in local health promotion. The Institute of Public Health at Rīga Stradiņš University in collaboration with the National Centre of Disease Prevention and Control carried out an online survey for studying the experiences and needs of health-promoting schools and municipalities in Latvia in implementing health promotion interventions. The aim of this study was to explore the context for developing new intervention model for a health behaviour change model within the Interreg BaltCityPrevention project "Innovative Lifestyle-Related Disease Prevention Model in the Baltic Sea Region." Totally 112 municipalities and 100 schools within the National Health Promoting Networks were targeted. Only schools and municipalities that had previous experiences in health promotion intervention development were included in the analysis. This study showed that school-aged children were the most prevalent target group for the both schools and municipality interventions. The interventions in physical activity and nutrition used to be the most common areas for health behaviour changes. Health promotion schools and municipalities reported several significant differences in experienced difficulties and expected future needs in implementing health promotion interventions, which should be considered for increasing the capacity for better health promotion for both schools and municipalities in Latvia.
Key words: health promotion, municipalities, schools.

1 Introduction

Health promotion is the key approach for reducing the burden of non-communicable diseases by changing the lifestyles. Creating supportive environments and strengthening community actions have been recognised as important health promotion actions in the World Health Organization (WHO) Ottawa Charter for Health Promotion [1]. Thus, at the local level, schools and municipalities may have a substantial potential in promoting health on both individual and community levels.

* Corresponding author: Inese.Gobina@rsu.lv

The existing evidence from systematic reviews suggests that the health promotion approach implemented at schools may contribute to improving child and adolescent health and promotion of healthy lifestyle factors, such as increasing physical activity and vegetable and fruit consumption and reducing tobacco use [2–5]. However, the description of information on interventional elements and the implementation context in the studies is limited, causing difficulties in understanding the interventional development process and understanding the learning experience [6].

Within the World Health Organisation's (WHO) Global School Health Initiative, a Health-Promoting School can be characterised as "a school constantly strengthening its capacity as a healthy setting for living, learning and working" [7]. In Latvia, the existing National Network of Health Promoting Schools was established in 2014, building upon the previous initiatives of Health Promoting Schools' Network started in 1993. The National Network of Health Promoting Schools in Latvia belongs to the larger European Network of Health Promoting Schools, Schools for Health in Europe (`https://www.schoolsforhealth.org`).

Studies of determinants of health promotion actions in local contexts are the focus of health promotion research and practice. Health promotion at municipal level involves essential components, such as community organisations, social planning, resource mobilisation, and cross-section collaboration [8–10]. The National Network of Health Promoting Municipalities in Latvia was started in 2013 with the aim of supporting municipalities for community health promotion at the local level by sharing the best practices and increasing the capacity of health promotion practices.

The health behaviour change, aimed at different behavioural domains, is the key instrument in health promotion practices. There are well-developed existing frameworks, including the behaviour change wheel [11], interventional mapping [12], and the PRECEDE-PROCEED model [13] for health intervention development. In all of those, several elements such as needs assessment of the target group, understanding the context of health behaviours, identifying determinants of health behaviours, and selecting appropriate tools for behaviour change are recognised as important for developing the interventions of health behaviour change.

This study was implemented within the InterregBaltCityPrevention project "Innovative Lifestyle-Related Disease Prevention Model in the Baltic Sea Region" (`www.https://baltcityprevention.eu/`). The project has several aims: 1) to strengthen the institutional capacity of the public health authorities (PHAs) by identifying their needs and requirements in order to better tailor interventions to user-specific groups and 2) to develop an innovative health-related lifestyle intervention model for PHAs. Thus, at first, the survey for PHAs in all participating project countries was planned.

In Latvia, the Institute of Public Health at Rīga Stradiņš University, the key partner of the BaltCityPrevention Project in collaboration with the Centre of Disease Prevention and Control (CDPC, Rīga) carried out an online survey among health-promoting schools and municipalities. This study aims to explore the experience and needs of health-promoting schools and municipalities in Latvia for implementing health-promotion interventions in order better to understand the context for developing the new intervention model.

2 Material and methods

Data were collected from December 2017 to January 2018 by using the online questionnaire, which was developed by the International Partners' Group of the BaltCityPrevention project.

The standardised questionnaire was developed with the aim of collecting information on the experience of developing and implementing health promotion interventions for public health authorities by covering the major issues: 1) major target groups for health promotion interventions; 2) tools used for situational assessment used for health promotion intervention planning; 3) major target areas; 4) main partners for health promotion interventions; 5) barriers and difficulties in planning and implementing health promotion interventions; and 6) needs in developing and implementing health promotion interventions.

The online questionnaire was anonymous, and the link for accessing the questionnairewas sent to all 100 municipal schools (91 schools and 9 pre-schools) and 112 municipalities registered in the National Health Promoting Networks by the CDPC representatives using Google Forms. One reminder was sent to all participants after two weeks. The response rate was 82% ($n = 82$) and 38% ($n = 42$) for schools and municipalities, respectively. A total of 76.8% of the responding schools and 92.9% of municipalities reported having experience in developing health-promotion interventions. Thus, only schools ($n = 63$) and municipalities ($n = 39$) with previous experience in health promotion intervention development were included in the data analysis for this study.

Descriptive and non-parametric statistics were used to calculate percentages with 95% confidence intervals for each of the response given and to examine the differences between health-promoting schools and municipalities. The statistical software IBM SPSS 22.0 was used for statistical analysis.

3 Results

In Table 1, the characteristics of the experiences in implementing health promotion interventions at health-promoting schools and municipalities in Latvia are summarised.

The main target groups for health interventions for schools were school-aged children and parents (79.4% and 58.7%, respectively), but for municipalities, the targets were school-aged children and seniors (76.9% and 74.4%, respectively). Adolescents, as a particular target group in health promotion interventions, were frequently reported for both schools and municipalities (58.7% and 61.5%, respectively).

Most (90.5%) of the studied schools ($n = 57$) and 82.1% ($n = 32$) of municipalities used specific methods for the situational assessment within the planning process of health promotion interventions. Of those having had the experience in conducting the situational assessment, the surveys using questionnaires, target groups or community meetings, and desk research (secondary data analysis) were significantly ($p < 0.05$) more frequently reported tools for the situational assessments for both schools and municipalities.

The primary general focus of health-promotion interventions for both schools and municipalities concerned health behaviour changes. For schools, physical activity, nutrition (81.0%), also injuries (69.8%) and violence (46.0%) were the most common focus areas for health-promotion interventions. In comparison with other specific target areas for health promotion interventions, the municipalities significantly and more frequently reported interventions on physical activity (94.9%) and nutrition (76.9%) ($p < 0.001$) followed by mental health and sexual health and family planning.

Health professionals (58.7%) and local municipalities (54.0%) were the principal collaborative partners for health interventions followed by non-governmental organisations (36.5%) and other schools or educational institutions (31.7%) within the schools. The municipalities mainly reported collaborations with schools or other educational institutions (89.7%) and non-governmental organisations (76.9%). Partnerships with target group representatives and non-governmental organisations in health-promotion interventions were significantly more prevalent for municipalities than schools ($p < 0.001$). As the collaborative

Table 1. Key characteristics of experience in health promotion interventions at health promotion schools and municipalities in Latvia (%)[a].

Studied areas	Health-Promoting Schools, % [95%CI]	Health-Promoting Municipalities, % [95%CI]
Target groups for health promotion interventions		
School-aged children	79.4 [67.8, 87.5]	76.9 [61.7, 87.3]
Pre-school children	36.5 [25.7, 48.9]	53.8 [38.6, 68.4]
Adolescents	58.7 [46.4, 70.0]	61.5 [45.9, 75.1]
Parents	58.7 [46.4,70.0]	48.7 [33.9, 63.8]
Adults	14.3 [7.7, 25.0]	66.7 [51.0, 79.4]
Seniors	1.6 [2.8, 8.5]	74.4 [59.0; 85.4]
Persons with special health needs	6.3 [2.5, 15.2]	33.3 [20.6, 49.0]
General population	1.6 [2.8, 8.5]	59.0 [43.4, 72.3]
Tools used for situational assessment		
Interviews	12.5 [5.0, 28.1]	19.3 [11.1, 31.3]
Questionnaires	66.7 [53.7, 77.5]	59.4 [42.3, 74.5]
Desk research (secondary data analysis)	42.1 [30.2, 55.0]	31.3 [18.0, 48.6]
Target group/community meetings	77.2 [65.0, 88.2]	62.5 [45.2, 77.1]
Target areas for health promotion interventions		
General		
Health behaviour change	87.3 [76.9, 93.4]	89.7 [76.4, 95.9]
Disease prevention	57.1 [44.9, 68.6]	71.8 [56.2, 83.5]
Specific		
Injuries	69.8 [57.6, 79.8]	10.3 [4.1, 23.5]
Violence	46.0 [34.3, 58.2]	23.1 [12.6, 38.3]
Mental health	27.0 [17.6, 39.0]	35.9 [22.7, 51.6]
Physical activity	81.0 [69.6, 88.7]	94.9 [83.1, 98.6]
Nutrition	81.0 [69.6, 88.7]	76.9 [61.7, 87.3]
Sleep	20.6 [12.5, 32.2]	0.0 [0.0, 9.0]
Sexual health & family planning	14.3 [7.7, 25.0]	35.9 [22.7, 51.6]
Collaborative partners for health promotion interventions		
Health professionals	58.7 [46.4, 70.0]	64.1 [48.4, 77.3]
Schools/educational institutions***	31.7 [21.6, 44.0]	89.7 [76.4, 95.9]
Non-governmental organizations***	36.5 [25.7, 48.8]	76.9 [61.7, 87.3]
Local municipalities	54.0 [41.8, 65.7]	51.3 [36.2, 66.1]
Target group representatives***	23.8 [15.0, 35.6]	59.0 [43.4, 72.9]
Public health institutions	34.9 [24.3, 47.2]	23.1 [12.6, 38.3]
Mass media	19.0 [11.2, 30.4]	35.9 [22.7, 51.6]
Small and medium enterprises*	7.9 [3.4, 17.3]	23.1 [12.6, 38.3]

[a] Level of statistical significance between schools and municipalities are shown with asterisks:
*$p < 0.05$; **$p < 0.01$; ***$p < 0.001$.

Table 2. Experienced difficulties and expected needs in implementing health promotion interventions at health promoting schools and municipalities in Latvia (%)[a].

Studied areas	Health-Promoting Schools, % [95% CI]	Health-Promoting Municipalities, % [95% CI]
Difficulties in planning and implementing health promotion interventions		
Resources		
Lack of funding**	63.5 [46.4, 70.0]	35.9 [22.7, 51.6]
Limited access to expertise and professionals	30.2 [20.2, 42.4]	48.7 [33.9, 63.8]
Lack of e-tools	12.8 [5.6, 26.7]	19.0 [11.2, 30.4]
Lack of specific health data***	11.1 [5.5, 21.2]	38.5 [24.9, 54.1]
Limited access to evidence-based research	15.4 [7.2, 29.7]	6.3 [2.5, 15.2]
Skills and competencies needed for health promotion interventions development		
Lack of knowledge about how to approach the target group**	28.6 [18.9, 40.1]	53.8 [38.6, 68.4]
Lack of knowledge about the effectiveness of health behaviour change interventions	9.5 [4.4, 19.3]	20.5 [10.8, 35.5]
Lack of competence to understand and translate research	3.2 [0.9, 10.9]	5.1 [1.4, 16.9]
Lack of intervention evaluation competencies	4.8 [1.6, 13.1]	5.1 [1.4, 16.9]
Leadership and partnership		
Lack of cooperation between public health authorities	27.0 [17.6, 39.0]	38.5 [24.9, 54.1]
Lack of inter-sectoral collaboration***	1.6 [0.3, 8.5]	20.5 [10.8, 35.5]
Limited autonomy to make decisions	6.3 [2.5, 15.2]	17.9 [9.0, 32.7]
Needs in developing and implementing health promotion interventions		
Resources		
Financial resources	74.6 [62.7, 83.7]	71.8 [56.2, 83.5]
Health needs assessment**	30.2 [20.2, 42.4]	56.4 [40.1, 70.7]
Expert opinions	30.2 [20.2, 42.4]	33.3 [20.6, 49.0]
E-tools	34.9 [24.3, 47.2]	28.2 [16.5, 43.8]
Skills and competencies needed for health promotion interventions development		
Sharing of best practices	69.8 [57.6, 79.8]	69.2 [53.6, 81.4]
Increased knowledge about the effectiveness of health promotion interventions	28.6 [18.9, 40.1]	30.1 [18.6, 46.4]
Leadership and partnership		
Political support***	4.8 [1.6, 13.1]	38.4 [24.9, 54.1]
Administrative support*	22.2 [13.7, 33.9]	43.6 [29.3, 59.0]
Stronger collaboration with small and medium enterprises	39.7 [28.5, 52.0]	25.6 [14.6, 41.1]

[a] Level of statistical significance between schools and municipalities are shown with asterisks:
*$p < 0.05$; **$p < 0.01$; ***$p < 0.001$.

partners for health promotion institutions, small/medium enterprises and universities were the least likely to be reported by both schools and municipalities.

When studying the previous difficulties in implementing health promotion interventions, the schools reported lack of funding (63.5%), access to expertise (30.2%), and the lack of knowledge about the methods for approaching different target groups (28.6%) as the main barriers for effective health promotion (Table 2). Twenty-seven percent of schools also reported previous difficulties in cooperating with public health authorities for health-promotion interventions.

Among the studied municipalities, the most common obstacles when implementing health interventions were the lack of knowledge aimed at reaching the target groups (53.8%) and limited access to expertise and professionals in the field (48.7%) followed by the lack of funding (35.9%), lack of specific health data (38.5%), and difficulties in cooperation among the public health authorities (38.5%). Some (20.5%) of the studied municipalities also reported lack of knowledge of the effectiveness of health behaviour change interventions and the lack of inter-section collaboration.

The proportion of those studied schools and municipalities that reported difficulties in translating evidence-based knowledge and evaluating the interventional outcomes was relatively low.

More than 70% of both schools and municipalities reported the for more health-promotion intervention funding. However, also sharing best practices used to be an important issue in almost 70% of the studied participants. The requirement for health needs assessment was more frequently reported among municipalities (56.4%) than schools (30.2%), and the need for better political and administrative support for developing health-promotion interventions was more relevant for municipalities than for schools.

4 Conclusions

To our knowledge, this is the first published study showing some of the key characteristics of the previous practices, difficulties, and needs in planning and implementing health promotion interventions of those schools and municipalities belonging to the Health Promotion Networks in Latvia.

In general, for both school and municipalities, the school-aged children were the most common target group in health-promotion interventions. As all of the studied schools were municipal schools, some mixing and overlapping may have occurred between the interventions implemented primarily by schools or municipalities. Parents and adolescents also were frequently reported to be target groups at the school-level health promotion interventions. Among the studied municipalities, the health promotion interventions were also often targeted to seniors and the general adult population. The smallest proportion of municipalities reported persons with special health needs as a special target group for health promotion interventions. Physical activity and nutrition were by far the most frequently reported specific target areas in previous interventions of health behaviour changes for both schools and municipalities.

The spectrum of collaborative partners for health promotion interventions is different for schools and municipalities, which may reflect the existing true differences in governance and leadership between schools and municipalities. The municipalities identified schools or other educational institutions as key partners for collaboration in health promotion the most. The proportion of schools reporting local municipalities and health professionals as the main collaborative partners for health promotion was the highest, and there were no significant differences between the two groups. Both schools and municipalities reported less frequent collaboration with small and medium enterprises. Some existing administrative barriers and legal aspects might explain this; however, these findings suggest that for building multi-section partnerships, focusing on closer collaboration with the private sector could be considered.

When studying previous difficulties and barriers in implementing health promotion interventions, lack of funding was significantly less frequently reported by health-promoting municipalities (36%) in comparison with schools (63%).There are fundamental differences in fund management and administration systems at municipal and school-levels. However, the relatively low percentage of those health promoting municipalities reporting the lack funding for health promotion might be also explained by more opportunities to receive funding for health promotion at the local level for municipalities in Latvia in the frame of the European Social Fund project "Measures for Health Promotion and Disease Prevention at Local Level" lasting from 2016 to 2022.

The proportion of those schools reporting collaborations with the target group in planning or implementing health promotion interventions was relatively low(24%), and only 29% of schools reported the lack of knowledge about how to approach the target group as the main challenge in implementing health-promotion interventions in the past. In comparison, almost 59% of the studied municipalities reported previous collaboration with the target groups' representatives in health-promotion when more than half acknowledged the lack of knowledge in better target group approach. These findings might suggest that the use of a participatory approach when the target group is being involved in the different phases of health promotion interventions should be much better promoted and supported.

This study provides an overview of experience in implementing health promotion interventions at health promoting schools and municipalities in Latvia. The findings of this study could be used for training and support system planning in order to increase the capacity for better health promotion. More focused and detailed studies might be important for increasing the awareness of both specific features and the larger context of developing health promotion interventions in local settings.

The authors are grateful to all study participants within the National Health Promoting Networks in Latvia for their response. The team of the BaltCityPrevention project from the Institute of Public Health at Rīga Stradiņš University thanks the members from the Department of Health Promotion at the Centre of Disease Prevention and Control for their support and collaboration during data collection process.

References

[1] WHO, The Ottawa Charter for Health Promotion (1986). Available at: `https://www.who.int/healthpromotion/conferences/previous/ottawa/en/` [viewed 2019. 21.01]

[2] M. Dobbins, H. Husson, K. DeCorby, R.L. LaRocca, *School-Based Physical Activity Programs for Promoting Physical Activity and Fitness in Children and Adolescents Aged 6 to 18*, Cochrane Database Syst. Rev., Cd007651 (2013)

[3] C.E. Evans, M.S. Christian, C.L. Cleghorn, D.C. Greenwood, J.E. Cade, *Systematic Review and Meta-Analysis of School-Based Interventions to Improve Daily Fruit and Vegetable Intake in Children Aged 5 to 12 Y*, Am. J. Clin. Nutr. **96**, 889 (2012)

[4] C. Knai, J. Pomerleau, K. Lock, M. McKee, *Getting Children to Eat More Fruit and Vegetables: A Systematic Review*, Prev. Med. **42**, 85 (2006)

[5] R.E. Thomas, J. McLellan, R. Perera, *School-Based Programmes for Preventing Smoking*, Cochrane Database Syst. Rev., Cd001293 (2013)

[6] R. Langford, C. Bonell, K. Komro, S. Murphy, D. Magnus, E. Waters, L. Gibbs, R. Campbell, *The Health Promoting Schools Framework: Known Unknowns and an Agenda for Future Research*, Health Educ. Behav. **44**, 463 (2017)

[7] WHO, *Global school health initiative* (1995). Available at: `http://www.who.int/school_youth_health/gshi/en/` [viewed 10.01.2019]

[8] M.K. Helgesen, E. Fosse, S. Hagen, *Capacity to Reduce Inequities in Health in Norwegian Municipalities*, Scand. J. Public Health **45**, 77 (2017)

[9] N. Simonsen-Rehn, R. Laamanen, M. Brommels, S. Suominen, *Determinants of Effective Health Promotion Actions in Local Contexts: A Study of the Perceptions of Municipal Politicians*, Int. J. Public Health **57**, 787 (2012)

[10] E.V. Jansson, P.E. Tillgren, *Health Promotion at Local Level: A Case Study of Content, Organization, and Development in Four Swedish Municipalities*, BMC Public Health **10**, 455 (2010)

[11] S. Michie, M.M. van Stralen, R. West, *The Behaviour Change Wheel: A New Method for Characterising and Designing Behaviour Change Interventions*, Implement Sci. **6**, 42 (2011)

[12] L.K. Bartholomew, C.M. Markham, R.A.C. Ruiter, M.E. Fernàndez, G. Kok, G.S. Parcel, *Planning health promotion programs: An intervention mapping approach* (4th ed.) (Hoboken, NJ: Wiley, 2016)

[13] L. Green, M.K. Kreuter, *Health program planning an educational and ecological approach.* (4th Ed.) (New York: McGraw Hill, 2005)

Health care accessibility and disparities in medical staff availability in Latvia

Ieva Bikava[1] and *Andris* Skride[2]

[1] Rīga Stradiņš University, Riga, Latvia
[2] Rīga Stradiņš University, Department of Internal Diseases, Riga, Latvia

Abstract. The Health care accessibility for inhabitants, especially in rural regions, is related to the problem of the medical staff availability in state and municipal hospitals and ambulatory institutions as well as for General Practitioners.

Despite the fact, that there are two state Universities – University of Latvia and Rīga Stradiņš University, as well as colleges, that educate medical staff – there is a lack of human resources in health care institutions in Latvia.

The Ministry of Health and municipalities had several initiatives to solve this problem – changes in regulations for Residency students, programme financed by European Funds with obtainable funding in the amount of 10 million euro, as well as different stimuli and programmes from municipalities.

The research explores the current situation with staff availability and, basing on interviews and outcomes, analyzes the effectiveness of the available programmes to attract human resources, and predict the impact and outcomes of those activities.

The results showed that all the programmes had some fragmented positive impact and partially solve the problem, but those activities will not provide sufficient outcomes and the problem of medical staff availability will still be the urgent topic in future.

Key words: health policy, regional disparities, medical specialist availability, human resources in health care.

1 Introduction

The slogan – to provide qualitative, affordable and accessible to inhabitants health care services – has been used for implementation of different reforms since the 1991. Countless changes have been planned and implemented in health care sector under this flag, but the problem with availability of health care service persists.

One of the indicators describing the results of achieved availability is the data on unmet needs of inhabitants (Table 1).

Analysis of reasons for unmet needs indicates that the leading obstacle is high costs for both urban and rural inhabitants. In the course of time, the main cause – "too expensive" – is decreasing, but other warning trend is arising – the share of people who postpone the medical care rises significantly, which can also be explained with financial issues.

"Too far to travel" is not one of the leading obstacles, but at the same time is twice higher for rural than urban population. All indicators show that the level of unmet needs is higher for rural than the urban population.

The deeper analysis of inhabitant behaviour showed, that there is another tendency in Latvia related to healthcare availability – the high rate of out-of-pocket payments for those,

Table 1. Reasons for unmet needs for medical care for persons 16 or over (%) [1].

Reason	Could not afford to (too expensive)		Wanted to wait and see if problem got better on its own		Too far to travel/ no means of transportation	
Year	Urban	Rural	Urban	Rural	Urban	Rural
2013	60.9	58.1	13.7	21.5	1.8	4.4
2014	53.7	56.1	15.1	27.1	1.7	3.6
2015	57.2	59.1	14.5	18.4	2.2	4.2
2016	41.7	51.2	16.4	20.9	2.9	6.1
2017	42.5	44.6	21.6	25.4	3.3	6.6

Table 2. Health Spending. Out of Pocket in percentage of health spending [2].

Year, %	2009	2010	2011	2012	2013	2014	2015	2016
Estonia	20.34	21.92	21.56	21.52	22.61	22.62	22.77	22.69
Latvia	38.78	37.19	34.3	37.82	38.47	39.11	42.07	44.56
Lithuania	26.82	27.59	28.22	31.80	32.82	31.49	31.84	32.34

who got the medical service they needed. The level of out-of-pocket payment in Latvia reached more than 40% rate and in comparison to both neighbour states is highest not only among Baltic States, but also is the highest in European Union (Table 2).

High rate of out-of-pocket payments are not associated with more expensive service price in Latvia, in comparison with other Baltic states, but is due to the shortage of state provided service volume. The deficit of state provided service push an inhabitant to choose receiving of the service in private sector paying the full price.

There are several services and specialists, but to receive the service there is a queue for several months to wait. At the same time, the service or specialist appointment can be received faster, if patients are ready to pay themselves.

The long waiting lists are related to the amount of funding that state can pay to service provider (limited amount that can be distracted to one service) as well as the capacity of service providers to supply the needed service. The limited price that state can pay to service provider known as tariff, for several services is not adequate or is insufficient, and does not cover the real expenses of the service therefore the service providers are not interested in supplying all the amount of state financed service.

In those circumstances, service suppliers overcharge the same service when it is provided to a single payer and thus, combining the amount that is paid by the state with the amount of service that is paid by a single payer, balance the profit and loss. Those improper mechanisms are the consequences of state settled under-priced tariffs [3].

When the same service is provided for a single payer for overcharged price and the undercharged price – when paid by the state, it leads to the situation that medical staff prefers to provide the service to single payers more than state paid services, because higher revenue can be earned.

Even in cases, when the tariffs are adequate or when the service provider can find the mechanisms how to motivate specialists to provide the services, there is the lack of specialists in several specialities. As acknowledged by the managers of the three Clinical Hospitals in Latvia, the lack of specialists leads to the situation when even if there is enough equipment and funding, the service cannot be provided in necessary extent due to deficiency of specialists.

Table 3. Human resources. Medical Doctors and Nurses per 1,000 inhabitants [4, 5].

Year	2010	2011	2012	2013	2014	2015	2016
Doctors – Estonia	3.24	3.29	3.28	3.30	3.36	3.41	3.46
Doctors – Latvia	3.11	3.13	3.14	3.19	3.22	3.20	3.21
Doctors – Lithuania	3.95	4.10	4.22	4.28	4.31	4.34	4.47
Nurses – Estonia	6.12	6.24	6.17	5.64	5.74	6.01	6.10
Nurses – Latvia	5.01	4.96	4.86	4.88	4.82	4.68	4.64
Nurses – Lithuania	7.37	7.54	7.59	7.55	7.60	7.66	7.70

Table 4. Human resources. Medical and Nurse graduates per 100,000 inhabitants [6, 7].

Year	2011	2012	2013	2014	2015	2016
Medical graduates – EST	9.42	10.28	9.71	10.57	10.11	11.02
Medical graduates – LAT	11.26	11.50	13.46	14.04	16.18	16.33
Medical graduates – LIT	13.44	13.19	14.81	14.63	16.39	16.11
Nursing graduates – EST	29.15	34.4	35.43	34.31	32.61	32.38
Nursing graduates – LAT	79.91	29.10	92.66	27.89	27.56	20.00
Nursing graduates – LIT	18.96	15.86	16.97	18.28	18.42	18.96

Comparing human resource data about three Baltic States (Table 3), the number of doctors per 1000 inhabitants in Latvia is increasing during the recent years, but is still lower than in other Baltic States. In comparison to the recommended level, or the average level in European Union, that is 3.5 to 1000 inhabitants, it is not critically low.

Comparison of the data about nurses shows that the number of nurses in Latvia is significantly lower than in Lithuania or Estonia. Comparing to average EU level, where the number of nurses is 8.4 per 1000 population, in Latvia the lack of nurses is significant because the rate of Latvia 4.6 is almost twice lower than average in EU 8.4.

Every year both Universities – University of Latvia and Rīga Stradiņš University, as well as other educational institutions train new specialists – new doctors, nurses and other specialists, who should replace seniors and provide service supply. According to OECD data, in comparison to Lithuania and Estonia, the level of Medical graduates in Latvia, is higher than in Estonia and quite similar with Lithuania (Table 4).

The average number of nursing graduates in Latvia for the last year does not significantly differ from the other Baltic States, but comparing data over a longer period, we see that in previous years from 2000 to 2010 the number of graduated nurses in Latvia was lower than in other states, so it leads to accumulative effects.

According to data obtained from the Association of nurses in Latvia, there is a large gap between registered and practicing nurses, which means that there are many nurses who have gained the education and practice but are not working as nurses (Fig. 1).

As was previously indicated, comparing data from OECD the rate of nurses per 1,000 inhabitants in Latvia and the European Union, the rate was 4.6 in Latvia to 8.4 in EU. As can be seen from the data from Association of Nurses in Latvia, if all registered nurses worked in the health care service, it would be possible to reach the average EU rate. However, the challenge is how to motivate those nurses working in other sectors to come into healthcare sector.

Another problem observed from the data of Association of nurses in Latvia is the aging problem in Nurses sector (Fig. 2).

It means that new nurses are not motivated to join medical institutions to work as nurses, but prefer work in other sectors.

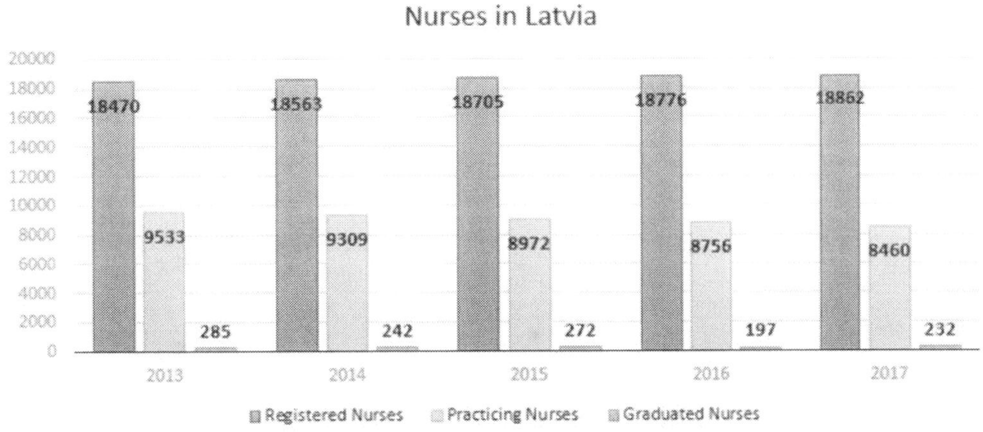

Fig. 1. Data from Medical Nurses register by year.

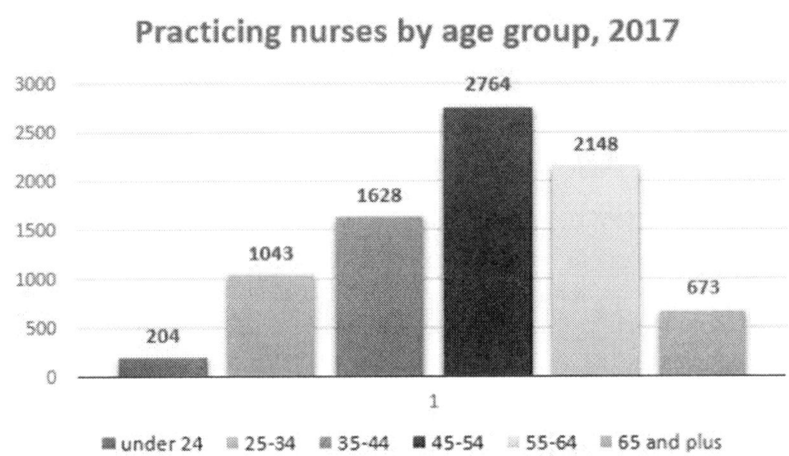

Fig. 2. Data from Medical Nurses register – Practicing nurses by age groups in 2017.

The accessibility to health care service also depends on availability of doctors. According to data on the main workplace of medical doctors – most of the doctors (62%) work in Riga city. In Riga region and in Latgale region – 9%, in Zemgale and Kurzeme – 7% and in Vidzeme only 6% of doctors work [8]. Of course, some specialists who are working in Riga, work also as guest specialists in regions, but they cannot provide all the necessary services in regions. Due to the lack of specialists in regions, there are long waiting lists for specialist consultations and services.

The problem with the availability of specialists is not only in hospital sector, but also in primary health care. In some regions, there are not General Practitioners (GP) available due to aging problem of GPs. For example, in Staicele district in Aloja municipality where, according to data from person registry 933 persons live in Staicele city and 565 in Staicele parish from 1st January 2019, the only GP practice has been closed because the doctor has retired and there is no new GP to replace the retired one. As the result, 899 registered patients, including 120 children are left without GP. The situation in Staicele is not unique, according to data from Rural GP association, of 1300 GP practicing in rural areas, about one third are

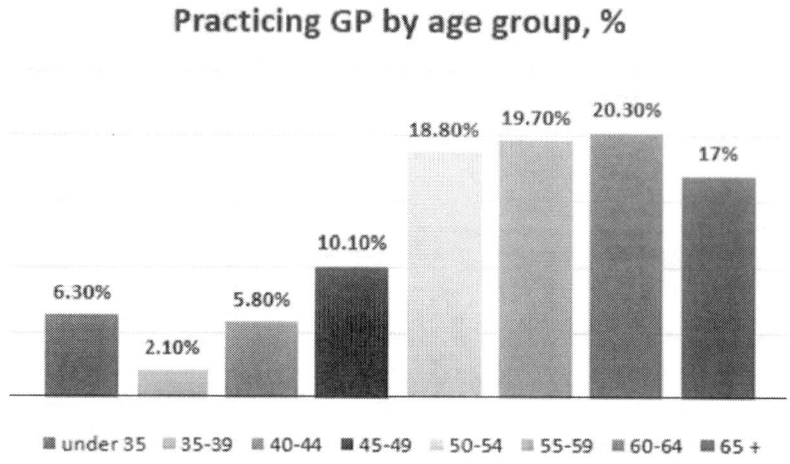

Fig. 3. Practicing GP by age group in 2017 [11].

at retirement age. In 2018, there were 380 GP, who had reached the retirement age 63, and some practicing GP were even 70 years old and older [9, 10].

According to data from The Centre for Disease Prevention and Control of Latvia, in 2017 more than 75% of all practicing GP had reached the age of 50 and more, and 17% of all practicing GP were of retirement age, more than 65 years old (Fig. 3).

Summarizing the above mentioned, there are significant problems with human resource in health care in Latvia, especially with nurses and GPs. The shortage can be observed in both sectors – rural and urban and is caused by the deficit of young nurses and GP entering the field in order to replace the persons at the retirement age.

In its turn, analysis of data on medical doctors does not indicate so pronounced lack of specialists in comparison to other countries, but shows significant regional disparities if comparing the number or practicing specialists in Riga and in regions.

Thus, there are two main problems in health care sector that should be addressed –how to attract more specialists – medical doctors to practice in regions, how to attract nurses to work in healthcare sector not in other sectors, and how to encourage young GP to take over practices of GPs of retirement age.

Evaluating health care in Latvia, the World Bank gave two recommendations for workforce deployment: 1) To develop the workforce in targeted specialities and geographical locations through education and training opportunities; 2) To develop multi – faceted strategies to attract and retain staff in rural areas involving career pathways, attractive living and working conditions, enhance remuneration, and recruitment of students from rural areas who will potentially return to their places of origin [12].

2 Analysis of reasons of human resources deficit

To get better understanding of the problem and find causes that provoke it, several interviews were carried out with representatives of health care sector – Minister of Health, Deputy Secretary of State on Health Policy Issues, managers of Project Management Division of Ministry of Health, Member of the Board of Rural GP association in Latvia, Member of the Board of Association of Nurses in Latvia, as well as Head of the Residency of Family Medicine of Rīga Stradiņš University.

During the discussions with representatives of Nurse association and the representatives from Rural GP associations, as well as representatives from the Ministry of Health it was pointed out that significant part of young nurses do not work in health sector. They prefer work in such sectors as beauty care, in pre-school institutions, in private care institutions, and in other sectors, that are not even close to health care – in offices, in supermarkets and even in gas stations. A big part of nurses is also working abroad – some as medical nurses, but many of them as care providers in social institutions. The main reasons for making such choices are – inadequate salary, paid in both state owned and even in private sector, the other reason is that duties are too heavy physically and emotionally, and the third – the low prestige of the speciality.

Discussions about the problem with GPs revealed that there are several reasons that hinder new GP to enter health care sector. In the rural areas, it is not only the small number of inhabitants that leads to small capitation money, but also the incapacity to provide the needed health care services to patients, especially seniors. From the other side, the senior GPs are not motivated to give up their practice to the new GPs, because there are no reimbursement mechanisms to support the elder GPs. This set of conditions lead to the situation where the elderly GPs cannot provide qualitative services, but the new GPs cannot enter the market.

The research on deficit of medical specialists – doctors in rural areas, illuminated several problems – the first is the amount of salary of a specialist practicing in rural area. If a specialist works in the capital city Riga or other big city, there is a possibility to combine work in more than one institution – for example, part of the day the specialist works in a state clinic, and part of the time in private consultation. It is also possible to combine several duties during one working shift. This is common especially in clinical hospitals – to work as a specialist, to work with the residency students and to work in clinical research. Many leading specialists also give lectures to medical students or work as lecturers in further education, which is paid by pharmacy industry. The combination of duties working in more than one workplace is common not only among doctors, but also among nurses. All those activities that medical staff are forced to execute are the result of small salaries.

Another stimulus to work in state university clinics, not in the rural clinic, is the opportunities of professional development. Rural hospitals cannot provide so much ordinary episodes and even more – rare and interesting cases than can provide the university clinic. The carrier ladders are more attractive in big state clinic than in smaller rural clinic.

As the third reason why the young doctors prefer work in Riga clinics was mentioned the possibility to work with contemporary equipment and participate in trainings and exchange of experience with foreign specialists.

In addition to abovementioned, the representatives of municipalities that try to attract new specialists also mentioned other factors that are significant for the new specialists in making decision where to work. They are the living conditions, the work possibilities for the spouse, the opportunities for children – preschool and school activities, as well as possibilities to enjoy cultural events.

Another factor that plays bilateral role is Latvia as a member of the European Union. On the one hand, this partnership provides possibilities to get better education, to have possibilities to practice in other EU countries. On the other hand, the more developed countries provide better reward system for specialists and the free market legislature provides more attractive working and living conditions in Germany, Norway, Sweden and even Estonia than in Latvia. That leads to the fact that new doctors, as well as experienced doctors and nurses, decide to move to work in other countries after getting highly valued medical education and practice in Latvia.

3 Evaluation of activities in human resource development

Several actions are taken by the state and municipalities to combat the deficit of medical staff, especially in rural areas, and to ensure better availability of service to inhabitants.

To attract young specialists to work in regions, from 2015 changes in regulation for state paid residency studies were made – giving priority to students who have accepted contract to work in regions after the residency for at least 3 years in state or municipality healthcare institutions. In the first study year 2015/2016 there were 20 percent or 43 of 230 new residency students who used this priority and signed contracts on work in regions after graduation [8].

This program is evaluated as successful and good references are available from regional healthcare institutions, especially from the larger regional clinics, Family Health Department and the Further Education Department in Riga Stradins University. Several articles are also available on the internet and in medical journals giving personal stories from young specialists who have residency practice in regional clinics and are planning to stay in regions for next decades.

Another group of actions are taken by municipalities with aim to attract both young and experienced specialists to work in their municipalities – medical specialists and GPs. The programmes that are performed by municipalities and the stimuli are broad – municipality pays residency in medicine, provides accommodation and special advantages to specialists and their families, gives discounts for municipality provided services, etc. Although these activities have some good results, unfortunately, the outcomes are not successful enough. Representatives of regional clinics mention some hindering factors in attraction of specialists to work in regions – environment or possibilities provided in social and culture sphere, difficulties for family members to find a job, insufficient facilities for children education, free time and sports activities. Another, very significant factor, mentioned by both state municipal representatives and by private sector is the carrier and professional development options. The small number of patients and lack of unusual cases do not give necessary challenge for young specialists. In addition, the new reforms with specialization of the rural clinics also have impact on long-term forecasts of further clinic development. Another obstacle is common to the whole of Health care sector – insufficient salary and the lack of human resources that make the existing staff overwork.

To solve the problem with medical staff availability in regions in 2017 the new state programme was initiated within the European Union Social Program "Investing in People" under the aim section "Growth and Employment". A specific program for specific support purpose – "SAM.9.2.5. – to improve health care accessibility in priority areas for inhabitants whose residency is outside Riga". The program provides special compensations or financial benefits to ensure activities necessary to move and start to work in other region. As priority specialities are mentioned – cardiovascular diseases, oncology, children (perinatal, neo-natal) and mental health. As benefit receivers can qualify – Medical doctors, doctor's assistants, nurses, nurse's assistants, midwives, physiotherapists and ergotherapists [13].

Another section of the programme is dedicated specially for GPs – to promote senior GPs to transmit their practice to young GP. Receivers on this section are senior GP of the retirement age who hands over their practice and the new GP who takes over this practice.

In the interviews with representatives of the health care sector all of them positively evaluated the program and characterised it as one of the good ways to improve the situation in health care, especially if the funding is used from EU Funds.

Representatives of the Ministry of Health, who are the program implementers, evaluated this program as very successful. The development of the program was prolonged and the program was conformed only from December 1, 2017. The participants could apply for the

Table 5. Applicants for program SAM 9.2.5. [Data from The Ministry of Health].

Indicator	10.09.2018	15.11.2018
Total applicants	149	264
Reviewed	78	181
Approved	72	163
Rejected	6	18
Applicants by speciality		
Doctor	14	29
Doctor's Assistant	40	58
Nurse	28	45
Nurse's assistant	60	124
Midwife	2	3
Physiotherapist	3	5
Ergotherapist	0	0
GP applicants – total	3	3
Reviewed	0	2
Approved	0	2
Rejected	0	0

programme since May, 2018 when the program was officially opened. During the first three months, till September 10, 2018, the Ministry of Health received 149 applications. At that time half of them – 78 were already reviewed and 72 applicants qualified to receive the support. The project team pointed out that the rate of interest is higher than it was predicted and the number of applicants is higher than the capacity to review. The project team is very satisfied with reached outcomes. The number of applicants and the status of application are represented on Table 5.

The outcome indicators of this programme till the end of the project in 31.12.2023 – is the number of medical specialists attracted to work in regions in priority specialities in amounts to 1420 (with intermediate goal on 31.12.2020 – 709). The number of GP practices that are undertaken by young GPs from retired GPs – 100.

As can be seen from the program results on Table 5, the total number of applications is high, and the overall goal can be reached even faster than planned, but if making deeper analysis of data, results are not satisfactory. According to human resource characterization, the main problem is human resource in Healthcare is the deficit of nurses and aging GPs.

Representatives of Medical Nurses association pointed out that the main problem with the deficit of nurses is not in rural clinics, but on the contrary – the deficit is in large state clinics in Riga and in some GPs practices. Therefore the interest rate in this category is not high and the program SAM 9.2.5. will not solve the problem. Of course, the fact that more nurses are attracted to work in health care is positive. However, the program support can be received only in case if the nurse is not already working in health care sector, and 45 nurses attracted to work in regions is not a significant accomplishment in the situation with nurse deficit of at least 800 in Latvia.

Representatives of association of Rural GPs of Latvia, commenting on results pointed to two reasons: the first – the amount of compensation that is paid to GP who gives away its practice is not adequate. The investments that GPs make to create and maintain the practice are more times higher than the compensation rate. In this situation, GPs are not motivated to give up their practice, and only in extreme situations, they agree on this program terms.

Therefore, the interest is so low. The other factor is that this programme is meant only for rural regions, but not for Riga city. But the majority of the retired GPs are the ones who practice in Riga. Based on this, representatives of GPs are very sceptical on the outcomes of this program.

4 Conclusions

Evaluation of the existing solutions that are provided by the State and municipalities to solve the problem of shortage of human resources in healthcare leads to conclusion that the state paid residency programme with priority to a candidate who commits themselves to work in regions for at least 3 years after graduation – is very successful.

The programmes developed and implemented by municipalities to attract workforce in regions have different outcomes, some are successful, some are less and there are a plenty of different factors that impact on those results. Basing on the results and feedback from graduates, municipality paid residency in medicine was highly evaluated. The other programmes and stimuli are not so attractive and are suitable for some specialities only, depending on regional institution specialization, demand and other factors.

The state run program SAM 9.2.5. giving a specific support for specialists in priority specialities and the compensation mechanisms has some good results and can be evaluated as successfully implemented, but according to evaluation of representatives of medical specialist associations, would not significantly impact on the results and solve the human resource problems.

That leads to conclusion that available programmes provided by the state, as well as municipalities – can solve only part of the problem. Short time reimbursement cannot solve low salary problem and lack of possibilities to earn more in long-term perspective.

As all of the interviewed representatives mentioned, while salaries in healthcare sector stay at the present level, working in other sectors and countries is more attractive than working in healthcare, and shortage of human resources will still exist.

At the end of 2018, the Parliament made changes in Healthcare financing law to raise the funding for healthcare domain by 20 percent averagely, but still after this increase the average salary in healthcare sector is lower than the average in the state. Considering the fact that deficit of human resources in health care is acute in both rural regions and in Riga there should be more comprehensive and better funded programmes implemented and run by the state.

References

[1] Central Statistic Bureau of Latvia, Social Processes. Health. Self-perceived Health Status (EU-SILC). *VPG072. Reasons for unmet need for medical care (except dental) for persons aged 16 or over (%).* Latest update 1/17/2019 [viewed 29.01.2019]

[2] OECD Data, Health Spending. *Out of pocket as percentage of Health spending.* Available at: `https://data.oecd.org/healthres/health-spending.htm#indicator-chart` [viewed 03.02.2019]

[3] Konkurences padome, *Medicīnas pakalpojumu tirgus uzraudzība. Noslēguma ziņojums* (Rīga, 2016)

[4] OECD Data. Health, *Doctors total per 1000 inhabitants.* Available at: `https://data.oecd.org/healthres/doctors.htm#indicator-chart` [viewed 03.02.2019]

[5] OECD Data, Health. *Nurses total per 1000 inhabitants*. Available at: `https://data.oecd.org/healthres/nurses.htm#indicator-chart` [viewed 03.02.2019]

[6] OECD Data, Health. *Medical graduates total per 100 000 inhabitants*. Available at: `https://data.oecd.org/healthres/medical-graduates.htm#indicator-chart` [viewed 03.02.2019]

[7] OECD Data, Health. *Nursing graduates total per 100 000 inhabitants*. Available at: `https://data.oecd.org/healthres/nursing-graduates.htm#indicator-chart` [viewed 03.02.2019]

[8] Ministru Kabinets, 07.08.2017 Rīkojums Nr.394. *Konceptuālais ziņojums "Par veselības aprūpes sistēmas reformu"*

[9] Pilsonības un Migrācijas lietu pārvalde, Iedzīvotāju registra statistika. Iedzīvotāju skaits pašvaldībās 01.07.2018. Available at: `https://www.pmlp.gov.lv/lv/assets/documents/statistika/Iedz%C4%ABvot%C4%81ju%20re%C4%A3istrs%20st.%20uz%2001072018/ISPV_Pasvaldibas_iedzivotaju_skaits_pagasti.pdf` [viewed 04.02.2019]

[10] LTV Ziņudienests, *Staicele palikusi bez ģimenes ārsta*. 26.01.2019 Available at: `https://www.lsm.lv/raksts/zinas/latvija/staicele-palikusi-bez-gimenes-arsta.a307404/` [viewed 04.02.2019]

[11] Slimību profilakses un kontroles centrs, Statistika, *Veselības aprūpes cilvēkresursi – Statistikas dati par veselības aprūpes cilvēkresursiem*. Available at: `https://www.spkc.gov.lv/upload/Veselibas%20aprupes%20statistika/Statistikas%20dati/2018/cilvekresursi_19122018.xlsx` [viewed 04.02.2019]

[12] World Bank, *Proscpects for Health Sector Reform in Latvia*. November 2016. Available at: `http://www.vmnvd.gov.lv/uploads/files/585933cb1d0d4.pdf` [viewed 04.02.2019]

[13] Veselības ministrija, *ESF Projekts "Ārstniecības un ārstniecības atbalsta personu pieejamības uzlabošanas ārpus Rīgas"*. Available at: `https://talakizglitiba.lv/projekti/kompensacijas-arstniecibas-personam-par-darbu-regionos` [viewed 09.02.2019]

Outpatient physiotherapy services in Latvian regional hospitals: Characteristics of receivers, the volume of services and source of funding

Ilze Heize[1] and *Daina* Smite[2]

[1] Rīga Stradiņš University, Latvia
[2] Rīga Stradiņš University, Department of Rehabilitation, Latvia

Abstract. *Objective:* To analyse the outpatient physiotherapy services provided in the regional hospitals of Latvia in 2017 and the profile of their receivers. *Methods:* In the research, 7 regional hospitals of Latvia participated, in which the analysis of statistical documentation of 2017 was carried out. The following data was collected about the receivers of outpatient physiotherapy services: age, gender, clinical diagnosis and place of residence; about the volume of service provided – the number of visits per patient; about the source of funding (state budget, private resources, insurance). *Results:* In the regional hospitals of Latvia, outpatient physiotherapy services in 2017 were received by 6957 patients, the most commonly adults of working age with basic clinical diagnosis of the musculoskeletal system and connective tissue diseases, and children up to the age of one year with diagnosis of certain conditions originating in the perinatal period. Among the regional hospitals of Latvia there are significant differences in the volume of outpatient physiotherapy services provided to one patient with a similar clinical diagnosis and age. Latvian regional hospitals still do not have electronic records of all outpatient physiotherapy services, which does not allow to judge about provided services and their receivers, which are financed by patients private resources.

Key words: outpatient physiotherapy services, physical therapists, regional hospitals.

1 Introduction

Outpatient care – health services provided to patients who are not confined to an institutional bed as inpatients during the time the services are rendered [1].

Physical therapists practise in a wide variety of settings that may include, but are not confined to, the following: hospitals, nursing homes, physical therapists' private offices, practices, clinics, rehabilitation centres, outpatient clinics [2]. It has been found that outpatient physiotherapy services have proven their effectiveness (both clinical and cost-effectiveness) when comparing different forms of physiotherapy services [3, 4]. Understanding the patient's profile is important not only for decision-making in the therapeutic process [5], but also for adjusting healthcare services to patient's changing needs [6].

Physiotherapists in Latvia provide their services in different environments providing outpatient, inpatient and home care physiotherapy services (services are provided both from the state budget and the private funds of the patient, including private insurance funds). Outpatient physiotherapy services in Latvia are in great demand, with different waiting time

Table 1. Data selection.

Hospital	Outpatient Physiotherapy patients (*n*)	The volume of service provided	Excluded patients (*n*)	Included patients (*n*)
Liepāja regional hospital	1672	8573	2	1670
Jēkabpils regional hospital	2540	7726	0	2540
Vidzeme hospital	1851	17000	177	1674
Daugavpils regional hospital	1161	11286(state)	418	743
Jelgava city hospital	71	171(charged)	0	71
North Kurzeme regional hospital	744	3842	485	259
Rēzekne hospital	3673	unknown	3673	0

for access in different institutions, but compared to the waiting time of other functional specialist services – it is the longest [7].

In Latvia, there are no statistics available about the provided outpatient physiotherapy services or about the receivers' characteristics. That would allow purposeful organizational activities for most commonly provided mono-professional outpatient rehabilitation service.

The objective of the research was to analyse the outpatient physiotherapy services provided in the regional hospitals of Latvia in 2017 and the profile of their receivers.

The tasks of the research were to collect and analyze the data on the characteristics of the receivers, the volume of services and source of funding of outpatient physiotherapy services provided by regional hospitals of Latvia in 2017.

2 Methods

2.1 Data sources and procedure

In the research, 7 (all) regional hospitals of Latvia participated, in which the analysis of statistical documentation of 2017 was carried out.

The study was carried out between 22.01.2018 – 31.03.2018. The design was quantitative, non-experimental, retrospective research. The approval for this study was obtained from the Ethics Committee of Rīga Stradiņš University and from the chairmen of the board of hospitals involved in the study. After obtaining permissions to obtain the necessary data, requests for the necessary information were sent electronically to the medical statistics units of the involved hospitals.

Data were collected about: the receivers of outpatient physiotherapy services (age, gender, clinical diagnosis and place of residence), the volume of service provided (the number of visits per patient); the source of funding (state budget, private resources, insurance). Descriptive data were analysed in general, by hospitals, as well as volume of service depending on the patient's clinical diagnosis, age, place of residence was analyzed in the total sample and in each hospital.

The study included only valid or usable (containing full information) data in electronic form, incomplete data were excluded (data selection is shown in Table 1). Due to various factors, in "Daugavpils regional hospital", "Vidzeme hospital" and in "North Kurzeme regional hospital" valid data were only available about the receivers and received services provided by state budget. "Rezekne hospital" does not perform electronic accounting of outpatient services provided, so data were excluded. Two patients from "Liepaja regional hospital" were excluded from the data analysis due to the large volume of services (number of visits per patient > 100).

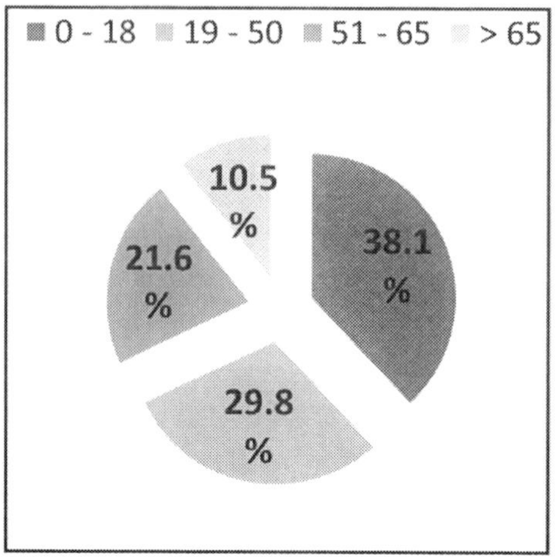

Fig. 1. Distribution of patients (%) by age (years).

2.2 Statistical analysis

The results were analyzed using Microsoft Excel, IBM SPSS Statistics 24.0 and StatPlus v.s.6.2.30 software. In statistical analysis descriptive statistics and analytical statistics were used. The statistically significant level was $p < 0.05$.

3 Results

3.1 Characteristics of receivers

3.1.1 Age and gender

In regional hospitals of Latvia in 2017, 6957 patients received outpatient physiotherapy services, of which 39.1% ($n = 2721$) were men and 60.9% ($n = 4236$) were women.

The mean age of patients was 33.5 ± 24.8 years (minimal age – 0, maximal age – 98), median was 35 years (25% percentile – 10 years, 75% percentile – 55 years). From the total number of patients 38.1% ($n = 2650$) were aged under 18 (children). From the total number of patients who had received physiotherapy services 51.4% ($n = 3574$) were at working age (19–65 years). Distribution of patients by age is illustrated in Fig. 1.

Differences were found between age, considering the gender ($p < 0.001$) – for women the median was 43 years (25% percentile – 14 years, 75% percentile – 58 years) and for men 20 years (25% percentile – 5 years, 75% percentile – 72.8 years).

3.1.2 Clinical diagnosis

Outpatient physiotherapy services provided by regional hospitals were received by patients whose primary clinical diagnosis corresponded to one of the 17 diagnostic code groups (by ICD-10).

From the total number of analyzed patients, 54.2% ($n = 3769$) were diagnosed with diseases of the musculoskeletal system and connective tissue (M00-M99), 12% were diagnosed with certain conditions originating in the perinatal period (P00-P96). Detailed distribution by diagnosis is shown in Fig. 2.

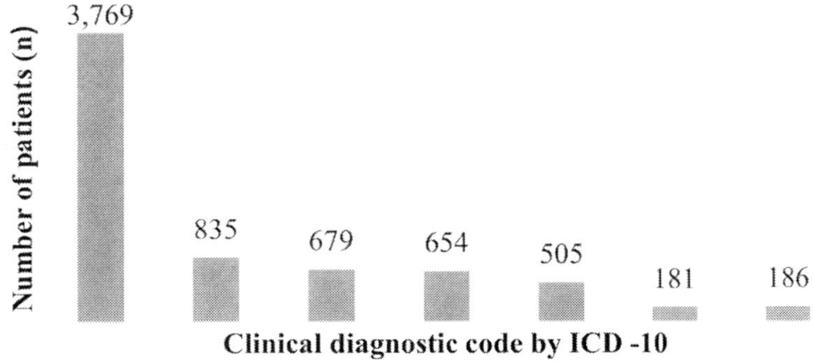

Fig. 2. Distribution of patients by diagnosis.

3.1.3 Place of residence

From the total number, 62.2% of patients lived in the town of hospital localization and only 37.8% lived in the region. Only The Vidzeme hospital has a higher proportion of patients with place of residence in regions – 59.9% of Vidzeme hospital patients. There were no gender or age differences by place of residence.

3.2 The volume of services and connection with the characteristics of the receivers

The total number of outpatient physiotherapy services, provided in regional hospitals in 2017, was 37 312 visits. Number of visits per patient varies from 1 to 91. The median volume of service per patient per year was 5 visits (25% percentile – 2 visits, 75% percentile – 10 visits).

It was found that patients with place of residence in hospital localization city received more services than those patients with place of residence in region.

Analyzing the data, it was found that the visits per patient were reliably related to the patients' diagnosis. Fewer visits per patient (median 2 and 3 visits per patient) had patients with diagnosis from P00-P96 and R00-R99 diagnostic codes compared to patients with diagnosis from M00-M99, S00-T98 and I00-I99 diagnostic codes (median – 5 visits per patient).

Statistically significant differences were found when analyzing the number of visits per patient in different hospitals ($p < 0.001$): the highest number of visits per patient was at Daugavpils regional hospital (median – 10 visits per patient), but the lowest number of visits per patient was at Jekabpils regional hospital (median – 3 visits per patient). At Liepaja regional hospital median number of visits per patient was 5 visits per patient, at North Kurzeme regional hospital median was 7 visits per patient.

The distribution of number of visits per patients by diagnosis (and by age groups in each hospital) is shown in Table 2, Table 3, Table 4 and in Table 5. Among the regional hospitals of Latvia there are significant differences in the volume of outpatient physiotherapy services, provided to one patient with a similar clinical diagnosis and age.

3.3 Source of funding

It was not possible to analyze data on the source of funding because of the lack of available, complete data on services provided by the patient's private or insurer's funds.

Table 2. The number of visits per child's age patient in different hospitals.

Hospital	Age		
	0–1	**2–6**	**7–18**
Liepaja regional hospital	6.0	7.2	6.3
Jekabpils regional hospital	1.9	3.0	3.2
Vidzeme hospital	6.4	8.4	10.2
Daugavpils regional hospital	19.7	19.4	23.6
North Kurzeme regional hospital	4.2	8.6	9.8

Table 3. The number of visits per patient in different hospitals. Patients with group M00-M99 diagnosis (by ICD-10).

Hospital	Age		
	19–50	**51–65**	**>65**
Liepaja regional hospital	5.4	4.7	5
Jekabpils regional hospital	3.2	3.3	3.4
Vidzeme hospital	9.5	10.8	9.8
Daugavpils regional hospital	12	14	14.7
North Kurzeme regional hospital	10.7	12.4	12.4

Table 4. The number of visits per patient in different hospitals. Patients with group S00-T98 diagnosis (by ICD – 10).

Hospital	Age		
	19–50	**51–65**	**>65**
Liepaja regional hospital	6.9	7.6	8.8
Jekabpils regional hospital	3.3	4.3	3.8
Vidzeme hospital	11.7	11.6	17.4
Daugavpils regional hospital	12.3	11.0	13.1
North Kurzeme regional hospital	16.8	14	4

Table 5. The number of visits per patient in different hospitals. Patients with group I00-I99 diagnosis (by ICD – 10).

Hospital	Age		
	19–50	**51–65**	**>65**
Liepaja regional hospital	8/10/20	5.8	3.5
Jekabpils regional hospital	3.6	5.3	4.3
Vidzeme hospital	6/16	7.1	7.7
Daugavpils regional hospital	14.5	20.3	15.1
North Kurzeme regional hospital	–	14	7/6

4 Discussion

The study revealed that there is a significant difference between the number of patients who had received outpatient physiotherapy services in different Latvian regional hospitals – the highest number of patients ($n = 2540$) was found at Jekabpils regional hospital, while the lowest at North Kurzeme regional hospital – only 259 patients. Perhaps the significant difference is affected by the activity of other service providers in the region (other healthcare facilities located in the regional hospital localization cities, as well as healthcare facilities in the regions that have a contract with the National Health Service and also provide outpatient physiotherapy services using state budget funds. According to National Health Service data, at least one outpatient health care centre is located in Valmiera, Daugavpils, Liepaja, Ventspils and Talsi which also provides outpatient physiotherapy services, while in Jēkabpils there is only one private practice which provides outpatient physiotherapy services by the state budget funds [8].

It was found that the 38.1% of outpatient physiotherapy receivers were children (most often infants) and 51.4% were adults at working age, but only 10.5% of receivers were over 65 years old. Attention should be paid to the low proportion of seniors. Literature shows that old-age individuals are often diagnosed with at least one or more chronic illnesses [9] and the aging process is characterized by the increase of functional limitations [10], which justifies the need of rehabilitation services.

It should be emphasized that only 2.6% ($n = 178$) of all patients who received outpatient physiotherapy had a primary clinical diagnostic code from I00-I99 (diseases of the circulatory system), including after stroke (I69.1, I69.3, I63). According to the data of the Central Statistical Bureau of Latvia, in 2016, disability was granted for the first time to 3906 persons suffering from diseases of the circulatory system, including stroke [11]. It is known that stroke has long-lasting consequences, most often having a negative impact on several body functions such as mobility, coordination and the ability to perform daily activities [12, 13]. For this reason, it is particularly important to provide qualitative rehabilitation services for these persons, it has been proven that outpatient, hospital-based stroke rehabilitation is effective in improving the physical functioning of stroke patients [14].

Among the regional hospitals of Latvia there are significant differences in the volume of outpatient physiotherapy services, provided to one patient with a similar clinical diagnosis and age. Unfortunately, the clinical diagnosis does not allow us to judge the severity of functional limitations, but there is no reason for argument that there are significant differences in functional limitations for patients from different regions. Such differences indicate that there are no established guidelines to define the algorithm how and how much to work with certain types of patients depending on their functional status and the purpose of the outpatient physiotherapy service.

5 Conclusions

The profile of receivers of outpatient physiotherapy services, provided by regional hospitals in Latvia, is characterized by adults of working age with basic clinical diagnosis of musculoskeletal system and connective tissue diseases (most commonly spondylosis and intervertebral disc disorders) and by children under the age of one, with diagnosis of certain conditions originating in the perinatal period (most commonly disorders of muscle tone of a newborn). It can be concluded that outside of the range of service receivers stay seniors, since the age structure of population, the increase of chronic diseases and functional limitations with aging, justify the need of rehabilitation services.

Among the regional hospitals of Latvia there are significant differences in the volume of outpatient physiotherapy services, provided to one patient with a similar clinical diagnosis and age. Given that there is no reason for argument that there are significantly different functional limitations for patients from different regions, it can be concluded that there is a difference in the content and planning of the service, and in cooperation model between specialists.

Latvian regional hospitals still do not have electronic records of all outpatient physiotherapy services, which does not allow judging about provided services and their receivers, which are financed by patients' private resources.

References

[1] World Health Organization, *Assessing health services delivery performance with hospitalizations for ambulatory care sensitive conditions* (2016)
[2] World Confederation for Physical Therapy (2017). Available at: https://www.wcpt.org/policy/ps-descriptionPT
[3] J.D. Childs, J.S. Harman, J.R. Rodeghero, M. Horn, S.Z. George, J. Orthop Sports Phys. Ther. **44**, 955 (2014)
[4] J. Zeidler, T. Mittendorf, G. Vahldiek, H. Zeidler, S. Merkesdal, Rheum. **47**, 1527 (2008)
[5] K. Lutfey, S.M. Campbell, M.R. Renfrew, L.D. Marceau, M. Roland, J.B. McKinlay, Soc. Sci. Med. **67**, 1391 (2008)
[6] Ensocare, *How Demographics Impact Healthcare Delivery* (2017), Available at: https://www.ensocare.com/resource-center/how-demographics-impact-health-care-delivery
[7] The National Health Service (2017), Available at: http://www.vmnvd.gov.lv/lv/veselibas-aprupes-pakalpojumi/ambulatoro-pakalpojumu-gaidisanas-rindas
[8] The National Health Service (2018), Available at: http://www.vmnvd.gov.lv/lv/503-ligumpartneriem/507-ligumiestazu-saraksts
[9] B.W. Ward, J.S. Schiller, R.A. Goodman, Prev. Chronic Dis. **11**, 1 (2014)
[10] J.H. Billson, I. Shaw, B.S. Shaw, African J. Phys. Health Edu. Recreat. Dance **18**, 672 (2012)
[11] Central Statistical Bureau of Latvia, *VAG13* (2017), Available at: http://data.csb.gov.lv/pxweb/lv/Sociala/Sociala__ikgad__veseliba/VA0130.px/table/tableViewLayout2/?rxid=cdcb978c-22b0-416a-aacc-aa650d3e2ce0
[12] R. Luengo-Fernandez, N.L.M. Paul, A.M. Gray, S.T. Pendlebury, L.M. Bull, S.J.V. Welch, F.C. Cuthbertson, P.M. Rothwell, Stroke **44**, 2854 (2013)
[13] M. Mollaoglu, T.K. Fertelli, F.O. Tuncay, Arch. Gerontol. Geriatr. **53**, e227 (2011)
[14] D. Rice, S. Janzen, A. McIntyre, J. Vermeer, E. Britt, R. Teasell, J. Stroke Cerebrovasc. Dis. **25**, 1158 (2016)

Palliative care in Latvia: Availability and the role of the family doctor

Baiba Strautmane, *Daiga* Behmane[1], and *Iveta* Ludviga[2]

[1] Rīga Stradiņš University, Faculty of Public Health and Social Welfare, Institute of Public Health, Riga, Latvia

[2] RISEBA University of Applied Sciences, Postgraduate Study Programme director, Riga, Latvia

Abstract. In line with rapid population aging, palliative care is becoming increasingly important. The purpose of the study is to evaluate the satisfaction with palliative care and its availability in Latvia, as well as identify factors which have a major impact on overall satisfaction with care. The paper describes the situation in Latvia by evaluating the satisfaction with the palliative care and the accessibility to it. The study relies on a quantitative methodology and primary data were obtained using structured online survey conducted during the 30-day period – in March and April 2018. In total 240 questionnaires were collected from which 131 (n = 131) were qualified for further analysis. The survey results indicated that the population of Latvia has insufficient information about the palliative care leading to low availability. The regression analysis showed that overall low assessment of the palliative care is influenced by two factors: 1) low satisfaction with the services provided by the primary care physicians (family doctors), 2) satisfaction with the received palliative care services and the availability of the services. The study indicated that there is no unified and comprehensive palliative care strategy and system in Latvia, however, some state-provided care is available for a limited number of patients. The results highlight the necessary improvements in palliative care in Latvia and may serve as a basis for health and social policymakers for the development of a unified and comprehensive population end-of-life or palliative care strategy.

Key words: palliative care in Latvia, barriers to palliative care, patient satisfaction, health care system, end-of-life care strategy.

1 Introduction

In the context of the rapid ageing and structural changes of population, Latvia must count on an increase in chronic non-communicable diseases. The aim of hospice care is to ensure the best possible quality of life not only for oncology patients and their family members, but also for all patients with chronic and non-communicable diseases up to the end of their life time. Palliative care is the prevention and relief of any kind – physical, psychological, social, or spiritual suffering.

Globally, it is estimated that palliative care is needed in 40–60% of all deaths [1]. Palliative care is required for a wide range of diseases. The majority of adults in need of palliative care have chronic diseases such as cardiovascular diseases (38.5%), cancer (34%), chronic respiratory diseases (10.3%), AIDS (5.7%) and diabetes (4.6%). Many other conditions may require palliative care, including kidney failure, chronic liver disease, multiple sclerosis, Parkinson's disease, rheumatoid arthritis, neurological disease, dementia, congenital anomalies, and drug-resistant tuberculosis [1].

According to OECD 2017 data people in OECD countries are living longer, but the burden of mental illness and chronic disease is emerging. Life expectancy at birth in 2015 was 80.6 years, on average, across OECD countries, in Latvia it was lower, reaching only 74.6 years. The main causes of mortality in Latvia encounter circulatory system diseases (more than 1/2 of all deaths) followed by cancer (~1/4 of all deaths), respiratory diseases and others. Longer life expectancy and declining fertility rates mean that older people make up an ever-increasing proportion of the populations of OECD countries. On average across OECD countries, the share of the population aged over 65 years increased from less than 9% in 1960 to 17% in 2015 and is expected to continue to increase, reaching 28% in 2050. In Latvia accordingly, the share of the population aged over 65 in 2015 and 2050 is respectively expected to increase from 20% to 28%, but population aged over 80% is expected to increase from 5% to 9% [2]. The above mentioned underlines the importance of considering issues related to the quality and availability of palliative care.

The aim of the study is to evaluate the satisfaction of patients and their family members with palliative care and its availability in Latvia, as well as identify factors which have a major impact on overall satisfaction with care. Two broad research questions are investigated: (1) which factors contribute to satisfaction of palliative care in Latvia and what the role of the family doctor is, and (2) what are the limitations to the availability of palliative care in Latvia.

The article is structured as follows. First, the essence and models of palliative care are discussed, followed by an empirical investigation in Latvia about factors affecting the satisfaction with palliative care and its availability. Further research findings are discussed, and conclusions presented.

2 Palliative care

Palliative care is explicitly recognised under the human right to health and becomes increasingly important with an aging population [3]. It should be provided through person-centred and integrated health services that pay special attention to the specific needs and preferences of individuals [1].

Pain is one of the most frequent and serious symptoms experienced by patients in need of palliative care. Opiate analgesics are essential for treating the pain associated with many advanced progressive conditions. For example, 80% of patients with AIDS or cancer, and 67% of patients with cardiovascular disease or chronic obstructive pulmonary disease experience moderate to severe pain at the end of their lives [1].

Other physical symptoms palliative care patients experience are anxiety, appetite loss, breathlessness, constipation, delirium, depression, diarrhoea, fatigue, nausea, pain and respiratory tract secretions, insomnia, etc. The most frequent and burdensome symptoms in patients with chronic, complex, life-limiting health problems tend to be the same across diseases [4].

Psychosocial support is another part of palliative care. Patients with life-threatening or terminal illness and their caregivers experience considerable stress, and health professionals need to be adequately trained or prepared to manage their stress. The health system and health facilities may need certain simple features to facilitate other end-of-life needs of a patient, such as spiritual needs, family support, legal support and a motivating physical environment [4].

Palliative care is not intended to hasten or postpone death but applies ethical principles, shared decision making and advanced care planning to identify patients' priorities and goals for their care at the end of life [4].

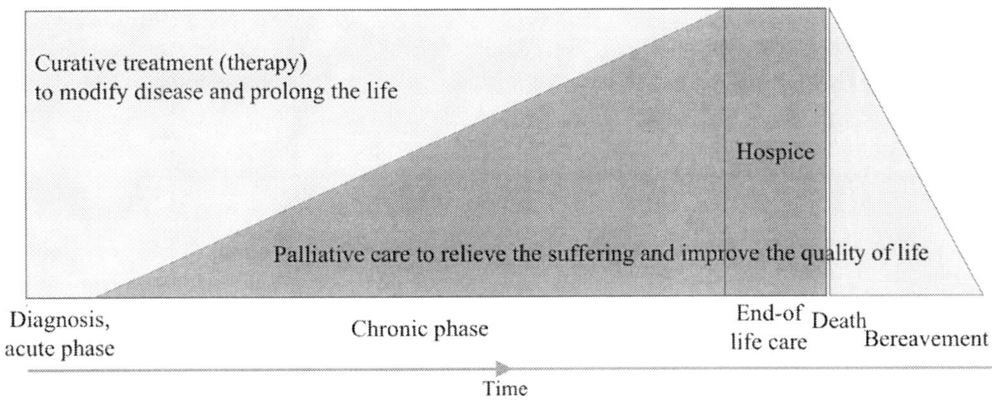

Fig. 1. Integrated palliative care. Source: adapted from WHO webpage `http://www.who.int/mediacentre/infographic/en/`.

Palliative care means planning, coordination and information sharing that includes identifying support and resources available, development and implementation of a care plan based on patient's needs, facilitating the availability and access to medication (especially opiates). Palliative care also identifies the psychosocial/spiritual needs of professionals providing care [4].

Palliative care also includes education and informing the patients and their families about the palliative care and progress of the disease, therefore, communication is a critical issue. Palliative care specialists have to communicate with the patient, family, and caregivers about the disease, prognosis, treatment, symptoms and their management. Other issues are related to care in the last days/weeks of life, identification and setting priorities with patient and family/caregivers and to providing information and guidance to patients and caregivers.

Studies suggest that in developed world palliative care more often is available already during the early stage or chronic phase of the disease (Fig. 1) in conjunction with other therapies that are intended to prolong life, such as chemotherapy or radiation therapy, and includes those investigations needed to better understand and manage distressing clinical complications [5].

Hospice like palliative care provides also comprehensive comfort care as well as support for the family, but, in hospice, attempts to cure the person's illness are exhausted. The hospice usually is provided for a person with a terminal illness whose doctor believes he or she has 6 months or less to live if the illness runs its natural course [6]. Hospice is an approach to care; it is not tied to a specific place. It can be offered in two types of settings – at home or in a facility such as a nursing home, hospital, or even in a separate hospice centre.

The general palliative care is usually provided by physicians and other healthcare professionals from all disciplines (such as family medicine, cardiology, and oncology). The practice of palliative care is an interdisciplinary area, covering doctors, nurses, social workers, chaplains, and others, is being actively taken over in the world [7]. Caregivers need regularly to help patients and their family members to understand changes in the patient's health as they relate to current and future care and treatment goals [8]. In addition, specialist palliative care (specialised or specialty palliative care) has grown substantially [9]. Specialist palliative care is provided by a specially trained team of doctors, nurse practitioners, nurses, social workers, and other health care professionals, who work together with a patient's primary care team to provide an extra layer of support for people with serious illness [10].

The World Health Assembly Resolution 2014 on Palliative Care [11] called for all member countries to incorporate palliative care provision into their health care systems to ensure access to palliative care for all patients in need. Need for systemic approach was acknowledged by researchers as having a huge impact on the quality of care and the effectiveness of the entire healthcare system [12].

But still there are identified several barriers to palliative care that include lack of awareness among policy-makers, health professionals and the public about what palliative care is, and the benefits it can offer to patients and health systems, cultural and social barriers [1, 5, 10]. There are misconceptions about palliative care, such as that it is only for patients with cancer, or for the last weeks of life and misconceptions that improving access to opiate analgesia will lead to increased substance abuse [1].

Other identified barriers to access the palliative care are lack of information and knowledge about palliative care in general and about particular care availability, service or specialist unavailability close to patient's place of living [5, 10], as well the waiting time to receive the care service and financial implications if family have to pay extra to receive palliative care services.

Palliative care requires financial investments, but, on the other side, offers considerable savings [1, 5] in other health care sectors, such as emergency medical care. Advanced health care settings have referred to this fact as an incentive to expand immediate palliative care, as this leads to economy in the growing health care budgets.

3 Method

This study relies on a quantitative methodology. Literature analysis on palliative care models and their organization elsewhere in the world was used, as well analysis of Latvian social and healthcare legislation and planning documents helped to design the survey questions. To obtain data about the palliative care availability and to analyse the impact of palliative care availability and system's components on a patient and their family member satisfaction with the overall palliative care received in Latvia, the primary data were obtained using structured online survey. The logic of the survey design is explained below.

3.1 Variables

Dependent variable: In line with the aim of the research overall satisfaction with palliative care in Latvia was designed as the dependent variable and measured with a single item statement where measured using 7-point Likert scale where 1 was assigned to 'care was not received' and 7 to 'high-quality care was received'.

Independent variables: The independent variables of the survey were chosen based on general palliative patient needs and services suggested by studies to meet these needs – general practitioners provided care (primary care), palliative care specialist consultations (secondary care), access to assistive technologies, pain relief medication, pain and symptom management, pain relief (algology), care in hospitals, care at home, psychological support and spiritual support. To understand the reasons behind the patient and their family satisfaction or dissatisfaction or barriers to access of palliative care services like service unavailability close to living place, lack of information, lack of financing and long waiting time to receive particular care where also taken into account. Thus, disease symptom strengths, the strength of need for care, restrictions to receive the care and finally satisfaction with the provided care were developed as independent variables.

Since in Latvia the main patient contact point in healthcare and also in palliative care is a general practitioner or family doctor, patients and their family members were asked to evaluate separately the service and care provided at primary health care level particularly from palliative care perspective.

Control variables: several factors which may have an impact on overall satisfaction with care such as disease, patient's place of residence and living conditions (whether the patient lives alone, with relatives or in care institution) as well as age and gender of the patient and that of the respondent. These factors were included as control variables.

3.2 Data

The survey was executed to obtain data from all statistical regions of Latvia to get an overview of overall palliative service availability across country. The survey included only respondents who had experience with severe illness and non-treatable patients in age above 18 (to exclude paediatric patients) during the last 24 months. The online survey was conducted during the 30-day period – in March and April 2018. In total 240 questionnaires were collected from which 109 were excluded either not having a relationship with severely ill patients or not having complete information on patient's needs and care received. The qualified respondents ($n = 131$) were both patients and their relatives, representing patients from all age groups (41% were 75 years and older), 64% were female, and represented all Latvia's statistical regions (53% patients were from Riga region). The most represented diseases were cancer (48%), cardiovascular diseases (15%), dementia and Alzheimer's disease (12%), cerebrovascular (11%) and musculoskeletal diseases (11%), the remaining 4% indicated other diseases. 75% of the sample patients were living together with relatives, 17% were living alone and 8% in care institutions.

Since data were gathered using a questionnaire and thus rely on self-reported measures, it may be subject of common method bias (CMB). Hartman one-factor test was used to address common method bias [13] and Factor 1 accounted for 41.2% of the variance indicating that CBM is unlikely to affect the data. All scales showed acceptable to good internal consistency reliability – Cronbach's Alfa coefficients ranged between 0.76 to 0.91 what is above 0.7 [14]. Multicollinearity test showed that variance inflation factor (VIF) range between 1.27 and 1.85, thus it is below 10, with an average being 1.63; and Tolerance from 0.54 to 0.75 what is well above 0.2, thus showing no multicollinearity [15].

4 Results and discussion

In overall only 45.7% of respondents can be accounted as satisfied with the palliative care services while 54,3% remain unsatisfied. Average evaluation of the overall satisfaction with palliative care was 3.51 in 7-point Likert scale with standard deviation 1.55, which shows low satisfaction rate with palliative care in Latvia. The detailed split of overall satisfaction is shown in Fig. 2.

To test the factor impact on overall satisfaction hierarchical regression analysis was used since it allows specifying a fixed order of entry for variables in order to test the effects of certain predictors independent of the influence of others (control variables). Table 1 summarises the results of the regressions: the estimated regression coefficients with their standard errors below them in are presented and the final six rows contain summary statistics. Model 1 to 5 present the effect of control variables and Model 6 and Model 7 present the effect of independent variable on the dependent variable (overall satisfaction with palliative care). Further the logic of the hierarchy is explained. Since variables are measured in different units of measurements, standardised coefficients are used for interpretation.

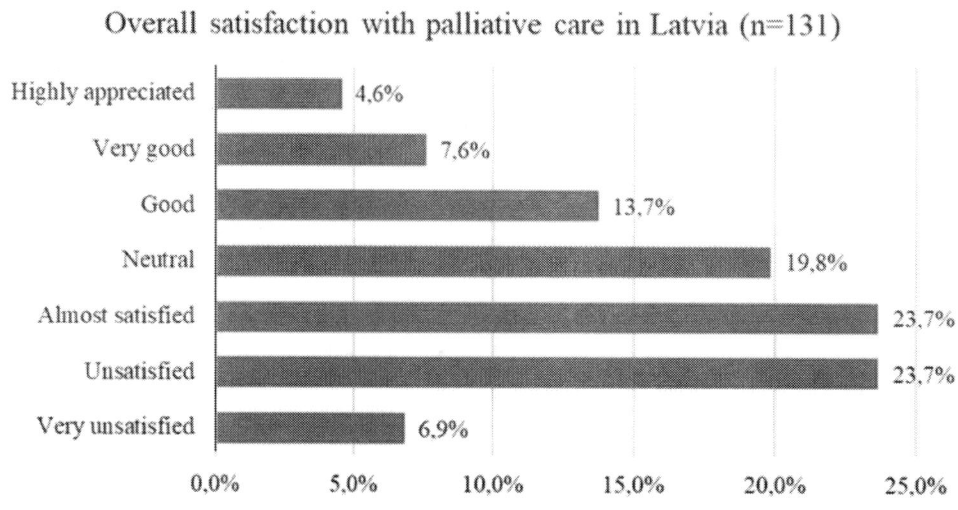

Fig. 2. Overall satisfaction with palliative care in Latvia (2018).

In the first step control variables (diagnosed diseases) were inserted (see Model 1). Disease dummies where created with 'other disease' as a baseline. All diseased have negative betas, as it might be expected still the model is not significant ($\bar{R}^2 = 0.04$).

In the second model, patient's place of residence was entered using 'Riga and district' as a baseline. Model 2 is not statistically significant ($\bar{R}^2 = 0.04$) showing that patients' place of residence has no impact on the dependent variable. Patient's living conditions using 'living alone' as a baseline (Model 3) added no explanatory power since model is not significant ($\bar{R}^2 = 0.04$). Patient's age and gender were entered in the fourth step (see Model 4). The model still remains not significant predictor ($\bar{R}^2 = 0.046$). Model 5 adds respondent's age and gender, still no predictive power ($\bar{R}^2 = 0.07$).

The results in Table 1 show that Models 1 to 5 are not statistically significant, thus, the type of the disease, place of residence and living condition, gender and age does not affect significantly the overall satisfaction with the care system. Therefore, the results can be equally related to all regions of Latvia, all diseases and all patients.

Further steps of hierarchical regression (Model 6 and Model 7) assess the impact of palliative care system's parameters and family doctor variable on the overall satisfaction of palliative care above the effect of control variables.

Model 6 shows the impact of the system's components excluding family doctor and shows statistically significant impact ($\bar{R}^2 = 0.332$; $p = 0.000$). Model adds 25% explanatory power and is statistically significant ($\Delta R^2 = 0.25^{***}$). Moreover, satisfaction with the care provided by palliative care specialists appears to be the only factor with statistically significant impact ($b = 0.71^{***}$ and standardised coefficient $\beta = 0.37^{***}$).

Figure 3 presents a detailed analysis of the variable 'satisfaction with the care provided by palliative care specialists. Figure 3 indicates that the most satisfied are people who received spiritual support and above average satisfied are people who could receive professional pain and other symptom management, pain medication, palliative specialist consultations, as well if palliative care is available at home or day centres. Thus, evidence in Latvia shows that more appropriate attention to the incurably ill patients should be paid by psychologists and professionals in hospitals, as well as the quality of assistive technologies provided should be increased.

Table 1. Hierarchical regression results.

Regression/ summary statistics	*Model 1*	*Model 2*	*Model 3*	*Model 4*	*Model 5*	*Model 6*	*Model 7*
Diagnose: cancer	−2.159* (0.904)	−1.984* (0.916)	−1.949* (0.919)	−1.877* (0.916)	−1.709* (0.907)	−1.167 (0.793)	−.191 (0.675)
Diagnose: heart or cardiovascular disease	−2.035* (0.950)	−1.912* (0.964)	−1.920* (0.964)	−1.775 (0.978)	−1.732* (0.964)	−1.060 (0.841)	−.122 (0.713)
Diagnose: Cerebrovascular	−2.600** (0.967)	−2.404* (0.976)	−2.406* (0.978)	−2.369* (0.991)	−2.137* (0.984)	−1.140 (0.869)	−.237 (0.735)
Diagnose: respiratory disease	−.667 (1.396)	−.588 (1.411)	−.765 (1.448)	−.663 (1.443)	.265 (1.506)	.020 (1.31)	.588 (1.094)
Diagnose: musculoskeletal diseases	−2.744** (0.98)	−2.643** (0.984)	−2.544* (0,989)	−2.659** (0.987)	−2.589** (0.975)	−1.179 (0.863)	−.236 (0.731)
Diagnose: Alzheimer's disease/ dementia	−1.979* (0.962)	−1.901 (0.971)	−1.993* (0.994)	−1.957* (1.02)	−1.825 (1.008)	−.895 (0.872)	−.175 (0.734)
Patient residence Kurzeme		−.429 (0.525)	−.340 (0.528)	−.519 (0.537)	−.528 (0.531)	−.263 (0.453)	−.211 (0.377)
Patient residence Latgale		−.961 (0.550)	−.951 (0.551)	−.979 (0.569)	−.981 (0.564)	−1.200* (0.511)	−.871* (0.428)
Patient residence Vidzeme		−.157 (0.363)	−.157 (0.364)	−.175 (0.367)	−.162 (0.364)	−.127 (0.311)	−.140 (0.259)
Patient residence Zemgale		−.196 (0.503)	−.189 (0.505)	−.215 (0.504)	.062 (0.515)	−.187 (0.446)	−.106 (0.372)
Lives with relatives			−.390 (0.364)	−.278 (0.37)	−.289 (0.368)	−.454 (0.315)	−.505 (0.262)
Lives in care centre			.223 (0.673)	.302 (0.674)	.239 (0.667)	−.058 (0.582)	.171 (0.486)
Patient's age				−.035 (0.107)	−.060 (0.110)	−.017 (0.095)	.010 (0.08)
Patients' gender				−.501 (0.297)	−.334 (0.307)	−.277 (0.265)	−.178 (0.221)
Respondent's age					.024 (0.013)	.013 (0.011)	.008 (0.009)
Respondent's gender					.729 (0.459)	.647 (0.39)	.589 (0.325)
Strength of the symptoms						.221 (0.147)	.072 (0.125)
Assessment of the needs						−.162 (0.133)	−.093 (0.095)
Palliative care received						.710*** (0.122)	.501*** (0.105)

Table 1. Continued.

Regression/ summary statistics	*Model 1*	*Model 2*	*Model 3*	*Model 4*	*Model 5*	*Model 6*	*Model 7*
Restrictions						−.077 (0.134)	.035 (0.113)
Family doctor							.582*** (0.083)
Intercept	5.667*** (0.883)	5.667*** (0.885)	5.926*** (0.917)	6.184*** (1.046)	3.596* (1.578)	1.417 (1.55)	−.604 (1.322)
SER	1.53	1.53	1.53	1.53	1.50	1.28	1.06
Model sig	.087	.155	.173	.145	.072	.000	.000
R Square	.084	.110	.125	.148	.186	.435***	.612***
R Square Change	.084	.026	.015	.023	.038	.249***	.177***
Adjusted R Square	.040	.036	.036	.046	.072	.332	.537
F Change	1.895	.867	1.027	1.590	2.661	12.095	49.591
Sig. F Change	.087	.486	.361	.208	.074	.000	.000

$^{*}p < 0.05$; $^{**}p < 0.01$; $^{***}p < 0.001$.

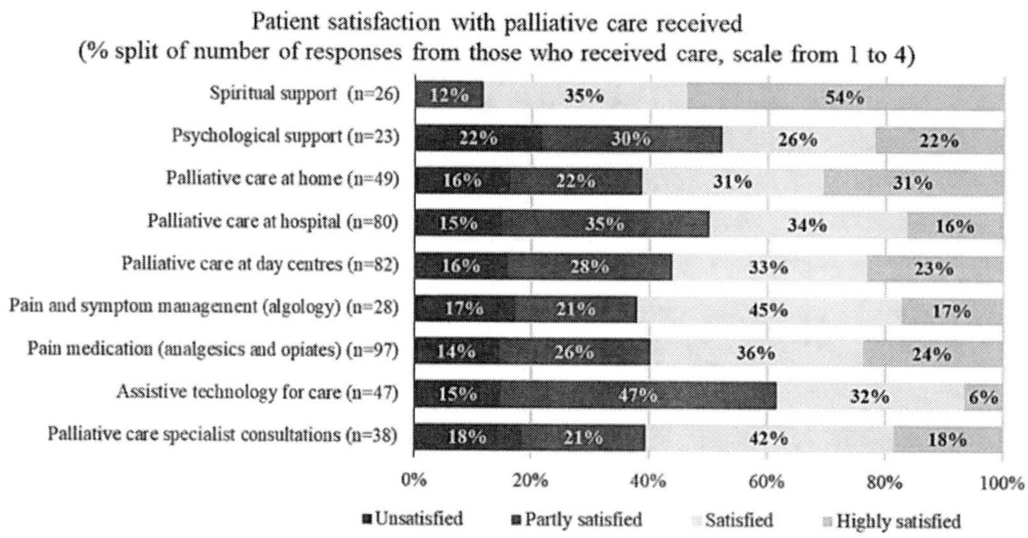

Fig. 3. Representation of patient satisfaction with care received in Latvia (2018).

The final model in Table 1 (Model 7) adds family doctor and it accounts for an additional 18% of overall satisfaction ($\Delta R^2 = 0.18^{***}$). Thus, the final model explains up to 61% of overall satisfaction ($R^2 = 0.61$ and $\bar{R}^2 = 0.54$). The effect of the family doctor ($b = 0.58^{***}$ and standardised coefficient $\beta = 0.49^{***}$) appears to be even larger than that of palliative care specialist ($b = 0.50^{***}$ and standardised coefficient $\beta = 0.37^{***}$). Thus, the results uncover the important role which family doctors have in ensuring satisfactory palliative care in Latvia.

This finding is in line with research by Ramanayake who found that family doctors have a major role to play in providing palliative care since they are closest to the community. Similarly, in US family doctors are significantly more likely to provide non–clinic-based

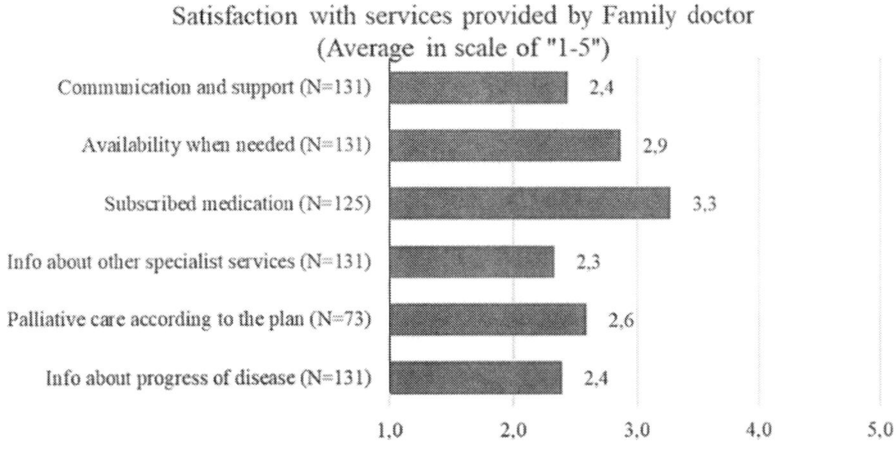

Fig. 4. Patient satisfaction with family doctor practices in Latvia (2018).

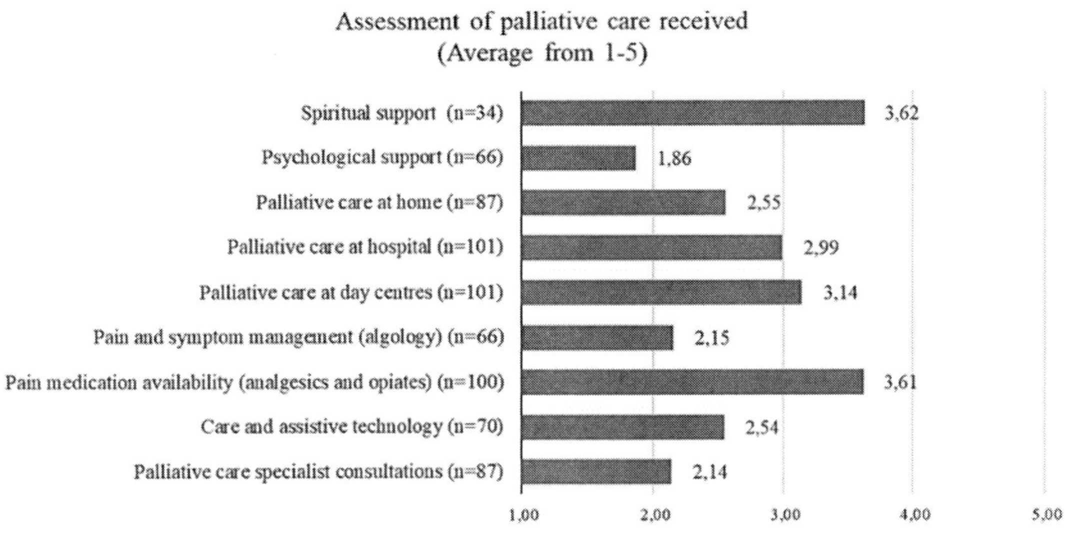

Fig. 5. Patient satisfaction with palliative care service received in Latvia (2018).

palliative care services [3]. Their role is to communicate and coordinate with other health care resources to address complex issues faced by patients [16]. Moreover, researchers have found that family doctors' tasks and roles are to ensure the integration of palliative care and collaboration [17].

The assessment of the care provided at primary level by family doctors in Latvia indicates the lack of adequate communication, insufficient information on the possibilities to receive the necessary care and services provided by other healthcare and social professionals (Fig. 4). Respondents also mentioned insufficient information delivery on the progress of the disease. Respondents also stated that care was not carried out according to a palliative care plan or it even did not exist in their opinion.

The survey also included questions about the actual availability of the service if it was needed. By accessing palliative care through special palliative services respondents mainly indicated insufficient availability of psychological support (Fig. 5 and Fig. 6), unavailability

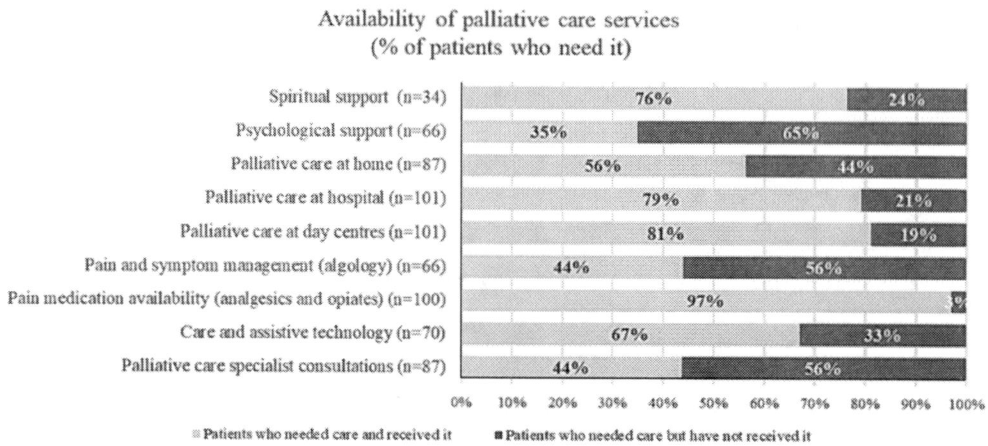

Fig. 6. Palliative care service availability in Latvia (2018).

of palliative specialist consultations and professional pain and symptom management. Other special care like spiritual support and palliative care received in day centres and hospitals were positively evaluated above the average. Highly evaluated was the availability of pain medication prescribed by family physicians.

One of the reasons of poor evaluation for special palliative care services is availability. Figure 5 shows that psychological support, palliative specialist consultations and adequate pain management is not available for more than half of the patients who desire it.

Thus, the results of the research show that particularly problematic aspect in Latvia is the psychological support. The palliative care is different from other medical fields exactly with the emphasis on the emotional support of the patient and his/her family members. It should be noted that terminal patients are particularly vulnerable to psychological disorders, existential problems, depression and suicide. The families of these patients need support to fight the loss and sadness after the family member loses the disease. This key role is exercised by different care team members, including psychologists, social workers and religious leaders (pastors, chaplains). Generally, in EU these specialists are not only responsible for the care of the patients and their families, but also for the welfare of the care team, which causes problems such as burnout [18].

Situation in Latvia shows that spiritual support is generally available and received, but the problem exists with access to psychological support.

Considering restrictions to receive palliative care, the survey data show that practically all worldwide recognised barriers regarding access to palliative care exist also in Latvia. The primary factors (Fig. 7 and Fig. 8) restricting reception of palliative care are the physical unavailability of palliative care services and too long waiting time followed by insufficient information and restricted financial coverage.

In the survey (Fig. 8) 73% of qualified respondents (n = 131) mentioned that there was at least one of restricting factors to receive appropriate palliative care and only 27% were not affected by any of barriers to receive appropriate palliative care.

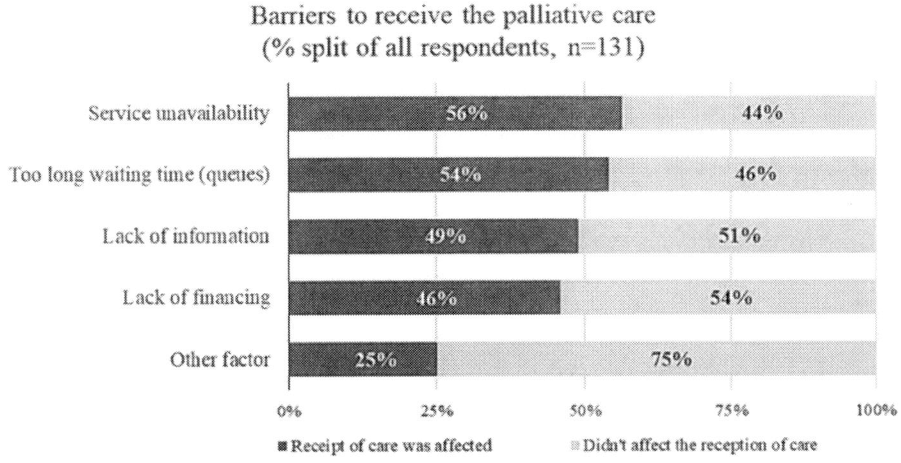

Fig. 7. Assessment of barriers to receive the palliative care in Latvia (2018).

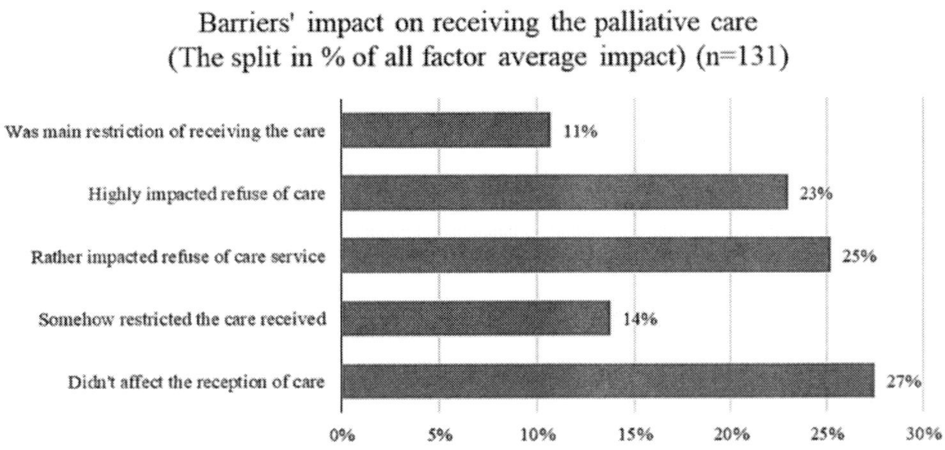

Fig. 8. Impact of barriers to receiving the palliative care in Latvia (2018).

5 Conclusions

The aim of this research was to evaluate the satisfaction with palliative care and its availability in Latvia, as well as identify factors which have a major impact on overall satisfaction with care.

According to the first research question (which factors contribute to satisfaction of palliative care in Latvia and what the role of the family doctor is), the survey analysis highlighted the crucial role of family doctors as the most important factor contributing to overall satisfaction with the care, and service provided by palliative care specialists being the second important factor. Results also show that the communication and exchange of information between those specialists and information provided to patients and their family members are crucial to ensure the patient needs and to improve the palliative care. Indeed, as stated by Ramanayake, developing palliative care models should include opportunities for family doctors to learn sound palliative care principles in order to meet the challenges of the future [16].

In Latvia, there is a need to establish an interdisciplinary team approach not only between the healthcare specialists but also social care specialists that should be involved in ensuring qualitative palliative care. According to the concept developed by Sisli Sounders, modern palliative care includes a holistic model of care provided by a multidisciplinary team. Moreover, modern palliative care is centred on general health care for patient care according to his needs. The patient and his/her relatives are not only recipients of care, but also active team members [19].

According to the second research question (what the restrictions to the availability of palliative care in Latvia are), the survey results provide an idea of the necessary improvements in palliative care in Latvia. There is a need to increase the availability of palliative care services across all regions of Latvia and to inform the population about the palliative care and how to access it. The situation in Latvia shows that while spiritual support is generally available and received when needed the problem exists with availability and access to psychological support. Consequently, there is also a need to increase the funding to finance the whole spectrum of palliative care services thus increasing access to them.

Palliative care should be available from the time of diagnosis to anyone who is experiencing a serious, chronic illness or other life-threatening or terminal illness. Such patients and their family members in addition to curative therapy need support and care provided by primary and secondary healthcare professionals, social workers, psychologists, and chaplain.

The study indicated that there is no unified and comprehensive palliative care strategy and system in Latvia, however, some publicly funded care is available for a limited amount of oncology patients. The world literature and models available may serve as a basis for Latvia policymakers to develop unified and comprehensive end-of-life or palliative care strategy in Latvia.

References

[1] World Health Organization, *Global Atlas of Palliative Care at the End of Life* (2014). Online: `http://www.who.int/nmh/Global_Atlas_of_Palliative_Care.pdf`

[2] World Health Organization, *Palliative care* (2018). Online: `https://www.who.int/en/news-room/fact-sheets/detail/palliative-care`

[3] OECD, *Health at a Glance 2017: OECD Indicators, OECD Publishing*, Paris (2017). Online: `http://dx.doi.org/10.1787/health_glance-2017-en`

[4] C.K. Ankuda, A. Jetty, A. Bazemore S. Petterson, J. Am. Board Fam. Med. **30**, 255 (2017)

[5] World Health Organization, *Planning and Implementing Palliative Care Services: A Guide for Programme Managers*, World Health Organization, Geneva (2016). Online: `http://www.who.int/iris/handle/10665/250584`

[6] National Institute of Aging, *What Are Palliative Care and Hospice Care?* (2018). Online: `https://www.nia.nih.gov/health/what-are-palliative-care-and-hospice-care`

[7] M. Pardo, R.D. Miller, *Palliative Care in Basics of Anesthesia*, edited by S. Gebauer, 7 ed. (Philadelphia, PA, Elsevier, Inc., 2017), p. 829

[8] S.B.E. Yennurajalingam, *Oxford American Handbook of Hospice and Palliative Medicine* (Oxford: Oxford University Press, 2011)

[9] M.A.A.P.A.M. Timothy, E. Quill, N. Engl. J. Med. **368**(13), 1173–5 (2013)

[10] P. Hawley, *Barriers to Access to Palliative Care*, Palliat Care **10**, 1–6 (2017)

[11] World Health Assembly, *Resolution WHA67.19*, Geneva: World Health Organization (2014)

[12] X. Gómez-Batiste, M. Martínez-Muñoz, C. Blay, J. Amblàs, L. Vila, X. Costa, J. Espaulella, J. Espinosa, C. Constante, G.K. Mitchell, Palliat Med. **28**, 302 (2014)
[13] P.M. Podsakoff, S.B. MacKenzie, N.P. Podsakoff, Annu. Rev. Psychol. **63**, 539 (2012)
[14] G. Darren, P. Mallery, *SPSS for windows Step by Step* (Pearson Education, 2003)
[15] A. Field, *Discovering Statistics using IBM SPSS Statistics* 5 ed., SAGE (2018)
[16] R. Ramanayake, G. Dilanka, L. Premasiri, J. Family Med. Prim. Care **5**, 234 (2016)
[17] K. Beernaert, L. Van der Block, K. Van Thienen, D. Devroey, K. Pardon, L. Delien, J. Cohen, Fam Pract. **6**, 694 (2015)
[18] J.M. Martin-Moreno, M. Harris, L. Gorgojo, D. Clark, C. Normand, C. Centeno, *Policy Department Economic and Scientific Policy, Palliative Care in the European Union*, European Parliament, (2008). Online: `http://www.europarl.europa.eu/RegData/etudes/etudes/join/2008/404899/IPOL-ENVI_ET%282008%29404899_EN.pdf`
[19] Laže, O. Gusakova, E. Bladiko, I. Mockus, G. Āmara, ESF, *Onkologiskās aprūpes Māsas papildspecialitāte*, Rīga, Nacionālais apgāds (2009)

The impact of diagnosis on the psychosocial and spiritual needs of the patients

Anna Kralova[1,*] and *Sona* Hlinkova[2]

[1] Central Military Hospital Ružomberok – Faculty Hospital, Ružomberok, Slovakia
[2] Catholic University in Ružomberok, Faculty of Health, Ružomberok, Slovakia

Abstract. *Introduction:* Holistic care focuses on the relationship between the body, mind, and spirit. Spiritual needs are among the basic needs of the individual. From the physical and spiritual dimension and the interaction of these two dimensions, the spiritual needs of the person are developed. *Aim:* The aim of our study was to analyze if the patient's knowledge regarding their diagnosis and possible proximity of death have significant impact on satisfaction of psychological and spiritual needs. *Material and Methods:* In our study we used standardized questionnaire The Spiritual Needs Assessment for Patient – SNAP constructed by Sharma Rashmi, applied for 113 respondents and processed by SPSS statistical program. *Results:* Reliability of questionnaire is given by Cronbach's Alpha 0.945 for the total SNAP. A statistically significant difference at the level $p < 0.01$ has been confirmed in all questions among respondents with fatal diagnosis and respondents with not life-threatening diagnosis. In domain of psychosocial needs, significant difference was not confirmed regarding to gender, age, education, income, religion and place of residency. In domain of spiritual needs, significant difference was partially confirmed regarding religion in 4 questions from 13. *Conclusion:* Faced with chronic or fatal diseases, many patients rely on dealing with spiritual and religious issues. In fact, spirituality/religiosity can be considered an important source of support and management of severe chronic diseases. It is possible to identify the four basic dimensions of spiritual need: interconnection, peace, meaning/purpose, and transcendence. Patients often have the problem addressing their needs for related psychosocial and physical problems such as physical disability, fatigue, sleep disorders, side effects of treatment, etc. Importance of satisfaction of psychosocial and spiritual needs is increased in time of threat, fatal diagnosis.

Key words: diagnosis, psychological needs, spiritual needs, religious needs, patient, holistic approach.

1 Introduction

The behaviour of the patient, acceptance of the diagnosis and coping with reality are important factors during treatment process. Therefore one of the greatest challenges for nursing is to provide the greatest possible comfort for the patient and it is necessary to focus nursing care to the whole personality of the patient. In recent years, scientific advances in health care have evolved to the detriment of the importance of human spirituality, particularly in terms of health and disease. The importance of spiritual needs is declared by authors Best,

* Corresponding author: `anna.kralova88@gmail.com`

Butow and Olver who state that patients require their doctors to reflect on their spiritual needs and facilitate their access to their satisfaction [1]. Holistic care focuses on the relationship between the body, mind, and spirit. Solutions to emotional, social and existential concerns can be realized through a team of interdisciplinary professionals. Spiritual needs may not always be associated with life satisfaction, but sometimes with anxiety, and can be interpreted as patient's desire for spiritual well-being. The need for peace, health and well-being is universal human need and is particularly important for patients with long-term illnesses [2]. It is possible to identify the four basic dimensions of spiritual need: interconnection, peace, meaning/purpose, and transcendence. Apart from addressing the physical needs of patients healthcare needs to address also emotional, social, existential and spiritual needs. Authors Mesquita et all identified eight groups of spiritual needs: finding the meaning and purpose of life; finding the meaning in experiencing the disease; being connected to other people, God and nature; having access to religious/spiritual practices; physical, psychological, social and spiritual wellbeing; talking about death and the experience of dying; making the best out of their time; being independent and being treated like a normal person [3]. Faced with chronic or fatal diseases, many patients rely on dealing with spiritual and religious issues. The patient in the terminal stage of the disease falls into palliative care, which is currently aimed at ensuring that the patient has a good quality of life during this period. But despite this effort, we are faced with situations when the dying is accompanied by pain, loneliness and loss of human dignity. The spiritual needs of the dying patient are related especially with his/her accompanying spiritual experience. Spiritual care can be perceived by such patient as a need for understanding their own lives, suffering and death [4].

Aim: The study was carried out as a pilot study in the project KEGA no. 007KU-4/2018 focused on professional applied ethics in the teaching process at the Faculty of Health in Ružomberok. The main aim of the study is to identify the psychosocial, spiritual and religious needs of the patients on the basis of a survey carried out in designated health care facilities in Slovakia. Based on our research, we tried to identify the components that affect patients' needs most, especially, how severity of diagnosis is influencing patient's needs.

2 Methods

2.1 Study design and setting

In our survey, we used the standardized questionnaire The Spiritual Needs Assessment for Patient – SNAP constructed by Sharma Rashmi [5]. This questionnaire is divided into three domains: psychological needs, spiritual needs and religious needs. The questionnaire has been filled on the basis of the Likert scale with values: 4 = very, 3 = little, 2 = not very, 1 = not at all. The patient – the respondent tells in the first domain how much help they need with the individual problems, in the second domain they talk about the problems and in the third domain what great benefits the actions for them are giving. Respondents can fill in the questionnaire independently – if their health status allows them, or data collector help them fill out by way of an interview. Responses were analyzed by SPSS statistical program with $p < 0.01$ and $p < 0.05$.

Research was carried out in 2017 and 2018 in several health care facilities and also in homecare agencies in regions of Ružomberok, Banská Bystrica and Liptovský Mikuláš.

2.2 Research sample and limitations

The final research sample consisted of 113 respondents: 44.2% men and 55.8% women. 69 respondents belong to Christian faith, 13 belong to another faith, 7 did not give religion

Table 1. Pearson correlations for questionnaire SNAP – Domain of Psychological Needs.

Q = questions of SNAP		Q 01	Q 02	Q 03	Q 04	Q 05
Gender	Pearson Correlation	.016	–.156	–.134	–.067	–.151
	Sig. (2-tailed)	.865	.098	.159	.478	.109
Age	Pearson Correlation	–.034	–.030	–.039	–.067	–.003
	Sig. (2-tailed)	.720	.753	.680	.483	.978
Education	Pearson Correlation	–.021	–.181	–.068	–.123	–.082
	Sig. (2-tailed)	.824	.055	.475	.196	.388
Income	Pearson Correlation	–.043	–.045	–.009	–.031	.055
	Sig. (2-tailed)	.652	.639	.928	.742	.566
Religion	Pearson Correlation	–.038	–.122	–.091	.007	.006
	Sig. (2-tailed)	.688	.198	.336	.940	.948
Residency	Pearson Correlation	–.023	–.002	.086	–.059	.038
	Sig. (2-tailed)	.813	.985	.364	.534	.689
	N	113	113	113	113	113

* Correlation is significant at the level $p < 0.05$ **Correlation is significant at the level $p < 0.01$.

and 24 identified themselves as unbelievers. A fatal (deadly) diagnosis was determined in 49.6% of respondents and not life-threatening in 50.4% of respondents. We addressed to 150 respondents, but 21 were excluded due to unfavourable mental state and 16 questionnaires were excluded due to incomplete filling. Limitation was set up as:

1. The first set of patients, hospitalized in standard hospital departments with not life-threatening diagnoses. The removing criteria are: severe cognitive deficits, sensory deficit, terminal status, and psychiatric illness according to patient's documentation. The age limit is from the age of 18 to 65 years.
2. The second group patients, hospitalized in the oncology department – incurably ill. The defining criteria are: severe cognitive deficits, sensory deficits, and the incidence of psychiatric illnesses as documented. The age limit is from the age of 18 to 65 years.

The division into groups was chosen in order to compare the satisfaction of psychosocial, spiritual and religious needs among groups. We wanted to find out if incurable diagnoses were influencing satisfaction the patient's spiritual needs.

3 Results and discussion

Reliability of questionnaire is given by Cronbach's Alpha 0.945 for the total SNAP. Authors Andrasi et al. achieved Cronbach's Alpha 0.921 for the total SNAP on the similar research with the sample of 433 respondents [4]. Authors of the questionnaire in research related to the validation of this instrument achieved Cronbach's alpha for the total SNAP 0.95, with the psychosocial subscale 0.74, the spiritual subscale 0.93, and the religious 0.86 [5]. This means that questionnaire is strongly reliable, valid and consistent.

In domain of psychosocial needs, significant difference was not confirmed in one question regarding to gender, age, education, income, religion and residency. The result can be seen in Table 1.

Table 2. Pearson correlations for questionnaire SNAP – Domain of Spiritual Needs.

Q = questions of SNAP		Q 06	Q 07	Q 08	Q 09	Q 10	Q 11	Q 12	Q 13	Q 14	Q15	Q16	Q 17	Q 18
Gender	Pearson Correlation	–.112	–.163	–.045	.017	.009	–.162	–.140	–.054	.223*	.185*	–.130	–.029	.065
	Sig. (2-tailed)	.238	.085	.640	.858	.923	.087	.138	.568	**.018**	**.050**	.170	.758	.497
Age	Pearson Correlation	.047	.117	.027	.041	–.091	.116	–.032	–.178	–.095	–.031	–.121	–.104	–.056
	Sig. (2-tailed)	.623	.217	.779	.666	.336	.220	.735	.059	.319	.743	.201	.272	.553
Education	Pearson Correlation	–.095	–.047	–.069	–.165	–.063	–.049	–.066	.259*	–.035	–.157	–.111	–.113	–.064
	Sig. (2-tailed)	.317	.618	.470	.080	.507	.603	.486	.006	.714	.096	.240	.233	.500
Income	Pearson Correlation	.063	.148	.020	.048	.003	.101	–.026	.186*	–.075	–.034	–.107	–.078	–.006
	Sig. (2-tailed)	.506	.118	.831	.610	.976	.289	.788	.049	.428	.723	.261	.409	.946
Religion	Pearson Correlation	–.061	–.036	–.032	.212*	.274*	–.115	–.016	.191*	–.062	–.103	–.136	.219*	–.174
	Sig. (2-tailed)	.518	.708	.738	**.024**	**.003**	.224	.867	**.042**	.513	.278	.150	**.020**	.066
Residency	Pearson Correlation	.097	.042	.101	.011	–.007	.168	.061	–.014	.078	–.011	.012	.070	–.014
	Sig. (2-tailed)	.308	.662	.286	.908	.943	.076	.522	.882	.411	.905	.901	.458	.887
	N	113	113	113	113	113	113	113	113	113	113	113	113	113

* Correlation is significant at the level $p < 0.05$ ** Correlation is significant at the level $p < 0.01$.

This means that there is not a direct dependence between severity of diagnosis and satisfaction of psychological needs of the patient. In contrast, authors Elizabeth Luth at Holly Prigerson in her study focused on the relationship between advance care planning and sadness and anxiety at the end of life. In the study, 315 respondents stated that surely there was some distinction in feelings and ways of coping with advance care planning [6]. After the patient first learns that they have diagnosis of an incurable disease, they are on the way of dealing with this. Their ability to cope with illness depends heavily on themselves, on their inner attitude to life. Often, however, this attitude affects the patient's belief. In many cases, we find that the dying is afraid of death.

In domain of spiritual needs, a significant difference was partially confirmed regarding religion and gender in 6 questions of 13. Detailed results are shown in Table 2.

Spiritual care is an important component of high-quality health care, especially for critically ill patients and their families. In addition, it is common that spiritual care resources that can improve both patient outcomes and family member experiences are underutilized [7]. As to gender, significant correlation at the level of 5% was in the question 14 = death and dying as well as in 15 = finding peace of mind. The difference is quelling from the divergence in attitudes and manifestation of feelings of men and women. Women are usually more willing to talk about their problems and inner feelings. Men are usually closed in time of suffering. They do not like to manifest anything that can be pointing at their weakness.

Table 3. Pearson correlations for questionnaire SNAP – Domain Religious Needs.

Q = questions of SNAP		Q 19	Q 20	Q 21	Q 22	Q 23
Gender	Pearson Correlation	–.143	–.068	–.011	–.203*	–.100
	Sig. (2-tailed)	.130	.476	.911	**.031**	.290
Age	Pearson Correlation	.073	–.066	–.102	–.035	–.179
	Sig. (2-tailed)	.440	.485	.283	.713	.057
Education	Pearson Correlation	–.038	–.126	–.219*	–.081	–.098
	Sig. (2-tailed)	.686	.185	**.020**	.393	.304
Income	Pearson Correlation	.098	–.053	–.105	–.032	–.168
	Sig. (2-tailed)	.304	.576	.269	.735	.075
Religion	Pearson Correlation	–.117	–.114	–.026	–.050	–.102
	Sig. (2-tailed)	.216	.231	.784	.600	.280
Residency	Pearson Correlation	.116	.045	.048	.177	.105
	Sig. (2-tailed)	.222	.632	.614	.060	.271
	N	113	113	113	113	113

* Correlation is significant at the level $p < 0.05$ **Correlation is significant at the level $p < 0.01$.

In domain of religious needs, significant correlation between gender groups in question 22 and between groups of different education in question 21 at the level 5% was confirmed.

Responses for questions 23: "What great benefits the actions for him/her give religious rituals such as chant, prayer, lighting candles or incense, anointing, or communion?" have emerged from the situation in Slovakia in the last 20 years. Looking into church ceremonies in Slovakia (no matter of a particular belief) you can see the remarkable predominance of the women who take part in them.

Responses for questions 22: "What great benefits the actions for him/her give visits from fellow members of your faith community?" are revealing the situation, when younger patients belong to different religious groups and movements based on strong interpersonal communication and attachment. The generation of people from 20 to 50 at the same time have had better opportunity to complete study at the university level. Nowadays in Slovakia about 45% of young population achieve university level of education. Authors Johnson, Engelberg, Nielsen et al. in their research focused on association of spiritual care provider's activities with family members' satisfaction with care. They detected a significant association between the discussions on patients' wishes for end-of life care and higher overall assessments experience. Possibly, the opportunity for family members to give voice to the patient's wishes and have this acknowledged by a spiritual care provider ensures some support [8]. Andrasi in his research that was conducted among paramedics pointed out that the spiritual needs of a dying person can sometimes be overlooked while dealing with physical care. However, for those experiencing it, spiritual distress is very real [9]. Role of spiritual provider is very important for believers, because the medical staff – doctors, nurses and paramedics are often too busy with medical solutions.

4 Conclusions

Even though it is possible to identify the four basic dimensions of spiritual need: interconnection, peace, meaning/purpose, and transcendence the majority of patients often have a problem addressing their needs related to psychosocial and spiritual domain because of dominant physical problems such as physical disability, fatigue, sleep disorders, side effects of treatment, etc. If we want to provide good and complex nursing care, we need to assess the presence of symptoms and their severity [10]. The importance of satisfaction of psychosocial and spiritual needs is increased in time of threat and fatal diagnosis, especially in women. Generally, we have not confirmed deep interdependence between the seriousness of the diagnosis on the one hand and gender, age, education, income, religion and residence on the other. Despite the benefits of increased longevity, it is widely acknowledged that enough has not been done to adequately address end-of-life care decisions at the crossroads between medical futility and quality of life [11]. The person with their needs remains essentially on the way they are brought up and how they were accustomed to act before the onset of the illness.

References

[1] M. Best, P. Butow, I. Olver, *Spiritual support of cancer patients and the role of the doctor*. Support Care Cancer. **22**(5), 1333–1339 (2013)
[2] L. Kober, *Satisfying the Spiritual needs in Long-Term Hospitalized Patients with End Stage Chronic Disease*. Rev. Int. Sci. Hum. Nat. **1**, 31–58 (2015)
[3] A.C. Mesquita, E.D.L Chaves, G.A.M. de Barros, *Spiritual needs of patients with cancer in paliative care: an integrative review*. Curr. Op. In Supp. And Pall. Care. **11**(4), 334–340 (2017)
[4] I. Andrasi, V. Littva, M. Sichman et al., *Psychosocial and Spiritual Needs of Dying Patients from the Perspective of the Nurse and Length of Professional Practice*, Clinical Social Work and Health Intervention **8**(4), 7–16 (2017)
[5] R.K. Sharma, A.B. Astrow, K. Texeira, D.P. Sulmasy, *The Spiritual Needs Assessment for Patients (SNAP): Development and Validation of a Comprehensive Instrument to Assess Unmet Spiritual Needs*. J. Pain and Symptom Manage. **44**(1), 44–51 (2012)
[6] E.A. Luth, H.G. Prigerson, *Psychological distress at the End of Life*. J. Pain Sympt. Manage. **56**(5), 752–759 (2018)
[7] J.Q. Ho, C.D. Nguyen, R. Lopes, et al., *Spiritual Care in Intensive Care unit: A Narrative Review*. J. Int. Care Med. **33**(5), 279–287 (2018)
[8] J.R. Johnson, R.A. Engelberg, E.L. Nielsen, et al., *The association of spiritual care providers' activities with family members' satisfaction with care after a death in the ICU*. Crit. Care Med. **42**(9), 1991–2000 (2014)
[9] I. Andrasi, *Spiritual needs of dying and emergency health care. SHS Web of Conferences SOCIETY.HEALTH.WELFARE. 2016* **40**, 03001, (2018)
[10] V. Littva, M. Šichman, *Troublesome symptoms in Palliative Care. Proceedings from the 7th International Conference of Hospice and Palliative Care*, 6–13, Solen (2015)
[11] K. Lutz, S.R. Rowniak, P. Sandhu, *A Contemporary Paradigm: Integrating Spirituality In Advance Care Planing*. J. Relig. & Health **57**(2), 662–671 (2018)

Comparison of antenatal care factors and pregnancy outcome in rural and urban context

Irisa Zile[1,*], *Dace* Rezeberga[2], *Gunta* Lazdane[3], and *Iveta* Gavare[1]

[1] The Centre for Disease Prevention and Control of Latvia, Riga, Latvia
[2] Rīga Stradiņš University, Department of Obstetrics and Gynaecology, Riga, Latvia
[3] Rīga Stradiņš University, Institute of Public Health, Riga, Latvia

Abstract. Antenatal care plays a basic role in the management of woman's health during pregnancy, and women who have not been registered to antenatal care are at increased risk of poor pregnancy outcome, including stillbirth. The aim of this study was to identify differences in maternal characteristics, antenatal care factors and perinatal mortality between urban and rural area in Latvia. The Medical Birth Register of Latvia (MBR) was used as a data source, all births in 2017 (n = 20,406) were included in the analysis. Mother's place of residence was categorized in 3 groups: Riga, other cities and rural area. Higher risk of smoking (OR = 1.4; $p < 0.001$), late first antenatal visit (after 12th gestation week) (OR = 1.2; $p < 0.05$) and incomplete antenatal care (including cases without care) (OR = 1.3; $p < 0.05$) were detected more in other regions (other cities and rural area) than in Riga. After adjustment by maternal age and education, just late first antenatal visit was significantly higher in other regions (ORadj = 1.2; $p < 0.05$) vs. Riga. New-born mortality in perinatal period is substantially higher in other regions vs. Riga. The results indicate health education gap and potential inequalities in receiving antenatal care and emergency obstetric care by regions.

Key words: antenatal care; rural; urban; pregnancy outcome.

1 Introduction

Antenatal care is defined as the routine care of pregnant women provided between conception and the onset of labour. Antenatal care is an opportunity to provide care for prevention and management of existing and potential causes of maternal and new-born mortality and morbidity [1].

Women who have not been registered to antenatal care are at increased risk of poor pregnancy outcomes, including stillbirth [1–4]. According to statistical data in the general population in 2017, 0.8% of live births occurred without antenatal care, but the rate of stillbirths that occurred without prenatal care was 8 times higher, at 6.9% [5].

Literature data show that the place of residence, urban or rural area, is also a factor affecting the use of antenatal care. There are differences between the urban and rural areas in knowledge, attitude, and practices of women towards antenatal care [6–8].

There is a lack of sufficient epidemiologically based data about this issue in Latvia. The use of antenatal care varies within the country, proportion of women with timely first antenatal visit (before the 12th gestational week) differs from 97% in Riga to 83% in Kurzeme region,

* Corresponding author: `irisa.zile@spkc.gov.lv`

an average in Latvia – 93% [5]. The aim of this study was to identify differences in maternal characteristics, antenatal care factors and perinatal mortality between urban and rural areas in Latvia.

2 Material and methods

The Medical Birth Register of Latvia (MBR) was used as data source. All births in 2017 ($n = 20,406$) were included in the analysis. The study analysed and compared antenatal care factors and pregnancy outcomes between urban and rural areas. Mother's place of residence was categorized into 3 groups: Riga, other Latvian biggest cities and rural areas (i.e., counties).

Information about the evaluation of antenatal care was based on the timing of the first antenatal visit and completeness. Antenatal care model in Latvia recommends that the first antenatal care visit takes place within the first trimester (i.e., gestational age of <12 weeks) and eight contacts with a gynaecologist throughout pregnancy [9]. Antenatal care is complete if several conditions are covered: first antenatal visit till 12th week of pregnancy and obligatory antenatal visits with all necessary antenatal checks and tests as per guidelines (blood, urine tests; ultrasound screening; genetic tests; etc.) and incomplete care was defined if one or more conditions are missing, e.g., late first antenatal visit or less antenatal visits or checks. Low birth weight (LBW) has been defined as the first weight recorded within hours of the birth of <2500 grams.

The descriptive statistics for all continuous variables are reported as mean and standard deviation. Categorical data are reported as percentages and 95% confidence interval (CI) and Chi-square test for comparing categorical variables were used. The odds ratios (OR) were calculated (adjusted and non-adjusted) and 95%CI were computed to measure the strength of the association between the outcome and the explanatory variables.

Place of residence was merged into 2 groups for OR analysis: Riga and other regions (other cities and rural regions).

3 Results

The study analysed data on 20,406 births in 2017. 34.5% (95CI 33.8–35.1) ($n = 7030$) birth took place in Riga, 15.3% (95%CI 14.8–15.8) ($n = 3121$) other cities and 50.3% (95%CI 49.6–51.0) ($n = 10,255$) in rural regions. The average mother's age in Riga was higher – 31 years than in other cities and rural area 29 ($p < 0.001$). Younger mothers (≤ 20) were more often in rural regions ($p < 0.001$) – 6.2% (95%CI 5.7–6.7) ($n = 634$) comparing with Riga 2.5% (95%CI 2.2–3.0) ($n = 179$). Higher education level was in Riga ($p < 0.001$), by 20 percentage points higher than in rural regions (65% to 44.4%). There are no differences between regions in average gestational age and birth weight of new-borns, although a higher proportion of new-borns with low birth weight (≤ 2499 g) was observed ($p < 0.01$) in rural regions 4.6% (95%CI 4.2–5.1) ($n = 476$) than in Riga and other cities. The characterization of parameters by region is shown in Table 1.

Higher proportion ($p < 0.05$) of mothers with incomplete antenatal care was outside Riga (5.4%; 95%CI 4.9–6.0; $n = 377$) – in other cities 6.4% (95%CI 5.6–7.3) ($n = 200$) and rural area 6.6% (95%CI 6.2–7.1) ($n = 681$) as well as with late first antenatal visit ($p < 0.001$) – rural regions 6.4% (95%CI 6.0–6.9) ($n = 653$) against 5.3% (95%CI 4.8–5.8) ($n = 370$) in Riga. US in the 1st trimester an average were 77.4% (95%CI 77.0–78.1) and in the 2nd trimester – 93.4% (95%CI 93.1–93.8), there are no differences between regions.

In relation to such lifestyle factors as smoking, differences were observed by regions – a higher proportion was in other cities and rural areas than in Riga, an average by 2 percentage

Table 1. Characterization of parameters by region.

Characteristics	Riga ($n = 7030$)	Other cities ($n = 3121$)	Rural regions (10255)	p value
Mean maternal age[1]	31 (±5.2)[†]	29 (±5.7)	29 (±5.7)	<0.01
Mean birth weight (g)[1]	3515.7 (±587)	3519.1 (±587)	3485.3 (±603)	NS
Maternal age group[2]:				
≤ 20 years	2.5 (179)[†]	5.9 (185)	6.2 (634)	<0.001
21–34 years	74.0 (5199)	75.6 (2359)	75.0 (7695)	NS
≥ 35 years	23.5 (1652)[†]	18.5 (577)	18.8 (1926)	<0.001
Maternal education[2]:				
elementary	7.5 (525)[†]	13.5 (422)	14.6 (1500)	<0.001
secondary	27.6 (1938)[†]	40.2 (1256)	41.0 (4204)	<0.001
higher	65.0 (4567)[†]	46.2 (1443)	44.4 (4551)	<0.001
Parity[2]:				
Primipara	44.4 (3120)[†]	37.8 (1181)	38.7 (3971)	<0.001
Low birth weight[2]	3.8 (270)	3.5 (110)	4.6 (476)[††]	<0.01
Maternal smoking[2]	5.7 (398)[†]	8.4 (263)	7.5 (766)	<0.001
Antenatal care[2]:				
Late fist antenatal visit	5.3 (370)	5.1 (159)	6.4 (653)[††]	<0.01
Incomplete antenatal care	5.4 (377)[†]	6.4 (200)	6.6 (681)	<0.01
Without antenatal care	0.6 (42)	0.4 (14)	1.1 (110)[††]	<0.001

[1] *Represents means ± standard deviation (SD) and T- test is used.*
[2] *Represents % (n) and Chi square test is used; NS: Not Significant.*
[†] *statistically significant difference between Riga and other groups (other cities and rural regions).*
[††] *statistically significant difference between rural regions and other groups (Riga and other cities).*

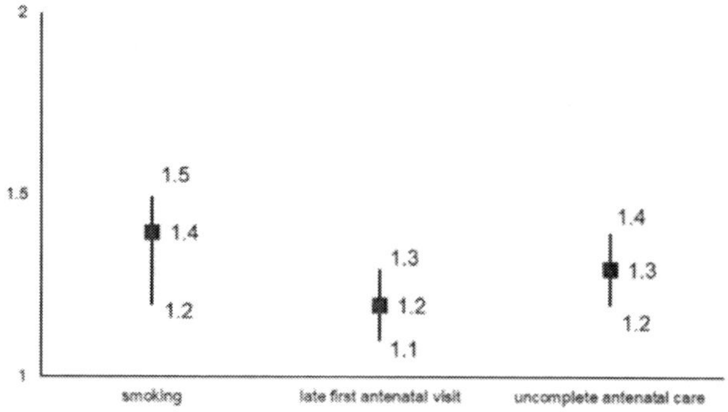

Figure 1. Odds ratio of different factors in other regions (other cities and rural area) vs. Riga, based on univariate analysis.

points ($p < 0.001$). An average obesity was detected in 3.5% (95%CI 3.3–3.8) ($n = 723$), there are no differences between regions.

Higher risk of smoking (OR = 1.4; $p < 0.001$), late first antenatal visit (after 12th gestation week) (OR = 1.2; $p < 0.05$) and incomplete antenatal care (including cases without care) (OR = 1.3; $p < 0.05$) were more detected in other regions (other cities and rural area) than in Riga (Fig. 1). After adjustment by maternal age and education, just late first antenatal visit was significantly higher in other regions (ORadj = 1.2; $p < 0.05$) vs. Riga.

The gestational week and the birth weight of a new-born are independently associated with an increased risk of perinatal mortality in all regions.

New-born mortality in the perinatal period is substantially higher in other regions than in Riga (ORadj = 1.5; $p < 0.05$). OR was adjusted by maternal age, gestational week, birth weight and parity. Adjusted by maternal education, perinatal mortality OR is still slightly higher (ORadj = 1.3; $p > 0.05$) in other regions vs. Riga but not more statistically significant. Mother education also as an independent factor was associated with increased perinatal mortality (OR = 2.1; $p < 0.01$).

4 Discussion

Antenatal care plays a basic role in the management of a woman's health during pregnancy. An early antenatal care visit allows providing screening and tests that are most effective in early stages of the pregnancy.

The procedure of antenatal care as well as during and after childbirth is determined by the Cabinet of Ministers Regulation No. 611 "Procedures for providing childbirth assistance" [9]. Prenatal care in Latvia is provided by a gynaecologist (obstetrician), a midwife or family doctor if the pregnancy is physiological. The health care services, which have agreed with the National Health Service (NHS), shall provide state-funded health care services. Pregnant women do not have to pay patient co-payments for health care services related to pregnancy and post-natal monitoring and pregnancy with a specialist contracted with NHS, as well as childbirth in a hospital with NHS is a state-funded service.

In this study, the proportion of timely first antenatal visit an average was 94.2%. For example, in the United States 77.1% of women who gave birth in 2016 initiated prenatal care in the first trimester of pregnancy [10].

The estimated worldwide coverage of early antenatal care visits increased from 40.9% in 1990 to 58.6% in 2013, corresponding to a 43.3% increase. Overall coverage in the developing regions was 48.1% in 2013 compared with 84.8% in the developed regions. In 2013, the estimated coverage of early antenatal care visits was 24.0% in low-income countries compared with 81.9% in high-income countries [1].

The Childbirth and Health Study in Iceland indicates that antenatal care can play an important role in preparing women for birth. The majority (87%) of the women want to be informed about birth in the antenatal phase of care, and 41% reported 5–6 months postpartum that too little time had been spent on this issue by health care professionals. Women who experienced their birth as difficult or very difficult were more likely to report that insufficient time had been spent on information than women who had experienced their birth as easy or very easy [11].

Many studies have identified socio-demographic predictors of late initiation of antenatal care [1, 6–8, 10, 12, 13]. This study also demonstrates variation in the timely and complete antenatal care receipt by place of residence. Higher risk of late first antenatal visit and incomplete antenatal care were observed in rural regions. An average 6% of received incomplete antenatal care. Study in the United States observed a higher proportion – 15% [10]. Survey data from England has found that the proportion of pregnant women having their booking appointment by 12 completed weeks of pregnancy is 86% and by 10 weeks of pregnancy is 63% [14]. In the United Kingdom, the timing of the antenatal booking appointment is used as an indicator of access to care in pregnancy and antenatal care ought to be promoted as woman-centred and as means to empower women to feel in control of their pregnancy [14].

To improve health outcomes and reduce disparities among pregnant women and new-borns in low- and middle-income countries more should be done to increase access to quality maternal health care services for women, especially for those from vulnerable populations [12]. This issue is also very important within the country because from our study results, we can assume that delayed care in rural regions could be related to geographic area and socio-economic factors. Distance to a healthcare facility could be an important barrier to the use of services, especially in rural areas.

The literature review identifies multiple socio-demographic, reproductive and access related factors that affect the utilization of antenatal care among pregnant women in different countries. Several studies, conducted in different countries, have shown that factors like maternal age, number of living children, education, socioeconomic status, previous bad obstetrical history, support from a spouse, quality of care and distance from health care facility are significantly associated with the use of antenatal care [7].

Another finding of this study is that a higher proportion of smoking during pregnancy was also outside the capital Riga, difference by 2–3 percentage points. Results of the Australian study showed that among pregnant women who smoked antenatal-care attendance was lower. The same was observed among women who were of younger age, higher parity and unemployed, and not partnered [15].

In some European countries, more than 10% of women smoke during their pregnancy. A preventive population approach is important to reduce the prevalence of smoking before pregnancy in the childbearing population. The EURO-PERISTAT network has pointed that smoking among pregnant women has declined in high-income countries, but it nonetheless continues to account for a substantial proportion of foetal and infant morbidity and mortality. Smoking before pregnancy and the likelihood of stopping smoking are associated with lower maternal educational level and poverty. Smoking thus contributes to the creation of social inequalities in perinatal health [16].

Literature data show that educated women tend to have a greater awareness of the existence of antenatal care services and the advantages of using such services. It is argued that educated women were more aware of health problems, know more about the availability of health care services, and utilize the information more effectively than non-educated women [7]. Other studies' results also showed that inadequate antenatal care has also been associated with stillbirth, women who had their first antenatal visit after 20 weeks' gestation were slightly more likely to have a stillborn baby than those who had earlier first visits (OR = 1.12) and odds more than tripled with no antenatal care [17].

Our study findings indicated that mother education as an independent factor associated with increased perinatal mortality (OR = 2.1) and it is substantially higher in other regions than in Riga (ORadj = 1.5).

Latvian perinatal deaths audit discovered that 19% of perinatal death cases were related to women who had not taken advantage of antenatal care. Non-attendees were more likely to be smokers ($p < 0.001$) and alcohol abusers ($p < 0.005$), above 35 years of age ($p < 0.005$) and had higher parity ($p < 0.001$). Non-attendees more often had systemic diseases and pregnancy complications. There was no difference in rates of preterm birth and low birth weight between attendees and non-attendees. One-fifth of mothers with perinatal death did not attend antenatal care, and just in some cases when women who attended antenatal care, lack of intervention was related to the perinatal death [18].

The strengths of this work include the Register-based study. Population-based data are essential for planning of health care and for determining temporal trends.

Health of a mother and child is one of the top priorities in Latvia as in worldwide. It should be a priority in the national level that all women have their first antenatal visit by 12 completed weeks of pregnancy, but it is not always of equal priority to women. Lower antenatal care coverage and poor pregnancy outcomes in rural regions detect discrepancies between regions and necessity to improve the equality aspects of receiving health care services regardless of the place of residence. The supply of health workers has changed significantly in Latvia and particularly in rural areas. In addition to the absolute shortage and skills imbalance, there is also a strong concentration of health workers in the capital Riga and there is evidence of severe shortages in key specialties (e.g. obstetricians).

Improving maternal and child health and reducing infant mortality are also objectives of the Latvian Policy document "Maternal and Child Health Improvement Plan 2018–2020", developed by the Ministry of Health [19]. The action plan provides for improvements also in screening and maternal and perinatal mortality audit. The action plan foresees changes in the legislative documents to include additional screenings for women and new-borns. Health care standards and guidelines are fundamental requirements for good quality assurance systems. They typically consist of comprehensive, evidence-based clinical guidelines and pathways, which serve as the criteria for quality monitoring activities.

Antenatal care is an opportunity to educate the woman about her health, pregnancy, and childbirth, recognizing danger signs, the benefits of good nutrition and exclusive breastfeeding, the harms of alcohol, tobacco and drugs, and other relevant issues. Latvia has implemented also Maternal Mortality Confidential Analysis conducted by the Expert Committee. The results of this study may deserve attention for policy implementation regarding strategies to improve antenatal and obstetric care and strengthen perinatal audits at the national level.

5 Conclusions

A higher proportion of younger mothers, lower education level, smoking during pregnancy and low birth weight are more likely to be in rural regions.

Considerable variation exists in the content of antenatal care by regions. Mothers with incomplete antenatal care, with the late first antenatal visit and smoking more often were observed in other regions vs. Riga, but after adjustment for maternal age and education there is no significant difference, just higher odds of late first antenatal visit in rural regions.

New-born mortality in the perinatal period is substantially higher in other regions vs. Riga.

The results indicate health education gap and potential inequalities in receiving antenatal care and emergency obstetric care in regions.

References

[1] A. Moller, M. Petzold, D. Chou, L. Say. Early antenatal care visit: a systematic analysis of regional and global levels and trends of coverage from 1990 to 2013. *Lancet*, **5**(10), 977–983 (2017)

[2] J.E. Lawn, H. Blencowe, P. Waiswa, A. Amouzou, C. Mathers, D. Hogan, V. Flenady, J.F. Frøen, Z.U. Qureshi, C. Calderwood, S. Shiekh, F.B. Jassir, D. You, E.M. McClure, M. Mathai, S.E. Cousens. Stillbirths: rates, risk factors, and acceleration towards 2030. *Lancent* **387**(10018), 587–603 (2016)

[3] E.M. McClure, S. Saleem, S.S. Goudar, J.L. Moore, A. Garces, F. Esamai, A. Patel, E. Chomba, F. Althabe, O. Pasha, B.S. Kodkany, C.L. Bose, M. Berreuta, E.A. Liechty, K.M. Hambidge, N.F. Krebs, R.J. Derman, P.L. Hibberd, P. Buekens, A. Manasyan, W.A. Carlo, D.D. Wallace, M. Koso-Thomas, R.L. Goldenberg, Stillbirth rates in low-middle income countries 2010–2013: a population-based, multi-country study from the Global Network. Reprod. Health **12**(2), S7, (2015)

[4] M. Aminu, R. Unkels, M. Mdegela, B. Utz, S. Adaji, N. van den Broek. Causes of and factors associated with stillbirth in low- and middle-income countries: a systematic literature review. BJOG **121**(4), 141–53 (2014)
[5] The Centre for Disease Prevention and Control of Latvia. Maternal and infant health care. *Statistical yearbook of health care in Latvia* **19**, 258–259 (2017)
[6] A.A. Ewunetie, A.M. Munea, B.T. Meselu, M.M. Simeneh, B.T. Meteku. Delay on first antenatal care visit and its associated factors among pregnant women in public health facilities of Debre Markos town, North West Ethiopia. BMC Pregnancy Childbirth **18**, 173 (2018)
[7] S.A. Ali, A.A. Dero, S.A. Ali, G.B. Ali. Factors affecting the utilization of antenatal care among pregnant women: A literature review. J. Preg. Neonatal. Med. **2**(2), 41–45 (2018)
[8] J. Amo-Adjei, K. Aduo-Adjei, C. Opoku-Nyamaah, C. Izugbara. Correction: Analysis of socioeconomic differences in the quality of antenatal services in low and middle-income countries (LMICs). PLoS ONE **13**(2), e0192513, (2018)
[9] Republic of Latvia Cabinet Regulation No 611 (25.07.2016.) Procedures for providing childbirth assistance
[10] A. Martin, B.E. Hamilton, M.J.K. Osterman, A.K. Driscoll, P. Drake. Births: Final Data for 2016. *National vital statistics reports, Division of Vital Statistic* **67**(1) (2018)
[11] H. Gottfredsdottir, Þ. Steingrímsdóttir, A. Björnsdóttir, E.Y. Guðmundsdóttir, H. Kristjánsdóttir. Content of antenatal care: Does it prepare women for birth? *Midwifery* **39**, 71–7 (2016)
[12] J. Sharma, M. O'Connor, R.R. Jolivet. Group antenatal care models in low- and middle-income countries: a systematic evidence synthesis. Reprod. Health **15**, 38 (2018)
[13] M. Redshaw, K. Heikkila. Delivered with care: a national survey of women's experience of maternity care 2010. *National Perinatal Epidemiology Unit, Oxford*, November 2010
[14] B. Hatherall, J. Morris, F. Jamal, L. Sweeney, M. Wiggins, I. Kaur, A. Renton, A. Harden. Timing of the initiation of antenatal care: An exploratory qualitative study of women and service providers in East London. Midwifery **36**, 1–7 (2016)
[15] C. Mittiga, K. Ettridge, K. Martin, G. Tucker, R. Dubyna, B. Catcheside, W. Scheil, L. Maksimovic. Sociodemographic correlates of smoking in pregnancy and antenatal-care attendance in Indigenous and non-Indigenous women in South Australia. Aust. J. Primary Health **22**(5), 452–460 (2015)
[16] Euro-Peristat Project. European Perinatal Health Report. Core indicators of the health and care of pregnant women and babies in Europe in 2015. November 2018, pp. 56–58
[17] V. Flenady, L. Koopmans, P. Middleton, J.F. Frøen, G.C. Smith, K. Gibbons, M. Coory, A. Gordon, D. Ellwood, H.D. McIntyre, R. Fretts, M. Ezzat. Major risk factors for stillbirth in high-income countries: a systematic review and meta-analysis. Lancet **377**, 1331–40 (2011)
[18] M. Jansone, G. Lindmark, J. Langhoff-Roos. Perinatal deaths and insufficient antenatal care in Latvia. Acta. Obstet Gynecol. Scand **80**(12), 1091–5 (2001)
[19] Republic of Latvia Cabinet Rules No. 259 (06.06.2018) Maternal and Child Health Improvement Plan 2018–2020

Readiness of independency in health care management for young adults with cerebral palsy

Zane Rozkalne[1,2,*], *Maksims* Mukans[1], and *Anita* Vetra[1,2]

[1] Rīga Stradiņš University, Riga, Latvia
[2] Children's Clinical University Hospital, Riga, Latvia

Abstract. Transition from paediatric to adult health care services is a crucial process for young adults with long-term medical conditions and well organized and coordinated transition process has been strongly highlighted. The aims of this research were to assess the young adults with cerebral palsy overall readiness for adult life, the independency in health care management and independency impacting factors. Eighty young adults (age range 16–21 years) with CP and normal or slightly decreased mental function participated. There were done "face-to-face" structured interviews using the Rotterdam Transition Profile and the Transition Readiness Assessment Questionnaire also the characteristics of participants were obtained. It was revealed that young adults with cerebral palsy show low levels of participation in almost all domains of transition to adult life, the level of attendance of rehabilitation services is not sufficient, and the level of independency for young adults with CP in health care management is between phases of preparation and action. The most impacting factor for independency in health care management was found to be the cognitive status.

Key words: cerebral palsy, transition age, health care, young adult.

1 Introduction

It has been long thought that cerebral palsy (CP) is a medical condition more attributed to paediatric care, even though most persons with CP now survive to adulthood [1]. The prevalence of CP in Europe has been estimated 2 per 1000 live births [2]. Although the neurological injury causing CP is defined as non-progressive, the musculoskeletal, neurological and mental conditions worsen in the course of time [3–5]. Adults with CP face health problems such as decreased mobility and self-care [6], hypoactivity [7], chronic pain [3, 8], sarcopenia [9, 10], spasticity [11, 12], musculoskeletal deformities (e.g., articular contractures [12, 13], osteopenia [14], degenerative arthritis [14, 15], deformities of hips [16, 17], knees [16, 18], ankles [12, 19], feet [16, 17] and spine [20–22], epilepsy [3], genitourinary problems [16, 23] and depression and anxiety [24]. It has been proven that as the age for persons with CP progresses also the use of certain medication (respectively, analgesic, antispastic, and psychotropic drugs in ambulatory patients and psychotropic, antispastic, antiepileptic, and digestive drugs in non-ambulatory patients)

* Corresponding author: Z_Rozkalne@inbox.lv

increases [3]. However, the physical medicine and rehabilitation doctor follow-up, the use of orthoses, and receiving rehabilitation decreases with age [3]. Young adults with CP have reported having too few information about their medical condition [25] and that there is a lack of real life opportunities of becoming independent [26]. Unsuccessful transition to adult life may lead to unnecessary lifelong dependency, unemployment and low quality of life [27, 28]. The transition from paediatric to adult health care is crucial process for persons with CP [29–34] and specific therapeutic approach is essential [35–38]. The aims of this research were to assess young adults with CP overall readiness for transition to adult life, the readiness of independency in health care management and the factors having impact on readiness for independency in health care management.

2 Material and methods

2.1 Study design

The design of the study was a cross-sectional, non-experimental research.

2.2 Study sample

Participants were searched through the data basis of Children's Clinical University Hospital and through open invitations in smaller rehabilitation centres. "Face-to-face" structured interviews using the assessment scales and questionnaires were done from October 2017 till June 2018. Permissions from Ethics Committee of Rīga Stradiņš University (August 2016) and from Children's Clinical University Hospital (July 2016) were received. Before signing the consent form, potential participants and their parents read the explanation form of the research and received answers to any questions concerning the participation. For participants under the age of 18 the consent form was signed by the legal representative (all cases – one of the parents). The inclusion criteria were: young adults aged 16–21 years, diagnosed with cerebral palsy (G80-G83, International Statistical Classification of Diseases and Related Health Problems,10th revision), with normal or slightly decreased cognitive status and able to understand the research procedure. Exclusion criteria were: not actually having cerebral palsy (having other cause of the paresis), severe overall physical and cognitive status (MMSE less than 24 points), not living in or coming from family (living in social institution), not living in Latvia, was not possible to gain contact information, not answering to phone call over three times, dead. Two hundred twenty – five potential participants were identified. Considering exclusion criteria and having the will to participate, a total of 80 persons participated.

2.3 Measures

For the characteristics of participants (age, gender, type of cerebral palsy, status of speech, vision, hearing, epilepsy, the need of assistant, place of living, level of income, parent status, number of siblings) we used a structured descriptive questionnaire. Type of cerebral palsy was detected through participants' medical records. Sever problems with speech, vision and hearing were detected if others often cannot understand the person's speech and/or the participant's health status was not fully compensated with medical assistive devices (e.g. glasses, hearing aids). Having epilepsy was approved if the person was using anti-epileptic drugs prescribed by doctor. Level of average income to a household was defined as 1017.60eurper month (index of year 2016) [39]. "Brought up by both parents" was marked as "yes" even if one of the parents was a step-parent and as "no" if only one parent raised the participant.

2.3.1 Functional status

The gross motor function was assessed using Gross Motor Function Classification System (GMFCS) [40], which is a five-level classification system, where the fifth level (person is transported in a manual wheelchair) is the most sever and first level (person walks without limitations) is the least. The manual abilities were assessed using Manual Ability Classification System (MACS) [41], which also is a five-level classification system, where the fifth level (person does not handle objects and has severely limited ability to perform even simple actions) is the most sever and the first (person handles objects easily and successfully) – the least. The cognitive status was assessed with Mini-Mental State Examination (MMSE) [42], in which the maximum score of points is 30 and the less points a participant receives the more cognitive impairments he/she might have, whereas 24–30 points indicate none or some/uncertain cognitive deficit.

2.3.2 Transition process to adult life, the Rotterdam Transition Profile

The readiness for transition to adult life was assessed using the Rotterdam Transition Profile (RTP) [43]. It is a ten transitional domain assessment tool. First seven domains are applied to measure participation and the last three – health care. The assessment of each domain is ranged in three or four phases: phase 0 – the young adult has no experience in a particular field, phase 1 – the young adult is dependent on adults, phase 2 – the young adult is experimenting and orientating towards a more independent life, phase 3 – the young adult has achieved more-or-less independence and autonomy. The total score of RTP is 30 points – the more points a participant has received the more independent and autonomous he or she is.

2.3.3 Readiness of independency in health care management, the Transition Readiness Assessment Questionnaire

The readiness of independency in health care management was assessed using the Transition Readiness Assessment Questionnaire (TRAQ) [44]. The TRAQ consists of 20 questions evaluating: medication managing, appointment keeping, tracking health issues, talking with providers and managing daily activities. Each question is scored from 1–5 points, describing the independency level of health care task managing: 1 – *No, I do not know how* (primary idea), 2 – *No, but I want to learn* (intent), 3 – *No, but I am learning to do this* (preparation), 4 – *Yes, I have started doing this* (action), 5 – *Yes, I always do this when I need to* (maintenance). The total score of TRAQ is 100 points – the bigger score, the better health care management.

2.4 Statistical analysis

Statistical analysis was conducted using SPSS software (IBM SPSS Statistics, v. 23.0, Chicago, IL, USA). The sample was tested for the normality of distribution of the measurements of scales and age by the means of Shapiro-Wilk test. As the sample did not reveal normal distribution, quantitative data was in median (Me) with inter-quartile range (IQR) and non -parametric statistics were performed. Mann-Whitney U test for interval data comparison, Wilcoxon test for time course of interval data, Spearman rho for correlation of interval data, Pearson chi square or Fisher's exact test for categorical data. TRAQ data was divided by median of $\leq$3.00 points and $\geq$3.01 points to find the impact of factors: characteristics of participants, functional status, and the readiness for transition to adult life. The correlation analysis was done with TRAQ groups to find the strongest association with the impacting factors. Correlation coefficient interpretation: up to $|\pm 0.19|$ very weak, $|\pm 0.20|$ – $|\pm 0.39|$ weak, $|\pm 0.40|$ – $|\pm 0.59|$ moderate, $|\pm 0.60|$ – $|\pm 0.79|$ strong, $|\pm 0.80|$ – $|\pm 0.999|$ very strong [45]. Adjusted residual was marked as significant with value more than $|\pm 1.96|$. Binary logistic regression analysis was done for separate impacting factor to find best (Wald criterion and percentage predicted probability) association with TRAQ data. Second time Binary logistic regression analysis was done for multivariate impacting factors to find most exact and strongest impacting factor. Criterion of choice was Wald, significance $p < 0.05$ and odds ratio. In all analyses $p < 0.05$ was significant. Graphs were done in SPSS software or MS office Excel (2010).

3 Results

3.1 Characteristics of the study sample

Table 1 presents the personal characteristics, type of cerebral palsy and the functional level of the 80 young adults aged 16 to 21 years with cerebral palsy who participated in the study. The median age was 18 (IQR 20-17) years and 41% percent were men. Thirty-four percent of the participants had quadriplegia, 74% had high level of gross motor function and 70% had high level of manual ability (GMFCS and MACS level I or II). Fifty percent of the participants had full or almost full score of the MMSE (29–30 points), indicating normal cognitive status and 16%had severe speech problems. The need of an assistant marked 52% of the participants. For 38% of the participants the place of living was town (other than capital city) and 61% had the income level below the average. Seventy-one percent of the participants were brought up by both parents and 66% had one sibling.

3.2 Transitional process to adult life

The outcomes of the RTP (see Fig. 1) show the general phases of the transition process of the participants.

3.3 Readiness of independency in health care management and the impacting factors

Table 2 demonstrates participants' median values of TRAQ, grouped in five health care management related themes.

The median value of the TRAQ score for women was 3.5 (IQR 4.4-3.0) and for men 3.3 (IQR 4.1-2.6). The median values in relation to participants' age, functional and cognitive level are demonstrated in Fig. 2.

To analyze the outcomes of the TRAQ and the outcome impacting factors the score was divided in two groups: ≤ 3.00 points, group I(the participant is rather in passive than in active phase of task managing) and ≥ 3.01 points, group II (the participant is rather in active than in passive phase of task managing).The statistical significance of the impacting factor on the participants' appearance in group I or group II was tested with Mann-Whitney U tests and for dichotomous parameters Pearson chi square tests. Person chi square was also done if the data was more demonstrative with this method (factors – Need of an assistant and Level of income) and if p-value was not significant, but in some categories adjusted residual was (factors- Level of income and Finance) (see Table 3).

Statistically significant were found to be the level of GMFCS, MACS, MMSE, whether the participant had or had not severe speech problems and the need of an assistant. When

Table 1. Characteristics of the study participants ($n = 80$).

Age in years, Me (IQR)	18 (20-17)		
Gender, n (%)			
Men	41 (51)		
Women	39 (49)		
Type of CP, n (%)		MMSE, n (%)	
Quadriplegia	27 (34)	30 points	17 (21)
Diplegia	24 (30)	29 points	23 (29)
Hemiplegia	16 (20)	28 points	8 (10)
Other	13 (16)	27 points and less	32 (40)
GMFCS, n (%)		MACS, n (%)	
Level I	35 (44)	Level I	23 (29)
Level II	24 (30)	Level II	33 (41)
Level III	13 (16)	Level III	16 (20)
Level IV	8 (10)	Level IV	8 (10)
Level V	0 (0)	Level V	0 (0)
Severe problems with speech, n (%)	no 67 (84) yes 13 (16)		
Severe problems with vision, n (%)	no 73 (91) yes 7 (9)		
Severe problems with hearing, n (%)	no 79 (99) yes 1 (1)		
Epilepsy, n (%)	no 72 (90) yes 8 (10)		
Need of an assistant, n (%)			
No	38 (48)		
Only for transportation	25 (31)		
All the time outside the home	17 (21)		
Place of living, n (%)			
Capital city	29 (36)		
Other town	30 (38)		
Rural area	21 (26)		
Level of income, n (%)			
Below average	49 (61)		
Average	18 (23)		
Above average	13 (16)		
Brought up by both parents, n (%)	no 23 (29) yes 57 (71)		
Number of siblings >1, n (%)	no 53 (66) yes 27 (34)		

Me: median. IQR: inter-quartile range.
CP: cerebral palsy, GMFCS: Gross Motor Function Classification System, MACS: Manual Ability Classification System, MMSE: Mini-Mental State Examination.

testing the impact of the scores of RTP on the participants' appearance in group I or group II, significant were found to be almost all the domains of RTP, except for Rehabilitation services. The correlation analysis was made to detect the most associated factors. For further binary logistic regression were chosen factors with $r_s > 0.39$ (see Table 4).

The impacting factors with the score of Wald >10.00 (MMSE 17.50, Transportation 15.92, Total score of RTP 14.13 and Leisure (social activities) 12.39) were analyzed in binary logistic regression model of equation (see Table 5).

The only impacting factor with $p < 0.01$ was the MMSE – level of participants' cognitive status (OR 2.27, 95% CI 1.36–3.79). Model correctly classified 85% of cases.

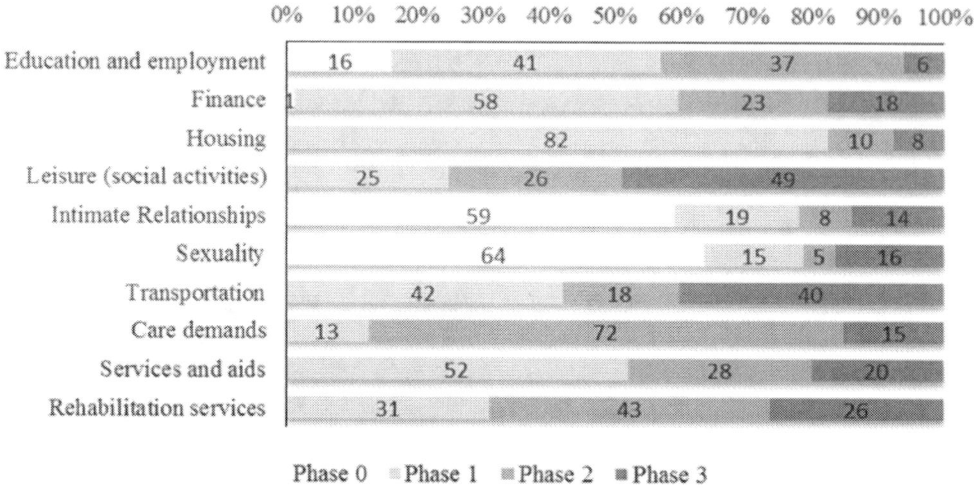

Figure 1. Percentage of participants in the transitional phases (the Rotterdam Transition Profile): 0 – no experience, 1 – dependent on parents, 2 – experimenting and orientating, 3 – independence and autonomy.

Table 2. Median values of TRAQ.

TRAQ themes	<18 y (IQR)	≥ 18 y (IQR)	All (IQR)	*p*-value*
Managing Medications	3.0 (3.8-2.0)	3.6 (4.8-2.8)	3.5 (4.3-2.5)	<0.01
Appointment Keeping	2.6 (3.0-2.0)	3.0 (4.0-2.4)	2.9 (3.9-2.2)	<0.05
Tracking Health Issues	3.0 (3.8-2.3)	3.3 (4.1-2.5)	3.3 (4.0-2.5)	0.23
Talking with Providers	4.5 (5.0-4.0)	5.0 (5.0-4.0)	5.0 (5.0-4.0)	0.25
Managing Daily Activities	4.7 (5.0-3.3)	4.7 (5.0-4.0)	4.7 (5.0-4.0)	0.63
TRAQ total	3.3 (3.6-2.8)	3.6 (4.4-3.0)	3.5 (4.2-2.9)	<0.05

*Median differences' significance associated with participants' age (minor vs. adult).

Y: years, IQR: interquartile range, TRAQ: Transition Readiness Assessment Questionnaire.

4 Discussion

This is the first study in Latvia that measures the readiness for transition to adult life and the impacting factors for independency in health care management for persons with CP.

We have demonstrated the levels of transitional phases according to the RTP for the young adults with CP at the process of becoming independent and autonomous. At the level of education and employment we found that almost one quarter of participants were not learning or having a job, and at the level of financial independency more than half were still dependent on parents/care takers. Less than one fifth of participants were orientating or already obtaining some independency in housing activities. Only half of the participants were able to spend time on leisure activities outside home having no restrictions on whether it is day time or late evening. More than half of the participants had completely none experience in intimate and sexual relationships. Also, more than half of the participants were fully or partially dependent on parents/care givers at the transportation domain. Seventy-two percent of participants claimed that they take part in decision-making about their health care management, but again – more than half of finding services and aids were done by parents/care givers. Even though persons with CP require throughout life health care, 43% had not received any rehabilitation

Table 3. Values of factors in comparison to TRAQ group I and group II.

Factors	p-value	Adj.res.	TRAQ	
			≤ 3.00	≥ 3.01
Gender		<1.96		
Age	0.44			
GMFCS, level 1–5 (IQR)	**<0.01**		2(3-2)	1(2-1)
MACS, level 1–5 (IQR)	**<0.01**	3 (3-2)	2(2-1)	
MMSE, points 30–24 (IQR)	**<0.01**		26 (28-25)	29 (30-27)
Severe problems with speech, n (%)		**>1.96**		
no			16(20)	51(64)
yes			8 (10)	5(6)
Severe problems with vision		<1.96		
Severe problems with hearing		<1.96		
Epilepsy		<1.96		
Need of an assistant, n (%)	**<0.01**	**>1.96**		
No			2(3)	36(45)
Only for transportation			16(20)	9 (11)
All the time outside the home			6(7) *	11(14) *
Place of living	0.88			
Level of income, n(%)	0.14	<1.96		
Below medium			17(21)*	32(40)*
Medium			6(8)*	12(15)*
Above medium			1(1)**	12(15)**
Brought up by both parents		<1.96		
Number of siblings		<1.96		
RTP, phases 0–3 (IQR)				
Education and employment	**0.02**		1 (2-0)	2 (2-1)
Finance, n(%)	0.13	**>1.96**		
Phase 0			0 (0)*	1(1)*
Phase 1			17(21)*	30(38)*
Phase 2			6(8)*	12(15)*
Phase 3			1(1)	13(16)
Housing	**0.01**		***	1 (2-1)
Leisure (social activities)	**<0.01**		2(2-1)	3(3-2)
Intimate Relationships	**<0.01**		0 (0-0)	1 (2-0)
Sexuality	**<0.01**		0 (0-0)	0 (2-0)
Transportation	**<0.01**		1 (1-1)	3(3-1)
Care demands	**<0.01**		2 (2-1)	2(2-2)
Services and aids	**<0.01**		1 (1-1)	2(3-1)
Rehabilitation services	0.12			
Total score of RTP, points 6–30 (IQR)	**<0.01**		11(14-9)	17(21-13)

*Adj. res. <1.96, ** adj.res. =1.9, ***constant when TRAQ ≤3.00.

Adj.res.: adjusted residual, IQR: interquartile range, RTP: Rotterdam Transition Profile, Phase 0 – no experience, Phase 1 – dependent on parents, Phase 2 – experimenting and orientating, Phase 3 – independence and autonomy (RTP).

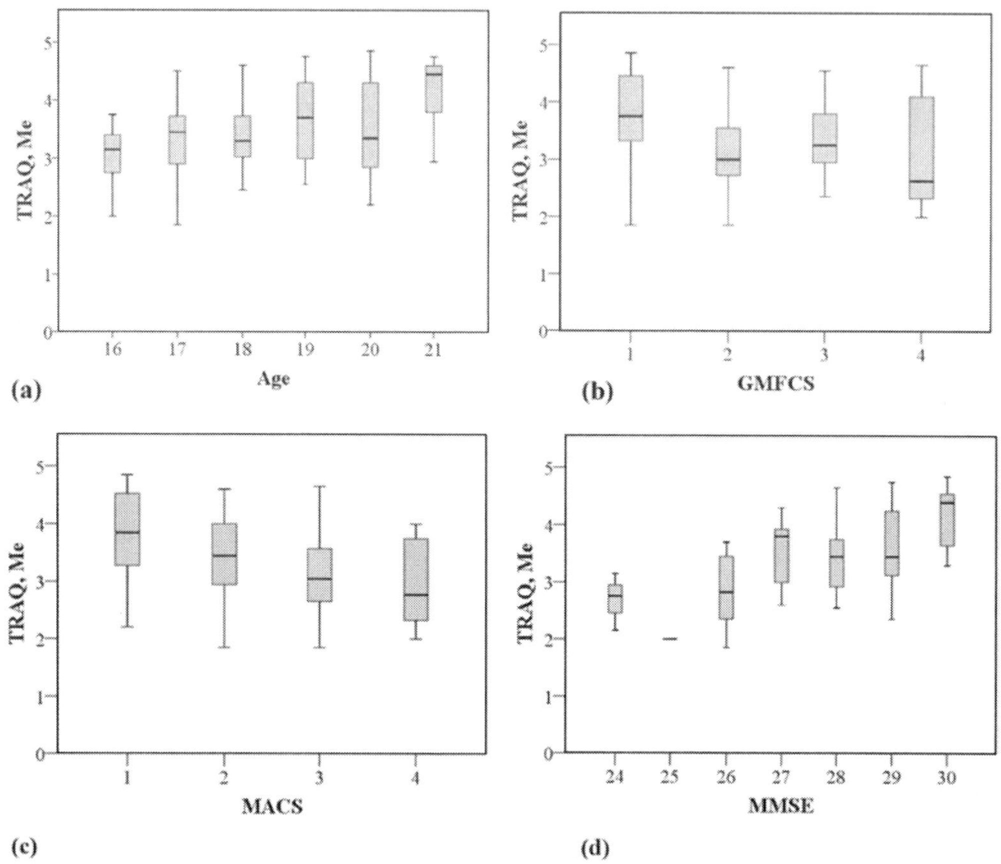

Figure 2. Median values of the Transition Readiness Assessment Questionnaire (TRAQ) total score impacted by: (a) age, (b) gross motor function (Gross Motor Function Classification System, GMFCS), (c)manual abilities (Manual Ability Classification System, MACS), and (d) cognitive status (Mini-Mental State Examination, MMSE).

services during the past year. These results are closely associated with findings in previous researches done in the Netherlands [6, 43].

In our research we analyzed the readiness of independency in health care management as the median scores of TRAQ. The lowest scores were found in thematic block "Appointment keeping" with median value 2.9 which indicates an inactive phase in making appointments, following-up on referral for test or check-ups, health change related call making to specialists etc. Thematic block "Talking with Providers" received the highest median score (Me 5.0). This was similarly seen also in the results of RTP – young adults with CP are being involved in discussions, but the real action is done by others (parents/care takers). It has been previously revealed that the assessment of health care management with generic and condition-specific measures can unfold targeted interventions and specific supports in achieving independence [46]. Our results for the total median value of TRAQ are found to be lower than those of previous research done in Latvia by L. Baranova and A. Vētra, 2016 [47] (Me 3.6 vs. 4.3, age 18–21 years). This could be explained by the fact that our research included only participants with CP and even those who had slightly decreased cognitive status. Research done by L. Baranova and A. Vētra included participants with functional limitations, not only with CP, although the cognitive status of participants was not described.

Table 4. Correlations and binary logistic regressions of the impacting factors.

	Spearman's rho	Wald	Predicted, %	OR (95% CI)
GMFCS	−0.41	9.98	73	0.43 (0.26–0.73)
MACS	−0.36			
MMSE	0.53	17.50	79	2.18 (1.51–3.13)
Severe problems with speech	−0.30			
Need of an assistant	−0.40	9.26	64	0.36 (0.19–0.70)
Education and employment	0.26			
Housing	0.30			
Leisure (social activities)	0.43	12.39	73	3.19 (1.67–6.09)
Intimate Relationships	0.36			
Sexuality	0.39			
Transportation	0.52	15.92	75	5.16 (2.31–11.56)
Care demands	0.34			
Services and aids	0.36			
Total score of RTP	0.53	14.13	76	1.43 (1.19–1.73)

$p < 0.05$.
OR: odds ratio, CI: confidence interval.

Table 5. Binary logistic regression equation* of the impacting factors.

	Wald	*p*-value	OR (95% CI)
MMSE	9.89	**<0.01**	2.27 (1.36–3.79)
Leisure (social activities)	0.38	0.54	1.37 (0.51–3.67)
Transportation	0.68	0.41	1.65 (0.50–5.41)
Total score of RTP	3.46	0.06	1.37 (0.98–1.91)

*85.0 percentage correct.

In our study it was found that the highest scores of TRAQ median were associated with the increase of participants' age, gross motor function, manual abilities and cognitive status. A linear increase was detected in the association with manual abilities. Other factors demonstrated non-linear increase of the TRAQ median score. It might be induced that a person with lower gross motor function have relatively high independency in health care management, and a person at an older age might not be as independent as age-matched peers because of other impacting factors. This highlights the idea that none of these factors can be taken as the only one when assessing young adults with CP readiness in health care management. It has been previously revealed that assessing the person, e.g., by age is not the only determinant of person' s readiness to be transited to adult health care [48].

To find out which of the factors have the most significant impact on independency in health care management we divided the TRAQ scores in two groups: ≤3.00 (rather passive phase, less independency) and ≥3.01 (rather active phase, more independency). It was detected that level of GMFCS, MACS, MMSE, having or not having severe speech problems or the need for an assistant and having income that exceeds the average have an impact on the participant's presents in one of the groups. The more capable the participant was in terms of gross motor or manual function, the less cognitive deficit and severe speech problems he/she had, and the more his/her income exceeded the average, the more he/she was likely to be present in group with TRAQ score ≥3.01. Almost all domains of RTP (except for Rehabilitation services) and the total score of RTP had an impact on the participants' presence in one of the groups. Respectively, the higher score of RTP domains and the total score of RTP the more likely for the participant to be in a rather active phase of independency in health care management. For the overall readiness to adult life for persons with CP and normal cognitive status it has been proven that there are associations with participant's age, gross motor functioning (GMFCS), manual abilities (MACS) and level of education [43]. We did not find the participants' gender, age, having or not severe problems with vision or hearing, having or not epilepsy, the place of living, being brought up by one or both parents, coming or not from a large family as an impacting factor on participants' presence in one of the groups.

The most correlating factors for the independency in health care management were found to be level of MMSE, level of leisure (social activities), independency in transportation and the total score of RTP. Suggesting that if the person has higher levels of cognitive status, ability to participate in leisure activities, arrange transportation and has higher level of overall readiness to adult life, he or she can achieve greater independency in health care management. After the binary logistic regression equation, the most impacting factor on whether the participant will receive TRAQ score $\geq$3.01, was detected to be the level of cognitive status (MMSE). Cognitive status has been found as an impacting factor also for other aspects of life for persons' with CP, e.g., employment [49], communication, receiving support and accessing services [50].

Limitations of this study include not having an able-bodied control group as a reference and the fact that RTP and TRAQ are similar in their aim of an assessment, therefore some results were overlapping. This is planned to be taken into account in the future research.

5 Conclusions

Young adults with cerebral palsy show low levels of participation in almost all domains of transition to adult life. Regardless of the fact that persons with CP need throughout life health care, the level of attendance of rehabilitation services is not sufficient. Even though young adults with CP seem to take part in decision-making and the increase of independency has been strongly highlighted, the level of independency for young adults with CP in health care management is between phases of preparation and action. The most impacting factor for independency in health care management was found to be the cognitive status, which should be considered when assessing person's readiness for transition to adult health care.

References

[1] E. Blair, L. Watson, N. Badawi, F.J. Stanley, *Life expectancy among people with cerebral palsy in Western Australia*. Dev. Med. Child. Neurol. **43**(8), 508–15 (2001)

[2] Prevalence and characteristics of children with cerebral palsy in Europe. Dev. Med. Child. Neurol. **44**(9), 633–40 (2002)

[3] M. Roquet, R. Garlantezec, O. Remy-Neris, E. Sacaze, P. Gallien, J. Ropars, et al., *From childhood to adulthood: health care use in individuals with cerebral palsy*. Dev. Med. Child. Neurol. **60**(12), 1271–7 (2018)

[4] M.E. Roebroeck, R. Jahnsen, C. Carona, R.M. Kent, M.A. Chamberlain, *Adult outcomes and lifespan issues for people with childhood-onset physical disability*. Dev. Med. Child. Neurol. **51**(8), 670–8 (2009)

[5] L.L. Tosi, N. Maher, D.W. Moore, M. Goldstein, M.L. Aisen, *Adult swith cerebral palsy: a workshop to define the challenges of treating and preventing secondary musculoskeletal and neuromuscular complications in this rapidly growing population.* Dev. Med. Child. Neurol. England. **52**(Suppl 4), 2–11 (2009)
[6] C. Nieuwenhuijsen, M. Donkervoort, W. Nieuwstraten, H.J. Stam, M.E. Roebroeck, *Experienced problems of young adults with cerebral palsy: targets for rehabilitation care*. Arch. Phys. Med. Rehabil. **90**(11), 1891–7 (2009)
[7] C. Nieuwenhuijsen, W.M. van der Slot, A. Beelen, J.H. Arendzen, M.E. Roebroeck, H.J. Stam, et al., *Inactive lifestyle in adults with bilateral spastic cerebral palsy*. J. Rehabil. Med. **41**(5), 375–81 (2009)
[8] R. Jahnsen, L. Villien, G. Aamodt, J.K. Stanghelle, I. Holm, *Musculoskeletal pain in adults with cerebral palsy compared with the general population*. J. Rehabil. Med. **36**(2), 78–84 (2004)
[9] A. Shortland, *Muscle deficits in cerebral palsy and early loss of mobility: can we learn something from our elders?* Dev. Med. Child. Neurol. **51**(Suppl 4), 59–63 (2009)
[10] M.D. Peterson, P.M. Gordon, E.A. Hurvitz, *Chronic disease risk among adults with cerebral palsy: the role of premature sarcopoenia, obesity and sedentary behaviour.* Obes. Rev. **14** (2), 171–82 (2013)
[11] *Surveillance of cerebral palsy in Europe: a collaboration of cerebral palsy surveys and registers. Surveillance of Cerebral Palsy in Europe (SCPE)*. Dev. Med. Child. Neurol. **42**(12), 816–24 (2000)
[12] K.P. Murphy, *The adult with cerebral palsy*. Orthop. Clin. North. Am. **41**(4), 595–605 (2010)
[13] T.D. Sanger, M.R. Delgado, D. Gaebler-Spira,M. Hallett, J.W. Mink, *Classification and definition of disorders causing hypertoniain childhood*. Pediatrics. **111**(1), e89–97 (2003)
[14] D.G. Whitney, E.A. Hurvitz, M.J. Devlin, M.S. Caird, Z.P. French, E.C. Ellenberg, et al., *Age trajectories of musculoskeletal morbidities in adults with cerebral palsy*. Bone. **114**, 285–91 (2018)
[15] K.P. Murphy, *Cerebral palsy life time care – four musculoskeletal conditions*. Dev. Med. Child. Neurol. **51**(Suppl 4), 30–7 (2009)
[16] L.B. Green, E.A. Hurvitz, *Cerebral palsy*. Phys. Med. Rehabil. Clin. N. Am. **18**(4), 859–82, vii (2007)
[17] D.S. Morrell, J.M. Pearson, D.D. Sauser, *Progressive bone and joint abnormalities of the spine and lower extremities in cerebral palsy*. Radiographics. **22**(2), 257–68 (2002)
[18] R.L. Lenhart, S.C. Brandon, C.R. Smith, T.F. Novacheck, M.H. Schwartz, D.G. Thelen, *Influence of patellar position on the knee extensor mechanism in normal and crouched walking*. J. Biomech. **51**, 1–7 (2017)
[19] S.H. Schless, F. Cenni, L. Bar-On, B. Hanssen, M. Goudriaan, E. Papageorgiou, et al., *Combining muscle morphology and neuromotor symptoms to explain abnormal gait at the ankle joint level in cerebral palsy*. Gait Posture. **68**, 531–7 (2018)
[20] K. Yoshida, I. Kajiura, T. Suzuki, H. Kawabata, *Natural history of scoliosis in cerebral palsy and risk factors for progression of scoliosis*. J. Orthop. Sci. **23**(4), 649–52 (2018)
[21] J.G. Thometz, S.R. Simon, *Progression of scoliosis after skeletal maturity in institutionalized adults who have cerebral palsy*. J. Bone Joint. Surg. Am. **70**(9), 1290–6 (1988)
[22] T. Sakai, H. Yamada, T. Nakamura, K. Nanamori, Y. Kawasaki, N. Hanaoka, et al., *Lumbar spinal disorders in patients with athetoid cerebral palsy: a clinical and biomechanical study*. Spine (Phila. Pa. 1976), **31**(3), E66–70 (2006)

[23] B. Samijn, E. VanLaecke, C. Renson, P. Hoebeke, F. Plasschaert, J. VandeWalle, et al., *Lower urinary tract symptoms and urodynamic findings in children and adults with cerebral palsy: A systematic review*. Neurourol. Urodyn. **36**(3), 541–9 (2017)

[24] K.J. Smith, M.D. Peterson, N.E. O'Connell, C. Victor, S. Liverani, N. Anokye, et al., *Risk of Depression and Anxiety in Adults With Cerebral Palsy*. JAMA Neurol. (2018)

[25] C. Nieuwenhuijsen, Y. van der Laar, M. Donkervoort, W. Nieuwstraten, M.E. Roebroeck, H.J. Stam, *Unmet needs and healthcare utilization in young adults with cerebral palsy*. Disabil. Rehabil. **30**(17), 1254–62 (2008)

[26] M. Freeman, D. Stewart, C.E. Cunningham, J.W. Gorter, *"If I had been given that information back then": An interpretive description exploring the information needs of adults with cerebral palsy looking back on their transition to adulthood*. Child. Care Health Dev. **44**(5), 689–96 (2018)

[27] M.A. Chamberlain, R.M. Kent, *The needs of young people with disabilities intransition from paediatric to adult services*. Eura. Medicophys. **41**(2), 111–23 (2005)

[28] J. Magill-Evans, L. Wiart, J. Darrah, M. Kratochvil, *Beginning the transition to adulthood: the experiences of six families with youths with cerebral palsy*. Phys. Occup. Ther. Pediatr. **25**(3), 19–36 (2005)

[29] E. Bjorquist, E. Nordmark, I. Hallstrom, *Living in transition – experiences of health and well-being and the needs of adolescents with cerebral palsy*. Child: care, health and development. **41**(2), 258–65 (2015)

[30] D. Lariviere-Bastien, E. Bell, A. Majnemer, M. Shevell, E. Racine, *Perspectives of young adults with cerebral palsy on transitioning from pediatric to adult health care systems*. Semin. Pediatr. Neurol. **20**(2), 154–9 (2013)

[31] E. Racine, D. Lariviere-Bastien, E. Bell, A. Majnemer, M. Shevell, *Respect for autonomy in the health care context: observations from a qualitative study of young adults with cerebral palsy*. Child Care Health Dev. **39**(6), 873–9 (2013)

[32] P.Y. Chu, G.R. Maslow, M. vonIsenburg, R.J. Chung, *Systematic Review of the Impact of Transition Interventions for Adolescents With Chronic Illnesson Transfer From Pediatric to Adult Healthcare*. J. Pediatr. Nurs. **30**(5), e19–27 (2015)

[33] K.S. Steinbeck, L. Brodie, S.J. Towns, *Transition in chronic illness: Who is going where?* J. Paediatr. Child Health. **44**(9), 478–82 (2008)

[34] N. Bagatell, D. Chan, K.K. Rauch, D. Thorpe, *"Thrust in to adulthood": Transition experiences of young adults with cerebral palsy*. Disability and Health Journal. **10**(1), 80–6 (2017)

[35] A. Bolger, J. Vargus-Adams, M. McMahon, *Transition of Care in Adolescents With Cerebral Palsy: A Survey of Current Practices*. Pm. r. **9**(3), 258–64 (2017)

[36] M.J. Okumura, M. Saunders, R.S. Rehm, *The Role of Health Advocacy in Transitions from Pediatric to Adult Care for Children with Special Health Care Needs: Bridging Families, Provider and CommunityServices*. J. Pediatr. Nurs. **30**(5), 714–23 (2015)

[37] R. Linroth, *Meeting the needs of young people and adults with childhood-onset conditions: Gillette Lifetime Specialty Healthcare*. Dev. Med. Child Neurol. **51**(Suppl 4), 174–7 (2009)

[38] R. Watson, J.R. Parr, C. Joyce, C. May, A.S. LeCouteur, *Models of transitional care for young people with complex health needs: a scoping review*. Child Care Health Dev. **37**(6), 780–91 (2011)

[39] *IIG080. Mājsaimniecību rīcībā esošie ienākumi pēc mājsaimniecības tipa (euro, mēnesī)*. Available at: `https://data1.csb.gov.lv/pxweb/lv/sociala/sociala__mb__ienakumi/IIG080.px/table/tableViewLayout1/?rxid=03d1d4fb-19c9-4c7f-b4f9-d8a532aca243`. Accessed 17.11.18

[40] R. Palisano, P. Rosenbaum, S. Walter, D. Russell, E. Wood, B. Galuppi, *Development and reliability of a system to classify gross motor function in children with cerebral palsy*. Dev. Med. Child Neurol. **39**(4), 214–23 (1997)

[41] A.C. Eliasson, L. Krumlinde-Sundholm, B. Rosblad, E. Beckung, M. Arner, A.M. Ohrvall, et al., *The Manual Ability Classification System (MACS) fo rchildren with cerebra lpalsy: scale development and evidence of validity and reliability*. Dev. Med. Child Neurol. **48**(7), 549–54 (2006)

[42] T.N. Tombaugh, N.J. McIntyre, *The mini-mental state examination: a comprehensive review*. J. Am. Geriatr. Soc. **40**(9), 922–35 (1992)

[43] M. Donkervoort, D.J. Wiegerink, J. vanMeeteren, H.J. Stam, M.E. Roebroeck, *Transition to adulthood: validation of the Rotterdam Transition Profile for young adults with cerebral palsy and normal intelligence*. Dev. Med. Child Neurol. **51**(1), 53–62 (2009)

[44] D.L. Wood, G.S. Sawicki, M.D. Miller, C. Smotherman, K. Lukens-Bull, W.C. Livingood, et al., *The Transition Readiness Assessment Questionnaire (TRAQ): its factor structure, reliability, and validity*. Acad. Pediatr. **14**(4), 415–22 (2014)

[45] *Spearman's Rank Correlation– Video Tutorial with Dr.Iain Weir.* Available at: `http://www.statstutor.ac.uk/resources/uploaded/SpearmanCorrelation/Tutorial/spearmans.htm`. Accessed 17.12.18

[46] S. Warschausky, J.N. Kaufman, W. Schutt, M. Evitts, E.A. Hurvitz, *Health self-management, transition readiness and adaptive behavior in persons with cerebral palsy or myelomeningocele*. Rehabil. Psychol. **62**(3), 268–75 (2017)

[47] L. Baranova, A. Vētra, *Funkcionāli ierobežotu jauniešu pāreja uz pieaugušo veselības aprūpes sistēmu: sagatavotība un pieredze*. RSU Scientific Conference: Riga Stradins University (2016)

[48] F. Campbell, K. Biggs, S.K. Aldiss, P.M. O'Neill, M. Clowes, J. McDonagh, et al., *Transition o fcare for adolescents from paediatric services to adult health services*. Cochrane Database Syst. Rev. **4**, Cd009794 (2016)

[49] S. Rutkowski, E. Riehle, *Access to employment and economic independence incerebral palsy*. Phys. Med. Rehabil. Clin. N. Am. **20**(3), 535–47 (2009)

[50] A. Ali, K. Scior, V. Ratti, A. Strydom, M. King, A. Hassiotis, *Discrimination and other barriers to accessing health care: perspectives of patients with mild and moderate intellectual disability and their carers*. PLoS. One. **8**(8), e70855 (2013)

Aerobic capacity for breast cancer survivors 2 to 3 years after breast surgery

Rudolfs Ceseiko[1,*], *Signe* Tomsone[1], *Laura* Majevska[1], and *Aivars* Vetra[1,2]

[1] Rīga Stradiņš University, Faculty of Rehabilitation, Riga, Latvia
[2] Riga Eastern University Hospital, Riga, Latvia

Abstract. Breast cancer treatment may cause unfavourable side effects that directly or indirectly influence the cardio-respiratory capacity among breast cancer survivors, thus affecting physical functioning and the quality of life. Mounting evidence supports that Maximal oxygen uptake (VO_{2max}) is a strong predictor of all-cause mortality and cardiovascular disease, and low VO_{2max} is associated with increased mortality from breast cancer. The aim of the study was to measure cardio-respiratory fitness in breast cancer survivors and determine to what extent VO_{2max} levels are lower compared with healthy counterparts from population. The study included 11 women with a history of breast cancer, the average time after breast surgery being 2–3 years. Aerobic capacity was determined by the Cardiopulmonary exercise test. HUNT3 study was used to define aerobic capacity for healthy women in population. Mean Age for participatns 50 ± 4 years ($M \pm SD$). The comparison of the results from Cardiopulmonary exercise test indicated that breast cancer survivors display VO_{2max} values significantly lower 27.1 ml/kg/min (27%) than healthy counterparts from population, i.e., 34.4 ml/kg/min ($P = 0.001$). Recognizing the increasing number of breast cancer patients worldwide and the physical reconditioning among the survivors, aerobic exercise training is suitable for improving cardio-respiratory health, functionality, and quality of life.

Key words: breast cancer, rehabilitation, aerobic capacity.

1 Introduction

Breast cancer (BC) is the most common type of cancer among women. In 2012, about 1.7 million new cases were diagnosed worldwide, thus representing 25.2% in morbidity rates and 14% in mortality rates from the overall cases of oncologic diseases among women [1]. Every year more than 1000 new cases of BC are diagnosed in Latvia. At the end of year 2016, more than 12,000 BC patients had been registered. After the diagnosis, about 89% of the patients live more than 5 years [2]. The current methods for treatment of oncologic diseases imply long combined local and/or systematic therapy. It initiates various acute and toxic manifestations leading towards increased risk of mortality and chronic diseases not primarily associated with the diagnosis of cancer. The decrease in the values of mortality and quality of life are mainly associated with cardiovascular and cardiopulmonary diseases [3] having huge effect on the overall participation in daily activities and increased rates

* Corresponding author: Rudolfs.Ceseiko@rsu.lv

of disability [4]. Aerobic capacity is the most objective indicator of the functional condition of the cardiovascular and respiratory systems. Aerobic capacity characterizes the possibilities of using oxygen transport in skeletal muscles. Reduced aerobic capacity rates can be explained by several variables, e.g., pulmonary, cardiovascular and peripheral. Aerobic capacity is determined by the results from the maximum or sub-maximal exercise test performed under laboratory conditions with appropriate equipment [5].

Long-term complications after cancer treatment are associated to the dysfunction of specific organ systems, resulting in various chronic diseases, functional restrictions (for instance, lymphodema), specific pain syndromes, neuropathy, secondary malignant tumours and aesthetic defects. All of the mentioned side effects are mostly typical to patients younger than 60 years and those having suffered from chronic diseases (diabetes, hypertension, etc.), or patients being exposed to infectious diseases or recent surgical interventions [4]. BC patients are exposed to nearly three times higher risk of developing one of the cardiovascular diseases in comparison to healthy counterparts [6]. In most of the postmenopausal BC patients, at least one of the risk factors for cardiovascular disease is identified already at the time of cancer diagnosis [7]. The cardio-toxic side effect derived from adjuvant therapy has the potential of significantly increasing mortality rates among BC patients diagnosed with cardiovascular diseases. Anthracycline medicine causes immediate damage of myocardial cells while clinically relevant changes can be noticed only after several months or years. Acute cardio-toxicity during treatment period is very rare, while sub-acute or chronic myocardial damage (within a year after treatment) is noticed in about 23% of cases. Similar side effects are caused also by hormone therapy, which increases triglyceride levels, thus leading towards increased risk of atherosclerosis [8]. Radiation therapy can cause local damage of myocardial walls, resulting in decrease in myocardial perfusion [9].

The cardiovascular system is especially sensitive to the reduction in physical activities, thus even periods of short-term physical inactivity or bed rest causes significant decrease in aerobic capacity for both physically active and inactive population [10]. The decrease in physical activities among BC patients is observed at the beginning of chemotherapy. The entire treatment process lasts for 9 months or longer; in this period physical activity is reduced to minimum or excluded completely. After 4 weeks of chemotherapy, VO_{2max} decreases for about 3–8% [11]. The studies show that VO_{2max} reaches 25 ml/kg/min in average prior to adjuvant therapy. That is by 17% lower than among healthy, yet physically inactive women. However, the maximum oxygen consumption after the adjuvant treatment is about 22.2 ml/kg/min, 25% lower than in healthy women of the same age [9]. To be of complete functional independence, the minimum VO_{2max} value has to be at least 15.4 ml/kg/min. It is surprising that 30% of the patients are below this value, which indicates huge effort needed to perform daily activities. As a result, higher energy is needed and the sense of fatigue becomes present, which is very typical for BC patients undergoing treatment [12]. Decrease in the value of VO_{2max} for BC patients is present even for approximately 7 years after treatment [9].

Thus, the aim of the study is to measure aerobic capacity for breast cancer patients having undergone breast surgery 2–3 years ago, and compare these results with the literature data about healthy and physically inactive women of the same age.

2 Methods

2.1 Participants

The selection of the subjects was based on the following inclusion/exclusion criteria. Inclusion criteria were: breast cancer stages I–III, breast surgery done 2–3 years ago,

40 to 55 years old, body mass index ranging from 22 to 28, not using beta blockers, were undergoing chemotherapy, radiotherapy and hormone therapy, at least a year has passed since the last chemotherapy, not engaged in regular, targeted physical activities on a daily basis. Exclusion criteria were: breast cancer recurrence 2–3 years after the surgery, Chronic Obstructive Pulmonary Disease (COPD), bronchial asthma, hypertension, arrhythmia, heart failure, ischemic heart disease or orthopaedic disease. The study was conducted in Riga Stradiņš University, Faculty of Rehabilitation, in corporation with research laboratory of Rehabilitation. The data was collected between January 9th and February 10th, 2018. The selection of research subjects was performed in corporation with cancer patient support society "Dzīvības koks" ("Life Tree"). Each participant received a letter with information about the study beforehand. Informed consent was obtained from all the participants. Before engaging in the study, the subjects were reached by phone, to explain testing details. When arriving to the assessment procedure, the subjects had to take with them comfortable clothing and footwear, as well as the statement from the medical treatment institution confirming information on the surgery and treatment received. The study was conducted in accordance with the ethical standards – the study was authorised by the Ethics Committee of Riga Stradiņš University and followed the principles set in Declaration of Helsinki.

To evaluate aerobic capacity, incremental treadmill test was applied to measure VO_{2max} [13]. The test was performed by the use of treadmill *Mercury (h/p/cosmos, Nussdorf-Traunstein, Germany)*, heart rate monitor *Polar (F7, Finland)* and spiroergometer *Cortex MetaMax 3B (Cortex Biophysik, Leipzig, Germany)*. On the day of the test, the spiroergometer was calibrated according to the manufacturer's instructions. A short history of diseases and the body weight of each participant were collected prior to the beginning of the testing. The participants were introduced with the study procedure. They were engaging in a 5-minute long warm-up session on a treadmill at a self-selected speed. A modified Balke protocol was used for measuring VO_{2max} [14]. The maximum expected heart rate was calculated on the basis of formula *220-age* (in years) [15]. The heart rate was monitored at the beginning of the test and during the last 30 seconds of it. Afterwards, incremental test on a treadmill was performed, starting with constant speed of 4.8 km/h and an incline of 0%. The intensity was gradually increased by lifting the treadmill slope by 2% every minute until 20% level is reached. The test was interrupted under the following conditions: 90% of the maximum heart rate is reached; VO_{2max} stops increasing (reaching plateau); limiting symptoms emerged and the participant demanded the test to be interrupted or failed to respond to repeated incentive to continue. The test was performed once for each participant and the whole testing procedure was supervised by the laboratory personnel.

The data on VO_{2max} value in healthy, physically inactive women was obtained from HUNT3 study – the most prominent collection of data on the results from maximum load test among the population of Europe. Totally 428 women at the age between 50 to 59 years were evaluated. The mean VO_{2max} for this group is 34.4 ± 5.7 ml/kg/min [16].

2.2 Statistics

The statistical processing of data was performed by computer programme *IBM SPSS*. *GraphPad Prism* was applied for creating graphics. The method of descriptive statistics was used for group characteristics (all values being expressed as mean $\pm$ standard deviation). One sample T test was applied for comparing VO_{2max} value for the group with the mean value for the population of the same age and gender as described in the literary sources. The result was considered statistically significant if $P < 0.05$.

Table 1. Participant characteristics. Breast cancer stage: number of patients with diagnosis. Therapy received: number of patients having received specific treatment.

	(N =11)
Age	50 ± 4
Weight, kg	75 ± 17
Height, cm	168 ± 6
Body mass index	26 ± 5
Stage of cancer (N)	
I	2
II	7
III	2
Treatment:	
Chemotherapy	10
Radiation	10
Hormone therapy	10

Data are presented as mean ± SD.

Table 2. Physiological parameters after VO_{2max} test.

	(n =11)
VO_{2max} ($mL \cdot kg^{-1} \cdot min^{-1}$)	27 ± 4
VO_{2max} ($L \cdot min^{-1}$)	1.9 ± 0.2
V_E ($L \cdot min^{-1}$)	59 ± 14
HR_{max}	167 ± 7

Data is presented as mean ±SD. VO_{2max}: maximal oxygen uptake; V_E: ventilation; RER: respiratory exchange ratio; HR_{max}: maximal heart rate.

3 Results

Basing on the letters sent, 21 participants applied for the study. According to the inclusion/exclusion criteria, 11 women were eligible. Patient's characteristics are presented in (Table 1). The rest of the women were not included in the study due to their age or too high body mass index.

Maximal oxygen uptake. The literary sources describing data on Cardiopulmonary exercise test among the population of Europe claim the maximal oxygen uptake for healthy women at the age of 50 years to be 34.4 ml/kg/min [16]. The incremental treadmill test, performed on the research subjects ($n = 11$) resulted in the mean VO_{2max}27.1 ml/kg/min (Table 2). As a result, the difference 7.3 ml/kg/min between the average value in patients and value existing in society is assumed to be statistically significant ($P = 0.001$) and (95% TI is from -10 to -5); thus, Cohen's d value shows extra-large difference ($d = 1.7$) (Fig. 1). The results obtained are by 27% lower in comparison to the data on healthy peers from the population.

4 Discussion

The main finding of the present study was that aerobic capacity of BC patients is by 27% lower than that of their counterparts (healthy, physically inactive women of the same age). Several studies investigating the aerobic capacity of 50-year old BC patients have demonstrated decrease in maximum oxygen consumption by approximately 22–30% after the treatment [17]. The data indicate the negative influence (which can persist even several years afterwards) of the treatment side effects on the body's aerobic capacity in a long run. Reduced aerobic capacity can remain present even up to 7 years after diagnosis [9].

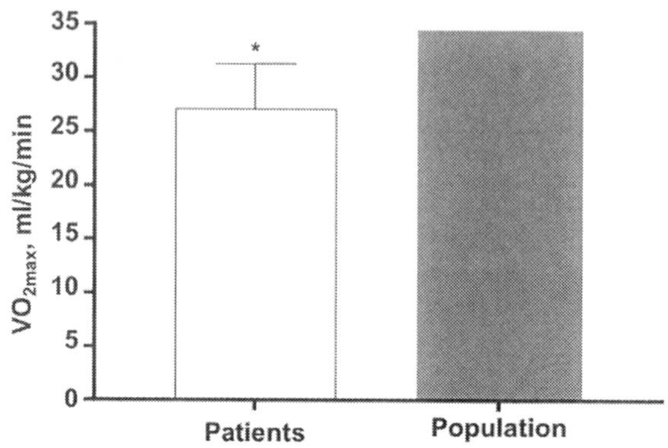

Figure 1. Maximal oxygen uptake for breast cancer patients compared with heathy counterparts from population. Data is presented as mean ±SD, $*P = 0.001$.

VO_{2max} indicators depend on BC stage and the treatment received. Aerobic capacity will be higher for patients in the stage I or II, while the capacity will be lower in stage III or IV of the cancer. A study conducted by Jones et al. (2012) investigated BC patients, who had undergone the last chemotherapy from 6 months to 3 years ago. The average VO_{2max} was 18.4 ml/kg/min, while the average VO_{2max} in forth (metastatic) stage was 16.3 ml/min/kg. It is by 33% lower than for healthy and physically inactive women of the same age [12].

Nine of the 11 participants in the present study have been diagnosed with BC at the stage I or stage II. In general, the results obtained are higher in comparison to other researches devoted to aerobic capacity. However, the results greatly depend also on the testing methods of aerobic capacity. Cardiopulmonary exercise testing (CPET) mostly is applied in 2 ways, either using treadmill or veloergometer. One of the techniques is by gradually increasing the load on a veloergometer or treadmill. According to the data of various studies, VO_{2max} rates on a treadmill for BC patients are slightly higher than on a veloenergometer. This phenomenon could be explained by the fact that bigger number of muscle groups is involved in walking activity. It leads towards higher blood supply to the congested muscles and need for more oxygen. Even though the Cardiopulmonary exercise test demands great physical effort, it is relatively safe for BC survivors when accompanied by appropriate screening techniques [17].

Reduced VO_{2max} rates are explained by multi-factor reasons, including several organ systems engaged in oxygen transport. The limitations of cardiovascular system are the ones described most prominently in the study literature devoted to BC patients. These limitations are mostly associated with the chemotherapy medicine applied during the treatment procedure and the cardio-toxic effect caused by this medicine in a long run. Also, medical exposure causes local damage of myocardial walls [9]. As 10 out of 11 participants had been exposed to chemotherapy, local radiation or hormone therapy, and none of the participants had been suffering from cardiopulmonary diseases, it can be concluded that the effect of the treatment has influenced the results of VO_{2max} obtained during the study. 9 of 11 participants acknowledged the failure of respiratory system being the limiting factor of the incremental test. The data indicate the main limiting factors of the aerobic capacity, including cardiovascular, respiratory and circulatory systems. Since gas diffusion in the lungs is hardly a limiting factor, it can be assumed that aerobic capacity is mostly affected by the volume of cardiac output and myocardial contractility. The medicine used during the therapy does not affect the functions of the skeletal muscles, yet they can be influenced by low physical activities. As a result, the muscle mass decreases while the mass of fat increases. It leads towards diminished maximum muscle strength significantly limiting the aerobic capacity [2].

The mean body mass index for the study group was 26 kg/m^2. It suggests that BC patients are slightly overweight. Gain in weight during therapy and after it, body mass index being above 25 kg/m^2, is associated with increased mortality risk [18]. Women with extra weight (BMI <25 kg/m^2) are at twice the risk of BC recurrence than those within the normal BMI [19].

Recently, the interest in physical activities for BC rehabilitation is increasing. The data available support the effectiveness of aerobic training for improving the physical condition of BC patients during and after the therapy. The studies suggest that the lowest physical activity among BC patients is just after finishing therapy. However, the situation stabilises after several weeks, but due to various treatment side effects and other diseases such condition can persist for a longer period. For instance, the mean VO_{2max} rate just after the adjuvant therapy is from 18.4 to 20.6 ± 6.7 ml/min/kg [2]. When experiencing severe decrease in aerobic capacity, patients find it difficult to engage in daily activities – maximum effort is required for going upstairs or working in a garden. Thus, moderate physical activities are crucial for BC patients in order to feel good, diminish the presence of the side effects of chemotherapy and limit the risk of additional diseases even before starting the therapy.

Physical activities can help in controlling body weight. It has been proved that gradual overweight reduction can contribute to better control of hypertension, hyper insularity, dyslipidaemia and pain in cancer patients [4]. Clinically relevant physical activities are those occurring after the diagnosis of BC [20]. In a study conducted by Coumeya and colleagues, VO_{2max} for BC patients increased by 14.5% after 17 weeks of aerobic training with a load intensity from 70% to 75% of the maximum heart rate. Also, after several weeks of training, the participants acknowledged such symptoms as fatigue, nausea and sleeping disorder having diminished [21]. According to a 24-week study performed by Scott et al. (2016) maximum oxygen uptake in BC patients increased by 32% due to aerobic training [22]. These studies verify that aerobic training in breast cancer patients do not have any negative effects on their health.

5 Conclusion

The present study demonstrates that aerobic capacity in breast cancer patients is greatly diminished just after the completion of therapy. Thus, effective aerobic training should be included as a part of rehabilitation procedure. However, the optimal aerobic training techniques (intensity, duration, frequency) to be performed by breast cancer patients just after the completion of the therapy have not been established yet. As the result, further clinical trials and the analysis of the obtained results are necessary for comparing different types of training methods.

References

[1] McGuire, S., *World Cancer Report 2014. Geneva, Switzerland: World Health Organization, International Agency for Research on Cancer, WHO Press, 2015.* Adv. Nutr. **7**(2), 418–9 (2016)
[2] Klassen, O., et al., *Cardiorespiratory fitness in breast cancer patients undergoing adjuvant therapy.* Acta Oncol **53**(10), 1356–65 (2014)
[3] Lakoski, S.G., et al., *Exercise rehabilitation in patients with cancer.* Nat. Rev. Clin. Oncol. **9**(5), 288–296 (2012)

[4] Aziz, N.M., *Cancer survivorship research: state of knowledge, challenges and opportunities.* Acta Oncol **46**(4), 417–32 (2007)
[5] Helgerud, J., et al., *Aerobic high-intensity intervals improve VO2max more than moderate training.* Med. Sci. Sports Exerc. **39**(4), 665–71 (2007)
[6] Ky, B., *Priorities in the Cardiovascular Care of Breast Cancer Survivors.* J. Oncol. Pract. **14**(4), 205–211 (2018)
[7] Park, N.J., et al., *Cardiovascular disease and mortality after breast cancer in postmenopausal women: Results from the Women's Health Initiative.* PLoS One **12**(9), e0184174 (2017)
[8] Bird, B.R., Swain S.M., *Cardiac toxicity in breast cancer survivors: review of potential cardiac problems.* Clin. Cancer Res. **14**(1), 14–24 (2008)
[9] Peel, A.B., et al., *Cardiorespiratory fitness in breast cancer patients: a call for normative values.* J. Am. Heart Assoc. **3**(1), e000432 (2014)
[10] Coyle, E.F., Hemmert, M.K., Coggan, A.R., *Effects of detraining on cardiovascular responses to exercise: role of blood volume.* J. Appl. Physiol. **60**(1), 95–9 (1985)
[11] Courneya, K.S., et al., *Effects of exercise dose and type during breast cancer chemotherapy: multicenter randomized trial.* J. Natl. Cancer Inst. **105**(23), 1821–32 (2013)
[12] Jones, L.W., et al., *Cardiopulmonary function and age-related decline across the breast cancer survivorship continuum.* J. Clin. Oncol. **30**(20), 2530–7 (2012)
[13] Herdy, A.H., et al., *Cardiopulmonary Exercise Test: Background, Applicability and Interpretation.* Arquivos brasileiros de cardiologia **107**(5), 467–481 (2016)
[14] Flo, G.L., et al., *Development and Safety of an Exercise Testing Protocol for Patients with an Implanted Cardioverter Defibrillator for Primary or Secondary Indication.* Cardiopulmonary Phys. Ther. J. **23**(3), 16–22 (2012)
[15] Tanaka, H., Monahan, K.D., Seals, D.R., *Age-predicted maximal heart rate revisited.* J. Am. Coll. Cardiol. **37**(1), 153–6 (2001)
[16] Loe, H., et al., *Aerobic capacity reference data in 3816 healthy men and women 20–90 years.* PLoS One **8**(5), e64319 (2013)
[17] Neil-Sztramko, S.E., et al., *Aerobic capacity and upper limb strength are reduced in women diagnosed with breast cancer: a systematic review.* J. Physiother. **60**(4), 189–200 (2014)
[18] Playdon, M.C., et al., *Weight Gain After Breast Cancer Diagnosis and All-Cause Mortality: Systematic Review and Meta-Analysis.* J. Natl. Cancer Inst. **107**(12), djv275 (2015)
[19] Peel, A.B., et al., *Cardiorespiratory fitness in breast cancer patients: a call for normative values.* J. Am. Heart Assoc. **3**(1), e000432–e000432 (2014)
[20] Ogunleye, A.A., Holmes, M.D., *Physical activity and breast cancer survival.* Breast cancer research : BCR **11**(5), 106–106 (2009)
[21] Courneya, K.S., *Exercise guidelines for cancer survivors: are fitness and quality-of-life benefits enough to change practice?* Curr. oncol. (Toronto, Ont.) **24**(1), 8–9 (2017)
[22] Scott, J.M., et al., *Cardiovascular Late Effects and Exercise Treatment in Breast Cancer: Current Evidence and Future Directions.* Can. J. Cardiol. **32**(7), 881–890 (2016)

Physical activity in healthy urban adults and its association with body composition parameters

Rudite Lagzdina* and *Maija* Rumaka

Rīga Stradiņš University, Department of Human Physiology and Biochemistry, Riga, Latvia

Abstract. An adequate amount of physical activity (PA) in adults improves cardio-respiratory fitness, metabolic health and helps maintain a normal body weight and composition. Here we examined sex- and occupation-related PA patterns and their association with body composition parameters.

Healthy volunteers ($n = 97$) aged 21–49 years completed a self-administered PA questionnaire and characterized their occupational duties depending on the volume of the physical work involved. A bio-impedance analyser was used to determine body composition. Men reported more metabolic equivalent of task (MET) minutes/week of vigorous PA than women ($p < 0.05$).

Participants with low occupational PA had less walking, moderate and total MET minutes/week, and more sitting time than those with moderate and vigorous occupational PA ($p < 0.05$), but their body composition parameters did not differ.

In males, negative correlations were found between body fat percentage, visceral fat range, segmental fat mass, and moderate MET minutes/week, whereas positive associations were observed between muscle mass in legs and vigorous MET minutes/week ($p < 0.05$). In females, negative correlations were found between visceral fat and vigorous MET minutes/week but positive – between lean body mass, trunk and leg muscle mass, and moderate PA ($p < 0.05$). These observed associations might be used to develop interventions to promote sex- and occupation-specific PA.

Key words: physical activity, body composition, healthy adults, occupational physical activity, IPAQ.

1 Introduction

An adequate amount of physical activity (PA) in adults improves their cardio-respiratory fitness and metabolic bone health and helps them maintain a normal body weight and composition. A randomized controlled study demonstrated that lifestyle changes including increased PA is the primary intervention for maintaining an appropriate body weight and reducing excess visceral and upper body fat, the presence of which has a strong correlation with metabolic syndrome [1, 2]. In addition to leisure time, domestic and transport-related activities, daily PA substantially depend on occupational PA (OPA). Many people in urban areas are employed in professions with negligible PA, which can be considered a potential risk factor for decreased total daily PA.

Changes in society such as mechanization and motorization have decreased the OPA, transportation PA, and home PA. A longitudinal study of a Finnish adult population revealed

* Corresponding author: `Rudite.Lagzdina@rsu.lv`

a decreased OPA and increased prevalence of leisure-time PA over a 30-year period [3]. However, data regarding long term changes in leisure time PA are inconsistent. Some population studies reported that leisure time PA remained at the same level, while others – that it increased, but the results varied among geographical regions due to age and education level [4]. A systematic review published in 2017 identified evidence of a positive relationship between living in urban areas and greater sitting time and total sitting behaviour [5].

The present study evaluated the pattern of PA in a typical week to assess aspects related to OPA and body composition parameters among healthy adults involved in professional duties in an urban environment such as office work, health, recreation, education, and service. A cross-sectional study design was used to collect self-reported PA data and objective measurements of body parameters.

2 Material and methods

Participants were recruited using a chain-referral sampling technique. The study included adult volunteers of both sexes who met the following inclusion criteria: healthy, age 20–50 years, not using any medication during study participation, and engaged in typical occupational routines (e.g., not during annual holiday or leave). The study was approved by the Ethics Committee of Rīga Stradiņš University and informed consent was provided by each participant upon study entry. The PA of the participants was assessed using the International Physical Activity Questionnaire, short form (IPAQ-SF). Participants reported information about their PA in the previous 7 days and four domains of PA were analysed – sitting, walking, and moderate and vigorous PA. In the further data processing, metabolic equivalent of task (MET) minutes for each activity were calculated. Self-reported PA data were handled according to the procedure described in the Guidelines for Data Processing and Analysis of the IPAQ [6]. In addition, participants were asked to characterize the type of their occupational duties depending on the volume of the physical work involved.

The body weight and height of each participant was measured to the nearest 0.01 kg and 1.0 cm, respectively, and body mass index (BMI) was calculated. A Tanita MC-180 MA (Tanita Corp., Tokyo, Japan) multi-frequency bio-impedance analyser was used to determine body mass and body composition, including fat and muscle mass segmental distribution in each extremity and the trunk. To obtain more precise results, the measurements were taken in the morning after an overnight fast with limited PA before the test.

The statistical analysis was performed using SPSS Statistics v. 22.0 (IBM Corporation, USA) with significance set at $p < 0.05$. The analysis included descriptive statistics for the entire sample stratified by sex and occupational groups in all cases. The results are presented as mean and standard deviation or median with interquartile range. Parameter differences between sexes and groups were assessed using an independent samples t-test in cases of Gaussian data distribution and the Mann-Whitney test in cases of non-Gaussian data distribution. Pearson chi-square and Spearman's tests were used to assess the relationship between variables.

3 Results

The study sample consisted of 97 participants and contained more women (55%) than men. The anthropometric characteristics of the participants are presented in Table 1. Mean age, height, body mass, and BMI were higher among men ($p < 0.05$, independent-samples t-test). According the answers about work-related PA, 64 (66%) of the subjects had low OPA and 33 (34%) had moderate-high OPA. There were no statistically significant differences in the mean values of the four mentioned anthropometric parameters among the OPA groups.

Table 1. Descriptive characteristics of the study participants.

	Male (n = 44) mean ± SD	Female (n = 53) mean ± SD
Age (years)	31.4 ± 5.4	29.0 ± 5.7
Height (cm)	183.5 ± 6.7	169.9 ± 5.9
Body mass (kg)	81.6 ± 10.6	63.7 ± 7.1
BMI (kg/m^2)	24.2 ± 2.7	22.1 ± 2.6

Table 2. Characteristics of physical activity assessed by the IPAQ-SF.

Parameter	Male median [IQR]	Female median [IQR]	p value*
Sitting in working day (min/day)	390 [240–480]	360 [240–495]	ns
Sitting in day off (min/day)	300 [191–360]	300 [180–360]	ns
Walking (MET-min/week)	660 [282–1386]	907 [479–2376]	ns
Moderate PA (MET-min/week)	480 [205–1110]	540 [260–960]	ns
Vigorous PA (MET-min/week)	840 [0–1890]	240 [0–1440]	0.025
Total PA (MET-min/week)	2625 [1541–4412]	2772 [1541–4166]	ns

* p values show gender differences, comparisons performed by Mann-Whitney U test. ns – $p > 0.05$.

The IPAQ-SF results are summarized in Table 2. The reported time of PA in each domain was characterized by MET minutes spent on them in 1 week. Walking MET minutes were computed as reported walking minutes in 1 week multiplied by the intensity coefficient 3.3. For moderate and vigorous MET minutes, coefficients 4.0 and 8.0, respectively, were used. The total number of MET minutes/week was defined as the sum of walking, moderate, and vigorous MET minutes.

The summary of PA expressed as MET minutes/week showed that men had significantly more MET minutes/week of vigorous PA than women, whereas no sex-based differences were observed in the sitting and moderate PA domains. The median values of walking and moderate MET minutes/week were higher in women, but the difference was not statistically significant. Walking and moderate PA in women and vigorous PA in men were the activities that also contributed the most to the total PA MET minutes/week.

As proposed by the IPAQ data processing and evaluation guidelines, PA results were evaluated as three-level categorical scores [6]. An equal number of participants (45.4%) met the criteria to be qualified as moderate or highly physically active, while 9.2% of participants were low active. Since there was no association between sex and occupational group (Pearson chi-square $\chi = 1.634$, $p = 0.201$), separate results of men and women in each occupational group were not analysed.

Analysis of the distribution of participants by OPA level between IPAQ PA categories revealed that, in the low PA category, all participants had a low OPA; in the moderate PA category, 36% had a moderate to high OPA; and in high PA category, 68% had a moderate to high OPA. The Pearson chi-square test results showed a statistically significant association between OPA group and PA score ($\chi = 13.67$, $p = 0.001$). The mean reported sitting time in working days (minutes per day) was significantly higher in subjects in the low OPA group

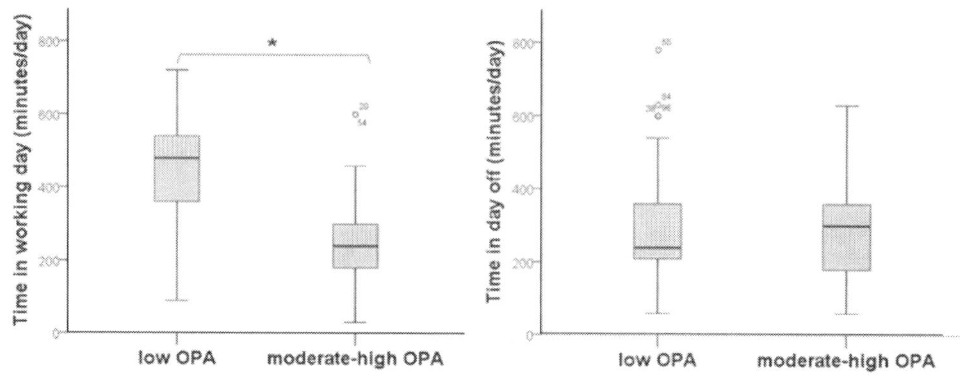

Figure 1. Sitting time by occupational physical activity (OPA) group in working day and in day off. $^{*}p < 0.05$, Mann-Whitney U test.

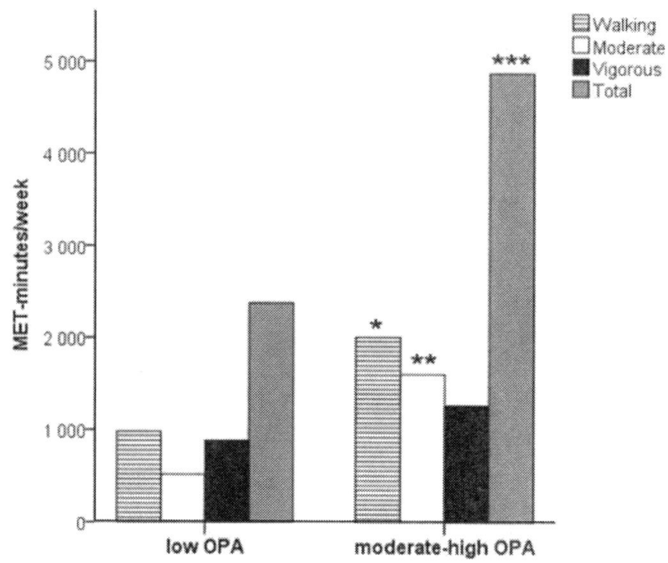

Figure 2. Physical activity domains of the occupational physical activity (OPA) groups. $^{*,**,***}p < 0.05$ *vs.* low OPA group, Mann-Whitney U test.

than in the moderate to high OPA group, whereas sitting time duration on days off was similar (Fig. 1).

Analysis of MET minutes of the separate PA domains in the occupational groups (Fig. 2) revealed that the subjects in the moderate to high OPA group had significantly higher results of walking, moderate, and total MET minutes/week than those in the low OPA group.

The relationship between PA domain and body composition parameters was assessed using the Spearman correlation test (Table 3). Since there were no statistically significant differences in analysed body parameters between the OPA groups, the data were stratified by sex only.

In men, a moderate MET minutes/week was inversely correlated with body fat, visceral fat range, and segmental fat mass of the trunk, arms, and legs, whereas a positive relationship was observed between muscle mass in the legs and vigorous activity MET minutes/week.

Table 3. Correlation coefficients (Spearman's rho) between MET-minutes in physical activity domains and body composition parameters.

	Male		Female	
	Moderate MET-min	Vigorous MET-min	Moderate MET-min	Vigorous MET-min
Visceral fat (range)	–0.379	–	–	–0.339
Fat mass (kg)	–0.355	–	–	–
Fat (% from body mass)	–0.359	–	–	–
Trunk fat mass (kg)	–0.348	–	–	–
Left arm fat mass (kg)	–0.373	–	–	–
Right arm fat mass (kg)	–0.316	–	–	–
Left leg fat mass (kg)	–0.329	–	–	–
Right leg fat mass (kg)	–0.314	–	–	–
Lean body mass (kg)	–	–	0.337	–
Trunk muscle mass (kg)	–	–	0.305	–
Left leg muscle mass (kg)	–	0.319	0.299	–
Right leg muscle mass (kg)	–	0.333	0.348	–

$p < 0.05$ for all presented correlation coefficients.

In women, an inverse correlation was noted between visceral fat range and vigorous activity MET minutes/week, whereas a positive correlation was found between lean body, trunk, and leg muscle mass and moderate PA.

4 Discussion

A slightly higher proportion of participants in the present study fulfilled the criteria for classification as adequately active (moderate or highly active) than that in the European Union population in general and inhabitants of Latvia in particular [7]. Our study included only participants under age 50 years in good health, which could have led to the recruitment of more active individuals who were more inclined towards and interested in a health-promoting lifestyle; thus, our PA data may have been improved compared to those in the Eurobarometer 2013 survey. The higher self-reported vigorous PA in men than in women in this study conforms to the Eurobarometer findings [8]. The comparably higher number of minutes of walking and moderate PA in women could be reflective of their lifestyle and exercise preferences; according to the results of another study, their PA level might be strongly influenced by availability of fitness and PA facilities in their immediate living area [9]. A similar number of men and women described their occupation as involving moderate or vigorous PA. However, in the same occupational group, the greater effort in men cannot be ignored. Additionally, moderate to vigorous non-OPA might be more common in men. This could be partially explained by the fact that enjoyment is the motivational factor that stimulates higher-intensity PA in men [10]. Analysis of a single week's data should consider that PA patterns can vary over time (week, seasons, etc.); however, individual activity behaviours of adults could be significantly influenced by occupational rhythms. The more time one spends sitting and less time one spends for walking, moderate, and total PA on working days increases the risk of a decreased weekly PA. There is no clear evidence of a pattern in which a high PA level on working days leads to inactivity during days off or vice versa. A large-scale national study in the United States revealed that in the most physically active groups of adults, highly active weekdays were followed by a significant decrease in activity over the weekend, whereas in adults with the lowest total volume of activity per week, activity levels were relatively stable [11]. Other studies reported that full-time employment in a mostly sedentary occupation increased the uncompensated sitting time in other activity domains (leisure, transportation, screen time) of the same individuals [12, 13].

A low OPA increases one's risk of a decreased weekly PA; however, it did not correlate with body composition in this group of relatively young and healthy adults. Since weight management depends on both energy expenditure and energy intake, it is possible that an adequate regulation of energy intake may sufficiently maintain normal body weight even if one's PA is low. In the present study, body parameters such as BMI, body lean mass, and fat tissue amount fat mass did not differ among OPA groups. BMI is routinely used to detect underweight or overweight status, but since it is not sensitive to particular tissue type changes, in this study, more specific obesity-related indicators such as body fat percentage and fat amount in different body segments were used to examine the relationship between body composition parameters and PA type.

Durations of moderate and vigorous PA influenced body composition parameters of both sexes. Men who reported more moderate PA had a lower visceral fat rating, body fat mass, fat percentage, and fat in all body segments, whereas vigorous PA correlated with higher leg muscle mass. In women, a higher amount of moderate PA correlated with a higher lean body mass and trunk and leg muscle mass, while vigorous PA inversely correlated with visceral fat range. Correlation differences due to sex can be attributed to reliance on different metabolic substrates for energy production depending on PA duration and intensity as well as sex-specific proportions of fat and muscle tissue and their placement in the body. Women's bodies normally contain more fat tissue, and there is evidence that over the wide range of exercise intensities, lipids remain a dominant source of fuel for oxidative metabolism in women versus men [14, 15].

Fat is the main energy source for moderate-intensity PA, in which oxygen uptake is <65% of maximum [16, 17]. Exercise duration determines the ratio of intra- and extramuscular fat input for fat oxidation. For exercises longer than 90–120 minutes, extramuscular fat delivery becomes more important [18], what might lead to decrease in fat mass not only in the exercising extremity but all over the body. Investigations in young non-obese men and women found greater susceptibility of the subcutaneous fat in the legs of men to epinephrine than that in the legs of women; epinephrine induces free fatty acid release in the arm fatty tissue comparably in both sexes. Men also show greater epinephrine release during exercise than women, which resulted in more significant fat breakdown in the legs during exercise for men than in women in the majority of studies using microdialysis or isotope tracers [19–21]. Moreover, free fatty acid release from subcutaneous fat in the upper body was greater in males than in females [22], which could lead to decreased fat mass in the arms of subjects with greater PA.

Meta-analysis of data from a large number of studies revealed that male sex is more receptive to exercise and diet interventions resulting in a larger reduction of total and visceral fat mass [23, 24].

Different types of physical loads have distinct effects on the skeletal muscles. Physical exercises with increased resistance stimulate protein synthesis in the muscle fibres, leading to muscle hypertrophy [25]. High and moderate resistance result in the greatest hypertrophy, while light resistance has a small effect [26]. Exercise duration is also important: If high-intensity exercise is performed for a short time or fewer repetitions, the hypertrophic effect is smaller than that of moderate intensity with longer duration or more repetitions.

Aerobic exercise increase can cause muscle hypertrophy comparable to that of resistance training [27, 28]. Activities that lead to hypertrophy should be intense (70–80% of the heart rate reserve) maintained for 45 minutes a day for at least 4 days a week. This may explain the greater muscle mass in the legs of men who spend more MET minutes in vigorous PA. In women, the lack of an association between vigorous MET minutes and muscle mass can be explained by reduced quantities of such activities in a typical week. Instead, they reported engaging in great amounts of moderate PA, which could lead to muscle mass increases, and we observed a correlation between moderate PA and higher lean and muscle mass in women.

5 Conclusions

The findings of this study indicate that, independent of OPA, men tend to engage in vigorous PA more than women. Adults who are involved in professional duties with insignificant PA have longer sitting time on their working days, less walking and moderate PA, and lower PA overall compared to those with moderate to vigorous OPA. A sex-dependent association was detected between amount of particular PA intensity and muscle or fat tissue distribution. In men, a greater duration of moderate PA was associated with lower amounts of total, visceral, and segmental body fat, whereas women who engaged more in moderate PA had higher lean body mass and segmental muscle masses of the trunk and legs. Greater time spent in vigorous PA was associated with higher leg muscle mass in men and lower visceral fat in women. These observed associations might be used to develop interventions to promote sex- and occupation-specific PA.

References

[1] S.M. Grundy, Trends Cardiovas. Med. **26**(4), 364–373 (2016)
[2] C.A. Slentz, B.D. Duscha, J.L. Johnson, K. Ketchum, L.B. Aiken, G.P. Samsa et al., Arch. Intern. Med. **164**(1), 31–39 (2004)
[3] K. Borodulin, T. Laatikainen, A. Juolevi, P. Jousilahti, Eur. J. Public Health **18**(3), 339–344 (2008)
[4] C.L. Craig, S.J. Russell, C. Cameron, A. Bauman, Can. J. Public Health; January/February 2004, **95**(1) (2004)
[5] S.A. Prince, J.L. Reed, C. McFetridge, M.S. Tremblay, R.D. Reid, Obes. Rev. **18**(8), 915–935 (2017)
[6] IPAQ Group, `http://www.ipaq.ki.se` (2015)
[7] V. Gerovasili, I.T. Agaku, C.I. Vardavas, F.T. Filippidis, Prev. Med. **81**, 87–91 (2015)
[8] European Commission `http://ec.europa.eu/commfrontoffice/publicopinion/archives/ebs/ebs_412_en.pdf` (2014)
[9] J.M. Oppert, M.A. Charles, H. Charreire, M. Menai, I. De Bourdeaudhuij, S. Brage, B. de Lauzon-Guillain, G. Fagherazzi, B. Balkau, Int. J. Environ. Res. Public Health **13**(8) (2016)
[10] T. Schwaneberg, F. Weymar, S. Ulbricht, M. Dörr, W. Hoffmann, N. van den Berg, Prev. Med. Rep. **7**, 162–168 (2017)
[11] K.R. Evenson, F. Wen, J.S. Metzger, A.H. Herring, Int. J. Behav. Nutr. Phys. Act. **12**, 20–20 (2015)
[12] M. Saidj, M. Menai, H. Charreire, C. Weber, C. Enaux, M. Aadahl, E. Kesse-Guyot, S. Hercberg, C. Simon J.M. Oppert, BMC Public Health **15**, 379 (2015)

[13] M.P. Jans, K.I. Proper, V.H. Hildebrandt, Am. J. Prev. Med. **33**(6), 450–454 (2007)
[14] M.C. Venables, J. Achten, A.E. Jeukendrup, J. Appl. Physiol. (1985) **98**(1), 160–167 (2005)
[15] S.G. Dasilva, L. Guidetti, C.F. Buzzachera, H.M. Elsangedy, K. Krinski, W. De Campos, F.L. Goss, C. Baldari, J. Strength Cond. Res. **25**(9), 2544–2551 (2011)
[16] J. Achten, M. Gleeson, A.E. Jeukendrup, Med. Sci. Sports Exerc. **34**(1), 92–97 (2002)
[17] L.L. Spriet, Sports Med. **44**(Suppl 1), 87–96 (2014)
[18] M.J. Watt, G.J.F. Heigenhauser, D.J. Dyck, L.L. Spriet, J. Physiol. **541** (Pt 3), 969–978 (2002)
[19] P. Arner, E. Kriegholm, P. Engfeldt, J. Bolinder, J. Clin. Invest. **85**(3), 893–898 (1990)
[20] S.L. Carter, C. Rennie, M.A. Tarnopolsky, Am. J. Physiol. Endocrinol. Metab. **280**(6), E898–E907 (2001)
[21] L. Hellstrom, E. Blaak, E. Hagstrom-Toft, Int. J. Sports. Med. **17**(6), 439–447 (1996)
[22] E. Blaak, Curr. Opin. Clin. Nutr. Metab. Care **4**(6), 499–502 (2001)
[23] R.J. Verheggen, M.F. Maessen, D.J. Green, A.R. Hermus, M.T. Hopman, D.H. Thijssen, Obes. Rev. **17**(8), 664–690 (2016)
[24] D. Vissers, W. Hens, J. Taeymans, J.P. Baeyens, J. Poortmans, L. Van Gaal, PLoS One **8**(2), e56415 (2013)
[25] H. Ozaki, T. Abe, A.E. Mikesky, A. Sakamoto, S. Machida, H. Naito, J. Phys. Fitness Sports Med. **4**(1), 43–51 (2015)
[26] G.E. Campos, T.J. Luecke, H.K. Wendeln, K. Toma, F.C. Hagerman, T.F. Murray, K.E. Ragg, N.A. Ratamess, W.J. Kraemer, R.S. Staron, Eur. J. Appl. Physiol. **88**(1), 50–60 (2002)
[27] J.D. Crane, L.G. Macneil, M.A. Tarnopolsky, J. Gerontol. A. Biol. Sci. Med. Sci. **68**(6), 631–638 (2013)
[28] A.R. Konopka, M.P. Harber, Exerc. Sport Sci. Rev. **42**(2), 53–61 (2014)

Fatness and thermoregulation of qualified rugby players

Inese Pontaga[1,*], *Jekaterina* Liepina[1], *Dzintra* Kazoka[2], and *Silvija* Umbrasko[2]

[1]Latvian Academy of Sports Education, Riga, Latvia
[2]Rīga Stradiņš University, Department of Morphology, Riga, Latvia

Abstract. A large body size and mass are advantages in rugby. The desire to gain weight can bring players to become overweight or obese. This can worsen their thermoregulation and health risks. The aim was to evaluate anthropometric characteristics and to determine the effect of additional body fat percentage on sweat loss during play-match in male rugby players. Nineteen qualified male rugby players were tested during play-match. The age, height, body mass, body mass index (BMI) and body fat percent of participants were: 29 ± 6 years, 183 ± 7 cm, 96.86 ± 12.88 kg, 29.07 ± 3.90 kg/m^2, 20.52 ± 5.64%, respectively. The skin fold thickness measurement was used to assess body fat percent. Body mass loss was detected by weighting. The mean BMI was 26.18 ± 2.37 (kg/m^2) and the body fat 15.87 ± 3.97% in backs. Forwards were significantly heavier and had BMI 31.18 ± 3.44 (kg/m^2) ($p = 0.002$) and the body fat 23.91 ± 4.02% ($p < 0.001$). The mean body mass decrease in the play-match was 1.83 ± 0.84%. The mean sweating intensity was 2.24 ± 1.07 l/h, but individual varied among players in very wide range (1.12–6.16 l/h). Relationships between the body fat percentage and sweating intensity was not determined ($p > 0.05$). Recommendation is to increase the volume of regular strength training, to correct the diet and liquid consumption.

Key words: overweight, dehydration, performance, health risks prevention, sport.

1 Introduction

Physical activity promotion and participation in organised sports is an important recommendation in the prevention and treatment of obesity, metabolic and cardiovascular risks reduction, health and fitness improvement [1]. Obesity is a complex, multi-factorial, and largely preventable disease, affecting, along with overweight, over a third of the world's population today [2, 3]. Investigation conducted by Kelly et al. [4] is in a good agreement with these data, they evaluated that 23.2% of the world's adult population in 2005 was overweight and 9.8% was obese. If the trends of body fatness increase continue, by 2030 an estimated 38% of the world's adult population will be overweight and 20% will be obese [4]. Therefore, increase of daily physical activity duration and participation in sports training must improve the body mass composition and health in population. However, sports participation not always improve the body mass composition in athletes. A large body size and mass is an advantage in some contact sports games (rugby, American football, team

* Corresponding author: inese.pontaga@lspa.lv

handball), especially in some positions' players. For example, Hill et al. [5] determined that international level rugby players' body mass has significantly increased since 1995: in 1955 the mean rugby players' body mass was 84.8 ± 8.2 kg but in 2015, it was 105.4 ± 12.1 kg, an increase of 24.3%. Between 1955 and 2015, the body mass of forwards increased steadily, whereas that of backs has mostly gone up since 1995. Large body size and mass can be achieved by increase of skeletal muscle mass due to resistance training, which is beneficially. Unfortunately, sometimes fatness of athletes occurs because high body mass is achieved by increase in body fat content. So, Malina et al. [6] observed a high prevalence of overweight and obesity (42.6%) among young American football players.

The Body Mass Index (BMI) is a commonly used measure of fatness by which individuals may be classified as lean (norm), overweight or obese. The BMI is calculated using the following Eq. (1):

$$BMI(kg/m^2) = body\,mass\ (kg)/height^2\ (m^2) \tag{1}$$

Individuals with a BMI of 25.0–29.9 kg/m^2 are classified as overweight, whilst those with BMI >30.0 kg/m^2 are classified as obese [7, 8].

Heo et al. [9] developed cut-offs of percentage of body fat on the basis of the relation between dual-energy X-ray absorptiometry-measured fat mass and BMI (in kg/m^2) stratified by sex, age, and race-ethnicity by using 1999–2004 NHANES data. They determined that in males, the body fat percentage corresponded to a BMI of 18.5, 25, 30, 35, and 40 across age and racial-ethnic groups ranged from 12.2% to 19.0%, 22.6% to 28.0%, 27.5% to 32.3%, 31.0% to 35.3%, and 33.6% to 37.6%, respectively.

High value of the body mass index (BMI) can be estimated as overweight in athletes with great skeletal muscles mass. This means that the reason of high BMI can be hypertrophy of skeletal muscles in athletes training in many sports specializations [10]. For example, BMI (25.1 ± 2.9 kg/m^2) is high in young Latvian handball players from Murjani Sports Gymnasium, but this is due to high lean body mass because the body fat percentage (13.6 ± 2.8%) is in norm in young male handball players [11]. The mean BMI (23.4 ± 1.4 kg/m^2) is lower and its value coincides with norm is Latvian qualified football players but the mean body fat percentage is greater (18.4 ± 3.7%) in comparison with the handball players [11]. Therefore, fitness rather than fatness has been proposed as the main marker of health. One rugby player can be overweight according the BMI without being fat and other athlete can also be overweight and fit [12].

The desire to gain weight can bring rugby players to become overweight or obese. This could worsen their thermoregulation and increase health risks in future. Therefore, the aim of the present study was to evaluate anthropometric characteristics and to determine the effect of additional body fat percent to sweat loss during play-match in qualified male rugby players.

2 Material and methods

2.1 Participants

Nineteen male rugby 15 players from the Latvian National team were tested during a play-match. They voluntary participated in our study (every athlete signed an informed consent form to take part in the investigation). The study was performed in accordance with the standards of the Ethics Committee of Latvian Academy of Sports Education and in accordance with the Declaration of Helsinki. The mean age of the participants was 29 ± 6 years. Their training experience in rugby 15 ranged from seven to fifteen years. These athletes trained four times per week (one and half to two hours per day) and played regularly on the weekends (the mean: 8.2 ± 3.4 training or playing hours per week).

2.2 Methods

The players' height was measured using an Ultrasound Height Measuring Unit MZ10020 (ADE, Hamburg). Body mass was measured in athletes wearing briefs using of special industrial scales Midrics1 (Sartorius, Germany) with precision of 0.01 kilogram. The body mass index (BMI) was calculated by using Eq. (1) [7, 8].

The mean anthropometric characteristics of all athletes and each position players (forwards and backs) separately were analysed. The method of skin fold thickness measurement in seven sites by Jackson/ Pollock with a calliper SH5020 (SAEHAN Ltd., Korea) was used to assess percentage of body fat (% BF) by equation for males (2) [13]:

$$BF(\%) = 495/[1.112 - (0.00043499 \cdot s) + (0.00000055 \cdot s \cdot s) - (0.00028826 \cdot a)] - 450 \quad (2)$$

Where: s = sum of seven skin fold (mm); a = age (years).

Weighting of every athlete wearing only briefs was repeated immediately after play-match using the same industrial scales Midrics1 (Sartorius, Germany). Body mass loss of every player was detected by calculating of the body mass difference of every player before and after the play-match.

Measurements were performed at October 1st, 2016 during the play-match with the Lithuanian Premium league team "Ažuolas". The air temperature was +13 °C and humidity 69%. The game took place at the Riga Technical College Sports Stadium. Duration of the play-match was 80 minutes (2 x 40 minutes). Time spent in the field by every player was measured. We asked the participants to consume the fluid as they usually do according to their subjective thirst sensation and not change their drinking habits before the play-match and during it. All water bottles of every player were weighed before and after training by using of industrial scales Midrics1 (Sartorius, Germany). The urine produced by athletes during the break and after the play-match was weighed with the same scales. Sweating intensity of every rugby players was calculated from the data of the body mass changes, consumed water mass, produced urine mass and the duration of playing on the field in this play-match, taking into account that 1kg of water = 1 litre of water (3) [14]:

$$\begin{aligned} &Sweat\ rate\ (l/h) = \{[pre - practice\ mass(kg) - post\ practice\ mass\ (kg)] - \\ &produced\ urine\ mass\ (kg) + fluids\ consumed\ during\ practice\ (kg)\} \\ &/duration\ of\ playing\ on\ the\ field\ in\ this\ play - match\ (h) \qquad (3) \end{aligned}$$

2.3 Statistical analysis

The data distribution was evaluated from the values of the skewness and kurtosis and skewness and kurtosis *t*- test analysis. The data of the athletes' age, height, body mass, body fat percent, body mass loss after play-match were normally distributed. Therefore, using parametric statistics was appropriate. Mean values and standard deviations for all characteristics were calculated. Student's independent t-test for unequal data groups was employed to determine differences between the mean characteristics of the different position players (forwards and backs). The differences were considered significant at $p < 0.05$. A correlation between body fat percent and sweating intensity during the play-match was evaluated. Microsoft Excel 2010 was used to perform all statistical procedures.

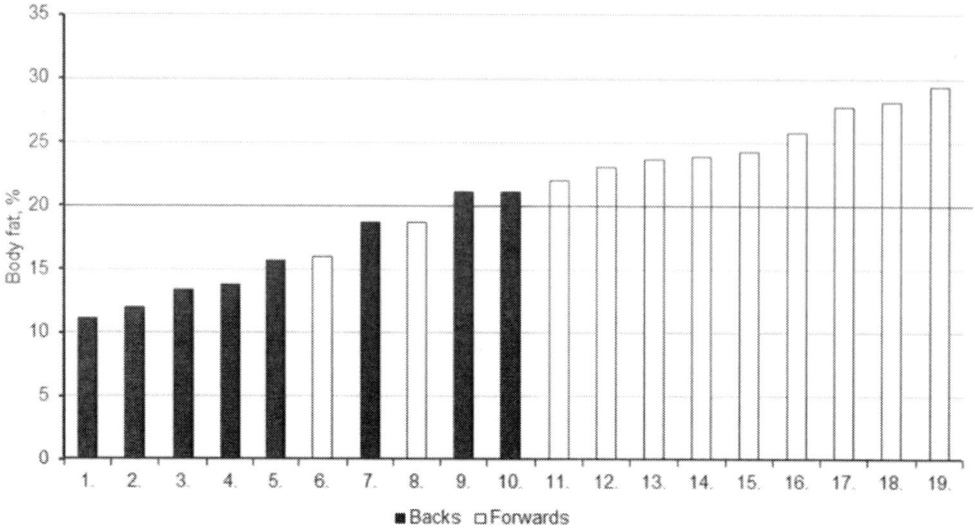

Fig. 1. Body fat percentage in every male rugby player in dependence on his position. Horizontal line at 20% indicates cut-off value for overweight of the athletes [9].

Table 1. The mean anthropometric characteristics of the rugby athletes in dependence on players' position.

Rugby players' position	**Body height, cm (± SD)**	**Body mass, kg (± SD)**	**Body mass index, kg/m² (± SD)**	**Body fat, % (± SD)**
Backs	180 ± 6	84.75 ± 6.38	26.18 ± 2.37	15.87 ± 3.97
Forwards	184 ± 7	105.94* ± 8.62	31.18* ± 3.44	23.91* ± 4.02
Significance of difference	$p = 0.09$	$p < 0.001$	$p = 0.002$	$p < 0.001$

3 Results

The mean height, body mass, body mass index (BMI) and body fat percent of all participants were: 183 ± 7 (cm), 96.86 ± 12.88 (kg), 29.07 ± 3.90 (kg/m^2) and 20.52 ± 5.64 (%), respectively. The mean BMI gives evidence about overweight (between 25 and 30 kg/m^2) in the players. The mean body fat percentage also is greater than the norm – above 19%. The body fat above 20%– overweight is observed in 11 players (nine forwards and two backs) from 19 (Fig. 1), but the first step of obesity in three forwards with body fat 27.8%, 28.2% and 29.4%, respectively (the borders from 27.5 to 32.3%, according to [9].

Comparison of the mean anthropometric characteristics of the backs and forwards are shown in the Table 1.

The mean height does not differ significantly in different position players, but the body mass, BMI and body fat percent is statistically significantly larger in forwards than in backs (Table 1). The mean BMI slightly exceed the highest value of norm and can be qualified like overweight in backs but their body fat content is in the norm (< 20%). The mean BMI of forwards (between 30.0 and 34.9 kg/m^2) coincides with the first step of obesity, but the body fat percentage: 23.91 ± 4.02% proves only overweight of them (22.6 to 28.0%).

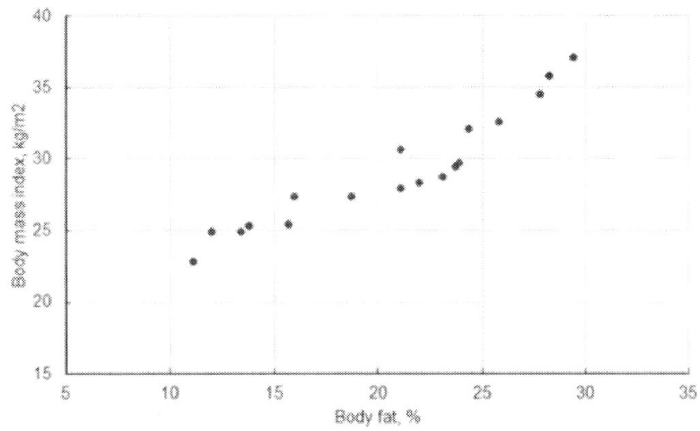

Fig. 2. Relationship between the body fat percentage and BMI in male rugby players ($r = 0.94$, $p = 0.001$).

The close linear positive relationship is determined between the body fat percentage and BMI, equation (4) (correlation coefficient $r = 0.94$; probability level $p = 0.001$):

$$BMI(kg/m^2) = 15.67 + 0.65 \cdot BF(\%) \qquad (4)$$

This confirms that the reason of high BMI is increased body fat content in our rugby players (Fig. 2).

The time spent by every player in the play-match is 58.74 $\pm$23.69 minutes (0.98 $\pm$ 0.39 hours). The mean water intake during the game is not high: 0.460 ± 0.334 kg. The mean body mass decrease of the player due to sweating in the play-match is 1.73 ± 0.71 kg or $1.83 \pm 0.84\%$ from the body mass at rest. This is not high and might not affect the performance [15]. Nevertheless, body mass loss exceeds 2% in six athletes: three backs and three forwards. The body mass decrease is very large (4.36%) in one back player. Therefore, the performance of these athletes should be worsened (Fig. 3).

The mean sweating intensity of all rugby players is 2.24 ± 1.07 litres (kg)/ h. Comparison of the mean body mass changes, sweating intensity, time spent on the field during the play-match and consumed water intake of the backs and forwards are shown in the Table 2. The playing time spent on the field and consumed amount of water do not differ significantly in backs and forwards (Table 2).

The mean body mass lost by backs is none significantly greater in comparison with the body mass decrease of forwards due to large variation of the weight loss and playing time on the field among different players. The mean sweating intensity does not differ significantly in backs and forwards but individual sweating intensity varies among players in a very wide range: from 1.12 to 3.28 l/h in the backs and from 1.45 to 6.16 l/h in forwards.

Significant relationships between the body fat percentage and sweating intensity are not determined in all rugby players ($p = 0.54$), as well as in the back players ($p = 0.21$) and forwards ($p = 0.70$).

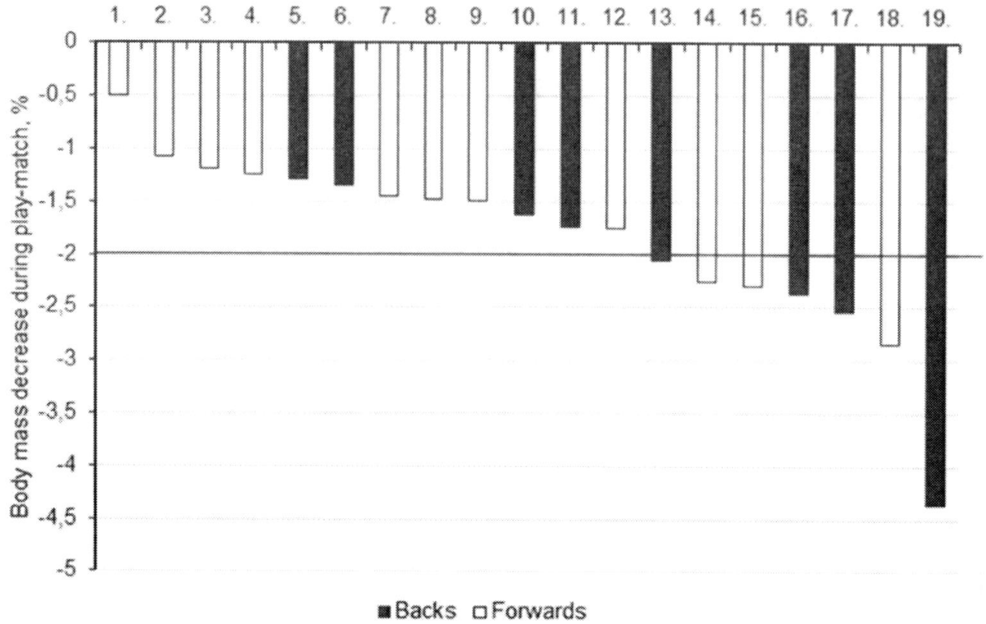

Fig. 3. Body mass decrease according to player's position during play-match. Horizontal line at –2% indicates cut-off value for decrease in physical performance [15].

Table 2. The mean duration of playing on the field, consumed water, lost body mass and sweating intensity of the male rugby athletes in dependence on players' position.

Rugby players' position	**Time spent on the field, hours**	**Consumed water, kg**	**Body mass decrease, kg**	**Body mass decrease, %**	**Sweating intensity, litres (or kg)/h**
Backs	1.14 ± 0.33	0.53 ± 0.42	1.82 ± 0.79	2.16 ± 1.00	2.15 ± 0.75
Forwards	0.86 ± 0.41	0.41 ± 0.26	1.66 ± 0.68	1.59 ± 0.66	2.76 ± 1.27
Significance of difference	$p = 0.06$	$p = 0.213$	$p = 0.327$	$p = 0.09$	$p = 0.10$

4 Discussion

The mean BMI of all players was 29.07 ± 3.90 (kg/m^2), which gives evidence about overweight (between 25 and 30 kg/m^2, [7, 8]) in the players, the mean body fat content $-20.52 \pm 5.64\%$ also is greater than the norm – above 19% (highest border in males) [9]. The reason of high BMI is increased body fat content in our rugby players which is confirmed by the close linear positive correlation between the BMI and body fat percent ($r = 0.94$; $p = 0.001$). Our data of the BMI in Latvian National male rugby team agree with the mean BMI 29.0 ± 2.3 (kg/m^2) of an Australian National Rugby League team male players at 2009 preseason [16]. Our results are in a good agreement with the data of King et al. [12] who evaluated BMI of international professional rugby players from the four teams involved in the semi-finals of the 2003 Rugby Union World Cup. According to the World Health Organisation [7, 8] values for BMI, 65% of the players were classified as overweight and 25% as obese.

Gavarry et al. [17] determined high prevalence of fatness in adolescent rugby players aged from nine to 14 years: 32.8% of young rugby players were considered as overweight and 12.8% as obese. They concluded that BMI can overestimate an excess of adiposity for an adolescent player having a high fat-free mass classified as obese or overweight: only 53% of young players classified as obese and overweight by BMI had an excess body fat.

All previous data demonstrate that high BMI is commonplace and a desirable attribute for sport performance in professional rugby players.

The mean BMI was 26.18 ± 2.37 (kg/m^2) and the body fat content $15.87 \pm 3.97\%$ in our back position players. This also gives evidence about overweight (BMI is between 25 and 30 kg/m^2, [7, 8], but body fat percent is in norm – below 19% [9]. The similar results in the back position rugby players from the A-League (the highest level of the local league in Japan, practicing six days per week, three hours per day) are detected by Nirengi et al. [18]: BMI 25.3 (kg/m^2) (23.4–26.9 kg/m^2) and the body fat content 15.2% (10.6–19.7%).

Our forwards were significantly heavier than backs and had BMI 31.18 ± 3.44 (kg/m^2) which coincides with obesity (BMI > 30 kg/m^2, [7, 8]. The body fat content $23.91 \pm 4.02\%$ in forwards coincides only with overweight value (22.6 to 28.0%) and does not shows obesity according to classification of Heo et al., [9]. This can be explained by large skeletal muscle mass in our forwards. Anthropometric characteristics observed in Japanese forwards: BMI 29.1 (kg/m^2) (27.3–32.8 kg/m^2) and the body fat content 21.6% (16.1–28.5%) observed by Nirengi et al. [18] also are close to our data.

Bell [19] estimated body mass composition in amateur male rugby players by skin fold thickness measurement method and determined that the mean body fat content was 19.5% in forwards (the highest border of the norm 19%, according to classification of Heo et al. [9], and 12.2% in backs (in norm: 12.2 to 19.0%, according to classification of Heo et al. [9]. Body fat percentage in both: backs and forwards are lower in comparison with our data, but significantly higher in forwards than in backs in both investigations. Morehen et al. [20] has assessed the body composition of elite European Super League rugby players from more than one club using DXA scan technology. They also reported that there are significant differences between players' positions, especially separating the forwards from the other positional groups, who have larger fat percentage in the body $16.3 \pm 4.3\%$ in comparison with players of other positions with the body fat content 12–14%.

The main characteristics necessary to backs for success are high power, speed and agility. Therefore additional body fat would worsen their performance. Our backs are leaner than forwards. The main task of forwards is the contact fight to get a ball, and therefore they need high strength and strength endurance to achieve success. High body mass and additional fat mass are important to increase of inertia forces in the contact fight and represent the unique positional demands of forwards, who are required to withstand high-speed physical collision on a more regular basis than the other position players [21]. Therefore our forwards have greater BMI and the fat content than backs. The performance worsening due to higher body fat percent in the forwards might be compensated by the fact that they have less game time than the other groups through tactical substitutions [22]. Increased body fat percentage may be a reason of chronic diseases. For example, Nirengi et al. [18] determined that the rugby players, especially those in the forwards position, are at higher risk of developing of non-alcoholic fatty liver disease. This can be explained by consuming of a diet to increase body mass.

The time spent by every player in the play-match is 58.74 ± 23.69 minutes (0.98 ± 0.39 hours), heavier forwards played shorter time on the field but the playing duration highly varied among the players. Therefore, the playing time does not differ significantly between the backs and forwards. The mean water intake during the game is small: <0.5 litres. Insufficient fluid intake for athletes training in cold and moderate temperatures can be explained by reduced thirst sensation [23]. The data of our present investigation are in good agreement with our previous results obtained in rugby seven players: fluid consumption according to the thirst sensation in cool conditions (the air temperature in that day of March: +4 °C, humidity 65%) was insufficient to prevent the body hydration status worsening and a decrease of the body mass for more than 1.5–2% in some players but the fluid consumption in warm environment (the air temperature in that day of June: +19 °C, humidity 35%) better allows preventing the body dehydration and keeping the body mass decrease below 1.5% [24]. The mean body mass decrease of the player due to sweating in the play-match is 1.73 ± 0.71 kg or $1.83 \pm 0.84\%$ from the body mass at rest, this is not high and might not affect the performance because is lower than 2% [15].

The mean sweating intensity of all rugby players is 2.24 ± 1.07 l/h. Individual sweating intensity varied among players in a very wide range: from 1.12 to 3.28 l/h in the backs and from 1.45 to 6.16 l/h in forwards. Therefore, significant relationships between the body fat percentage and sweating intensity is not determined ($p < 0.05$). Differences of sweating rate of an individual are determined by activated sweat gland density and sweat output of the activated glands, which are observed between tropical and temperate natives. These differences should be explained by acclimatisation and genetic reasons [25]. This proves that additional attention must be paid to rehydration strategy of some our rugby players with large sweating intensity (more than 3l/h) and body mass loss for $3-4\%$ from the rest value [15] to maintain their performance to the end of play-matches.

Jones et al. [26] compared body mass composition in professional and semi-professional rugby players using dual-energy X-ray absorptiometry (DXA) and did not observed significant differences in age, stature or body mass. They detected relatively less soft tissue fat in professional players in comparison with semi-professional ($17.5 \pm 3.7\%$ and $14.8 \pm 3.6\%$ in backs; $21.4 \pm 4.3\%$ and $20.8 \pm 3.8\%$ in forwards, respectively) and greater trunk lean mass. Jones et al. [26] concluded that to reduce this discrepancy, some semi-professional players should reduce fat mass and increase lean mass, which may be of benefit for the next season and beyond.

Significant anthropometric changes were observed in players across a professional rugby league playing season, including an overall loss of lean body mass and skeletal muscle mass which regain with preseason training [16]. These findings are in agreement with the observations of Duthie et al. [27], who determined substantial individual variation in lean body mass in qualified male rugby players within and between seasons. They confirmed the lean body mass and skin folds thickness measurement is a potentially useful tool for assessing body composition of athletes. Crewther et al. [28] proved that different equal-volume resistance training protocols (full-body or split-body) four – week application improves the strength and slightly also body mass composition by decrease of body fat percent in male rugby union players.

Limitation of the study is investigation of small number of rugby players during only one play-match. Further research is necessary to determine possible relationships among body fat percentage, sweating intensity and body mass decrease during training and play-matches in dependence on rugby player's position and environmental conditions.

5 Conclusion

The mean BMI 29.07 ± 3.90 (kg/m^2) and body fat of $20.52 \pm 5.64\%$ of all the National team players prove overweight. The mean BMI 31.18 ± 3.44 kg/m^2 and body fat $23.91 \pm 4.02\%$ of our forwards are significantly higher than these characteristics in the backs: BMI 26.18 ± 2.37 (kg/m^2) and body fat 15.87 ± 3.97%, $p \leq 0.002$. The first step of obesity is observed in three forwards with body fat 27.8%, 28.2% and 29.4%. Significant relationships between the body fat percent and sweating intensity is not determined ($p > 0.05$) due to individual sweating intensity variation in wide range among the players.

Therefore, large body mass in rugby players would be better achieved by additional regular strength training to gain skeletal muscle mass and decrease the body fat percent.

The athletes should be educated about nutrition and liquid consumption during training to prevent dehydration and performance decrease.

The authors would like to sincerely thank all the rugby players who participated in the study, and their coach who encouraged the athletes to do that.

References

[1] A.P. Hills, L.B. Andersen, N.M. Byrne, Br. J. Sports Med. **45**, 866–87 (2011)
[2] M.N. Fleming, M. Robinson, B. Thomson, et al., Lancet. **384**, 766–781 (2014)
[3] G.A. Stevens, G.M. Singh, Y. Lu, G. Danaei G, J.K. Lin, M.M. Finucane, A.N. Bahalim, R.K. McIntire, H.R. Gutierrez, M. Cowan, Ch.J. Paciorek, F. Farzadfar, L. Riley, M. Ezzati, Popul. Health Metr. **10**, 22 (2012)
[4] T. Kelly, W. Yang, C.S. Chen, K. Reynolds, J. He, Int. J. Obes. **32**, 1431–7 (2008)
[5] N.E. Hill, S. Rilstone, M.J. Stacey, D. Amiras, S. Chew, D. Flatman, N.S. Oliver, BMJ Open Sport Exerc. Med. **4**, e000459 (2018)
[6] R.M. Malina, P.J. Morano, M. Barron, S.J. Miller, S.P. Cumming, A.P. Kontos, B.B. Little, J. Pediatr. **151**, 378–82 (2007)
[7] *World Health Organization. Obesity: preventing and managing the global epidemic: report of a WHO consultation* (World Health Organization, Geneva, 2000), pp. 1–253
[8] D. Gallagher, S.B. Heymsfield, M. Heo, S.A. Jebb, P.R. Murgatroyd, Y. Sakamoto, Am. J. Clin. Nutr. **72**, 694–701 (2000)
[9] M. Heo, M.S. Faith, A. Pietrobelli, S.B. Heymsfield, Am. J. Clin. Nutr. **95**, 594–602 (2012)
[10] J.J. Ode, J.M. Pivarnik, M.J. Reeves, J.L. Knous, Med. Sci. Sports & Exerc. **39**, 403–409 (2007)
[11] I. Pontaga, J. Zidens, LASE J. Sport Sci. **2**, 33 (2011)
[12] N.A. King, A.P. Hills, J.E. Blundell, Eur. J. Sport Sci. **5**, 73–75 (2005)
[13] A.S. Jackson, M.L. Pollock, Br. J. Nutrit. **40**, 497–504 (1978)
[14] F.S. Godek, A.R. Bartolozzi, J.J. Godek, Br. J. Sports Med. **39**, 205–211 (2005)
[15] S. Shirreffs, Scand. J. Med. Sci. Sports. **20**, 90–94 (2010)
[16] E.C. Georgeso, B.K. Weeks, C. McLellan, B.R. Beck, BMJ Open **2**, e001400 (2012)
[17] O. Gavarry, G. Lentin, P. Pezery, A. Delextrat, G. Chaumet, A. Boussuges, J. Piscione, Sports Medicine – Open **4**, 17 (2018)
[18] S. Nirengi, M. Fujibayashi, S. Furuno, A. Uchibe, Y. Kawase, S. Sukino, Y. Kawaguchi, S. Minato, K. Kotani, N. Sakane, Frontiers in Endocrinology. **9**, 341 (2018)
[19] W. Bell, Br. J. Sports Med. **13**, 19–23 (1979)
[20] J.C. Morehen, H.E. Routledge, C. Twist, J.P. Morton, G.L. Close, Eur. J. Sport Sci. **15**, 523–529 (2015)
[21] C. Twist, M. Waldron, J. Highton, D. Burt, M. Daniels, J. Sports Sci. **30**, 359–367 (2012)

[22] C. Twist, J. Highton M. Waldron, E. Edwards, D. Austin, T.J. Gabbett, Int. J. Sports Physiol. Perform. **9**, 925–930 (2014)

[23] R.W. Kenefick, M.P. Hazzard, N.V. Mahood, Med. Sci. Sports Exerc. **10**, 1528–1534 (2004)

[24] I. Pontaga, L. Voitkeviča, J. Liepina, *Proceedings of the 1st World Congress on Health and Martial Arts in Interdisciplinary Approach* (Archives of Budo, Warsaw, 2015), pp. 155–164

[25] J. Lee, Y. Shin, Auton. Neurosci. **205**, 67–71 (2017)

[26] B. Jones, K. Till, M. Barlow, M. Lees, J.P. O'Hara, K. Hind, PLOS ONE **10**, e0133188 (2015)

[27] G.M. Duthie, D.B. Pyne, W.G. Hopkins, S. Livingstone, S.L. Hooper, Br. J. Sports Med. **40**, 202–207 (2006)

[28] B.T. Crewther, T.O.L. Heke, J.W.L. Keogh, Biol. Sport. **33**, 111–116 (2016)

Readiness of military personnel for high intensity combat training course

Liana Plavina, *Silva* Smagare, *Andris* Cakstins, and *Silvija* Umbrasko

Rīga Stradiņš University, Department of Morphology, Institute of Anatomy and Anthropology, Riga, Latvia

Abstract. High standards of physical ability and health are essential for Military personnel and its career. Training of military personnel in field conditions is an essential process. The aim of the study was to evaluate readiness of cadets to high physical load during the combat training course (CTC). We carried out the following: evaluated posture status, analysed anthropometric characteristics, controlled parameters of musculoskeletal system such as muscle tone, muscle strength characteristics, and provided Nordic Questionnaire list containing a figure of Human Body, where a respondent marked the regions of musculoskeletal disorder (pain). All subjects gave their informed consent to the protocol approved by the local Medical Ethics Committee of Rīga Stradiņš University for biomedical research. Combat training course has duration of ten days and includes various physical, tactical, and psychological activities in military conditions that are a compulsory part of study process in National Defence Academy of Latvia. Cadets are facing complex challenges in tactical situations where they can approve their physical abilities, endurance and psychological persistence. They have dietary and sleeping deprivation during the course. We provided standardized Nordic Questionnaire for analysis and evaluation of musculoskeletal disorders in the study group and indicated main problematic regions with musculoskeletal symptoms. The general health status of participants was checked by medical specialists in military Medical Centre and advanced examination was done by Sports medicine doctor. The aim of the study is to analyse the posture status, foot status, to make muscle functional tests that allow managing pre-courses training program for CTC participants and optimizing adaptation of participants of the course to field condition of military training. Evaluating musculoskeletal symptoms cadets indicated that main problematical regions were lower back and knee. In 77% (42/54) of cases cadets with lower limb hyper-tonus had asymmetric posture in frontal plane. Analysis of anthropometric characteristics showed that body mass index changes were in the interval from 21.34 to 33.24. BMI value for 32.2% of participants corresponded to the standards of WHO recommendations. Analysis of posture and foot status was important for provision of preventive measures and reduction of the risk of health disorders related to high physical load in military environment. Persons with expressed asymmetry of posture and foot during dynamic load and static load can have overstress with high risk of health disorders, musculoskeletal pathology and psychological overload.

Key words: anthropometric characteristics of military personnel, posture disorders, muscular functional tests data.

1 Introduction

High standards of physical fitness and health are essential for Military personnel and its career. Training of military personnel in field conditions is an essential process. We have carried out evaluation of health status of participants of combat training course (CTC) that is a compulsory part of study process in National Defence Academy of Latvia. We made evaluation of posture analysis, foot status analysis, muscle function testing, which can help to manage pre-courses training program for CTC participants and optimize adaptation of participants of CTC to field conditions of military training. Military operational stress in combination with physical exertion, sleep restriction and dietary insufficiency during the CTC demanded participants' adaptation and adjustment to a changing military environment. Combat training course has duration of ten days and includes various physical, tactical, psychological activities in military conditions. Military instructors ensure control, evaluation and training the readiness and resiliency of participants of CTC. Cadets face challenges and train their physical fitness and endurance, as well as psychological approach to the tactical situation. They have dietary and sleeping deprivation during the CTC. Military exercise outcomes depend on physical fitness, aerobic fitness, muscular strength, health capacity, knowledge and practice [1]. Physical training, physical preparedness and endurance have influence on individual ability to withstand stressors [2–4]. Environmental factors have impact on stress tolerance and injury risk regarding the gender [5]. Risk of musculoskeletal injuries also depends on physical fitness [6, 7]. Sleep restriction and limitation of adequate rest have negative impacts on cadet's reaction time and decision making, which may result in making mistakes [8, 9]. Cold stress or extreme heat in combination with physical exercises can have effect on cardiovascular and endocrine function and decrease tactical performance [10, 11]. Musculoskeletal disorders and injuries are the primary health problems for military personnel during military training [12, 13]. Programs for prevention of musculoskeletal injuries are organized in armed forces of other countries [14–16]. Anthropometric parameters, such as increased body mass, are one of the risk factors for musculoskeletal disorders [17–19]. High physical load has influence on various health parameters [20]. High intensity exercises have influence on fat mass decreasing it. High physical load induces insulin sensitivity [21, 22]. Anthropometric parameters like body mass, body mass index (BMI) are associated with health problems such as cardiovascular pathology, diabetes, cancer [23]. Changes of muscle characteristics are related to muscle strength [24]. BMI can be used for personnel selection according to physical fitness abilities, but in some cases BMI does not relate with military performance [25, 26]. The aim of our study is to provide evaluation of posture analysis, foot status analysis, muscle function testing and assessment of anthropometric characteristics, which allows managing pre-courses training program for CTC participants to optimize adaptation to field conditions of military training.

2 Material and methods

Study group included 59 participants of Combat Training Course of both genders (47 males and 12 females) at the age of 22–30 years. We assessed readiness of cadets to high physical load during the CTC. According to the aim of the study, we carried out the following tasks: evaluation of posture status, analysis of anthropometric characteristics, checked such parameters of musculoskeletal system as muscle tone, muscle strength characteristics, and used Nordic Questionnaire list with a figure of Human Body, where a respondent marked the regions of musculoskeletal disorder (pain). All subjects gave their informed consent to the protocol approved by the local Medical Ethics Committee of the Riga Stradins University for biomedical research. The Combat training course lasts for ten days and includes various

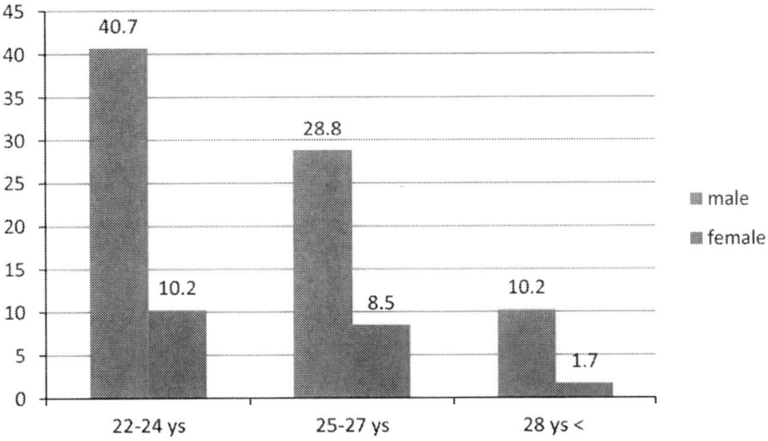

Fig. 1. Distribution of participants of Combat training course according to age and gender.

physical, tactical, and psychological activities in military conditions with equipment. Military training during the CTC was in combination with physical exertion, sleep restriction and dietary energy insufficiency. Cadets of CTC with full military equipment (22.5 kg) performed tactical exercises that included loaded road marches with obstacle course elements that were implemented during the CTC. Participants of CTC underwent medical examination programme by medical specialist in Medical Centre during the period of May–June, 2017. Sports medical doctor provided examination of posture status asymmetry (in frontal and sagittal planes), foot status analysis, muscle function testing (balance, coordination, elasticity tests) and assessment of anthropometric characteristics. The height was checked by anthropometry with an accuracy of 0.001 m. The body weight was checked with scales with an accuracy of 0.01 kg. The body mass indexes (BMI) were calculated as the quotient of body mass (kg) and the square of height (m^2). Evaluation of health status of CTC participants allows managing pre-courses training program for CTC participants. We used standardized Nordic Questionnaire [27] for analysis and evaluation of musculoskeletal disorders in study group. The whole statistical analysis was performed in IBM SPSS Statistics 23 software. The Pearson's Chi-Square test was applied to examine the differences between the groups.

3 Results

Study group included 59 participants of both genders (47 males and 12 females) aged 22 to 30 years. Distribution of participants according to the age and genders are presented on (Fig. 1).

We used standardized Nordic Questionnaire for analysis and evaluation of musculoskeletal disorders in study group. Cadets indicated main problematic regions with musculoskeletal symptoms. The most problematic region was the back: 42.3% of respondents in study group indicated musculoskeletal symptoms in the lower back, 15.4% of respondents in the upper back and 21.1% of respondents – in the neck region. The next problematic region was lower extremities: 48.0% of respondents noted musculoskeletal symptoms in the knee region, and 11.5% of respondents indicated musculoskeletal disorders in the hip region and 17.3% of respondents – disorders in the foot region. Musculoskeletal symptoms in the upper extremities also were shown in standardized Nordic questionnaire data analysis: 34.6% of respondents mentioned musculoskeletal symptoms in the shoulder region, 15.4% of respondents noted musculoskeletal symptoms in the elbow region and 11.5% of respondents –

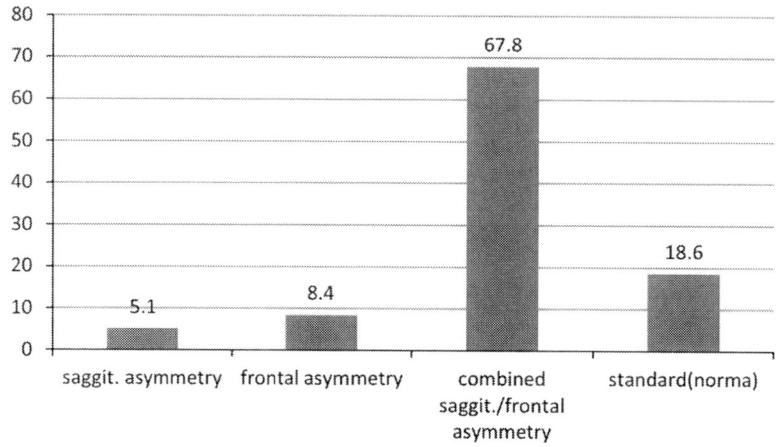

Fig. 2. Posture state assessment for participants of Combat training course.

in the wrist region. These musculoskeletal disorders were related with previous sports trauma, physical load and leisure time sports activities.

Posture analysis of participants was performed by Sports medicine doctor of Military Medical Centre in order to characterize functional status of musculoskeletal system, to reveal muscle over-stress and overload as well as posture disorders that can result in performance reduction during Combat training course.

Analysis revealed also posture problems: the combined asymmetric posture (in sagittal plane and frontal plane) was found in 67.8% (40/59) cadets. The asymmetric posture in one plane – sagittal plane was revealed in 5.1% (3/59) of cadets and the symmetric posture in frontal plane in 8.4% (5/59) of cadets of study group. 18.6% (11/59) of cadets of study group had a normal standard posture.

Musculoskeletal complaints during the physical load were indicated by 25.4% (15/59) of participants of study group during Sports doctor's examination. Assessment of the foot status revealed foot asymmetry in 28.8% (17/59) of participants and deviation from standard weight distribution was found in 3.4% (2/59) of participants.

Increased level of muscle tone – hyper-tonus was fixed in 91.5% (54/59) of participants during examination tests. In 77% (42/54) of cases, cadets with lower limb hyper-tone had asymmetric posture in frontal plane, but data between both groups are not statistically significant ($P > 0.05$). Examination of the coordination functional status of participants revealed that 32.2% (19/59) of respondents had dysfunctions and 25.4% (15/59) of participants had decreased level of elasticity. Evaluation of the balance status showed that 82% (37/45) of cadets with asymmetric posture in frontal plane had a balance disorder. Results of Sports doctor's examinations corresponded to standards (norm) and the rest part of participants of study group had dysfunction of balance characteristics.

Examination of the foot status showed the foot asymmetry in 28.8% (17/59) and deviation from standard weight distribution in 3.4% (2/59) of respondents.

Analysis of anthropometric characteristics showed that BMI changes were in the interval from 21.34 to 33.24. Anthropometric body mass and height parameters were used for calculation of the value of BMI. 30.5% (14 males and 4 females) of participants had BMI value that corresponded to the standards (20–25 for males and 19–24 for females). 38.9% (20 males and 3 females) of participants had BMI value that exceeded standards by 10% (25–28 for males and 24–26 for females). 20.3% (8 males and 3 females) of participants had BMI

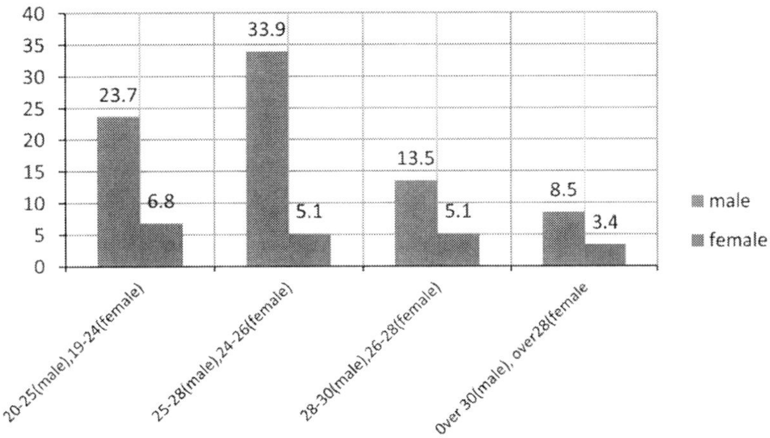

Fig. 3. Distribution of body mass index value for participants of CTC according to values and gender.

value over the standards by 20% (28–30 for males and 26–28 for females). For 11.9% (5 males and 2 females) of respondents BMI value was over the standards by 30% (over 30 for males and over 28 for females) (Fig. 3).

4 Discussion

Musculoskeletal disorders formed high morbidity rate among military personnel. According to data of Central Medical Expertise Commission, each third retired military person had musculoskeletal pathology that was acquired during military service. Military medical specialists should pay more attention to organizing the preventive measurement programme for diagnostics of early musculoskeletal dysfunctions. It allows reducing injury consequences and morbidity of military personnel. Standardized Nordic questionnaire for analysis of musculoskeletal symptoms revealed that in participants of the study group the most problematic regions were the back region (the lower back, the upper back and the neck region), then followed the lower extremity region (the knee, the hip and the foot region), and lastly the upper extremity (the shoulder, the elbow, the wrist region). Only 11.9% (7/59) of respondents in study group did not indicate any musculoskeletal symptoms. We have organized advanced medical examinations for participants of CTC, where participants faced high physical loads, and severe psychological stress conditions in military environment. During medical examination Sports medicine doctors found health problems and that will help to develop preventive procedures and special physical exercise program three months before starting Combat Training Course. 18.6% (11/59) of participants of study group had posture status corresponding to standard. There were many participants with posture disorders in one plane (frontal and sagittal) and combined planes. The combined asymmetric posture (in sagittal plane and frontal plane) was found for 67.8% (40/59) of participants, the asymmetric posture in one plane – sagittal plane – was found for 5.1% (3/59) and the symmetric posture in frontal plane in 8.4% (5/59) of participants. 25.4% (15/59) of respondents indicated musculoskeletal complaints during the physical load. Early diagnostics is important for providing special sports exercise program for training of deep muscle groups with the aim to reduce musculoskeletal symptoms during high physical load for large number of participants with posture asymmetry. Assessment of functional tests revealed an increased level of lower limb muscle tone in 91.5% (54/59) cases. In 42 out of 52 cases cadets with lower limb hyper-tone had an asymmetric posture in frontal plane. Deviation of posture status from standards was the main risk factor in functional tests. Examination of the coordination status of participants of study group revealed that 32.2% (19/59) of respondents had dysfunctions. 25.4% (15/59) of participants had decreased level of elasticity. Evaluation of the balance status of participants revealed that 18.6% (11/54) of respondents had balance disorders. We have found balance dysfunction for individuals with frontal posture asymmetry. Examination of the foot status showed the foot asymmetry for 28.8% (17/59) and deviation from standard weight distribution for 3.4% (2/59) of respondents. Early diagnosis of foot disorders requires specialist's consultations, which would allow increasing elasticity and stability of foot during military tactical exercises. Assessment of anthropometric parameters, calculation of BMI showed that 30.5% (14 males and 4 females) of participants of study group had BMI value that corresponded to the standards (20–25 for males and 19–24 for females). 38.9% (20 males and 3 females) of participants of study group had BMI value that exceeded standards by 10% (25–28 for males and 24–26 for females). 20.3% (8 males and 3 females) of participants had BMI value over the standards by 20% (28–30 for male and 26–28 for female). In 11.9% (5 males and 2 females) of respondents of study group BMI value was over standard by 30% (over 30 for males and over 28 for females). Increased values of BMI for military persons aged 22–30 did not have impact on military performance results, all participants of study group started Combat training course and all of them finished the course.

5 Conclusion

Analysis of posture and foot status was important for providing preventive measures and reduction of the risk of health disorders related to high physical load in military environment. Persons with expressed posture and foot asymmetry during dynamic load and static load can suffer from overstress with high risk of health disorders, musculoskeletal pathology and psychological overload.

References

[1] Hauschild V.D., DeGroot D.W., Hall S.M., Grier T.L., Deaver K.D., Hauret K.G., Jones B.H., Fitness tests and occupational tasks of military interest: a systematic review of correlations. Occup. Environ. Med. **74**(2), 144–153 (2017) (http://dx.doi.org/10.1136/oemed-2016-103684)

[2] Nindl B.C, Billing D.C., Drain J.R., Beckner M. E., Greeves J., Groeller H., Teien H.K., Marcora S., Moffitt A., Reilly T., N.A.S., Young A.J., Friedl K.E., Perspectives on resilience for military readiness and preparedness: Report of an international military physiology roundtable. J. Sci. Med. Sport **21**, 1116–1124 (2018) (https://doi.org/10.1016/j.jsams.2018.05.005)

[3] Burley S.D., Drain J.R., Sampson J.A., Groeller H., Positive, limited and negative responders: The variability in physical fitness adaptation to basic military training. J. Sci. Med. Sport **21**, 1168–1172 (2018) (https://doi.org/10.1016/j.jsams.2018.06.018)

[4] Blue M.N.M., Smith-Ryan A.E., Trexler E.T., Hirsch K.R., The effects of high intensity interval training on muscle size and quality in overweight and obese adults. J. Sci. Med. Sport. **21**, 207–212 (2018) (https://doi.org/10.1016/j.jsams.2017.06.001)

[5] Knapik J.J., Sharp M.A., Canham-Chervak M., Risk factors for training related injuries among mem and women in basic combat training. Med. Sci. Sports & Exercise **33**(6), 946–954 (2001)

[6] Jones B.H. Bovee M.W., Harris J.M., Intrinsic risk factors for exercise-related injuries among male and female army trainees. Am. J. Sports Med. **21**(5), 705–710 (1993) (`https://doi.org/10.1177/036354659302100512`)

[7] Roy T.C., Knapik J.J., Ritland B.M., Risk factors for musculoskeletal injuries for soldiers deployed to Afghanistan. Aviation, Space and Environ. Med. **83**(11), 1060–1066 (2012) (`https://doi.org/10.3357/ASEM.3341.2012`)

[8] Hynynen E., Uusitalo A., Kontinen N., Cardiac autonomic responses to standing up and cognitive task in overstrained athletes. Internat. J. Sports Med. **2997**, 552–558 (2008) (`https://doi.org/10.1055/s-2007-989286`)

[9] Smith C.D., Cooper A.D., Merullo D.J., Sleep restriction and cognitive load after performance on a simulated marksman ship task. J. Sleep Res. (2017) (`http://dx.doi.org/10.1111/jsr.12637`)

[10] Liberman H.R., Castellani J.W., Young A.J., Cognitive function and mood during acute cold stress after extended military training and recovery. Aviation, Space and Environ. Med. **80**(7), 629–636 (2009)

[11] Sawka M.N., Leon L.R., Montain S.J., Integrated physiological mechanisms of exercise performance, adaptation and maladaptation to heat stress. Comprehensive Physiology **1**(4), 1883–1928 (2011)

[12] Wardle L.S., Greeves J.P., Mitigating the risk of musculoskeletal injury: A systematic review of the most effective injury prevention strategies for military personnel. Journal of Science and Medicine in Sport **20**, S3–S10 (2017) (`https://doi.org/10.1016/j.jsams.2017.09.014`)

[13] Cameron K.L., Owens B.D., The burden and management of sports-related musculoskeletal injuries and conditions within the US military. Clin. Sports Med. **33**(4), 573–589 (2014)

[14] Zambraski E.J., Yancosek K.E., Prevention and rehabilitation of musculoskeletal injuries during military operations and training. J. Strength and Conditioning Res. **26**(2), S101–S106 (2012)

[15] Nindl B.C., Williams T.J., Deuster P.A., Strategies for optimizing military physical readiness and preventing musculoskeletal injuries in the 21st century. US Army Medical Department Journal, pp. 5–23 (2013)

[16] Nindl B.C., Jaffin D.P., Dretsch M.N., Human performance optimization metrics: consensus findings, gaps, and recommendations for future research. J. Strength and Conditioning Res. **29**(11), S221–S245 (2015) (`https://doi.org/10.1519/JSC.0000000000001114`)

[17] Kyröläinen H., Pihlainen K., Vaara J.P., Ojanen T., Santtila M., Optimising training adaptations and performance in military environment. J. Sci. Med. Sport **21**, 1131–1138 (2018) (`https://doi.org/10.1016/j.jsams.2017.11.019`)

[18] Santtila M., Häkkinnen K., K. Karavirta I., Changes in cardiovascular performance during an 8-week military basic training period combined with added endurance or strength training. Military Med. **173**(12), 1173–1179 (2008)

[19] Knapik J.J., Harman E.A., Steelman R.A., A systematic review of the effects of physical training on load carriage performance. J. Strength and Conditioning Res. **26**(2), 585–597 (2012) (`https://doi.org/ 10.1519/JSC.0b013e3182429853`)

[20] Boutcher S.H., Dunn S.L., Factors that may impede the weight loss response to exercise – based intervention. Obesity Reviews **10**(6), 671–680 (2009) (`https://doi.org/10.1111/j.1467-789X.2009.00621.x`)

[21] Boudou P., Sobngwi E., Mauvais-Jarvis F. Absence of exercise-induced variations in adiponectin levels despite decreased abdominal adiposity and improved insulin sensitivity in type 2 diabetic men, Eur. J. Endocrinol. **1149**(5), 421–424 (2003) (https://doi.org/10.1038/sj.ijo.0803781)

[22] Trapp E., Heydari M., Freund J., The effects of high intensity intermittent exercise training on fat loss and fasting insulin levels of young women. Internal J. Obesity **32**(4), 684–691 (2008) (https://doi.org/10.1038/sj.ijo.0803781)

[23] Bradshaw P.T., Monda K.L., Stevens J., Metabolic syndrome in healthy obese, overweight, and normal weight individuals: the arteriosclerosis risk in communities' study. Obesity **21**(1), 203–210 (2012) (https://doi.org/10.1002/oby.20248)

[24] Cadore E.L., Gonzalez-Izal M., Pallares J.G., Muscle conduction velocity, strength, neural activity, and morphological changes after eccentric and concentric training. Scandinavian J. Med. Sci. Sports **24**(5), e343–e352 (2014) (https://doi.org/10.1111/sms.12186)

[25] Sanderson P.W., Stacy A.C., Friedl K.E., Biddle S.J.H., The association between obesity related health risk and fitness test results in British Army personnel J. Sci. Med. Sport **21**, 1173–1177 (2018) (https://doi.org/10.1016/j.jsams.2018.08.003)

[26] Pierce J.R., DeGroot D.W., Grier T.L., Hauret K.G., Nindl B.C., East W.B., McGurk M.S., Jones B.H., Body mass index predicts selected physical fitness attributes but is not associated with performance on military relevant tasks in US. J. Sci. Med. Sport **20**, S79–S84 (2017) (https://doi.org/10.1016/j.jsams.2017.08.021)

[27] Kuorinka I., Jonsson B., Kilbom A., Vinterberg H., Biering-Søorensen F., Andersson G., Jøorgensen K., Standardised Nordic questionnaires for the analysis of musculoskeletal symptoms. Applied Ergonomics **18**(3), 233–237 (1987) (https://doi.org/10.1016/0003-6870(87)90010-X)

[28] Maze R., Cavallaro G., Battling bureaucracy: the way forward requires modernizing the modernization process. Army Magazine **682018**, 36–38 (2016)

Smoking and alcohol abuse – predictive factors in oropharyngeal squamous cell carcinoma: A retrospective study

Andrejs Lifsics[1], *Elza* Rate[2], *Juris* Tars[2], *Modra* Murovska[3], and *Valerija* Groma[4]

[1] Rīga Stradinš University, Department of Otolaryngology, Riga, Latvia
[2] Riga East Clinical University Hospital, Oncology Centre of Latvia, Department of Head and Neck Surgery, Riga, Latvia
[3] Rīga Stradiņš University, Institute of Microbiology and Virology, Riga, Latvia
[4] Rīga Stradiņš University, Institute of Anatomy and Anthropology, Riga, Latvia

Abstract. Oropharynx is one of the most common localizations for malignant neoplasms in the head and neck region. HPV-positive OPSCCs are the most common subtypes in Western countries with tobacco and alcohol use, remaining the major risk factors for cancer development in this region. The study aims to address the significance of smoking and alcohol abuse in OPSCC cases and perform a survival analysis of the patients with OPSCC treated over a 10-year period in a single hospital in Latvia, correlating these data with patients' hazardous habits (smoking and alcohol abuse). Out of 247 patients, the majority were smokers (75.95%, $n = 180$), approximately one third were habitual drinkers (35.19%, $n = 82$) or were exposed to both aforementioned major risk factors (31.47%, $n = 73$). Kaplan-Meier survival analysis showed significantly better survival for the patients without the aforementioned hazardous habits. Cox regression analysis proved a higher early death rate when both factors (smoking and alcohol abuse) were present. It seems that smoking has a higher predictive value than alcohol abuse.

Key words: smoking and alcohol use, oropharyngeal squamous cell carcinoma (OPSCC), oropharynx, pharyngeal cancer, hazardous habits.

1 Introduction

Oropharynx is one of the most common localizations for malignant neoplasms in the head and neck region. The GLOBOCAN data (2012) confirm over 140 000 new cases of pharyngeal cancer worldwide and age-standardized incidence of 1.9 per 100,000, whereas – 34 000 new cases and age-standardized incidence of 2.9 in Europe [1].

Nowadays, oropharyngeal squamous cell carcinoma (OPSCC) is divided into two major subgroups according to Human Papilloma Virus (HPV) status (HPV-positive and HPV-negative). HPV-negative OPSCCs are believed to be associated with tobacco and alcohol use [2]. However, these hazardous habits can appear as combined risk factors affecting the development of the neoplasm.

For all mentioned, HPV-positive OPSCCs are the most common subtypes in Western countries with tobacco and alcohol use, remaining the major risk factors for cancer development in this region [3].

Furthermore, two etiological pathways for head and neck squamous cell carcinoma have been suggested, one driven by smoking and alcohol abuse and the other by HPV infection [4, 5].

Therefore, worth the discussion is the fact about the presence of the synergic effect of smoking/alcohol use and HPV infection in the development of HNSCC however some researchers suggest that there is none [6].

A dose-dependent carcinogenic effect on OPSCC development was recognized [7] evidencing that the risk increases 6–7-fold in individuals overusing tobacco or alcohol and as much as 15-fold in subjects who both smoke and drink alcohol [8].

The world incidence of OPSCC varies and estimated differences in the incidence and survival are generally related to the distinct risk and socioeconomic factors, environmental agents, public health awareness and accessibility of health services as well as advances in diagnostics and therapy.

Therefore, we wanted to address the ever-actual question of the significance of smoking and alcohol abuse in OPSCC cases. The aim was to perform a survival analysis (overall and disease-specific) of the patients with OPSCC correlating these data with subjects' hazardous habits (smoking and alcohol abuse). To achieve the aim, we have conducted a retrospective study using a group of patients with OPSCC treated over a 10-year period in a single hospital in Latvia.

2 Material and methods

We carried out a retrospective study of 247 patients diagnosed with OPSCC, staged following the TNM classification of the International Union against Cancer (6th edition) for oropharyngeal carcinoma and treated in Riga Eastern Clinical University Hospital Stationary Oncology Centre of Latvia between January 1st 2000 and December 31st 2010. Patients are admitted to this hospital from all over the country, which has an estimated population of 1.91 million. The patients' data were collected from the Hospital Archive and The Centre for Disease Prevention and Control. Analysis of the patients' data was performed when the diagnosis of OPSCC was confirmed histologically.

Statistical analysis of correlation of aforementioned covariates with survival, and mean overall survival time after diagnosis was performed. Statistically significant differences between the study groups were assessed using Pearson's chi-squared and Fisher's exact test (depending on the size of the group), and the value of $P < 0.05$ was considered as significant. Cramer's V was used to measure an association between two nominal variables. Kruskal-Wallis test and Mann-Whitney test (depending on the number of groups) were used to assess the correlation between nominal variables and mean survival time after an established diagnosis of OPSCC.

The data collected was processed to calculate the overall and disease-specific 3 and 5-year survival rates for all patients. The Kaplan-Meier survival analysis was used for statistical data analysis. Statistical testing for differences in survival rates was performed using the log-rank and Breslow tests. Finally, a Cox regression method was used to estimate the hazard ratio. Age, sex, T stage, N status, risk factors (smoking and alcohol abuse), therapy modality (radiotherapy (RT), surgery (OP), chemotherapy (ChT), symptomatic therapy and combinations of aforementioned, primary tumour location, and histo-pathologic variant of tumour were included as covariates in the survival model. Cox regression methods for two-stage randomization designs were applied when conducting univariate, and secondly, multivariate analysis (see Table 1 for the list of covariates).

3 Results

Less than one-tenth of the patients presented with the stage I and II – 3 (1.22%) and 19 (7.72%) patients, accordingly, whereas a major portion – 224 (91.6%) revealed advanced

Table 1. List of covariates analyzed in multivariate analysis.

Variables	Variables
N status (N0 >N+)	Sex (female >male)
Alcohol abuse and/or smoking	Age group
None	<55 years old
1 of the aforementioned	55 – 64 years old
Both	>64 years old
T grade	Primary tumour location
1	Palatine tonsil
2	The base of the tongue
3	Pharyngeal wall
4	Soft palate
Therapy	Histological variant
RT	Squamous cell carcinoma, keratinizing
OP	Squamous cell carcinoma, non-keratinizing
RT + OP	Carcinoma, undifferentiated (Epit)
RT + ChT (Cetuximab) +/−OP	Squamous cell carcinoma, BCN (unspecified)
RT + ChT (Platinum) +/−OP	
Sympt (symptomatic)	

Table 2. Patients' breakdown by alcohol abuse and disease outcome (14 patients censored).

Alcohol abuse	Number of patients (count, %)	Result of the disease (therapy)	Number of patients (count, %)	Statistical analysis between groups	
				$P\chi$	V_1
No	151 (64, 81)	Positive (survived)	20 (13, 42)	0.12	0.11
		Negative (deceased)	129 (86, 58)		
Yes	82 (35, 19)	Positive (survived)	5 (6, 33)		
		Negative (deceased)	74 (93, 67)		

disease stage. Only 8.10% ($n = 20$) of all reviewed patients were female and 91.90% ($n = 227$) – male. The mean patient age was 60 years, median – 60.20 years.

The majority of the patients were regular smokers (75.95%, $n = 180$), habitual drinkers (35.19%, n=82) or were exposed to both aforementioned major risk factors (31.47%, $n = 73$). Most of the smokers were male ($n = 171$). Median pack-years were 30.0 ($SD = 17.9$).

We found that cancer death rates are not very much different when the distribution of patients by the total life/disease outcome (without distinguishing between patients whose death is due to cancer) was correlated to alcohol exposure (Table 2).

Different survival rates (in months) were found when estimating the mean overall survival calculated from the time of established diagnosis and correlated to patient alcohol abuse (Table 3), but differences were not statistically significant between the study groups.

Kaplan-Meier survival analysis (Table 4) showed a statistically significant difference between the overall ($P_L = 3.04 \times 10^{-2}$) survival (OS) curves reflecting variations in alcohol exposures – a higher survival rate was estimated in patients without this characteristic.

No statistically significant differences between the results deciphering disease-specific survival (DSS) when including and excluding those patients, whose cause of death is not cancer, were found over time (P value increased with each time period).

Table 3. Correlation between survival and alcohol abuse.

Alcohol abuse	Time, months			Statistical analysis	
	Mean	SD	CI 95%	P_{MV}	η
No	30.18	39.33	23.81 ... 36.55	0.08	0.14
Yes	19.68	26.50	13.74 ... 25.61		

Table 4. Kaplan-Meier analysis of hazardous habits for DSS, OS.

Variable	3-year Kaplan-Meier estimate %		5-year Kaplan-Meier estimate %	
	Overall survival (OS)	Disease-specific survival (DSS)	Overall survival (OS)	Disease-specific survival (DSS)
Smoking	32.1%	30.8%	25%	26.9%
• No	16.5%	14.6%	11.4%	8.7%
• Yes	***P*** **=0.0035**	***P*** **=0.005**	***P*** **=0.0035**	***P*** **=0.005**
Alcohol abuse	24.8%	22.2%	16.7%	16.3%
• No	13.9%	13.9%	11.4%	8.9%
• Yes	***P*** **=0.03**	*P* =0.08	***P*** **=0.03**	*P* =0.08
Alcohol abuse and smoking:				
• Neither	34%	31.8%	23.4%	25%
• 1 factor	22.7%	20.4%	16.4%	14.3%
• Both	11.4%	10.9%	7.1%	6.3%
	P **=0.002**	***P*** **=0.008**	***P*** **=0.002**	***P*** **=0.008**

Table 5. Patients' breakdown by smoking and disease outcome (10 patients censored).

Smoking	Number of patients (count, %)	Result of the disease (therapy)	Number of patients (count, %)	Statistical analysis between groups	
				P_{χ}	V_1
No	57 (24,05)	Positive (survived)	12 (21,43)	1.77×10^{-3}	0.21
		Negative (deceased)	44 (78,57)		
Yes	180 (75,95)	Positive (survived)	12 (6,82)		
		Negative (deceased)	164 (93,18)		

Table 6. Correlation between survival and smoking.

Smoking	Time, months			Statistical analysis	
	Mean	SD	CI 95%	P_{MV}	η
No	36.63	42.01	25.38 ... 47.88	1.44×10^{-2}	0.17
Yes	22.63	32.45	17.81 ... 27.46		

Regarding tobacco abuse, we found that the number of cancer-related deaths was higher for smokers when comparing the distribution of patients by the total life/disease outcome depending on smoking (Table 5). Moreover, this difference between the analyzed groups was statistically significant ($P_{\chi} = 1.77 \times 10^{-3}$) revealing a moderately strong correlation between survival and smoking – $V_1 = 0.21$.

We demonstrated that patients who don't smoke have a higher mean overall survival time. The difference between the analyzed groups was statistically significant ($P_{MV} = 1.44 \times 10^{-2}$), whereas the correlation between survival and smoking was weak – $\eta = 0.17$ (Table 6).

Table 7. Patients' breakdown by alcohol abuse and/or smoking and disease outcome (15 patients censored).

Alcohol abuse and/or smoking	Number of patients (count, %)	Result of the disease (therapy)	Number of patients (count, %)	Statistical analysis between groups	
				Pχ	V
Neither	48 (20,69)	Positive (survived)	11 (23,40)	6.00×10^{-3}	0.22
		Negative (deceased)	36 (76,60)		
Alcohol abuse or smoking	111 (47,84)	Positive (survived)	10 (9,09)		
		Negative (deceased)	100 (90,91)		
Alcohol abuse and smoking	73 (31,47)	Positive (survived)	3 (4,29)		
		Negative (deceased)	67 (95,71)		

Table 8. Correlation between survival and alcohol abuse and/or smoking.

Alcohol abuse dose-dependent smoking	Time, months			Statistical analysis	
	Mean	SD	CI 95%	P_{MV}	η
Neither	35.66	41.45	23.49 . . . 47.84	2.12×10^{-2}	0.21
Alcohol abuse **or** smoking	28.96	39.12	21.57 . . . 36.36		
Alcohol abuse **and** smoking	16.12	20.91	11.13 . . . 21.10		

When comparing smokers and non-smokers, a statistically significant difference in overall survival was found ($P_L = 3.50 \times 10^{-3}$; $P_B = 1.45 \times 10^{-2}$) suggesting that the proportion of survived patients in non-smokers group is higher at all times. Similar observations were made when estimating disease-specific survival.

Comparing the distribution of patients by the total life/disease outcome depending on alcohol consumption and/or smoking (Table 7), we found that the number of cancer-related deaths was lower in the group of patients who do not smoke and do not use alcohol. The difference between the analyzed groups was statistically significant ($P\chi = 6.00 \times 10^{-3}$), whereas the correlation between survival and alcohol abuse and/or smoking was weak – $V = 0.22$.

A higher mean total survival time was calculated for patients without hazardous habits (patients who do not use alcohol and do not smoke) when analyzing the mean overall survival time in cancer patients exposed and not exposed to alcohol consumption and/or smoking (Table 8). The difference between these groups was statistically significant ($P_{MV} = 2.12 \times 10^{-2}$), whereas the correlation between survival and presence of hazardous habits was not considered significant as it was weak – $\eta = 0.21$.

A significantly lower OS and DSS were demonstrated in OPSCC patients when both risk factors (alcohol abuse and smoking) were detected (Fig. 1).

Finally, when applying Cox regression model, we found that the presence of one of the hazardous habits (smoking or alcohol abuse) has an apparent effect and significantly elevates the risk of death.

4 Discussion

The main goal of this study was to collect and analyze data on cigarette smoking and alcohol consumption for patients with OPSCC to understand the relevance of these factors, used when assessing survival rates in case of cancer.

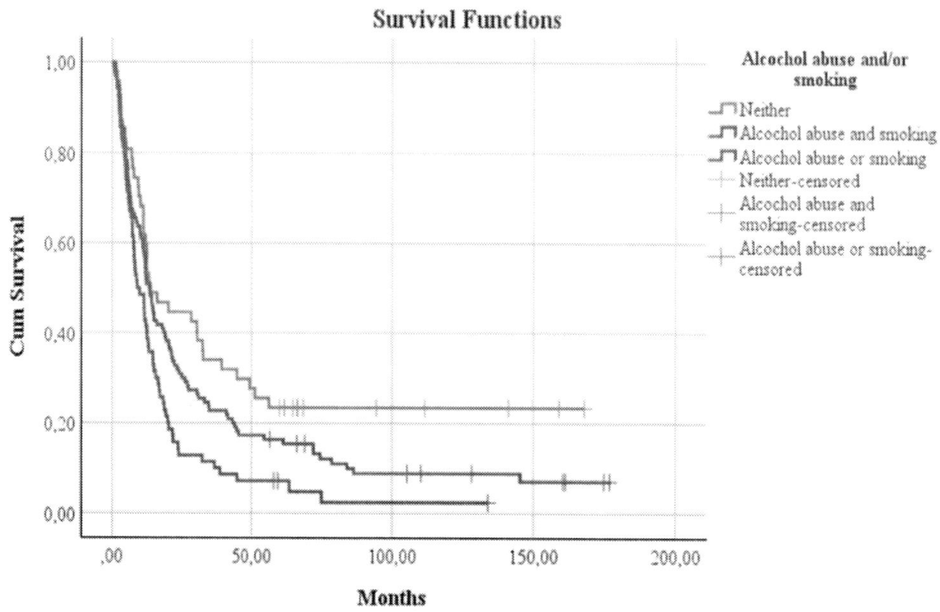

Fig. 1. Kaplan-Meier disease specific survival plot according to hazardous habits.

The results of this study revealed that a vast majority of patients were smokers (76%) whereas one third (35%) had drinking problems. We found that these habits independently predict OPSCC prognosis contributing to both OS and DSS, with smoking, playing a more important role in survival, especially DSS.

Furthermore, the combination of these two factors significantly decreases indicators of survival (DSS and OS). Supportive pieces of evidence were obtained by using Cox hazard model (multivariate analysis), which showed a higher risk of early death when at least one of the factors – smoking and alcohol abuse was present. These results seem to be similar to the contention made by other scientists [9–12].

Unfortunately, it was not possible to perform exact dose-dependent analysis of the aforementioned risk factors when studying these retrospectively, and it sets certain limitations of the given study.

Undoubtedly, hazardous habits (smoking, alcohol abuse) proved to be strong predictive factors when assessing survival in OPSCC. These should be considered when selecting the appropriate treatment for the patient.

Even with declining smoking rates reported by American physicians in 2018, assessment of the main risk factors for OPSCC – tobacco and alcohol consumption appear of great relevance when making survival-related suggestions [13].

A growing body of evidence accumulated during the last two decades demonstrates that the incidence of OPSCC increases and the role of HPV in the pathogenesis of OPSCC was highlighted. Many studies show that HPV positive OPSCC has a better prognosis than HPV negative [4, 13–15] therefore the value of the prognostic role of the HPV status appears to be in a focus of interest in some future survival-related studies.

5 Conclusion

Hazardous habits as smoking and alcohol abuse are main, independently acting, risk factors recognized and assessed in patients with OPSCC. Collectively, the given study suggested that patients without hazardous habits had better 3 and 5-year OS and DSS and lower hazard ratio. Furthermore, it appears that smoking has a higher predictive value than alcohol abuse when assessing survival rates in patients with OPSCC.

References

[1] M. De Camargo Cancela, D.L. de Souza, M.P. Curado, *International incidence of oropharyngeal cancer: a population-based study.* Oral Oncol. **48**(6), 484–490 (2012)
[2] B. Solomon, R.J. Young, D. Rischin, *Head and neck squamous cell carcinoma: Genomics and emerging biomarkers for immunomodulatory cancer treatments.* Semin. Cancer Biol. **52**(Pt 2), 228–240 (2018)
[3] A.K. Chaturvedi, G. D'Souza, M.L. Gillison, H.A. Katki, *Burden of HPV-positive oropharynx cancers among ever and never smokers in the U.S. population.* Oral Oncol. **60**, 61–67 (2016)
[4] M.L. Gillison, et al., *Evidence for a causal association between human papillomavirus and a subset of head and neck cancers.* J. Natl. Cancer Inst. **92**, 709–720 (2000)
[5] M.L. Gillison, et al., *Distinct risk factor profiles for human papillomavirus type 16-positive and human papillomavirus type 16-negative head and neck cancers.* J. Natl. Cancer Inst. **100**, 407–420 (2008)
[6] R. Herrero, et al., *IARC Multicenter Oral Cancer Study Group. Human papillomavirus and oral cancer: The International Agency for Research on Cancer multicenter study.* J. Natl. Cancer Inst. **95**, 1772–1783 (2003)
[7] D.M. Cohan, S. Popat, S.E. Kaplan, et al., *Oropharyngeal cancer: current understanding and management.* Current Opinion in Otolaryngology & Head and Neck Surgery. **17**, 88–94 (2009)
[8] B. Szybiak, P. Trzeciak, W. Golusiński, *Role of extended histological examination in the assessment of local recurrence of tongue and floor of the mouth cancer.* Reports of Pract. Onc. and Radiotherapy. **17**(6), 319–323 (2012)
[9] C.A. Benhamou, N. Laraqui, M. Touhami, et al., *Tobacco and cancer of the larynx: a prospective survey of 58 patients.* Rev. Laryngol. Otol. Rhinol. (Bord). **113**(4), 285–8 (1992)
[10] W. Winkelstein, *Smoking and cervical cancer—current status: a review*, Am. J. Epidemiol. **131**, 945–957 (1990)
[11] H. Kuper, P. Boffetta, H.O. Adami, *Tobacco use and cancer causation: association by tumour type.* J. Intern. Med. **252**, 206–224 (2002)
[12] N.J. Farsi, M.C. Rousseau, N. Schlecht, et al., *Aetiological heterogeneity of head and neck squamous cell carcinomas: the role of human papillomavirus infections, smoking and alcohol.* Carcinogenesis. **38**(12), 1188–1195 (2017)
[13] K. Panesar, *Oropharyngeal Squamous Cell Carcinoma.* US Pharm. **43**(8), HS2-HS6 (2018)
[14] F. Dayyani, et al., *Meta-analysis of the impact of human papillomavirus (HPV) on cancer risk and overall survival in head and neck squamous cell carcinomas (HNSCC).* Head Neck Oncol. **2**, 15 (2010)
[15] K.K. Ang, J. Harris, R. Wheeler, et al., *Human papillomavirus and survival of patients with oropharyngeal cancer.* N. Engl. J. Med. **363**, 24–35 (2010)

Discourse through Photovoice: "Lived experience" of epilepsy

Jurgita Tuitaitė[1] and *Polina* Šedienė[2]

[1] Vilniaus Kolegija/University of applied sciences, Vytautas Magnus University, Vilnius, Lithuania
[2] Lithuanian University of Health Sciences, Kaunas, Lithuania

Abstract. The purpose of this article is to help medical students during the interaction of people with epilepsy, to learn more about "lived experience", and find out more about patients' voices. People with epilepsy sometimes disincline to talk about their problems because of stigmatization. Stigmatization of epilepsy determines diminished self-esteem, depression, anxiety for the future and low quality of life [1]. **Methodology**—photovoice is a qualitative method that combines pictures and narratives in exploring "ived experience" of people with epilepsy, patients' and students' voices. Photovoice project was conducted in processing steps considering Luo (2016). The research members included people with epilepsy and medical students who were divided into 8 groups. The duration of the project was 6 weeks and after the project exhibition was arranged. **Results.** Voices of patients with epilepsy allow students to understand the discourse between insiders and outsiders, the differences between professional language and conversation among professionals.

Key words: photovoice, discourse, epilepsy.

1 Introduction

System of medical education is generally understood as training of professionals at an educational institution and practical classes. Different environment, people, and health problems allow medical students to gain practical experience. Students employ their theoretical background and professionals' help to find the solution in real practical workplace. Historically, doctors had been teachers for many years, but they had little knowledge how to teach students. Erlich and Shayghnessy (2014) distinguish two key elements of success. First, interleaving sessions devoted to didactic educational principles where practical teaching experience with actual students in a real course, and second, direct faculty guidance in developing knowledge, skills, and attitudes of teaching. Peer teaching is one of the suggested methods in accordance with this concept. Currently students' teaching models, such as peer teaching, are very effective because "...they use accessible language to describe concepts and relate to the feeling of being novice". Peer teacher personnel from similar social groups, somebody who is not a professional teacher helps others to learn and learn themselves by teaching [2]. Through the teaching process experiences of people with epilepsy remain unknown because the teacher observes students only during sessions and the teacher uses professional language which could be evaluated. People with epilepsy often have cognitive limitations and special needs and a doctor has to transform people's language into the professional language. How do students transform nonprofessional language? What is the impact of voices of people with epilepsy on students' professional language?

Photovoice is one of the methods allowing hearing the voice of people with epilepsy. Through photos and stories community members can identify important issues and acquire knowledge of what is meaningful for the community, but not for teachers [3, 4]. Photovoice often is used to draw attention to society otherwise the following method can change opinion of the community members and environment. Yamani et al. (2016) described that application of active teaching strategy, meaningful learning and integration of conceptual and practical knowledge are the core of curriculum. In this context the core curriculum is understood as a focal point for training teachers and as the basis for describing specific objectives. Practical knowledge of problems of people with epilepsy was explained by the teacher, but about the symptoms of the illness and arising social problems students heard from people with epilepsy. Students indicated the project as an innovative teaching method, without having previous experience. Innovations at all levels of medical education allow developing a curriculum and doing revisions in medical education [5]. It is important to remember that innovations could be evolutionary and revolutionary. Students noted that this project enabled them to change their opinion and they started to think more critically. Innovative education maintains learners' autonomy, self-efficacy, critical thinking and creativity [6]. Hjelm (2013) described learning through discourse analysis like directing a critical gaze at learning with discourse analysis process itself [7].

Photovoice is one of the methods that allow hearing the voice of people with epilepsy. Through the photos and stories community members can identify important issues and acquire knowledge of what is meaningful for the community [3, 4]. Photovoice is often used to draw attention to society otherwise the following method can change opinion of the community members and environment.

2 The Photovoice method

Photovoice is a community-based participatory research method which enables combining visuals and narrative to disclose what is meaningful for the community [8]. The project members included people with epilepsy and medical students who were divided into 8 groups. Project members decided to demonstrate "lived experience" of epilepsy. It was all members' decision to expose that epilepsy is not an illness but life with it. 5 workshops were organized during the project. The project was conducted in processing steps considering Luo (2016): determining the problem to be explored (1), identifying the community members to participate in the project (2), introductory meeting (3), training (4), photo taking (5), facilitated discussion (6), analysis (7), exhibition (8), and debrief (9). The first step in this project was to identify the problem and to focus only on the problems concerning epilepsy.

Each project member had to present his / her opinion on the idea of the project and to plan next workshops. Project members named the project "I Am as You Are". It is very important that not only people with epilepsy, but each student could talk and express themselves. Discussions were organized after each workshop and project members prepared reflections in written form. The project members combined visuals and stories to show social aspects of epilepsy, stigmatization problems, epilepsy and work, need of social assistance. The duration of the project took 6 weeks and after the project an exhibition was arranged.

3 Data collection procedures

All of the project members decided to choose 3 photos and to write stories. Peer work was divided into two stages: individual and group work. At first, students were outsiders and tried to discover the difference between professional language and conversation among people with epilepsy.

During the research ethnographic focus on examining what is insider or emic was used. Agar (2006) argues that culture is realized through language and a newcomer brings his/her own language culture [9]. This logic enquiry let us understand that every group member defines their situation during interactions [10]. Insiders' knowledge is unprofessional, and it depends on "lived experience", social constructs were made not only during insiders' interactions. Outsiders, professionals, joined the group and interacted with insiders: whole group tried to make notes and choose photos. During the discussions students clarified all the information they acquired from people with epilepsy and the amount of information they did not hear. The teacher was waiting for each answer as long as possible. Ingram et al. (2014) argues that extending the waiting time or not, is a part of toolkit for improving the quality of a dialogue. The teacher usually had to wait at least three seconds for an answer and it is not a productive strategy [11]. Language begins with the start of thinking, but sometimes people pronounce different words depending on the context. Language helps to construct social situations. This situation changed social construction of narrative stories. Outsiders used professional language within their group and tried to understand insiders' "lived experience" [10].

Reflective writing is one of the methods used for helping students to enhance self-awareness. It helps gaining an understanding of the perspectives of others. Reflection helps to explore experiences and these experiences help students to enhance their psychological well-being and their abilities to empathize with patients and themselves by enabling them to access and accept their feelings. Language changes could be the result in deepened critical thinking [12].

Language helps to translate internal thoughts and emotions to others and reflective writing could show us how students use language to construct their narratives and effective communication [12]. Lin et al. (2016) suggested that teachers enable students to see and hear what happens in the interactions of patients and their family members and interpreted the training purposes: "...medical students should be encouraged to enhance their empathetic understanding of the psychosocial issues in patients and illness by becoming aware of expressing their affective experience...". Common learning experience is essential for medical students to achieve clinical competencies [5]. These educational programs help students to analyze different psychological processes and this is challenging for all participants. Hjelm et al. (2013) explained that learning through discourse gives possibility for students and teachers to engage in critical reflection of the process while doing analysis. The discourse of classroom interaction constructs knowledge [7]. Students during this project obtained knowledge of themselves and of teamwork and group reflection. Medical students' learning process is individual and not only directed to knowledge about people with epilepsy.

Photovoice method enables students to see and hear what happens during interactions between them and patients. In order to construct epilepsy-narrative in a comprehensive way, which includes medical and non-medical understanding and experiencing of illness, it is essential to enhance their empathetic interaction and transform professional outsiders' language into understandable language for insiders (patients).

4 Findings

People with epilepsy and students discussed a range of challenges of living with epilepsy. Two key themes emerged from the analysis: coping with an illness strategies and portrait of a person with epilepsy.

Fig. 1. Everyone is similar and different.

4.1 Coping with illness strategies

People with epilepsy showed that self-realization, understanding of feelings and insight are the strategies in coping with an illness.

Self-Realization. People with epilepsy spoke of their purpose of life and their wishes to feel useful in community, but it is hard to control life because of unexpected seizure. People with epilepsy called these feelings "shadow of anxiety" because they feel unsafe and therefore it is hard to live and to do things in life, as the next excerpt illustrates:

Aistė:	I am sick. I go further to help other.
Romualdas:	I am sick. I'm taking care of my elderly mom.
Neringa:	I am sick. I bake muffin for my friends.
Kasparas:	I am sick. I do not forget my close friends.
Povilas:	I am sick. I am very active in social projects.

People with epilepsy fought with illness expressing themselves through the help to other people, participating in public life and realizing their abilities. Many people with epilepsy explained that the daily worries and intolerance of others – did not stop them achieving their goals. Emphasis is placed on self-realization for the purposes of life, regardless of public attitudes and veto (Fig. 1).

Photovoice, as peer teaching method, was experienced by medical students as innovative, creative and effective method which allowed students to understand epilepsy in bio-psychosocial consequences [8]. Visual narratives enabled us to see how and what was learned about epilepsy by the project members. People act and interact, learn from each other in communities and students learned that patients were source of inspiration for them—they showed how to live a full life and not to give up. Acquaintance and communication with sick people can inspire healthy ones to achieve their goals, and help people with epilepsy to realize them.

Understanding of feelings and insight. Efforts, skills and help are needed to overcome emotional abundance and confusion (Fig. 2). The study revealed feelings of epilepsy and many persons with epilepsy mentioned that fear about unexpected seizure lead to tension, shame that makes them close up inside, and disengage from the outside world. Group members noticed that if persons with illness manage to 'tease' the disease, it does not necessarily make life easier and people with anxiety, distrust and unsafeness got help from other people with their interest, understanding, and listening, as the next excerpt illustrates:

We should encourage the people with epilepsy to share their feelings and difficulties. When people talk—the problem is reduced by half. In addition, those who do not suffer need to know about other people's worries. There is only one way—to solve them together.

Fig. 2. Diversity of feelings.

Fig. 3. Not everyone has the opportunity to go through the door.

The results of the study showed that understanding, expressing feelings and listening were helpful for community members, but it is an important condition for cohabitation and satisfaction with communion.

The walls do not stand by themselves. In society, ignorance, misunderstanding, and mystification of epilepsy prevail. People with epilepsy are often stigmatized. In this way, an "invisible wall" is created between people with epilepsy and other people (Fig. 3).

Analysis of research data revealed that it is not impossible to break down the borders that limit people's with epilepsy ability to participate in society without public support, as the next excerpt illustrates:

We would like to emphasize that the unemployed should put more effort into integrating the people with epilepsy into community. First of all, educate members of the community about other members' problems–reduce exclusion. Exclusion leads to ignorance and fear. In our opinion, everything depends on the approach to epilepsy.

The Photovoice method helped all group members to fight with the consequences of the disease.

Fig. 4. Posture of people with epilepsy.

4.2 Portrait of a person with epilepsy

Medical students and people with epilepsy expressed themselves differently, as the next excerpt illustrates:

The first photo, with a person who has raised his hands up, symbolizes a self-confident and resolute personality, who knows of the despair of the disease, but he keeps his hands up and strive for the heights of life or inner peace; the second photo shows how person is crushed, lowered hands and lost the meaning of life; also, part of the patients with epilepsy lock in, this situation is symbolized by the third photo; in the fourth picture we see a person who asks for help because he understands that there is a way out and may be better.

Many people with epilepsy and students said that posture is very important in relation to the disease which is revealed by overcoming the psychosocial consequences of the disease (Fig. 4).

People with epilepsy said that it is very important to be realistic and to ask help from others. Epilepsy seizures can be unpredictable and sometimes people with epilepsy need help after seizure.

M*eaningfulness of disease.* According to Franklin, "man is looking for meaning". It was very important to see the meaning of suffering in the disease, as the next excerpt illustrates:

Just glance at it, it may seem that all the girls look alike, but looking closer, one can notice that one girl is distinguished by the fact that her hair is tied.

Maybe I wouldn't be who I was, maybe I wouldn't focus on selfishness and development, I wouldn't have time to spend with my daughter if I didn't get it.

The disease helped me to develop such qualities as patience, stubbornness to my goals, help and sensitivity to others. I think it's important, it strengthens self-esteem.

The disease gave significance to life for people with epilepsy, because they put effort into self-improvement, were devoted to their loved ones, developed self-esteem and sensitivity to other people.

5 Conclusion

Project members learned about people with epilepsy from live stories. Students acquired knowledge about discrimination, surrounding environment, insensibility for people with epilepsy. During the study process students learned about different types of epilepsy. Through working in groups like outsiders and after the project they broaden the understanding that each case is unique, different, and unpredictable. Students learned how to initiate the contact and how to maintain this close professional contact with people having epilepsy. People with epilepsy sometimes disincline to talk about their problems because of power balance. It is very important for students to collaborate and to seek for the aim together. Photovoice method helped to work as a team and to seek aims of the people with epilepsy.

References

[1] C. Soria, S. Escolano, S. El Sabbagh, S. Chmura, Ch. Bulteau, C. Chiron, G. Dellatolas, Child Neuropsychol. **18**(3), 209–227 (2012)
[2] R. D. Erlich, F. A. Shayghnessy, Medical Teacher **36**, 322–332 (2014)
[3] M. Teti, L. Johnson, D. Binson, JERHRE **7**(4), 34–43 (2012)
[4] T. Heyns, G., Van Rensburg, T. Botma, I. Coetzee, Gender and Behaviour **14**(1), 7028–7039 (2016)
[5] N. Yamani, M. Rahimi, Res. Dev. Med. Education **5**(2), 50–54 (2016)
[6] P. Serdyukov, JRIT&L **1**(10), 4–33 (2017)
[7] T. Hjelm, Teaching in Higher Ed. **8**(18), 871–882 (2013)
[8] L. Luo. L. Library Hi Tech. **1**(35), 179–185 (2016)
[9] M. Agar, FQS **7**(4), 36 (2006)
[10] N. Markee, *The Handbook of Classroom Discourse and Interaction* (West Sussex: John Wiley & Sons Inc., 2015)
[11] J. Ingram, V. Elliot, Cambridge J. Ed. **46**(1), 37–53 (2014)
[12] Chi-Wei. Lin, M. Lin, C. Wen, S. Chu, Med. Education **21**, 29522 (2016)

3D dissection tools in Anatomage supported interactive human anatomy teaching and learning

Dzintra Kazoka* and *Mara* Pilmane

Rīga Stradiņš University, Department of Morphology, Kronvalda 9, Riga, Latvia

Abstract. The main aim of this study is to present the usage and importance of 3D dissection tools in the teaching and learning of Anatomy and to describe and explain our experience with Anatomage Table in Human Anatomy studies at Rīga Stradiņš University. In 2017–2018 two 3D dissection tools (scalpels) were used every week in work with Anatomage Table during the practical classes. As methods for collecting data were used discussions between students and teachers. Together 200 students of the Faculty of Medicine and Dentistry were involved in this study. It was possible to create incisions and cuts in order to remove and uncover different layers of organic tissues, to move deep inside step by step and find out which structures it was necessary to look for. Afterwards students showed that they were able to place the organs back and reattach the bones, muscles, blood vessels in the body and put the skin back on. Students enjoyed virtual tools in the practical classes and learned the material better. Virtual tools helped students and tutors to easily understand and memorize different anatomy structures. 3D scalpels were useful for different education activities, but the learning experience may be suitable further for the study of real materials.

Key words: dissection tools, Anatomage, human anatomy, education

1 Introduction

Human Anatomy is one of the oldest subjects in medical education and there are different methods that are used for teaching and learning of it around the world [1–3]. At present, we still continue study process of human anatomy based on cadaveric dissections, but together with traditional methods, several combinations of modern methods and technologies are used. It is very important to find correct directions and reach the most effective ways in the upraising of knowledge of our students in Human Anatomy.

It is clear that teaching and learning of human body take time and may involve the use of different books, cadaver dissections, plastic models and special interactive 3D images in practical classes [4, 5]. These things help students to form a creative and productive approach, which can give them a lot of benefits for learning or self-education process [6]. New technologies and educational software are shaking up the role of educators in medical fields [7]. The educators of today should support and guide students' activities. The advantages of the technologies and different methods can be very helpful for teachers to maximize their

* Corresponding author: Dzintra.Kazoka@rsu.lv

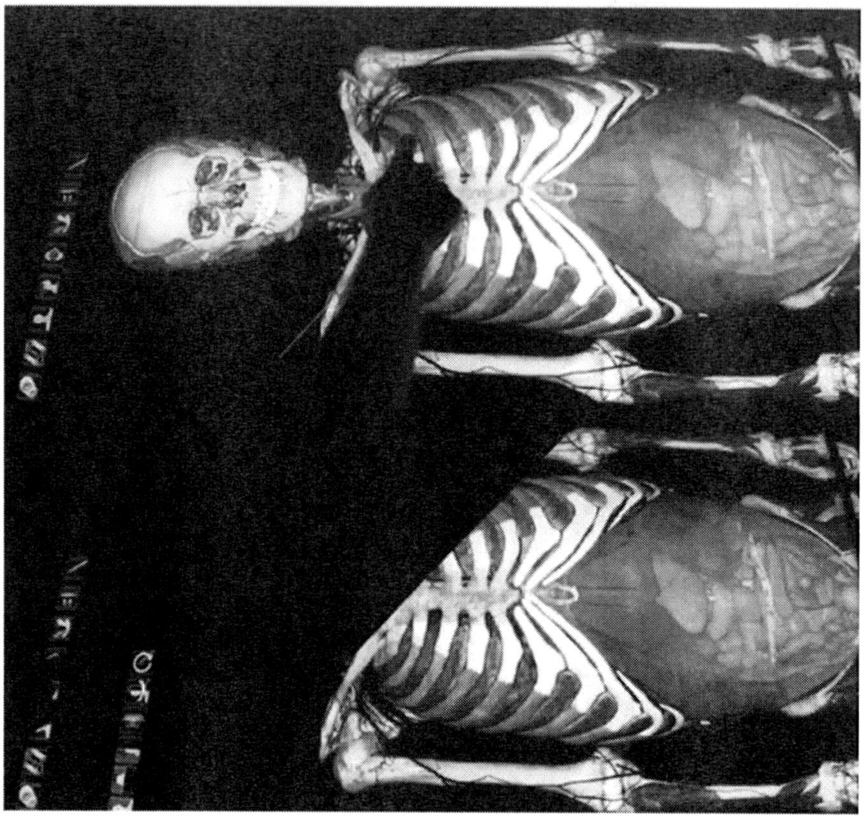

Fig. 1. The application of scalpel for straight line area – step 1 and step 2.

efforts in teaching [8–11]. It is very important that students are learning not only basics, but also how to use technologies that they have.

The Anatomage Table is life-size 3D interactive table and includes a segmented real human anatomy system. There are included virtual tools for performing an anatomical dissection and surgical approaches. It is known that the incorrect usage of scalpels is very dangerous to the patients and the surgeons [12], but in dissection the role of an incision with precision plays a very important role and it must be provided by strictly followed steps.

The main aim of this study is to present the usage and importance of 3D dissection tools in the teaching and learning of Anatomy and to describe and explain our experience with Anatomage table in Human Anatomy studies at Rīga Stradiņš University.

2 Material and methods

In academic years 2017-2018 two 3D dissection tools were used every week in work with Anatomage Table (Table Application software from Anatomage, Inc. (Table EDU 4.0)) during the practical classes.

We offered and validated the usage of two virtual scalpels. The first scalpel allowed cutting according to the straight line position (Fig. 1). The second scalpel was used for freehand dissection in any area (Fig. 2).

Students were divided into two groups. The first group consisted of 100 students of the Faculty of Dentistry and 100 students of the Faculty of Medicine. All students were also

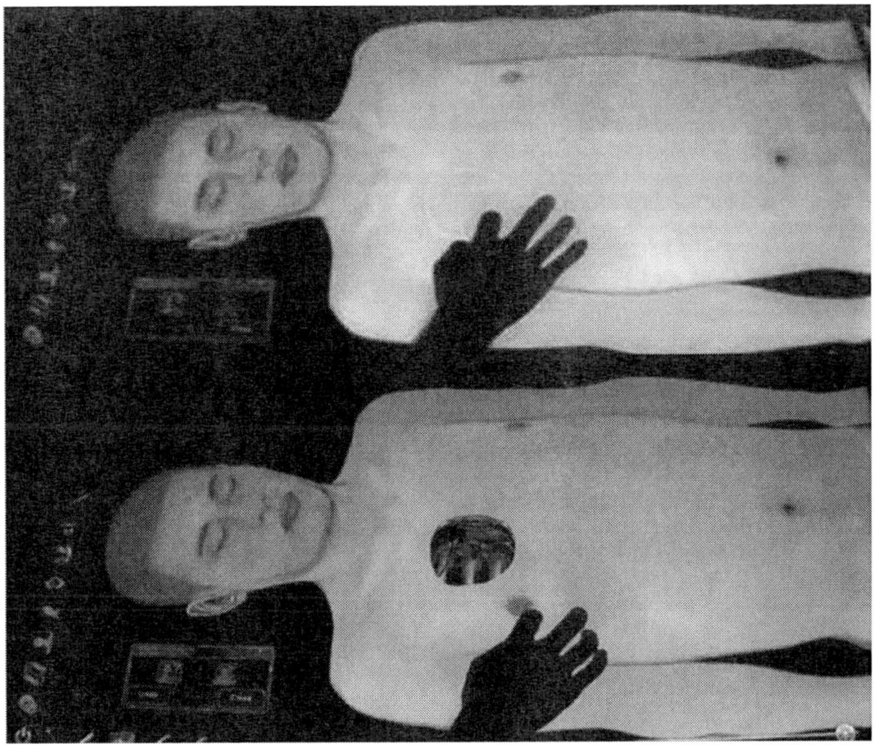

Fig. 2. The application of scalpel for freehand area – step 1 and step 2.

asked if they had ever worked with the virtual scalpels as part of any prior or present course works.

At the beginning of the practical class the teacher explained students the general elements necessary to create incisions of tissue and demonstrated details of all steps involved in creating an ideal incision. Before beginning an incision, the teacher and students planned their approaches, landmarks, distances and techniques.

The students were instructed to complete their virtual dissection with discussions which contained the answers to the questions about the usage of 3D dissection tools. The questions were divided into the following categories: advantages and disadvantages of the used virtual scalpels of the Anatomage Table. All free answers were used for collecting of data.

3 Results

The students started each practical class of the dissection by repeating previous theoretical materials, skills and following the instructions. Students from Faculty of Medicine focused more on different organs systems, but students from Faculty of Dentistry focused more on the head and neck. None of the 200 students had ever worked with the virtual tools as part of prior or present courses. First two practical classes were started by students with slow incorporation of the virtual tools. The Anatomage Table offered access to life-size cross sections in different directions and allowed students to do a cross-section on the spot, using "scalpels" tools. In some cases it was a very fast process which could be done during only one anatomy class.

During the virtual cutting students trained in controlling of the directions, length and depth of all performed incisions. Virtual scalpels were utilized in a variety of ways to better

Table 1. Some advantages and disadvantages of the virtual dissection tools in Human Anatomy.

Advantages	Disadvantages
"Hands-on" dissections, procedures and manipulations Safety and controlled settings Good training models or virtual, full human body Possibilities to train and make mistakes, learn from them No risks to cause problems to the cadavers or to the virtual patients Preparation for the unexpected situations Knowledge and learned skills transfer to a real life cases More details, zooming, highlighting and labelling of virtual images in dissections, procedures and manipulations Teams, group work and/or only individual sessions Real environment Training different scenarios and experience Testing of some very important procedures Effective time management Costs reduction of some procedures and manipulations Development of digital skills Education of students in medicine and dentistry Education of people to make positive changes about their health Improvement of quality of teaching and learning Combination between normal and clinical subjects	Quite expensive technology Functionality issues, technical difficulties Software problems Impossible to restore dissections, procedures manipulations when all is lost Lack of flexibility Difficult to simulate variations or pathologies Unable to simulate some important procedures and manipulations Challenging in large student groups

demonstrate layers and learn Topographical Anatomy using 3D format. For some students, watching a procedure performed by the teacher and afterwards performing the dissection by themselves with a scalpel were two very different things and experiences.

After some practical training, students became more comfortable with using virtual tools. It was possible to cut from the body surface to the inner body using a scalpel, as well as to watch images of 3D sections in the three planes. Majority of students realized that with just a click of a button they were given several options to study different parts and structures of human body. The general structures were labelled and further separated into next indentified elements. Our results of the study showed that the advantages of an adding the virtual scalpels to Human Anatomy course outweighed the disadvantages (Table 1).

During discussions students' answers as well as their feedback were used to explore details of an effectiveness of virtual tools in study process (Table 2). Students of both groups noted these activities to be very beneficial in classes.

Students performed several digital procedures on life-size male or female virtual cadavers that were viewed from different angles and tissue levels. Different types of cuts were used by students to look into the body. It was possible to create incisions and to remove and uncover different layers of tissues, to move deep inside step by step and to understand which structures it was necessary to look for. Afterwards students showed that they were able to place the organs back and reattach the bones, muscles, blood vessels in the body and put the skin back on. They increased or decreased areas of interest from superficial until deep levels of structures. Students enjoyed the virtual tools in the practical classes and learned theoretical material better. Majority of students performed work with the first type or straight scalpel because it was more useful for dissection or slicing.

All students highlighted or removed from the view direct or specific body parts. According to these procedures, the variations and/or abnormalities were compared with normal examples. Possibilities of virtual dissection allowed students to limit or to reduce any mistakes immediately. At the end of the practical class students performed reset of virtual body to its initial position and original state.

Table 2. The most common answers of students obtained after discussions about virtual tools at the end of practical classes.

Nr.	Free text answers
1	Resources, possibilities and virtual content of Anatomage Table increase interest in study course
2	Virtual tools help to understand and study better different anatomical topics
3	Virtual images and tools help to learn faster and remember longer
4	Possibilities give great experience and ways to learn Human Anatomy
5	Some images and tools give a clearer image of human body, parts, organs, structures and topography
6	Virtual dissection takes less time than traditional dissection
7	Procedures and manipulations are useful for training
8	No risks to perform something incorrect
9	Testing of level of knowledge and preparing for tests, colloquiums and exams in Human Anatomy
10	Fast access to different layers of human body

Virtual dissection was excellent for slicing with virtual scalpels, identification of special anatomical structures and the understanding of spatial relationships among organs. It was a new direction to see and interact with the same anatomy, as well as the access to a collection of normal and abnormal cases. Scalpels and their usage helped and allowed students to see structures from different positions and levels.

4 Discussion

Many of the world's medical schools and Universities use new visualization systems or technologies for medical education [13, 14]. Virtual 3D Anatomage Table is one of technology that can teach students different dissection procedures and can determine their levels of knowledge and competencies before they make real manipulations [15]. It allows students to repeat and return to the same tasks or steps of procedures or topics several times. Virtual 3D screen and possibilities can duplicate dissection or operative fields, support training and reduce some needs for expensive models in study process [16–18].

There is very limited information in the literature in regard to how the virtual scalpels of the Anatomage Table impact on students' teaching and learning. Several authors presented students' learning in different ways [19] and relationships between learning-teaching of anatomy [20]. Our students liked the ability to rotate, dissect and manipulate the virtual cadaver to better understand topography and cross-sections in Human Anatomy. The use of the "reset" button and possibility to return to previous steps allowed the students to repeat, train and make dissections of different regions and parts of the body several times. The tutors served as guides for students throughout the teaching and learning process. Students worked together in pairs or teams, thus forming a reality that was more similar to surgical environment [21].

The most significant challenge for students in learning Human Anatomy is direct understanding of body in dimensions and how different structures fit together [22]. There are many ways to improve the field of education with Anatomage Table using 3D dissection tools.

These new virtual things give possibility to make Anatomage supported training experiences available for students and tutors.

Can all students learn through the use of virtual tools in dissection of the human body? One way to determine this is to introduce the classes taught completely via virtual dissection and compare the results with those gained when the same classes are taught in a traditional way.

Several authors have shown that many technologies and virtual simulations can be used in medical education [23, 24]. But, definitely, they should not replace direct communication or face-to-face learning.

Our experience shows that virtual instruments can develop real learning and teaching environment for students and tutors. It is clear that technologies and virtual simulators will play a great and an important role in future, and quantitative measurements of competencies would have to be part of the education system [25–27]. Some dissection procedures or manipulations consist of a series of tasks and steps and different virtual instruments can be used.

In some cases virtual scalpels can be used for faster procedures, manipulations and dissection of the human body. Digital images can show some details of structures in more precise ways [28].

Students can remove away different layers, structures and understand precisely connections and relationships between structures. Virtual tools and demonstrations of visual images of performed manipulations can be included into study course creating a self-training environment. Students can perform not only individual sessions, but they can learn together and these possibilities are some of the most effective ways to learn Human Anatomy [29, 30]. Virtual scalpels also offer students the possibilities to practice responding to unexpected situations or scenarios [31].

We believe that the teaching and learning of Human Anatomy through the use of 3D virtual tools can be rather positive and productive accomplishment. One future area of this study will be to look at quantifiable data, comparing tests or exam scores from the students of previous years who had the virtual tools and dissection in their courses. Further studies are suggested to determine if a virtual experience can be a substitute for real dissections, or may serve as an enhancement to them.

5 Conclusion

Virtual dissections offered students the opportunity to prepare for and follow up on their human anatomy course. Virtual tools help students and teachers to understand and memorize better different anatomy structures. 3D dissection scalpels are useful for different teaching and learning activities, but the learning experience may be further enhanced by providing opportunity for the study of real materials.

References

[1] P. Benly, J. Pharm. Sci. Res. **6**, 242–43 (2014)
[2] A.W. Keedy, J.C. Durack, P. Sandhu, E.M. Chen, P.S.O' Sullivan, R.S. Breiman, Anat. Sci. Educ. **4**, 84–91 (2011)
[3] D.S. Barry, F. Marzouk, K. Chulak-Oglu, D. Bennett, P. Tierney, Anat. Sci. Educ. **9**(1), 90–96 (2016)
[4] J. Kerby, Z.N. Shukur, J. Shalhoub, Clin. Anat. **24**, 489–97 (2011)

[5] J.R. Fredieu, J. Kerbo, M. Herron, R. Klatte, M. Cooke, Medical Sci. Educator **25**, 183–194 (2015)
[6] P.K. Ganguly, Open Med. Educ. J. **3**, 5–10 (2010)
[7] K. Sugand, P. Abrahams, A. Khurana, Anat. Sci. Educ. **3**, 83–93 (2010)
[8] J. Collins, BMJ **337**, 665 (2008)
[9] L. Donnelly, D. Patten, P. White, G. Finn, Med. Teach. **31**, 553 (2009)
[10] K. Yammine, C. Violato, Anat. Sci. Educ. **8**, 525–538 (2015)
[11] T. Lewis, B. Burnett, R. Tunstall, P. Abrahams, Clin. Anat. **27**(3), 313–20 (2014)
[12] P. Ruisoto Palomera, J.A. Juanes Méndez, A. Prats Galino, Comput. Hum. Behav. **31**, 446 (2014)
[13] M. Hackett, M. Proctor, J. Sci. Educ. Technol. **25**(4), 641–54 (2016)
[14] H. Hoffman, D. Vu, Acad. Med. **72**(12), 1076–1081 (1997)
[15] C.F. Smith, H.S. Clin. Anat. **24**, 113–19 (2011)
[16] M. Zilverschoon, K.L. Vincken, R.L. A.W. Bleys, J. Biomed. Inf. **65**, 58–75 (2017)
[17] J. Sweitzer, Sci. **36**(2), 41–44 (1996)
[18] N. Hoyek, Ch. Collet, F. Rienzo, M. Almeida, A. Guillot, Anat. Sci. Educ. **7**(6), 430–437 (2014)
[19] D.I. Newble, N.J. Entwistle, Med. Educ. **20**(3), 162–75 (1986)
[20] K.M. Patel, B.J. Moxham, Clin. Anat. **21**, 182–89 (2008)
[21] M. Estai, S. Bunt, Ann. Anat. **208**, 151–57 (2016)
[22] R. Jaiswal, S. Sathe, V. Gajbhiye, R. Sathe, Int. J. Anat. Res. **3**(2), 1103–08 (2015)
[23] P. Bradley, Med. Educ. **40**(3), 254–62 (2006)
[24] J.P. Akpan, T. Andre, J. Comput. Math. Sci. Teaching **19**(3), 297–313 (2000)
[25] K. Wilkinson, P. Barter, J. Pedagogic Development **6**(1), 14–23 (2016)
[26] J. Qualter, F. Sculli, A. Oliker, Z. Napier, S. Lee, J. Garcia, S. Frenkel, V. Harbik, M. Triola, Stud. Health Technol. Inf. **173**, 359–61 (2012)
[27] T. Greenhalgh, BMJ **322**(7277), 40–44 (2001)
[28] R. Trelease, A. Rosset, Anat. Sci. Educ. **1**(2), 50–5 (2008)
[29] B. Preim, P. Saalfeld, Computers and Graphics **71**, 132–53 (2018)
[30] Z. Merchant, E.T. Goetz, L. Cifuentes, W. Keeney-Kennicutt, T.J. Davis, Comput. Education **70**, 29–40 (2014)
[31] M. Graafland, J.M. Schraagen, M.P. Schijven, Br. J. Surg. **99**(10), 1322–30 (2012)

Evaluation of results of social rehabilitation programs for young people with behavioural problems

Martins Moors[1,2] and *Juris* Osis[2]

[1] Rīga Stradiņš University, Social Work Department, Riga, Latvia
[2] Riga City Council Welfare Department, Riga, Latvia

Abstract. New targeted social service – social rehabilitation program for young people with behavioural problems is implemented in Riga since 2016. The program beneficiaries are young people from social risk families with multiple social problems – learning difficulties, delinquency, low communication skills, difficulties in problem and conflict resolution, addictions, social isolation, etc. The program is provided by two organizations – "The Samaritan Association of Latvia" and the "Social Services Agency". The duration of the programs is 6 months and 8 months respectively, each Program provides services for 15 young people aged 13 to 18 years. The study analyses results of the Program implemented by the "Social Service Agency" for 15 young people. The main objective was to determine if the implementation of the Program improves the social functioning of young people having behavioural problems.

Key words: social rehabilitation, teenage youth, behavioural problems, negative social consequences.

1 Introduction

One of the problems that are becoming more and more common nowadays is the behavioural disorder of teenage youth, which leads to different kinds of negative social consequences. There are many early school leavers. Several studies [1–3] indicate that lack of motivation for learning and problems in the family (lack of parental participation in youth life) are key factors in early school leaving.

1.1 Topicality of the problem

According to the data **of the Central Statistical Bureau of Latvia** (CSB), in Latvia in three quarters of 2017, the average proportion of NEET youth in the total number of young people (NEET level) aged 15–24 was 11.0% (12.0% in 2014), in the age group 25–29 – 15.8% (19.9% in 2014). Teachers point out that more and more of younger children are being over-aggressive or child-insensitive, which is a no less dangerous trend.[1]

According to the reports of the State Children's Rights Protection Inspectorate on the work of Orphan's Courts in 2015–2017 [4], the number of families in which the child's

[1] Arvien jaunākiem bērniem novēro agresivitāti un uzvedības traucējumus: `https://www.lsm.lv/raksts/dzive--stils/veseliba/arvien-jaunakiem-berniem-novero-agresivitati-uzvedibas-traucejumus.a242676/`

development and upbringing is not adequately ensured and for which the Orphan's Court has informed the municipal social service or other responsible institution during the reporting year, remains more than 1,500 annually. Moreover, in 2017 this was an increasing trend. The number of children in these families is approaching 3,000 each year.

The number of families with children in which there have been identified problems of social functioning in Riga has remained high during the last three years, i.e. in 2015, 2016, 2017, the number of families receiving social work services has remained at the previous year's level.

In 2017, the social work service was provided to 3 084 families with total family members of 12 405, of which 5 151 were children. Social workers of the Riga Social Service have identified and worked also with child behaviour problems (roaming, begging, etc.). According to the data of Riga Crisis Centres, more and more teenagers with serious behavioural problems are coming to Riga crisis centres in recent years.

Also, providers of Support Program for foster families, guardians and adoptive parents [5], in 2012–2015 have found that a large proportion of adolescent from foster families who have participated in the program are experiencing behavioural problems, school non-attendance and communication problems with authorities. Adolescents have experienced these problems for a longer time.

1.2 Characteristics of the problem and target group

Observations and experiences from teachers and social workers show that behavioural disorders can most often be observed in students in 5–9 grade. However, when they already appear as school non-attendance and other externally visible manifestations, the first signs have usually been three to four years earlier. The first signs appear as indifference, apathy when the child cannot be "moved" either by positive or negative reinforcements. Specialists point out that behavioural disorders appear as non-social behaviour – tiny thefts, school conflicts with adults, small hooliganism, fights in school and on the street, non-fulfilment of requirements and disrespect for authority. Often young people due to the behaviour mentioned above have taken notice of the police. The use of addictive substances, as well as dependence on smart devices, is also observed. Behavioural disorders also include roaming and periodic overnight stay out of the home, staying outside usual living area.

Young people have defensive mechanisms such as aggression, non-compliance, challenging action, gross abuse, which is a protective position because, in reality, the teenager is unhappy. Children with auto-aggression and pseudo-suicidal behaviour are more common than before [6] – hand and foot scratching, suicidal talk that professionals need to see as a "red" warning. Specialists point out that this is more common in girls.

The paradigm of internal and external forces is trying to explain why people sometimes act strange and unpredictable. The basic assumption in this paradigm is that there are forces that occur both in the person and in the environment, and that are the cause of certain types of behaviour. It is important to note that such forces exist and act upon a person, creating certain behaviour. Most often, the external forces put pressure on the young person who has internal forces that are usually weak [7].

Therefore, the causes of behavioural problems can be found not only in the young person, but also in his/her social environment and, first and foremost, in the family and the school environment, which do not meet the specific needs of the young person.

The reasons for the social environment are such as the parents' long-lasting conflicts in relationships, such as divorce, or parental violence when the child is forced to defend the mother. Then the child reflects this conflict outside – in school, on the street. Often there is only one parent, usually a mother, so the daily regime is not respected, and the discipline that is usually associated with the male role in the family is missing. But there is often no quality relationship with at least one parent. This means that in a lifetime when the family should help to meet the three main development tasks of the young person– develop autonomy, resolve parent-adolescent conflict, and ensure parent-adolescent attachment [6], the parents and the family do not provide that.

Other causes in the social environment mentioned by social workers are the use of alcohol by parents, which leads them to not be a resource for young people. Also overcrowding, inadequate living conditions when young people do not have their "place", sometimes also authoritarian, too strict parents. Then, when children reach the age of adolescence, they protest in different forms. In these circumstances, children, parents and employees of institutions often lack the skills to deal with conflict situations.

The more difficult the family situation is, the earlier the child must make decisions that are irrelevant to their age – to lie, to defend and not to "betray" parents, to manipulate "helping" adults and so on. Because of the above-mentioned situations, young people cannot follow learning material, and this leads to a lack of interest in learning. Children who have grown up in their class age (left for another year) are ashamed to learn with younger children. Without the support of adults, young people themselves have no internal capacity to change this situation. Children develop low self-esteem, insecurity and isolation – they don't tell their feelings to anyone and nobody knows about them. Depression and depressive symptoms are observed – children do not want anything, show grief, grief gets externally, in clothing. In theoretical literature it is acknowledged that mental health problems are most common among adolescents, youth who have been abused, youth whose caregivers have poor parenting skills. Maltreated children often have difficulties with self-regulation and social–emotional processing that may contribute to the development of mental health problems [8].

Some of the youngsters have attachment disorders. They do not trust the relationship because they are so-called "spotted" children, for example, having brought by their grandmother in the countryside (without formally prescribed custody) and alternately with both parents.

Around the age of eight to nine, there is a fracture point when children get into a mental crisis because they realize that parents' words do not match with their actions, that parents lie. But for about thirteen years you can notice the indifference when you offer what you want, the child does not accept it. Therefore, at this age, if the above-mentioned serious behavioural disorders are observed, long time and a special program are needed to change the behavioural pattern of the young person.

From the foregoing, it can be concluded that the most common serious behavioural disorder among young people starts to appear in adolescence, typically characterized as a period of time between about 10–12 years of age and about 21–22 years of age. Particularly important during this period is peer recognition and acceptance [6]. During this time, the focus is transferred from parents to peers and short-term intense emotional relationships with friends. If a young person has behavioural and attachment disorders, the conflict between his pursuit of autonomy and the lack of mature psychological competences increases. This leads to unrealistic ambitions, even though interest in learning is completely disappearing [9].

1.3 Characteristics of the social program

To address the above issues new targeted social service – social rehabilitation program for young people with behavioural problems (hereinafter – the Program) is implemented in Riga since 2016. The Program beneficiaries are young people from social risk families with multiple social problems – learning difficulties, delinquency, low communication skills, difficulties in problem and conflict resolution, addictions, social isolation, etc. The Program is provided by two organizations – "The Samaritan Association of Latvia" and the "Social Services Agency". The duration of the programs is 6 months and 8 months respectively, each Program provides services for 15 young people aged 13 to 18 years. The study analyzes the results of the Program, implemented by the "Social Service Agency" for 15 young people.

It is recognized that developmentally appropriate treatment for adolescents are often most concerned with forming and maintaining peer relationships. Their emotion regulatory deficits may be most salient in peer contexts, and thus building social competence with peers should be a priority [8]. That is why the new program was group based.

1.4 Theoretical perspective of the program

The service is based on Resilience's psychological intervention, the community's creative approach and adventure pedagogy. The resilience of people denotes their ability not only to overcome difficulties, but also to gain new strength after the stressful experience. Resilience is the personality ability and vitality that helps to cope with the trauma and the understanding that this power exists in every person.

Resilience is "a positive adaptation in extreme conditions after a trauma that develops the individual's functional and cognitive abilities" [10–12].

The focal of resilience is the concept that people's abilities, their capacity to cope, are greater than believed, and the task is to help people mobilize their resources to cope with the problem. It is possible to train and develop these abilities to deal with difficulties. There are protective factors in one's life and that can help explain why one young person may cope better with adverse life events than another. Some young people are more intrinsically resilient than others because of a whole range of these protective factors [13]. This approach recognises that although it may not always be possible to protect young people from further adversity and that while it may not always be possible to provide an ideal environment for them, boosting their resilience should enhance the likelihood of a better long-term outcome. A resilience-based approach focuses on maximising the likelihood of a better outcome for young people by building a protective network around them. The concept of resilience increasingly offers an alternative framework for intervention, the focus being on the assessment of potential areas of strength within the young person's whole system. This approach enables a move away from an assumption that a parent or alternative placement will provide all that the young person needs. Instead the emphasis is on building a network of support from the resources available and adding to them with professional support where necessary. It also emphasises the importance of building on the potential areas of resilience within the young person by maximising opportunities for engaging in hobbies, associating with friends, experiencing success, contributing and so on [13].

By providing help, the focus is not on the problem but on the human resources and strengths. Resilience allows perceiving and treating a person who has suffered from a trauma or a serious living condition, as a personality, as a human being, not a victim. Consequently, resilience is based on the strengthening of resources, emphasizing of positive aspects, and personality acceptance, separating the personality from behaviour.

The Community's creative approach is geared toward teenagers' social and behavioural changes. Development of creative community involves not only helping professionals (social workers, psychologists, psychotherapists, educators) but also representatives of other professions from the community (working in a team) – artists, coaches, architects, cultural workers, musicians, managers, etc. Working in a united team, the emphasis is on education and building new experiences. The Community's creative approach is process-oriented and resource finding-oriented. Tasks for this approach are:

- Helping the target group to decide, plan, act and identify their own needs,
- Develop and mobilize individual and environmental / community resources,
- Develop a creative and innovative approach by acting more efficiently and interactively,
- Build closer links between service providers and target groups,
- To change conditions and environmental conditions to create new quality needs.

There is an assumption in the Community's creative approach – any experience is a lesson. Each lesson is personal and unique to everyone. The dynamics of experience form a process of change that promotes personality openness, critical thinking, the need to acquire new knowledge and skills, and encourages a change of attitude towards life.

Adventure Pedagogy is about gaining experience by living outside your usual environment or travelling and participating in various exciting activities. Adventure therapy has a positive effect on the self-confidence and self-esteem of the participants. It promotes mutual trust and responsibility, develops socially acceptable behaviour and teaches to seek help.

All these theoretical concepts were integrated in the new service – social rehabilitation program for young people with behavioural problems (hereinafter – the Program).

1.5 Content of the social program

Aim of the program is to promote the social integration of adolescents/young people at risk of social exclusion in Riga and to reduce the risks of social exclusion by providing social rehabilitation and support programs for social risk persons aged 13–18 years.

The program is holistic in composition and includes physical training, games and recreation, career development, relationship development, nutrition, relaxation training, and a social network analysis where each youth contracts with a "village" of adult supportive mentors.

During the implementation of the program, six themes are updated, which correspond to the main components of resilience:

1. Identity (self-esteem, personal goals, values – the aim is to strengthen authenticity, specific resources, look for ways the group can help strengthen each participant).
2. A sense of belonging, relationships with peers, parents (family goals, goals and initiatives of the group).
3. Structure and consistency (self-discipline, rights and obligations, responsibility).
4. Challenges (personal, group and family difficulties, overcoming them, risks and mitigating them).
5. Security and Support (awareness of external resources, experience to use them) – the aim is to learn the skills to ask and provide support.
6. Realistic optimism (future goals are hopeful and optimistic, but they are based on the reality of the current situation).

These six themes, or sections of the program, take the form of activities such as:

- **Group relationship building camp** (3 days),
- **Creative group activities** (Interactive communication/resilience psychology; non-violence philosophy/survival skills; contemporary dance/freedom of movement; creative self-expression/music/drama/cinema/art) (256 hours–128 lessons/2 hours),
- **Psychotherapy sessions** (240 sessions/16 for adolescence and support person),
- **Social worker counselling** (300/20 for adolescence and support person),

- **Specialist consultations** (narcologist, gynaecologist, career counsellor, specialist teacher, etc.) as needed (75 counselling/5 adolescents),
- **Mentoring** (600 hours/40 mentor hours per 1 teenager), Inter-institutional cooperation (regular inter-institutional meetings during the provision of the service) 120 hours per program/8 on average for adolescents,
- **Lessons for adolescents with parents** 96 hours/32 lessons × 3 hours.

All program activities focus on changes in thinking, behaviour and attitudes, focus on things that succeed and regular positive feedback (building self-confidence), creating new and positive experiences for young people, development of awareness of actions and consequences. Program activities are regular and intense.

2 Material and methods

The main objective of the research was to determine if the implementation of the program improves the social functioning of young people with behavioural problems in Riga municipality.

An analysis was conducted on the grounds of the self-assessment of young people involved in the implementation of the program and their parents, in accordance with the 360-degree Competence assessment method adapted by the service provider.

The self-assessment of young people involved in the implementation of the program and their parents was conducted before and after participation in the program using the self-assessment questionnaire. According to the 360-degree Competence assessment method adapted by the service provider, self-assessment questionnaire comprised 20 questions, which were organized into four categories each with corresponding indicators: Motivation (6 indicators – School attendance, Diligence, Motivation, Reflection of the initiative, Vision of career, Persistence of interests), Emotional stability (7 indicators – Emotional balance, Skills to deal with conflict situations, Responsibility for actions, Ability to provide support, Ability to express constructive criticism, Respect for adults, Tolerance for peers), Social skills (4 indicators – Personal hygiene and self-care, Ability to follow the rules, Behavior, Ability to work in team) and Self-image (3 indicators – Self-esteem, Self-criticism, Creative self-expression). Each of 20 questions asked to assess the level of each indicator on the scale from 0 (the lowest assessment level) to 10 (the highest assessment level).

Descriptive statistics were used to calculate the mean values for data processing. The non-parametric Wilcoxon signed rank test for two dependent samples was used to determine service efficiency. The size of the effect was calculated using the formula:

$$r = z/\sqrt{N} \tag{1}$$

where r is the size of the effect, z is the calculated value of the Wilcoxon signed rank test, and N is the total number of participants from both samples. In statistics, the size of the effect is a quantitative indicator of the magnitude of the phenomena. Effect sizes complement the testing of statistical hypotheses and play an important role in Power analysis, sample size planning and meta-analysis. In order to determine how successful the implementation of the service will be in Riga municipality, Power analysis was used.

3 Results

The Wilcoxon signed rank test showedthat by the self-assessment of participants of the program all criteria of assessment have positive improvement (see Table 1). Assessment of

Table 1. Mean values of assessment before and after the program.

	Parents		Youngsters	
	Before	**After**	**Before**	**After**
Behavior	5	7	5	7
Diligence	4.5	7	3.5	5.5
School attendance	5	7	5	8
Personal hygiene and self-care	4.5	7.5	6.5	8
Responsibility for actions	3.5	6	6.5	8.5
Motivation	4.5	6.5	5.5	8
Self-esteem	4.5	6.5	5.5	8
Emotional balance	2	6	6	7.5
Ability to work in team	5	6	5	7
Skills to deal with conflict situations	4.5	6.5	6	8.5
Reflection of the initiative	5	7	5	7
Self-criticism	4	5.5	5	6
Ability to express constructive criticism	4	7	5.5	7
Ability to provide support	5.5	6.5	5	8
Respect for adults	5	6.5	5.5	8.5
Tolerance for peers	4	7	5.5	8
Vision of career	3	6	4.5	6.5
Persistence of interests	5.5	6	5	7.5
Ability to follow the rules	2.5	4.5	4	6.5
Creative self-expression	5.5	6.5	4.5	6.5
Total	4.5	6.5	5	7.5

parents showed positive improvement as well except for two criteria: School attendance and Ability to provide support. For example, in the criteria Ability to provide support parents did not see improvement, whereas, children found that they had the best results in this criterion.

The most remarkable improvements were recorded for following criteria: Personal hygiene and self-care, Emotional balance, Ability to express constructive criticism, Tolerance for peers, Vision of career, School attendance, Ability to provide support, Respect for adults.

The lowest improvements in assessment were observed for such indicators as Self-esteem, Self-criticism, Creative self-expression, which all 3 are indicators comprising Self-image category of indicators.

In general, parents evaluate the social functioning of their children more critical than the children themselves do. Statistically significant differences between parents and youngsters assessment were observed for such indicators as Responsibility for actions, Motivation, Emotional balance, Skills to deal with conflict situations, Respect for adults, and Ability to follow the rules – the self-assessment of youngsters for all 6 mentioned indicators was significantly higher than assessment given by the parents.

At the same time statistical analysis did not showed any differences in the self-assessment by the gender so it can be concluded that the program works equally efficiently for both boys and girls. On average, participants' social functioning improved by 2–

2.5 units. The calculated size of the program effect is $r = 0.5$ (average) – this means that the program not only statistically but also practically significantly affect the youngsters' social functioning.

When performing Power analysis, it was found that with the existing sample we can only be 75% sure that the program will work at the same level of effectiveness if introduced in the whole Riga municipality. The minimum level of confidence for this study should be 80%, and 95% is recommended. At present, data suggests that in order to be 95% sure that program will be effective if implemented in Riga, the experimental group should be increased to at least 27 children.

4 Conclusions

1. It was found that the implementation of the social rehabilitation program for young people with behavioural problems improves the social functioning of young people with behavioural problems from social risk families with multiple social problems in Riga municipality.
2. 360-degree Competence assessment method and tool proved to be effective approach for self-assessment of young people involved in the implementation of the program and their parents as it can give general assessment of the program as the whole as well as be more specific regarding various dimensions of the changes of the young person's behaviour.
3. Overall assessment of the program shows positive improvement in youngsters' social functioning. On average, social functioning of young people has improved by 2–2.5 units. The service effect is $r = 0.5$ (average) – it means that the service not only statistically but also practically substantially affects the social functioning of young people.
4. Youngsters' self-assessment and parents' assessment does not significantly differ in the majority of indicators of social functioning, but parents are more critical of the social functioning of their children than children themselves. Especially critical parents are for two criteria: School attendance and Ability to provide support– indicators the youngsters themselves estimate they have the best results. This difference in assessment is an important indicator for social workers when planning future case work with the family as a system.
5. There are no significant differences in the evaluation between girls and boys which means that program works equally well for both genders.
6. The greatest improvements are in the category "Emotional Stability": Emotional balance increased by 4 units, Ability to provide support, Ability to express constructive criticism, Respect for adults and Tolerance for peers increased by 3 units. The lowest improvements are in the category "Self-image", which include: self-esteem, self-criticism and creative self-expression.
7. It can be concluded that the activities of the program give the most significant contribution to the social skills related to the provision of peer support, the ability to interact with other adults and peers. In turn, the smallest impact of program activities is on the psychological state of the young person, which suggests that this factor should be given more attention to in the future implementation of the program.
8. In turn, to argue that such result would also be achieved by the introduction of a service throughout the municipality of Riga, it is necessary to collect additional data with at least one group of youth – in other words the experimental group

should be increased at least up to 27 participants to obtain 95% level of confidence of results.

9. Results of the assessment of the program prove that the combination of resilience's psychological intervention, the community's creative approach and adventure pedagogy is the successful foundation of the social rehabilitation program for young people with behavioural problems.

References

[1] Izglītības un zinātnes ministrija, Pētījums par priekšlaicīgas mācību pamešanas iemesliem un riskiem jauniešiem vecuma grupā no 13 līdz 18 gadiem [Ministry of Education and Science. Study on causes and risks of early school leaving of youth aged 13 to 18] (2014). Available at: `www.izm.gov.lv/images/statistika/petijumi/41.pdf`

[2] Maslo I., Jaunu izaicinājumu un to risināšanas iespēju identificēšana un analīze, kas ietekmē pieaugušo (18–24 gadi) iesaistīšanos (atgriešanos) mūžamācīšanās procesā [Identification and analysis of new challenges and opportunities influencing involvement (return) of adults (18–24 years) in lifelong learning] (2015). Available at: `http://sf.viaa.gov.lv/library/files/original/Informativais_zinojums_p_t_jumu_kops.pdf`

[3] Izglītības un zinātnes ministrija, Pētījums par politikas alternatīvu veidošanu priekšlaicīgas mācību pārtraukšanas problēmas risināšanai [Ministry of Education and Science. Study on development of policy alternatives for tackling early school leaving (2015). Available at: `http://www.izm.gov.lv/images/statistika/petijumi/IZM_PMP_Gala_zinojums_AptaujuAptauj_ExcoloLatvia_2015.pdf`

[4] Valsts bērnu tiesību aizsardzības inspekcija, Pārskati par bāriņtiesu darbu 2015–2017, gadā [The State Inspectorate For Protection Of Children's Rights. Reports on Orphan's Court Work years 2015–2017]. Available at: `http://www.bti.gov.lv/lat/barintiesas/statistika/`

[5] Rīgas domes Labklājības departaments, Rīgas domes Labklājības departamenta gadagrāmata "Sociālā sistēma un veselības aprūpe 2017. gadā" [Riga City Council Welfare Department. Riga City Council Welfare Department Yearbook "Social System and Health Care in 2017"] (2018), 55–56. lpp Available at: `http://www.ld.riga.lv/files/Gadagramatas/SOCIALA_SISTEMA_UN_VESELIBAS_APRUPE_RIGA_2017_GADA.pdf`

[6] J.B. Ashford, C.W. LeCroy, K.L. Lortie, *Human Behavior in the Social Environment: A Multidimensional Perspective* (Pacific Grove, CA: Brooks/Cole Pub. Co., 1997), p. 330, 333, 581

[7] R.A. Skidmore, G. Milton, O. Thackeray, O.W. Farley, L.L. Smith, S.W. Boyle, *Introduction to social work* (Boston, MA: Allyn and Bacon, 2000), 404

[8] *Neuroscience for Social Work: Current Research and Practice*, edited by Holly Matto, et al. (Springer Publishing Company, 2013), p. 112, 127

[9] Н.П. Рюгаард, *Детиснарушениемпривязанности: указаниякоргани зациитерапевтическойсредыдляребёнка: теория, причины, симптомы, лечение* [*Children with a violation of affection: instructions for organizing a therapeutic environment for a child: theory, causes, symptoms, treatment*] (Keila: IgaleLapselePere, 2016), p. 312

[10] M. Rutter, Implications of Resilience Concepts for Scientific Understanding, Ann. N.Y. Acad. Sci. **1094**, 1–12 (2006) doi:10.1196/annals.1376.002
[11] N. Garmezy, Stress-Resistant Children: The Search for Protective Factors. In J.E. Stevenson (Ed.), Recent Research in Developmental Psychopathology: Book Suppl J. Child Psychol. Psychiatr. **4**, 213–233 (1985)
[12] A.S. Masten, J. Coatsworth, The Development of Competence in Favorable and Unfavorable Environments: Lessons from Research on Successful Children. Am. Psychol. **53**, 205–20 (1998)
[13] D. Brigid, S.Wassell, *Adolescence: Assessing and Promoting Resilience in Vulnerable Children* (Jessica Kingsley Publishers, 2002), p. 173

Challenges and their overcoming when implementing the restructuring of children's care homes in Lithuania

Vida Gudzinskiene and *Rita* Raudeliunaite

Mykolas Romeris University, Vilnius, Lithuania

Abstract. The aim of the study is to reveal challenges and the ways to overcome them in the context of the restructuring of childcare, based on the experience of social workers who work in children's care homes, which participate in the restructuring. Qualitative research type was chosen for the study. The method of a semi-structured interview with social workers who work in care institutions, that take part in the deinstitutionalisation process. The challenges arising to social workers in the context of the restructuring, which came to light during the study, are related to the resistance and anxiety of social workers regarding their future, the lack of information on the restructuring, the lack of engagement of social workers in the restructuring processes, the lack of the dissemination of good practices, the need to improve professional qualifications and the possibility to choose training courses of qualification improvement which correspond to their needs. The unpreparedness of the community for the restructuring, as well as the finding of foster parents and adoptive parents for children or the reunification with their family and the anxiety of the children about their future were also identified. In order to overcome the challenges arising during the restructuring it is necessary: to have a restructuring plan which is clear and understandable to all, the preparation of the employees for changes, the increase of dissemination of information in the communityand society, the rallying of the community, inter-institutional cooperation and the preparation of parents/ adoptive parents/foster parents.

Key words: deinstitutionalisation, care homes, family and community-based care, children, parents, adoptive parents, foster parents, social worker.

1 Introduction

In 2016, according to the data of State Child Rights Protection and Adoption Service [1], 518 201 children under 18 years of age were living in Lithuania. Out of them, 9106 children were in care: 65% of all the children in care were growing in the families of their foster parents and 35% were growing in care homes. Although the number of the children in care homes is decreasing, a significant proportion of children, who were placed in care, are growing in care homes.

The studies conducted in Lithuania and Europe demonstrated that institutional care always result lower quality of life than quality services provided in the community and often result lifelong social exclusion [2–7]. The scientific studies on early child development required by the World Health Organisation and the United Nations demonstrate that, in the case of very little children, even relatively brief period of a stay in the institution might adversely affect the child's development and result in lifelong negative consequences for emotional well-being and behaviour ofchildren [8]. The studies demonstrate that mental development of children, who live in care homes, slows down due to the lack of quality relationships and meaningful communication with the adults [9]. The scientists also draw attention to the fact that living conditions in care homes can cause different problems in children within a field of their health, cognition, emotions and morality [10–12]. Care homes are not designed for a child's positive socialisation, because these institutions, although they have sufficiently good financial resources, are not comparable with family environment, it is impossible to safeguard child's individual needs, and in particular quality relationships with the adults. C.M. Connell, R.J. Prinz [13] claim that the child, who has close relationships with parents, has well developed social skills, whereas the child, who has not had any such relationships in the family, is characterised as lacking social skills, avoiding people, being aggressive and not able to control himself. Institutionalisation is increasingly recognised as a vicious policy and a violation of human rights.

The deinstitutionalisation of child care system has recently been one of the priority areas in Lithuanian social policy. The beginnings of the deinstitutionalisation of child care in Lithuanian social policy can be noticed in the Measures for Implementation of the Programme of the Government for 2001–2004 when it was foreseen to bring about the reform of child care system, giving priority to child care in the family and improving a support system for orphans and children deprived of parental care. The aspiration to reform childcare system was established in the Concept of Child Welfare [14], the Regulations on the organisation of child care [15] and in the Reorganisation strategy of Child Care Systemand the Plan for Its Implementation for 2007–2012 [16]. The most recent documents, which examine the issues of the deinstitutionalisation of child care system [17, 18], further emphasise the necessity for the transition from institutional to community-based social services, in order to safeguard harmonious environment and conditions for children deprived of parental care to grow with foster parents or adoptive parents and receive assistance in the community. Since 2008 Lithuania has started to implement foster parent preparation programme PRIDE (Parent Resources for Information, Development, and Education), which is carried out by certified social workers. The aspirations for the restructuring set out in the documents and proposals are rational and profound, however, as practice has shown, they are difficult to implement: there is still the lack of social services for families, the forms of alternative care, the legal regulation of child care forms has shortcomings, care forms being in force, are too strictly defined, temporary care is not flexible enough, legal framework, which would regulate emergency aid to the family having a crisis situation, has not been established yet [19].

The transition from institutional to family and community-based care is gradual. Political basis for that is important, however, not only authorities, but also care institutions themselves have to be involved in that. Therefore, while the restructuring is underway it is relevant to investigate what challenges social workers working in children's care homes, which participate in the restructuring, encounter, and while analysing the findings of the studies, to model strategies for overcoming arising problems, in order that the objectives set out in the restructuring would be achieved.

The aim of the study is to reveal challenges and the ways to overcome them in the context of the restructuring of childcare, based on the experience of social workers who work in children's care homes, which participate in the restructuring.

Problematic questions of the study: 1) what challenges arise during the restructuring of children's care homes; 2) how is it possible to overcome challenges related to the restructuring of care homes?

2 Research methodology

Qualitative research type was chosen for the study. The method of a semi-structured interview with social workers who work in care institutions that take part in the deinstitutionalisation process was used. The presented article overviews two problematic questions of the study: 1) what challenges arise during the restructuring of children's care homes; 2) how is it possible to overcome challenges related to the restructuring of care homes?

The obtained data were analysed using the content analysis method. Qualitative content analysis was carried out in accordance with the inductive, study data based and categories composed logic. Content analysis is a technique which, having examined the specificities of the text, allows, objectively and systematically, draw reliable conclusions [20]. The qualitative content analysis was made regarding the following sequence [20]: repeated reading of the content of transcript interview texts, distinction of meaning elements in the text analysed, the grouping of the distinguished meaning elements into categories and sub-categories, the integration of the categories/sub-categories into the context of the phenomenon analysed and the description of their analysis.

Criteria-based sample was used in the study. The informants were chosen according to three criteria: 1) social workers who have a degree in social work, 2) social workers who work in care institutions that take part in restructuring, which consists of the transition from institutional care to services, which are provided to children, who have lost parental care, in a family and community; 3) social workers whose length of their working time in children's care homes is at least 5 years.

The study was conducted in the September-October of 2015 in 10 children's care homes chosen for the restructuring. 14 social workers (all female) participated in the study.

The permission of the heads of children's care homes to conduct the research in their organisation has been obtained. Personal permission of the informants to take part in the research was sought. The researchers obliged themselves to the research participants not to divulge the information related to the organisation and the informant. The principles of anonymity, volunteerism and benevolence were followed. The social workers were acquainted with the goal of the research, the interview questions, and the importance of justified reflection on their experience.

3 Analysis of research results

3.1 Challenges arising during the deinstitutionalisation of children's care homes

During the study social workers, who work in children's care homes, which participate in the restructuring, were asked to express their opinion on challenges arising during the restructuring of children's care homes. While analysing the answers obtained from the informants, three categories were distinguished: challenges arising to social workers, challenges which are related to society and community and challenges which are related to children (Table 1).

The success of the restructuring of institutional care depends on the appropriate preparation of social workers for the restructuring. Based on the opinion of the informants, not all of the employees of care are positively disposed to the restructuring: a portion of employees feel anxiety about their future, they are afraid to lose their job and social guarantees. According to the informants, the anxiety of social workers about the restructuring and resistance to it arises due to the lack of information on restructuring. The informants underlined that the originators and the implementers of the reforms "do not consult with social workers who work in care homes". Social workers resist and are anxious about the effectiveness of the reform, whether transition from institutional care to family and community-based services will be really implemented properly. They doubt as regards the declared changes due to different reasons: the projects which failed earlier, the lack of

Table 1. Challenges in the context of the restructuring children's care homes.

Category	Subcategory
Challenges arising to social workers	The resistance of social workers and their anxiety about the future
	The lack of information on restructuring
	Insufficient involvement of social workers in the restructuring processes
	The lack of the dissemination of good practices
	Necessity to improve professional qualifications
	Challenges related to the supply and demand of the themes and trainings of qualification improvement
Challenges related to society and community	Negative approach of the community towards the educatees of children's care homes and resistance to the restructuring
	The unpreparedness of the community for the restructuring
Challenges related to children	The finding of foster parents, adoptive parents for children or the reunification with their family
	The anxiety of children about their future

information and negative personal experiences. The distrust of employees is basically related to systemic problems. Similar tendencies were also revealed by the studies of other countries, in which the deinstitucionalization of care homes is carried out [21–24], showing the fears of employees to lose their jobs, limited sustainability of the deinstitutionalization projects, because some projects, which were carried out and are being carried out, are based on abstract ideas, were insufficiently considered and discussed at various levels.

On the other hand, the lack of the dissemination of good practices came to light, which is associated by social workers with the desire to share good practices when carrying out the restructuring of childcare institutions. The informants state that it is important to share good practices, in order that mistakes, misunderstandings and tensions would be avoided. Other studies [23–25] demonstrate that the employees of care institutions lack information about specific decisions, how their further work will be organised and how posts will change. During the reform process there is the lack of discussions in care homes themselves and with originators of the restructuring and/or the representatives of the authorities. Internal discussions within the staff of care institutions and the discussions with the representatives of the authorities could be a good opportunity to engage employees in the process more and thus decrease their anxiety and fears, and sometimes scepticism or even resistance to the restructuring. The Common European Guidelines on the Transition from Institutional to Community Based Care [8] emphasise that the sufficiency of information and the dissemination of good practices are among the most important factors that guarantee the success of deinstitucionalisation.

When analysing the findings of the study the need of social workers to improve their professional qualifications in the context of changes came to light, because during the transition to community services the existing competences of social workers are not sufficient to provide quality services. It should be noted that social workers encounter the problem regarding the themes of qualification improvement and the supply and demand of trainings. The informants noted that the content of teaching programs does not always match the teaching theme. Trainings for the social worker of children care homes are organized together with other social groups, therefore the knowledge cannot be applied for work with children.

In addition, the trainings lack the examples of solutions to their practical problems and case studies. The informants underlined that "theoreticians discuss an urgent problem superficially without taking a close look at it". It should be noted that part of social workers, who work in care institutions that take part in deinstitutionalisation process, plan to be foster parents and participate in the trainings under the PRIDE programme. Trainings last for about 3 months. During the trainings, individual meetings are held in the institution and the homes of future foster and adoptive parents and group trainings are done (one meeting, which lasts about 3 hours, per week). 10 themes are discussed during the meetings: introduction; teamwork; a sincere conversation with an adopted or foster child; the satisfaction of developmental needs; the sense of self-worth; upbringing; the strengthening of constant relationships; preparation for changes; making conscious decisions. Social workers, who work in care institutions that take part in deinstitutionalisation process, and will take part in group foster care in the community, would like to have systematic trainings under homogeneous programmes similar to PRIDE. Based on the findings of the study, it can be maintained that, for social workers, who participate in the restructuring, qualification improvement is one of the most urgent challenges and requires immediate solutions. As it is stated in the Common European Guidelines on the Transition from Institutional to Community Based Care [8], a systematic and coordinated teaching program is a condition for the creation of services in the community, this ensures that employees would be adequately trained.

The study revealed that among the challenges, which arise in the context of the restructuring, are the negative approach of the community towards the children of care homes and resistance to the restructuring. It is noteworthy that the media also influence a negative approach of the community towards the restructuring of child care institutions. According to the informants, "the signs of labelling, prejudice could be seen in the television report when after closing one of children's care home the residents were screaming to the journalist, "What is going on here? Picture that, children from a care home were accommodated in our house. Nobody is going to ask us. How are we going to live in the neighbourhood now?" That shows the unpreparedness of the community for the restructuring, which was also underlined by the social workers. According to the informants, municipality institutions, which are responsible for the restructuring of children's care homes, do not prepare the community for the restructuring processes. On the other hand, social workers, in the absence of a suitable preparation for the restructuring, are also not able to contribute to the preparation of the community for the restructuring with expertise and quality. It is noteworthy that only properly communicating and maintaining relationships with community at different stages of the restructuring, it is possible to succeed in reducing its resistance and to safeguard child well-being [8].

When analysing the findings of the study challenges related to the finding of foster parents and adoptive parents for children or the return to families and the anxiety of children about their future were identified. Social workers feel the greatest anxiety regarding the finding of foster/adoptive parents and their preparation to receive children to families, as well as the preparation of children to live in their new family. On the other hand, the informants note that there is the lack of preparation of parents, whose parental rights were restricted, for the return of their children to the family.

3.2 Ways to overcome challenges

When analysing the findings of the study, the ways, proposed by the social workers to overcome challenges in the context of the restructuring, were identified: a restructuring plan which is clear and understandable, the preparation of the employees for changes, the increase of dissemination of information in the community and society, the rallying of the community, inter-institutional cooperation, the preparation of foster parents and children (Table 2).

During the study all of the informants noted that one of the most important challenges, which urgently need to be overcome, is the restructuring plan which is clear and understandable. It is noteworthy that, according to the regional plans of the restructuring, the restructuring plans and their course must be discussed in detailand it has to be agreed what particular community services and what infrastructure will be needed for the family and children. The informants noted that the course of the restructuring of care institutions should be understood not only by a care institution, but by the community and society as well. Therefore, the provision of information at various levels – micro-, macro- and meso-level is important.

The increase of the dissemination of information in the community and society is an important challenge in the context of the restructuring. It is noteworthy that greater regard should be given to awareness-raising/advertising on the restructuring itself to the communities in order that they would understand how and why the restructuring is carried out. In addition, there is a need for the wider dissemination of good practices, the publicity of the activity of foster parents, in order that community services would develop and a more favourable approach of the society towards the reform would form. Society needs to understand that the restructuring of child care system is very important and not only the Ministry of Social Security and Labour or other institutions or municipal institutions need it, but it is important in order to safeguard children's well-being. When the community rallies and receives into their midst children deprived of parental care, not only will children's rights be less violated, but the effectiveness of the social protection system will increase as well.

Another way to overcome challenges of the restructuring, which came to light during the study, is the preparation of the employees for changes. The participants of the study noted that, for them, trainings on the preparation of the child for independent living, assistance to persons with addictions, the development of personal and social skills in children, the solution of behavioural and psychological problems in children (ways to tackle conflictual and aggressive behaviour and their management, the work with children with behaviourial and emotional disorders, the encouragement of positive behaviour, sexuality education, the solution of problems) and the preparation of the child and employees for the restructuring are of particular relevance. The need for the supervisions of social workers, which, as is believed, will help them "cope with their anxiety and maybe unfounded fears", also came to light. The informants drew attention to the need for the organisation of trainings oriented towards practical applicability and the necessity for specific methodologies how to work with children of different agesand who have undergone different experiences. Social workers, who work in care institutions, that take part in deinstitutionalisation process, and will take group foster care in the community, as mentioned above, would like systematic trainings under homogeneous programmes.

One of the prerequisites for the successful implementation of the restructuring is inter-institutional cooperation. According to the informants, "it is important that the social workers of children care homes would receive information and assistance from the centres of social services, departments of the protection of children's rights, social workers of elderships, the representatives of schools and in particular social pedagogues of schools". Joint agreement of the parties, which are interested in the restructuring (care institutions, authorities, communities), is important to ensure children's well-being and to represent their interests.

One of the main challenges, which also guarantees the success of the restructuring, is the preparation of adoptive parents/foster parents. Social workers note that "the institute of

Table 2. Ways to overcome challenges.

Subcategory	**Confirming statements**
A restructuring plan which is clear and understandable	"According to regional plans of the restructuring, there is a necessity to discuss and to agree and to know what particular community services and what infrastructures will be needed for the family and children..." (D), "The plan and the course of the restructuring should be understood not only by a care institution, but also by the community and the entire society through wider publicity..." (M)
The increase of dissemination of information in the community and society	"More advertising on the restructuring itself should be created, the members of the community should be provided more information, in order that they would imagine how everything will be in reality" (N), "Society needs to be informed, prepared and a more favourable attitude towards the reform has to be promoted". (E)
The preparation of the employees for changes	"More training on urgent topics need to be organised: the preparation of the child for independent living, assistance for persons with addictions, the development of personal and social skills. . . " (L), ". . . very relevant themes are the solution for behavioural and psychological problems in children (ways to tackle conflictual and aggressive behaviour and their management, work with children with behaviourial and emotional disorders, the encouragement of positive behaviour, sexuality education, the solution of problems)..." (K), "The supervisions of social workers are necessary when assisting them to cope with their anxiety and maybe unfounded fears". (I)
The rallying of the community	(B), "Only when community rallies and renders assistance to foster parents children's rights will be less violated, ... the effectiveness of the restructuring will increase" (L), "The rallying of the community is very important when receiving a child and providing a sincere support". (G)
Inter-institutional cooperation	"It is important that the social workers of children care homes receive information and assistance from the centres of social services, departments of the protection of children's rights, social workers of elderships and social pedagogues of schools" (H), "Joint cooperation of the parties, which are interested in the restructuring, when responding to the child's interests is the most important thing". (J)
The preparation of foster parents and children	"To prepare foster parents... the institute of foster parents is one of priorities the restructuring of institutional care, because it is thanks to it that children will be able to grow in the environment which resembles the family environment" (I), "It is important to enable families to take their children from institution". (N)

foster parents is one of the priorities of the restructuring of institutional care, because it is thanks to it that children will be able to grow in the environment which resembles the family environment". The informants are concerned regarding the preparedness of adoptive parents/foster parents to receive children into their family: "We have been having problems with the preparation of foster parents, and it is on their preparedness to receive a child that the well-being of a child will depend very much". Although future foster parents participate in trainings, pass a selection test, in Lithuania, every year foster parents are suspended from their functions, because they perform them improperly (from 31 in 2011 to 23 in 2015). In 2015, when implementing the general programme of the preparation of foster parents and adoptive parents, 56 certified social workers, who work in 21 institutions, acted and provided the services of foster parents in Lithuania, whereas there are 60 towns and district municipalities in Lithuania. On average, there are 52 foster families, which have completed introductory and continuous trainings, and almost 100 existing foster families, which are periodically given guidance and various forms of assistance, per one certified social worker per year [26]. Thus, part of foster parents lack professional guidance and support. Recently, while the deinstitutionalisation of children's care homes is underway, more certified social workers, who are able to train and give guidance and provide continuous training modules for adoptive and foster families, are being prepared.

4 Conclusion

The challenges, which came to light during the study, arising for social workers in the context of the restructuring are associated with the resistance of social workers and the anxiety about their future, the lack of information on the restructuring, the lack of engagement of social workers in the restructuring processes, the lack of the dissemination of good practices, the need to improve professional qualifications and the possibility to choose training courses on qualification improvement corresponding to their needs. The negative approach of the community towards the children of care homes and resistance to the restructuring and the unpreparedness of the community for the restructuring, as well as the finding of foster parents, adoptive parents for children or the return to their family and the anxiety of children about their future have also been identified.

In order to overcome the challenges arising during the restructuring, among the immediate actions, which urgently need to be taken, are the detailed discussion of the plans of the restructuring and its course at micro-, macro- and meso-levels and agreements what community services and what infrastructure will be needed for the family and children in the particular community. In order that the community would take a more positive stance towards the restructuring of children care institutions which is being carried out, it is necessary to step up the dissemination of information in the community and society on the course of the restructuring and to disseminate good practices of deinstitutionalisation.

The preparation of the employees for changes is associated with the organisation of relevant trainings on the preparation of the child for independent living, assistance for persons with addictions, the development of personal and social skills in children, the solution of behavioural and psychological problems in children (ways to tackle conflictual and aggressive behaviour and their management, work with children with behaviourial and emotional disorders, the encouragement of positive behaviour, sexuality education, the solution of problems) oriented towards practical applicability and the need for specific methodologies how to work with children of different ages and children who have undergone different experiences.

In order that the restructuring of child care institutions would be successful, inter-institutional cooperation is important in order to safeguard the child's well-being and to represent his interests. The preparation of parents/adoptive/foster parents for the reception/return of children to the family is one of the most important problems requiring urgent decisions.

References

[1] *Valstybės Vaiko teisių apsaugos ir įvaikinimo tarnybos prie Socialinės Apsaugosir Darbo Ministerijos 2016 metų veiklo sataskaita* (Vilnius, 2017)
[2] I. Hojer, Y. Sjoblom, Child FamSoc Work **15**, 118–127 (2010)
[3] P.A. Fisher, M. Stoolmiller, A.M. Mannering, A. Takahashi, P. Chamberlain, J. Consult. Clin. Psychol. **79**(4), 481–487 (2011)
[4] E. Keymolen, D. Broeders, Br. J. Soc. Work **43**(1), 41–63 (2013)
[5] D. Pūras, *Institucinė Globa – vaikų iki trejų metų teisių pažeidimas* (ROE OHCHR, 2013)
[6] K. Samašonok, Pedagogika **110**, 103–114 (2013)
[7] D. Snieškienė, I. Tamutienė, *Institucijose Ir Šeimose globojamų vaikų žmogaus teisių užtikrinimo lyginamoji analizė* (2014)
[8] *Common European Guidelines on the Transition from Institutional to Community Based Care* (2012)
[9] M. Pileckaitė-Markovienė, T. Lazdauskas, *Šeima Ir Tėvų Globos Netekusio Vaiko Raida* (2007)
[10] K. Browne, *Adopt Foster* **29**(4), 23–33 (2005)
[11] A. Zbarauskaitė, *Globos namuose gyvenančių paauglių tapatumo išgyvenimas: biografiniopasa-kojimointerpretacinėanalizė* (VU, 2009)
[12] K. Wheten, J. Osterman, B.W. Pence, R.A. Whetten, L.C. Messer, S. Ariely, K. O'Donnell, A.I. Wasonga, V. Vann, D. Itemba, M. Eticha, I. Madan, N.M. Thielman, PLOS ONE **9**(8), 1–10 (2014)
[13] C.M. Connell, R.J. Prinz, J. Sch. Psychol. **40**, 177–193 (2002)
[14] Vaiko Gerovės Valstybės Politikos Koncepcija, Žin **52–2316** (2003)
[15] *Vaiko Globos Organizavimo Nuostatai,* Žin **35–1275** (2002)
[16] *Vaiko Globos (rūpybos) sistemos reorganizavimo strategijos ir jos įgyvendinimo priemonių 2007–2012 metų planas,* Žin **118–4817** (2007)
[17] *Vaiko Gerovės 2013–2018 metų programa*, Žin **140–7208** (2013)
[18] *Perėjimo nuo institucinės globos prie šeimoje ir bendruomenėje teikiamų paslaugų neįgaliesiems ir likusiems be tėvų globos vaikams 2014–2020 metų veiksmų planas* (2014)
[19] L. Gvaldaitė, S. Šimkonytės, STEPP **12**(12), 55–73 (2016)
[20] J.W. Creswell, *Research Design: Qualitative, Quantitative, and Mixed Methods Approaches* (SAGE Publications, Inc, 2009)
[21] V. Schmidt, SW&S **7**(1), 58–69 (2009)
[22] R. Carter, *Family matters: A study of institutional child care in central and eastern Europe and the former Soviet Union* (EveryChild, 2005)
[23] N. Lyalina, A.N. Severinsson, *Analysis of the progress and remaining challenges in Child Care System Reform: Azerbaijan, Kazakhstan, Kyrgyzstan, Tajikistan, Turkey, Turkmenistan and Uzbekistan* (UNICEF CEE/CIS, 2009a)

[24] N. Lyalina, A.N. Severinsson, *Analysis of the progress, remaining challenges and trends in Child Care System Reform: Armenia, Belarus, Georgia, Moldova and Ukraine* (UNICEF CEE/CIS, 2009b)
[25] A.A. Dunn, *Independent assessment of the deinstitutionalisation process in Georgia* (UNICEF, 2011)
[26] *Valstybinio audito ataskaita, Ar vaikų globos sistema atitinka globojamo vaiko geriausius interesus?* (2014)

The specifics in the work of social workers when rendering social services at the person's home

Rita Raudeliunaite

Mykolas Romeris University, Vilnius, Lithuania

Abstract. The objective of the study is to highlight the specifics of the work of a social worker when rendering social services at the person's home on the basis of the experience of social workers. Qualitative research strategy was used, including semi-structured interview.

The study revealed a wide spectrum of the activities of the social worker when organising the provision of social services at home: the identification and assessment of the need for a service, the planning, provision, coordination and assessment of the assistance or care at the person's home, the involvement of the service recipient in decision-making, the involvement of close relatives of the service recipient, teamwork, cooperation with other institutions, and improvement of skills of the employees.

The study revealed the benefit of social services at the person's home to the recipients of the services: living at their homes, they do not feel so lonely, they feel safe, needful, capable of benefiting from the needed assistance.

Difficulties, which arise when cooperating with the recipients of social services at home, are the following: people provide false information when identifying the need for a service, mental disorders of the service recipient, dissatisfaction with the rotation of the visiting care workers, complaints regarding the lack of communication. Organisational limitations of bodies providing services are linked to formalism, shortage of the staff, time restrictions for services, lack of transport, unavailability of services due to the limited financial resources of service recipients.

Key words: social services, home, elderly people, persons with disabilities, social worker.

1 Introduction

Recently, while improving social services system, ways are sought how to bring them closer to the person, so that the persons in need of social assistance would receive necessary services at their home or community thus remaining, as long as possible, in the social environment they are accustomed to [1]. Social services in the community and at the person's home are appreciated due to the increasing social security and social cohesion and reduced impact of stigmatisation [2]. It is recognised that social services in the community and at the person's home enable the person to live with dignity and maximal autonomy at their home rather than be separated from their community [3, 4].

In Lithuanian social policy, social services at home are identified as a priority. The Law on Social Services [5] emphasises that social services to elderly people should be provided creating conditions that enable the person to live at home with their family, to handle household activities independently as long as it is possible. It means social service should organize assistance which enables the person in need to retain social relationships with his family, the closest people and society, i.e. to reduce social exclusion.

Social services to the person with a severe disability must be provided while securing a safe and healthy environment, assistance adequate to the person's dignity which is harmonised with the person's healthcare, permanent care and permanent assistance. The Provisions of the Expansion Directions of Social Services at Home and the Enhancing of Efficiency in Residential Care [6] maintain that one of the most important social services in the community has to be social services at home, because they improve the life quality of people who are unable to take care of themselves, assist a person in remaining in a domestic environment, provide an opportunity to assess individual needs of the person and to provide him with the most necessary social services, and they are the most cost efficient kind of social services. The provisions underline that while expanding the system of social services at home it is necessary to provide various kinds of social services at the person's home, to create an integrated system of assistance at home by merging the services of primary health care in the community with the provision of social services at home, as well as by involving non-government organisations and non-formal service providers in the provision of social services at home. The experience in the European Union countries shows that providers of assistance at home, ensuring conditions to people with disabilities and the elderly to live in their homes as long as it is possible, must provide complex assistance [7]. Only by developing a unified structure of health care and of the network of social service it is possible to effectively create integrated assistance compatible with the needs of the people who are taken care of at home. Various services provided at home may improve the life quality of the elderly and persons with disabilities, to save state funds allocated to residential care of these persons.

In provision of social services at person's home an important place is taken by home help and social day care services at home. Home help is services provided at person's home assisting a person (family) in managing household activities and participating in the life of society [8]. It is a service which does not require continuous care of specialists. Social day care at the person's home is complex assistance requiring continuous care of specialists provided to a person during the day at their home.

Every year the increase of the recipients of social services at home is noticeable. According to the data of The Lithuanian Department of Statistics, in 2016, home help and social care services at the person's home were provided to 18.6 thousand people, or to 4.5 per cent more people than in 2015 [9]. There is a tendency observed that the majority (82.2 per cent) of persons, who received social services at home, were retired, 16.6 per cent of them were working-age persons with disabilities, 1.2 per cent of them were children with disabilities [9].

In accordance with the Law on Social Services [5] municipalities are responsible for the provision of social services at home. Municipalities take care of the development of the infrastructure of social services, provide for the mechanism of supply of social services at the person's home, allocate financial means for social services in their budget, and assess the needs of persons for social services and assign relevant services for a person. The studies have shown that the establishment of the system of social services at home and its development is relevant and necessary. Annual Audit Report [10] states that, although the demand for social services at home is rising, the present infrastructure of social services in the municipalities does not satisfy an increasing demand for social services at the person's home. Due to undeveloped infrastructure of social services provided at the person's home and the lack of appropriate selection criteria for a service when assigning a social care service for a person, services giving a person the possibility to stay at home as long as possible are not regarded as a priority. Consequently, the longest waiting lists in social care institution are for the services of long-term social care. The report states that there is the lack of qualified personnel providing social care and day care services at the person's home, which is having a negative effect not only on accessibility to services, but also on the quality of provided services. This shows that the problem of the provision of social services at home is topical and requiring systematic research that helps reveal the experience of the provision of services at home. This would help to identify the positive aspects of social services provided at home and emerging problems and possible approaches for solving them assisting in improving the system of organising and the provision of services at home in order to ensure the quality of social services at home.

It should be pointed out that, although social services at person's home are given priority, in Lithuania, research on the subject has not been conducted. The analysis of scientific literature revealed that scholars have drawn a great deal of attention to the problem of the quality of services [11–14], the management of social services and the effectiveness of services and the problem of the cost reduction of services [2, 16–18], the expression of empowerment in the context of social services [19, 20]. In the context of the aforementioned studies, the problem of social services at home, as one of the priority areas of change in social services system is highlighted. One of the forms of social services at home – home help – was investigated by V. Indrašienė and A. Katkonienė [21] more in-depth. The accessibility of social services at home was revealed by the of study ofI.Tamutienė and R. Naujanienė [22]. R. Raudeliūnaitė and A. Smalcer [23] analysed the assessment of the provision of social services at home based on the experience of service recipients. A substantial part of studies has been focused on the problem of integrated health care and social services at home [7, 24–26].

The study brings up a scientific problem which is formulated by the following question: What social work specifics and emerging problems when providing home help and social day care services at the person's home are identified by social workers?

The objective of the study is to highlight the particularities of the activity of a social worker when providing home help and social day care services at the person's home based on the experience of social workers.

2 Research methods

The strategy of qualitative research was used. Semi-structured interviews were conducted. The analysis of the results was performed by applying the method of content analysis: the text was read and essential aspects, which were reflected in phrases, sentences and words directly related to phenomenon under the investigation, were identified; the identification of notional elements in the read text; the breakdown of notional elements into categories and subcategories; the integration of categories into the context of the analyzed phenomenon and the description of the content [27].

The research sample. The targeted selection type of the researched ones was chosen. The main selection criteria for the research participants were no less than three years experience of social work providing home help and social day care services at the person's home. 10 social workers took part in the research.

Research ethics. The research was conducted without violating the rights of the research participants and in conformity with the principles of research ethics: the participation of the informants was voluntary; the research participants were sensitized to the goal of the research, the method of data collection; their confidentiality was ensured. In order to ensure the confidentiality of the research participants, their names were encrypted by assigning a letter to them.

3 Research results analysis

During the study we tried to reveal the particularities of the activity of social workers when providing home help and social day care services at person's home. When analysing the findings four qualitative categories became apparent: the characteristic of the service recipient, activity areas of a social worker when organising the provision of services at the person's home, the benefits of provided services to service recipients and difficulties arising when providing services at person's home.

Category "The characteristics of the service recipient". Social workers, when characterising the recipient of social services at home, pointed out that most often they are lonely elderly persons in a frail state of health, who have difficulty taking care of their own life independently, they need the assistance of other people to live independently at home: "the most typical recipient of services at home is an elderly person, who due to health problems is unable to take care of themselves and needs the assistance of other person" (A). Such portrait of the recipient of social services is attributable to changes linked to demographic ageing: the population of older people is growing, especially the number of the elderly people who are older than 80 [26]. On the basis of the forecast of the Statistical Office of the European Union (Eurostat), it is considered that, in the beginning of 2060, almost 37 per cent of Lithuanian population will consist of the elderly (65 years old and older). Consequently, the demand for social services will also rise, and that is a significant challenge to the organising system of social services of the state – it must adjust to the increasing demand for the services, to reorganise the present one or to develop a new infrastructure of services.

The informants pointed out that social services at person's home are also provided to other groups of persons, as provided for by legislation: adult persons with disabilities, children with disabilities and other persons, who have lost their independence due to an illness or other reasons and to whom the need for a service was identified.

The category "Activity areas of the social worker when organising the provision of social services at home" is further specified by the content of the distinguished subcategories: need identification and assessment; assistance planning; the provision of assistance, the assessment of support, the involvement of the service recipient in decision-making, promotion of activity of the service recipient, the involvement of close relatives of the service recipient, teamwork, cooperation with other institutions and organisations, and improving of skills of the employees.

The social workers emphasised that they, while providing social services at person's home, are guided by the procedure descriptions and service specifications regarding home help and social day care at person's home approved by municipalities. The informants pointed out that, in order to receive home help or social day care services at the person's home, a person or his guardian has to apply at the municipality's centre of social services, to complete an application for social services and to provide necessary documents. The participants of the study also pointed out that "as provided for by legislation, other persons are also able to contact respective institutions requesting the provision of home help or day care, to point out the reason why a person himself is not able to do that, a question regarding services is also discussed without the person's request when information is received on the necessity of social services for a person from the police, health care, social service institutions and other institutions" (E).

Social workers perform the identification and assessment of a person's need for services. The informants pointed out that the need for social services at home is identified and assessed individually and in a complex way, taking into account the person's health condition, independence, life and household conditions and the possibilities of the family members to help: "We evaluate what the client can do, how their environment is able to assist, whether there are relatives, what they can do, whether they have none of them and the like (A)". In the opinion of the participants of the study, the identification and assessment of needs ensures that the recipient at home receives services adequate to their condition. The informants told that, while assessing the person's needs, they sought to obtain the largest possible amount of information on the recipient of social services at home: "I am never in a hurry and pursue the detailed examination of every detail and triviality related to the client" (D). Exhaustive and comprehensive assessment enables a social worker to know the service recipient's situation better, to meet their needs. It is noteworthy that the social worker reviews the person's need for social services during the provision of social services, in the event of changed circumstances or the condition of the person's health.

A decision regarding the assignment of social services is made basing on the application of a social worker who identified the person's need for social services in line with the procedures established by the institution of the municipality. After the decision to assign social services has been made, a contract of service provision with a service recipient is concluded. The further activity of the social worker must be associated with the provision of services at the person's home. It is noteworthy that services, provided at person's home, in accordance with the laws of the Republic of Lithuania, are paid, free or partly paid. Every municipality approves the price list of such social services and payment mechanisms for services. Financial capacity of the person (family members) is assessed by social workers simultaneously with assessment of person's need for home help or social day care.

The social workers pointed out that service specifications of home help and social care at person's home foresee a wide spectrum of services: the provision of information and consulting, mediation and representation, communication, organising of catering, assistance with housework and domestic activities, organising of personal hygiene services, organising of health care, the development and maintenance of daily life skills and other services. Assistance is provided in accordance with the approved list of home help services in which service specifications are provided, the duration of their provision and prices, a visiting care specialist is appointed to provide services at home. The informants pointed out that, in accordance with the description of social day care provision at person's home approved by the municipality of their town, day care is provided from 2 to 8 hours per day up to 7 times per week. Those persons, whose adult children, adult grandchildren or heirs of the property live in that town or district, are provided with day care from 2 to 8 hours per day up to 5 times per week. In accordance with the Social Service Catalogue [8], the duration of social home help is up to 5 times per week.

The assessment of provided services is an important part of the provision process of social services at person's home. Assessment, according to the social workers, helps identify whether the person is satisfied with the provided service, what problems they encounter, what changes have occurred in their situation: "We see that the person's health is deteriorating, the situation is changing, and they are already not able to do some things, then we reassess their need (B)". The social workers noted that they endeavour to maintain permanent communication with a service recipient, to receive feedback from them: they phone, visit their clients, provide information, consult, discusses how the foreseen services are provided, what questions, uncertainties, problems arise to them. According to the informant, "there is a continuous communication, most often over the telephone. We inform clients regarding everything: changing schedules, service staff, time changes regarding service provision, what we can do in order that they would be satisfied, what he can do in order that any miscommunication, misunderstandings and the like would no longer occur" (A). Meetings and individual conversations with visiting care specialists are organised while discussing various aspects of service provision, problems which have arisen: "We talk to the staff. Every morning we organise short meetings. We start with the discussion on the previous day, what happened, what observations are there, what clients asked for. We talk individually with the employees" (B). According to L.C. Johnson [28], assessment is a recurrent part of the social work process and the means to ascertain whether desired objectives and tasks have been achieved.

According to the contemporary approach to social work, the recipient of social services is not only a passive user of social services, but also an active participant. Consequently, it is important to develop collaborative relationships with the client by involving them in the provision of the service. Social services are inclusive only if an institution providing services and its employees (providers) create conditions to a service recipient, the members of his family or guardians to get actively involved in the planning, decision-making and assessment of the provided services. Such participation in the provision process of social services enables the recipients of social services to recognise and identify their needs on their own and to control personal changes [20]. The study revealed that service recipients actively participate and express their opinion, make decisions at all stages of service provision starting from the assessment and identification of the need for a service and finishing with service assessment.

Activeness promotion in a service recipient is equally important. When analysing the findings of the study, it came to light that service recipients, while taking into account the level of their independence and health condition, are encouraged and involved in all possible activities: cooking, the tidying of their flat and the like. When providing services it is sought that a service recipient would not be passive, would not be rendered powerless: "there are clients who are passive, for example, when asked what they want to buy in the shop today, they respond – "open the fridge and see what is missing". So we ask that there would be no such things. That the person would get ready beforehand and would think over what he needs. In so doing we encourage him to take care of himself" (A).

Social workers cooperate with the family members of the recipient of social services or other relatives. The participants of the study revealed that the involvement of close relatives of the service recipient and cooperation with them takes place from the very first visit of the social worker at home: "When identifying the need I always contact the children. If there is a possibility, I even invite when identifying the need. Then we come to a joint conclusion and make decisions together. After that there is a continuous cooperation with the client's relatives, and I encourage them to respond very much (E)". According to M. Berg-Weger [29], family members have the greatest knowledge of their relative and perform the most important role – they take care of him. The more people a person has in their close social network and the more diverse roles they perform, the higher the likelihood for support.

In the opinion of the social workers, the provision of social services at home is the result of teamwork. According to the participants of the study, teamwork is based on the pursuit of a common objective and responsible implementation of the tasks entrusted to each one, consultation, trust, the provision of assistance to each other, the development of positive mutual relationships. Teamwork, according to the informants, "offers new ways of looking at a problem and allows us to help a person in a more diverse way" (H). In addition, in a team it is possible "to make faster and better decisions, to improve the client's situation in the most appropriate manner" (F). Researchers say that a successful cooperation is the result of good relationships between team members, mutual trust, respect and understanding. Such relationships create an important and valuable assistance network for service recipients [30]. The social workers pointed out that team meetings, during which "everyone has an opportunity express themselves, to say what they think" (C), are organised. The study revealed that team meetings take place in various ways. Shorter meetings take place every day, most often at the beginning of a working day. During them, current matters are discussed, information is shared, problems, which have arisen, are discussed. The main team meetings usually take place every two weeks, if need be, even more often. During them, the main work matters are discussed, more complicated events are analysed. The social workers clarified that, during meetings, not only problems are solved, but also their prevention is discussed: "We discuss it openly in order not to repeat the same mistakes again, because the employees have to go to work being safe and knowing what to do, this is my daily task (D)".

The participants of the study pointed out that when organising and providing services at home they cooperate with health care establishments, state institutions and non-governmental organisations, or communities. The informants maintained that the vast majority of all the recipients of social services at home are elderly people with disabilities and various health problems, therefore, cooperation with medical institutions, while helping them to solve health problems, is very important. The social workers also mentioned that they have to cooperate with such organisations like Caritas, the assistance service of the Order of Malta and charity and support fund The Food Bank: "We have clients with a very small income, and then the assistance of these organisations is badly needed (E)". The social workers noted that a community also plays an important role in the assistance network when small and big problems have to be solved or daily human concerns addressed: "When the person is lonely we can count on their neighbours or the chairman of the block of flats/community of the block of flats or the chairman of a local community (H)".

The social workers identified the competence of workers as one of the factors determining the provision of quality services: "The whole quality of work depends on the educational level of the worker, their abilities and knowledge, practical experience, continuous professional development" (D). Consequently, the informants deem the improvement of professional qualifications an important area of their professional activity. Social work is a multidisciplinary profession requiring significant knowledge, abilities and skills. In order to ensure professional assistance, the development of the workers of social services in different areas is necessary. They must continuously update their knowledge, skills, apply new work methods, and develop their need to continuously improve the existing competences. The findings of the study revealed that social workers improve their professional qualifications in various ways: by participating in trainings, courses, conferences, sharing good practice and novelties with their colleagues, consulting with other specialists.

When analysing the findings obtained during the study, the category "the benefits of the provision of social services at home to recipients" and four subcategories: a person lives at their own home, life quality improved, they feel safer, not so lonely – were highlighted.

The social workers emphasised the benefit of social services at home to services recipients. They noted that services provided at home enable service recipients to live at their home. The informants maintained that the majority of service recipients, "although their health is frail, and they have retained a small degree of independence, but they do not want to set off from their home", "...home is home, they feel best there, familiar environment, people they are familiar with" (A). According to E.L. Siegler, S.D. Lama, M.G. Knight, E. Laureano, M.C. Reid [4], when providing social services at home, conditions, which are created for the recipients to remain at their home, in their communities as long as possible, enable them to avoid institutionalisation and to reduce their social exclusion.

The social workers, who participated in the study, pointed out that services which are provided at home, make the life of service recipients easier, improve the quality of their life, because due to their precarious health, and reduced independence they find it difficult to satisfy their basic needs, to live a full-fledged life. The informants noted that, while receiving services at home, service recipients feel less lonely, have a sense of safety that others are in need of them, have awareness that they are able to receive assistance they need.

The social workers, organising the provision of social services at home, encounter various kinds of problems in their daily activity. When analysing the findings of the study, two qualitative categories came to light: difficulties arising when cooperating with recipients of social services at home and organisational limitedness of providing institutions.

The category "Difficulties arising when cooperating with recipients of social services at home" is made more concrete by the following distinguished subcategories: the provision of false information when identifying the need for a service, health problems of a service recipient, dissatisfaction with the rotation of visiting care workers, complaints regarding the lack of communication.

The social workers, who participated in the study, revealed that there are cases when the person applying for social services at home provides false information for the assessment of the need for social services in order to receive a service: "when assessing a need certain difficulties arise, because future clients make significant efforts to conceal the truth, the majority of them manipulate that they cannot do this or that, and when the contract is concluded and when we already start to provide services, it turns out that they can do both this and that (G)". The informants maintained that, if it turns out that a person provided false information, the provision of a service is discontinued until, taking into consideration new circumstances which came to light, the need for a service is re-identified.

The participants of the study mentioned health problems of a recipient of social services as reasons impeding collaboration between workers and service recipients. The participants of the study explained that most often difficulties arise in the case of people with mental and memory disorders. Difficulties with people with mental disabilities arise due to insufficient knowledge of workers how to work with these persons, this often leads to various misunderstandings and conflicts. The informants maintained that memory disorders of a client also impede work, for example, "A person does not remember where they put their money, how much of it they gave to workers, because of that disagreements and complaints occur that attempts are made to deceive them and the like." (E).

When analysing the information, which was obtained during the interviews, it has been observed that all the social workers, who participated in the study, noted that services recipients often complain of the lack of communication. The independence level of the recipients of social services has declined, therefore, they spend the majority of their time at home, and they would like that the workers of visitation care would spend more time to communicate with them. The social workers associate the lack of communication with service recipients with heavy workloads of the workers of visitation care and inability to give more time to communicate with a client.

The social workers pointed out that the recipients of services at home are dissatisfied with frequent changes of workers providing services to them. The participants of the study explained that frequent turnover of workers causes inconveniences for service recipients, because they need to explain their desires and wishes to a new worker every time. Another reason for the dissatisfaction of service recipients regarding the turnover of workers is the accustomedness of service recipients to one person and the appreciation of relationships with him. The informants maintained that it happens that after the change of a worker some clients begin as if to mutiny, to continuously complain that "the worker, who has newly come, does everything in a different manner than the previous one, therefore, they are dissatisfied (D)". Scholars revealed similar tendencies when full satisfaction of the recipients of social services at home with the provided services is impeded by frequent turnover of workers providing services [31].

When analysing the findings of the study a category "Organisational limitedness of service providing institutions" and four subcategories were distinguished: formalism, shortage of staff, restriction on the service time, the lack of transport means, services are not affordable due to limited financial resources of service recipients.

The social workers have noticed that, although all the documents, which they have to fill in are important, according to their words "red tape", burdens their work, because it is very time consuming and less time is available to communicate with their clients: "We spend loads of unnecessary time on paperwork, to enter data into one or another database and the like, I would rather visit people to fill that gap of their communication" (B).

The social workers mentioned that difficulties also arise due to staff shortage: "It is sometimes hard when in organising services there is a staff problem, when only three out of five work, because of illness, whereas we cannot tell a client that somebody is ill, and we will not come" (H). Due to staffing problems, social workers are compelled to increase the workload of visitation care workers.

Another obstacle identified by the social workers is that when services are provided at home limited time is allocated to provide them per week and the time is not sufficient to meet the needs of services recipients. Because of that, according to the informants, they must "decrease the amount of some services somewhere not because they are not necessary for a person, but because their time is limited" (I).

The social workers pointed out that difficulties arise when organising transport services. According to them, to serve the whole large town, only three automobiles were given, therefore, the recipients of social services at home must order transport a few weeks in advance: "There is a shortage of transport services, because it depends on the finance supplied. We have 3 cars to serve our town, those cars break, drivers get ill, without mentioning fuel..." (A).

The study revealed that services are not affordable due to limited financial resources of the service recipients, which is one of the obstacles due to which the recipients of services at home refuse a service they need: "due to financial possibilities they even restrict themselves to food fetching once per week. To tell the truth, they would use many of services, because they are necessary, but they can hardly afford that for themselves (C)". Similar tendencies were revealed by other studies [22] which demonstrate that services at home, which are necessary to the recipients of social services, are unaffordable due to financial difficulties. Therefore, the recipients of services are prone to refuse services or to choose only the most necessary services.

4 Conclusions

Social services at person's home are one of the most progressive forms of social services. An important role, when rendering social services at person's home, is played by domestic assistance and social day care services at person's home. Domestic assistance is the entirety of services by which the person is provided with integrated assistance which does not require continuous oversight of the professionals. Social day care at person's home is the entirety of services by which the person is provided with integrated assistance, which require continuous oversight of the professionals during the day. Social workers play an important role when organising the rendering the services of domestic assistance and social care at person's home. The main recipients of domestic assistance and social day care at person's home are elderly people and persons with disabilities.

The study revealed a wide spectrum of the activities of the social worker when organising the provision of social services at home: the identification and assessment of the need for a service, the planning, provision, coordination and assessment of the assistance or care at person's home. The social worker, while organising the provision of the services at person's home, engages the recipient of a service in the decision making when identifying and assessing the person's need for the services, foreseeing and assessing the services, making a decision regarding the continuity of a service. When providing services at person's home the social worker promotes the activity of the recipient of a service, cooperates with the closest people of the recipient of a service, the employees who provide services, various bodies and organisations, communities. Social workers improve their professional qualification by participating in trainings, professional development training courses, conferences, exchanging good practices and innovations with their colleagues, consulting with other specialists.

The services, which are provided at person's home, allow the service recipients to live at their homes, they do not feel so lonely, they feel safe, needful, capable of benefiting from the assistance they need.

Social workers, who organise the provision of social services at home, in their daily activity encounter difficulties, arising when cooperating with the recipients of social services at home and organisational limitations of bodies providing the services. Difficulties, which arise when cooperating with the recipients of social services at home, are the following: people provide incorrect information when identifying the need for a service, mental disorders of the service recipient, dissatisfaction with the rotation of the workers of visiting care, complaints regarding the lack of communication. Organisational limitations of bodies providing services are linked to formalism, lack of personnel, time restrictions on services, lack of transport, services are not affordable due to the limited financial resources of service recipients.

References

[1] *Common European Guidelines on the Transition from Institutional to Community Based Care* (2012). Retrieved from: `http://deinstitutionalisationguide.eu/`

[2] A. Guogis, Socialinių paslaugų ekonominiai svertai Lietuvos savivaldybėse. Socialinis darbas **4**(1), 66–73 (2005)

[3] C. Xie, J. Hughes, C. Sutcliffe, H. Chester, D. Challis, Promoting personalization in social care services for older people. Journal of gerontological social work **55**(3), 218–32 (2012)

[4] E.L. Siegler, S.D. Lama, M.G. Knight, E. Laureano, M.C. Reid, M.C., Community-Based Supports and Services for Older Adults: A Primer for Clinicians. J. geriatrics **2015** (2015)

[5] Lietuvos Respublikos socialinių paslaugų įstatymas (2006), Retrieved from: `https://www.e-tar.lt/portal/lt/legalAct/TAR.91609F53E29E`

[6] Lietuvos Respublikos socialinės apsaugos ir darbo ministro įsakymas "Dėl socialinių paslaugų namuose plėtojimo krypčių ir stacionarių globos įstaigų darbo efektyvumo didinimo nuostatų patvirtinimo" (1998) Retrieved from: `https://www.e-tar.lt/portal/lt/legalAct/TAR.4E5C43FF9A42`

[7] R. Kudukytė-Gasperė, D. Jankauskienė, K. Štaras, Sveikatos ir socialinių paslaugų integracija. Atvejo analizė VšĮ Centro poliklinikoje. Sveikatos politika ir valdymas, **1**(4), 127–146 (2012)

[8] Socialinių paslaugų katalogas, Valstybės žinios **43–1570** (2006)

[9] Statistikos departamento informacinis pranešimas. Socialinės paslaugos (2016) Retrieved from: `https://osp.stat.gov.lt/informaciniai-pranesimai?articleId=5178560`

[10] Valstybinio audito ataskaita: Ar teikiamos socialinės paslaugos tenkina didėjančius senyvo amžiaus asmenų poreikius (2015)

[11] L. Žalimienė, Nauji iššūkiai socialinių paslaugų plėtrai – mišrios socialinių paslaugų rinkos kūrimas. STEPP: socialinė teorija, empirija, politika ir praktika **3**, 139–147 (2006)

[12] L. Žalimienė, Socialinės globos paslaugų pagyvenusiems žmonėms standartizavimas: Lietuvos praktika ir užsienio šalių patirtis. Gerontologija **8**(1), 44–54 (2007)

[13] G. Kondrataitė, Evaluation of the Quality of Public Services in Lithuanian Municipalities. Intellectual Economics **3**(15), 393–411 (2012)

[14] L. Žalimienė, D. Skučienė, J. Junevičienė, A. Gataulinas, *Profesinė gerovė socialinio darbo paslaugų sektoriuje* (Lietuvos socialinių tyrimų institutas, 2013)

[15] I. Adomaitytė-Subačienė, Socialininų paslaugų kokybės standartizavimas Lietuvoje. STEPP: socialinė teorija, empirija, politika ir praktika **11**, 54–65 (2015)

[16] L. Žalimienė, *Socialinės paslaugos (*VU specialiosios psichologijos laboratorija, 2003)

[17] L. Bagdonienė, R. Hopenienė, *Paslaugų marketingas ir vadyba* (Kaunas: Technolologija, 2005)

[18] M. Išoraitė, *Socialinių paslaugų administravimas* (Vilnius: UAB Saulė, 2007)

[19] J. Ruškus, N. Mažeikienė, R. Naujanienė, R. Motečienė, D. Dvarionas, Įgalinimo samprata socialinių paslaugų kontekste. Socialinis darbas: patirtis ir metodai **12**(2), 9–43 (2013)

[20] D. Dvarionas, R. Motečienė, J. Ruškus, N. Mažeikienė, R. Naujanienė, Įgalinančių socialinių paslaugų modelis socialinės gerovės politikos kontekste. Filosofija. Sociaologija **25**(2), 89–97 (2014)

[21] V. Indrašienė, A. Katkonienė, Pagalbos namuose paslaugų teikimo vertinimas. Socialinis darbas **10**(2), 267–278 (2011)

[22] I. Tamutienė, R. Naujanienė, Senyvo amžiaus asmenų socialinių paslaugų prieinamumo namuose patirtys. Tiltai **4**, 63–78 (2013)

[23] R. Raudeliūnaitė, A. Smalcer, Socialinių paslaugų į namus teikimo vertinimas paslaugų gavėjų požiūriu. Socialinisdarbas **15**(2), 24–41 (2017)

[24] R. Ulianskienė, O. Vitkūnienė, L. Hitaitė, *Integruotų sveikatos priežiūros ir socialinių paslaugų poreikis ir plėtra* (2006)

[25] R. Jurkuvienė, S. Audronytė, Lėtinėmis ligomis sergančių senų ir pagyvenusių žmoniųintegruotų ilgalaikių paslaugų namuose poreikis. Sveikatos mokslai **5**, 1191–1197 (2007)

[26] R. Kudukytė-Gasperė, K. Štaras, Integruotų slaugos ir psichosocialinių paslaugų poreikis namuose slaugomiems sunkios būklės pacientams ir jų šeimos nariams. Atvejo analizė všĮ Centro poliklinikoje. Sveikatos mokslai **25**(1), 15–23 (2015)

[27] J.W. Creswell, *Research design: Qualitative, Quantative and mixed methods approaches* (London: Sage, 2009)

[28] L.C. Johnsosn, *Socialinio darbo praktika: bendras požiūris* (VU Specialiosio psichologijos laboratorija, 2003)

[29] M. Berg – Weger, *Social work and social welfare* (London: Routledge, 2010)

[30] V. Coppock, B. Dunn, *Understanding social work and mental health* (London: Sage, 2010)

[31] A. Guogis, D. Gudelis, Socialinių paslaugų sektoriaus plėtros galimybės Lietuvoje. Viešoji politika ir administravimas **12**(1), 77–85 (2005)

Experience of persons with epilepsy in labour market

Jolanta Pivoriene[1,*] and *Polina* Sediene[2]

[1] Mykolas Romeris University, Vilnius, Lithuania
[2] Lithuanian University of Health Sciences, Kaunas, Lithuania

Abstract. The goal of the article is to reveal how the right to work is implemented regarding persons with epilepsy by investigating their subjective experience. The article is based on qualitative research methodology. By using content analysis three categories with subcategories were formulated: motivation (high motivation for employment, demotivating factors), experience in labour market (negative emotions during job interview, hiding the illness, consequences of hiding the illness, stress for possible seizure at work), relations with employers (expectations toward employers, stigmatized attitude of employers, stereotypical thinking). Research data shows that persons with epilepsy have high motivation in searching job and attempting to stay in labour market, because they want financial stability, safety and inclusion in social life. The main demotivating factors are painful experience during job interviews; not adjusted working conditions and environment. Persons with epilepsy experience negative emotions during job interviews, they hide their illness from employers as long as it is possible, but even after illness is discovered they constantly feel stress for possible seizure. Research participants have clear and specific expectations, they feel stigmatized, stereotypical attitudes to them, and want individualized approach to the situation, and adjustment of working conditions to their abilities.

Key words: right to work, persons with epilepsy, labour market.

1 Introduction

Epilepsy is one of the most prevalent neurological disorders. It could affect any person no matter their age, gender, origin or other factors. According to Health Information Centre at Hygiene Institute data the number of people with epilepsy has raised by 13% since 2010. However, there is no exact statistical data as number of persons with epilepsy includes those only who have disability determined by medical authorities [1]. There is no necessary assessment, rehabilitation model, support system for persons with epilepsy [2]. This illness is surrounded by fear and mystery; persons with epilepsy find difficulties in leading meaningful social life. Job is one of the most important elements in human life leading to self-realization, safety and stability. A working person is valuable first of all for him/herself and then for his/her near environment, community and society [3]. The right to work is inscribed in the main national [4] and international documents [5]. However, there is very little data about how this right is actually implemented. Majority of researches on epilepsy are conducted in the field of medicine [6], there are only few investigations in social life of persons with

[*] Corresponding author: `jolantapiv@mruni.eu`

epilepsy. Sediene and Pivoriene have analysed theoretical aspects of epilepsy in the light of classical sociological [7] and existential [8] theories; Butvilas, Krisciunas, Endziniene, Jurkeviciene and Svedaitė-Sakalauske [9] questioned about fullness of life; Krutuliene, Bagdonas, Jakutiene and Zdanaviciute [10] researched psychosocial situation; Jurkuvenas, Mameniskiene and Bagdonas [11] – personal characteristics; Krutuliene [12] – quality of life; Ambramaviciens ir Aleksiene [13] – social inclusions using music therapy; Krutuliene [14] – traits of political, cultural and social participation.

The researches on situation of persons with epilepsy in the labour market are even more limited [15, 16]. Given the relevance of the problem and the lack of research the goal of the study is to find out how the right to work for people with epilepsy is exercised, investigating their subjective work experience. Object – work experience of persons with epilepsy. The article is based on qualitative research methodology using interview as data collecting technique. Selection criteria for research participants were: diagnosis of epilepsy, illness seizures, working age, and experience in labour market. Data of five research participants is included in qualitative content analysis. Three categories with subcategories are formulated: motivation (high motivation for employment, demotivating factors), experience in labour market (negative emotions during job interview, hiding the illness, consequences of hiding the illness, stress for possible seizure at work), relations with employees (expectations toward employees, stigmatized attitude of employees, stereotypical thinking). For ethical reasons depersonalised data is presented in the article. The main limitation of the research is that due to interview translations from Lithuania to English language the validity could be lessen.

2 Psychosocial definition of epilepsy and possibilities for coping with psychosocial consequences of epilepsy

Classical sociological theories analyse changing social roles of persons with illness, interactions between society and a person with new roles, stigmatization processes, acceptance or rejections. In the general context of health and illness, chronic illness is investigated as a specific research object. Epilepsy is one of the chronic illnesses which are in the field of sociological interest. Epilepsy in the light of classical sociological theories is analysed by describing changes of social roles, role of the society in forming social identity of a person with epilepsy, labelling and stigmatization. Big emphasis is put on how symbolic meanings are related to epilepsy and what are social consequences of the illness. A person and a society attach such meanings to epilepsy which enforce labelling and stigmatisation leading to stronger negative social consequences of the illness. Present and future social interactions are weakened and inclusion of persons with epilepsy to the society is diminished [17].

Understanding of epilepsy and its psychosocial consequences is complex and ambiguous. Attitudes toward persons with epilepsy depend on historical, psychosocial and medical understanding of the illness. This illness is invisible therefore is mystified and full of myth, which raises tension in dealing with persons with epilepsy. Historical cultural context of epilepsy presupposes attitude that epileptic seizure is "an awful" event, encounter with death, which reminds finality of human, event which causes fear, panic, rejection of surrenders and which restrains social life of person with epilepsy, his/her self-perception. Therefore people with epilepsy very often confine themselves in a sick person's role, isolate themselves from society, and become dependent on others [18].

Bio-psychosocial consequences of the illness have no direct relation with the process of illness in the current moment, but rather are connected with fear that a seizure could happen in the future. Negative attitudes of the society also interfere with lives of persons with epilepsy [19, 20].

It is proven that epilepsy affects all areas of person's life: personal development, interpersonal relations, friendships, marriage, family life, education, employment, etc. [21]. Persons with epilepsy very often lead an excluded life, have low self-esteem, they are stigmatized, prone to depression, have fears, are not motivated to study or work, have communication difficulties [22]. Education level of persons with epilepsy is lower in comparison with persons without epilepsy due to medical (side effects of medicines, repeated seizures, memory disorders) and psychosocial (disappointment, fears, hyper-care, restrictions in environment) reasons [23]. Research data shows that employment is one of the biggest problems for persons with epilepsy; every third respondent defined the problem as very important [24]. Very often these persons do not work, not because of health problems or incapacity to complete the tasks at work, but because of low motivation and negative attitudes to the illness by employers and persons with epilepsy themselves [25]. Krutuliene [26] agrees that employment is one of the most challenging problems and also it is one of the most stigmatized areas in the life of persons with epilepsy. Stigma could be diminished by providing more information on epilepsy, seizures, and suggestions how to behave during seizures [27]. It is necessary to reduce negative attitudes to persons with epilepsy by constantly educating society [28].

3 Experience of persons with epilepsy

In the category "Motivation" the following subcategories were formulated: high motivation for employment and demotivating factors.

Motivation very often is related with financial and social challenges. Research participants revealed what motivates them to search for a job:

"It is very difficult emotionally. You know that you live not in your own flat; you know that you need to put your children in bed at nine o'clock and can't even read or do anything because there is only one room and light would trouble children. I fall asleep only at twelve, so everything is burning in my head" (I interview).

"I wish, I really could cope with all these elementary things. Job. Family. Everything like that" (V interview).

Most of persons with epilepsy are not working and live only on social allowance which is not enough for living. Such situation motivates them for searching the job.

"For example, if my mother dies. Really, one day she will pass away and if I wouldn't get a job and will not get any money, only from state... and if for example my washing machine breaks, I will not have cash money. It is awful" (V interview).

In order to solve such problems persons with epilepsy are trying to get a job and to stay in labour market at any cost.

"When I was looking for a job, I didn't like it. I knew that I needed money because I was renting a flat and needed to pay taxis. I didn't know people there. Anything. It was a very big problem" (III interview).

A job is needed for leading as much as possible regular independent life, for possibility to make social connections, not to stay for all time at home.

"I was sitting at home all my life: home-hospital, home – hospital, home – sanatorium... I wanted to go out to real normal world" (II interview).

Research data also revealed demotivating factors for an employment. Research participants said that in some cases for them it is not financially beneficial to work. They have calculated that minimal salary is smaller than social allowances which persons with epilepsy get if they have defined disability.

"Very recently I was in Labour Exchange office. Everyone is laughing at the situation, people calculated that they get bigger social allowances than salary so they don't want

to work; my situation is exactly the same. When I told that I have disability and I have restrictions for work, they suggested me to take half-time position and it was very small money. I calculated how much I will spend for travel, and all compensations for heating disappear because I get salary, and I decided not to work. The employment would be absurd for me." (IV interview)

Another demotivating factor is difficulties to find a job which would fit to special needs of persons with epilepsy. A research participant has a profession and could work as a tiler, but there restrictions for work arise:

"Again, how will you carry instruments, you need a car. What car??? It is forbidden by law to have driving license for people with epilepsy" (IV interview).

Earlier employment experiences are painful and demotivating as well

"If I think about my painful employment story... " (IV interview);

"It is one of my most terrible experiences in my life. It is a tragedy when you feel rejected, you feel not important to anybody" (V interview);

"If they are telling that they will call you later, you know – they will never call" (I interview);

"Labour Exchange office gave me a contact to call for cleaner's position. The first question was what kind of disability I have. I asked, what the difference is, if I am looking for a job, it means that I can work" (V interview).

Summarizing this category, it could be said that persons with epilepsy tend to look for a job and stay in labour market because of financial stability and possibility to have social life. However, they face difficulties in finding job which would correspond to their needs. Persons with epilepsy who have disabilities do not have financial motivation to work as they get bigger social allowances than minimal salary. Other demotivating factors such as painful previous employment experience, not adjusted working places or conditions impede wish to have a job.

In the category "Employment experience" the following subcategories were formulated: negative emotions during job interviews, hiding the illness and consequences of it, stress for possible seizure at work.

Research participants talked about negative emotions during job interviews, especially emphasising employers' reaction after illness was discovered.

"If he sees everything in CV, if illness has been included, the beginning is still very difficult, he sees that you have illness but choose you. If CV is nice, clean, the most difficult moment comes when you say about illness and see his reaction" (IV interview);

"Sometimes job interview is so tense that you start crying, begging to give a job because you need money, that you need to live, you start telling all story. I have learnt to beg, to humble" (V interview).

Research data showed experience of hiding the illness during job interview and later. Persons with epilepsy know that it is easier to get a job if you do not reveal the fact:

"Yes, I tried. I tried not to tell that I am ill. I decided to work as long as I can. Anyway, I have to work" (I interview).

"They always tell, we have to discuss and we will inform you. And I started to be afraid. If you want a job and you tell about illness, no employee wants to take a risk that something could happen in working place. Then I started to hide. Someone told me to do like that, said that in this case you have a job at least for a while. I have no choice" (III interview).

Hiding the illness becomes like a tool for better adaptation to the reality, however new dilemmas rise:

„Of course, I felt bad. It is important information for an employer, you feel very frustrated and the interviewer feels that. Sometimes they ask 'have you told everything?" (I interview).

"I had to tell when they sent me to clean staircase. Cold water, you have to go to the basement, nobody is next to you. And I told. I told that I have epilepsy, and that sometimes I have seizure *but hope that nothing will happen during the time when I will be cleaning staircase. And nothing has happened. I went cleaning when I felt really well" (I interview).*

Epilepsy is episodic, not a visible illness which is easy to hide:

"If I felt worst, I just went away and waited. Don't know for how long. I came back when felt better. But I was working very hard. It was too hard for me. After seizure I felt weak, dizzy. Of course, they notice, were asking, am I ok. I replied that everything is ok" (I interview).

Illness seizures at work make impossible to hide them anymore:

"I was afraid. I was afraid that no one will know how to help me. Then I told that it could happen" (III interview).

The feeling of fear is the most prevailing emotion:

"With this illness you are always in fear. All the time. It is out of control. You wake up, dress yourself, but don't go anywhere because you are afraid of seizure, you always think – what if – and that – what if starts ruling all your life" (V interview);

"I was diagnosed. Oh god, I was afraid to cross the street. It was tremendous fear" (III interview).

Summarizing this category, it could be said that persons with epilepsy feel negative emotions during job interviews. Hiding the illness makes easier to get a job and to stay in it, however, even after illness revealing they fell a permanent fear of seizure, worry that nobody will help them and that employee could fire.

In the category "Relations with employees" these subcategories were formulated: expectations toward employers, stigmatized attitude of employers, stereotypical thinking.

Research participants talked about expectations toward employers. People with epilepsy would like that they would be interested to know more about the illness:

"They have to know the epilepsy, like everyone knows cancer" (V interview).

Also they expect to be accepted as all other people, looking for their capacities and readiness to work:

"Just to look at me like at normal person. Yes, I am ill, but I am controlling the illness, I take medicine. I can work. So, why not take me" (I interview).

Research participants feel stigmatizing attitudes of employers, not willingness to understand, rigidity:

"He made me stop working. He knew that I have a small child and scheduled inconvenient working hours. I signed" (I interview).

Lack of information is one of the reasons for stigmatizing attitudes. Very often epilepsy is understood as psychic illness or mental disability.

"If only an employer knew more. He doesn't know anything and doesn't want to know anything about me. Not even once, he talked with me about the illness" (IV interview);

"They think that epilepsy is a disaster, tragedy, that you are silly, retarded" (V interview).

Summarizing this category, it could be said that persons with epilepsy have clear expectations to employers but meet stereotypical attitudes and discriminating behaviour. Employers lack knowledge and understanding and want to avoid risk at working places.

4 Conclusions

Persons with epilepsy are motivated to have a job as they want to have financial stability and richer social life. Job gives possibilities for self-realization, allows feeling better emotionally and physically, provides meaning for the life. However, despite high motivation, persons with epilepsy find difficulties in getting a job or are demotivated by various aspects of economic and social systems which do not provide reasonable opportunities to be included in labour market. Job interview is one of the most complicated stages in job seeking. Most of experiences are very painful and force to hide the illness in order to get a job. Hiding the illness has reverse negative consequences for persons with epilepsy as they always feel tension of possibility to have a seizure at work. They also fear that employees sooner or later will know their diagnosis and fire or will start to devaluate their abilities.

Employers are not ready to hire persons with epilepsy as they do not know specifics of the illness and do not give opportunity to start a job and show their capacities. Employers lack knowledge and support from social system in employing and keeping at work places persons with epilepsy. Persons with epilepsy have to know and control their illness and be able to explain individual aspects of the illness to employers during a job interview and later in working process for co-workers. It could help for better inclusion in labour market.

Based on classical sociological theories, people with illness and society members relate such meanings to the illness which support and enforce stigmatized attitude and behaviour towards the person with illness, strengthens negative social aspects of epilepsy, which restrict present and new developed social interactions, and full engagement into society.

References

[1] LR Socialinės apsaugos ir darbo ministerija, *Socialinės integracijos kompleksinių (integruotų) paslaugų modelio neįgaliesiems, sergantiems epilepsija, įgyvendinimo 2015–2020 metų veiksmų planas* (2015)
[2] D. Butvilas, *Veiksniai, įtakojantys jaunų asmenų, sergančių epilepsija, gyvenimo visavertiškumą (*Kauno medicinos universitetas, 2010)
[3] Tereškinas, J. Bučaitė-Vilkė, *Socialinė atskirtis ir geras gyvenimas Lietuvoje* (Vilnius: Versus aureus, 2015)
[4] *Lietuvos Respublikos Konstitucija* (1992)
[5] *Neįgaliųjų teisių konvencija* (Autentiškas vertimas. Ministro Pirmininko tarnybos Organizacinių reikalų valdyba, 2009)
[6] S. Krutulienė,Socialinis darbas **14**(1), 53–68 (2016)
[7] P. Šedienė, J. Pivorienė, Socialinis darbas **14**(2), 188–198 (2016)
[8] J. Pivorienė, P. Šedienė, Socialinis darbas **16**(1), 35–47 (2018)
[9] D. Butvilas, A. Kriščiūnas, M. Endzinienė, G. Jurkevičienė, B. Švedaitė-Sakalauskė, Neurologijos seminarai **14**(44), 86–92 (2010)
[10] S. Krutulienė, A. Bagdonas, V. Jakutienė, G. Zdanavičiūtė, Specialusis ugdymas **1**(26), 29–44 (2012)
[11] V. Jurkuvėnas, R. Mameniškienė, A Bagdonas, Neurologijos seminarai **17**(55), 17–28 (2013)
[12] S. Krutulienė,Socialinis darbas **14**(1), 53–68 (2016)
[13] Z. Abramavičiūtė, V. Aleksienė, Acta Pedagogica Vilensia **35**, 113–131 (2015)
[14] S. Krutulienė, Socialinis darbas **14**(1), 53–68 (2016)
[15] S. Krutulienė, Socialinis darbas **14**(1), 53–68 (2016)
[16] P. Šedienė, J. Pivorienė, E. Laskevičiūtė, Eurika, Socialinis darbas: patirtis ir metodai **20**(2), 99–115 (2017)
[17] P. Šedienė, J. Pivorienė, Socialinis darbas **14**(2), 188–198 (2016)
[18] P. Šedienė, Socialinis darbas: patirtis ir metodai **17**(1), 117–139 (2016)
[19] A. Baker, Epilepsia **43**(6), 26–30 (2002)
[20] A. Jacoby, Epilepsy behav. **3**(6S2), 10–20 (2002)

[21] S. Krutulienė, A. Bagdonas, V. Jakutienė, G. Zdanavičiūtė, Specialusis ugdymas **1**(26), 29–44 (2012)
[22] LR Socialinės apsaugos ir darbo ministerija, *Socialinės integracijos kompleksinių (integruotų) paslaugų modelio neįgaliesiems, sergantiems epilepsija, įgyvendinimo 2015–2020 metų veiksmų planas* (2015)
[23] P. Smith, S. Wallace, *Clinician's guide to epilepsy* (Londonas: Arnold, 2001)
[24] A. Baker, Epilepsia **43**(6), 26–30 (2002)
[25] S. Krutulienė, A. Bagdonas, V. Jakutienė, G. Zdanavičiūtė, Specialusis ugdymas **1**(26), 29–44 (2012)
[26] S. Krutulienė, Socialinis darbas **14**(1), 53–68 (2016)
[27] P. Šedienė, Socialinis darbas: patirtis ir metodai **17**(1), 117–139 (2016)
[28] H. Baier, D. Dennig, M. Geiger-Riess, F. Kerling, U. Specht, R. Thorbecke, *MOSES, Modulinė epilepsijos mokymo programa. Darbo knyga* (Kaunas: Lietuvos Caritas, 2017)

SHS Web of Conferences **68**, 03005 (2019) https://doi.org/10.1051/shsconf/20196803005
Int. Conf. SOCIETY. HEALTH. WELFARE. 2018

Competences of older people to use smartphones as a factor of wellbeing in contemporary society

Irena Zemaitaityte and *Agata* Katkoniene

Mykolas Romeris University, Educational Science and Social Work Institute, Vilnius, Lithuania

Abstract. One of the key reflective aspects of modern society is the application of information communication technologies. According to SEB Bank and the information and communication technology company TEO [1] more people aged 65 and over are discovering the Internet and smart technology services to manage finances, communicate and spend their leisure time. Technological development has a large group of supporters even among seniors; however, many older adults have no skills and neither the intention nor the possibility to use digital devices which are critical for successful functioning in society. Escape from technological innovations in no way makes life easier. Modern smartphones can make life more enjoyable for seniors and to benefit society. The aim of this paper is to investigate seniors' opinions about the importance of increasing competence in smartphone usage in their daily lives. The investigation was conducted in two stages: 2018 April and 2018 September/October. During stage two, groups were former of 10–12 seniors. During and after the training, the participants were interviewed in focus groups about their opinions. The study revealed that older people's interest in new technologies activates their participation in the learning process, promotes interest in innovations and creates conditions for wellbeing in society.

Key words: information communication technologies, technological development, seniors, competence in smartphone usage, older people's education.

1 Introduction

The age structure of the population of Lithuania has undergone significant changes during the last five decades. Proportions of all age groups in the total number of the population have undergone changes; however, these changes went in different directions. The proportion of the youngest population has decreased almost twice – from 27 percent in 1959 to 15 percent in 2010, and the proportion of those over 65 has doubled – from 8 percent in 1959 to 16 percent in 2010. Mikulionienė [2] writes that the ageing of the Lithuanian population is seen in the quantitative data (growth of the absolute number of older people) and proportionally (when the proportion of children shrinks, the proportion of older people grows correspondingly). In order to measure the Lithuanian population's age structure in a wider context, it is useful to compare it to other EU member states. As Eurostat data indicates, the average growth of ageing population in European member states was 1.9 percentile. It is the percentage of 65-year-olds and over in comparison to the number of the rest of population [3]. In the future, according to Eurostat population projections (baseline scenario), the proportion of the youngest population (aged 0–14) will decrease in Lithuania up to 13.7 percent in 2050.

According to Statistics Lithuania, in 2003 over 20 percent of the population of Lithuania was 60 years old and older, and it is predicted that by 2030 the share of the population that is 60 years old and older will have reached more than 27 percent. As scientists have pointed out, a longer average life expectancy has caused changes in many domains of human life. Longer lives are accompanied by changing behaviour, norms and values, changing of family institution (family relationships, increasing duration of family members' co-survival), and changing education careers (the need for life-long learning arises along with the increase in lifespan and accelerated social changes). Therefore, it is recognized that, with the ageing of society, more attention should be paid to the quality of life for older people, because it is this group of people that will determine the general standard of living and societal stability of the country. One of the key reflective aspects of modern society is the application of information communication technologies (ICT). Though a spectrum of technologies has found its way to Lithuania relatively recently, they were first introduced in Lithuania 20–30 years ago. Many older adults have no skills and neither the intention nor the possibility to use ICT, which are critical for successful functioning of society.

Learning of older adults also has the immense potential to improve the levels of productive ageing by ensuring that older workers have the necessary skills required by potential employment opportunities [4, 5]. Since older adults tend to learn in non-formal rather than formal settings, a key question arises – how far ICTs actually influence their daily lives and whether this is, in fact, perceived as a learning area [6].

The aim of this paper is to investigate seniors' opinions about the importance of increasing competence in smartphones usage in their daily lives.

2 Older people's education

Older people's education is rapidly becoming one of the key elements in determining the development and advancement of the educational system of a country. By creating opportunities for personal fulfilment and for meeting the requirements of the labour market, the education of older people assists humanity in dealing with challenges springing from the globalization processes, demographical changes and development of knowledge and information technologies. Older people's education within the context of lifelong learning has established itself as a norm in developed societies. The challenges of world economies, social life and ecology demand not only restructuring of the economy but also restructuring the systems of social wellbeing and education. The purpose of such changes is to highlight the importance of the idea of sustainability for society, and this, in its turn, calls for changes in educational and learning aims. Adult education may provide the required conditions and meet the needs of learners for the positive processes of this change. For this reason, the system of adult education including older people's education is part of key changes in implementing new ideas and fostering society's capacity for change [7].

In the changing environment, one of the basic principles of the modern society is democracy. In practice this translates as the accessibility of diverse educational services for all who are willing avail themselves of these services [8]. Another feature is the activeness of the learners. According to Mikulionienė [2], the changing structure of the population from society with the dominance of young people to a society where all age groups are proportionally represented is and will continue to be a challenge for the social, economic and cultural development of society. Due to the scope and irreversibility of the phenomenon, demographic ageing of the population is altering. The structure of population needs and simultaneously challenges the social institutions whose function is to meet those needs. A growing number of older people encourage revisions and adaptation of the newly emerging needs in human resources in the areas of social security, economy, health protection and education. The growing average estimated age brings alterations to the individual's development and to the development of a family institute and education, since the societal growth brings forward the need for constant learning and upgrading. Social inclusion of older people presupposes opening up accessibility and participation in learning [9]. According to Gray [10], social inclusion occurs, when every society member actively participates in social life, and no activity, meaningful for societal development, is denied. In Peters' and Armstrong's [11] opinion learning should be seen as the central function in life. Speedy change of the learning contents, methods of teaching, and information transmission media introduce amendments to the conditions of the changing life and work. Experiential learning is indispensable, and this kind of learning is seen as self-guided learning, while the responsibility for the process stays with the learner. Learning is oriented to the achievement of social reform goals: it can upgrade the social and economic status of the individual, and it can solve social inequality and other societal problems [11].

The ageing of the population and the consequences of this phenomenon create problems that draw researchers' attention worldwide. The focus is on many aspects of this situation: changes in workforce supply, the effects of the ageing of the population on the social welfare systems, age influence on quality and productivity of work. Lithuania mostly focuses on the capabilities of the elderly to participate in the labour market and on the demographic and social effects of the ageing of the population. To deal with this problem, European Union member states with regard to the Lisbon Strategy [12] seek to implement measures which benefit the employability and social inclusion of the elderly. Today the key measure in the state policy for the elderly is the National Strategy of Overcoming Consequences of Ageing by which Lithuania started consideration of the problem on the national level: "Though many a measure is implemented for provision of active personal, public, professional and cultural lives for senior citizens, these measures do not meet completely all the required alterations for the welfare of the elderly, therefore, there still remains the need to combine efforts to build environment in which the elderly would live active personal, public, professional and cultural life" [13].

Although competence development is probably the most accepted goal of all forms of education, this is not always the case in older adult learning. This stems from the conjecture that older persons do not need new skills or knowledge, or that they will have no opportunity to use their new competencies [14].

3 Competences in using technologies

The spread of technologies is occurring regardless of the age group; therefore, all age groups will have to adapt to be able to function appropriately in modern society. New technologies are cable and satellite television, computerised communication, personal computers, and new bureau technologies, specifically online information services [15, 16]. Tomczyk [16] defines information technologies as a totality of information processing methods and tools. Researchers identify that the effect of new technologies on the humanity is multiple: it is related to building material and spiritual values and social-cultural environment [17–19]. The stress falls on the characteristics that the skills to use ICT are indispensable for all social groups and layers. They are instrumental for the successful participation in societal life. Modern ICT are applied to many areas of societal life; therefore, the abilities and skills to use them are of importance to all groups of society. Tolutienė and Puškorius [20] highlight the fact that the main goal of ICT use in education is to modernise the process of education and to provide the conditions to seek the modern goals of education: 1) to develop important societal skills for a complete life in the knowledge society (abilities to learn, critical thinking, creativity, information technology skills); 2) to update teaching content and to integrate diverse data; 3) to apply new teaching methods; 4) to form new learning and life culture. These goals are oriented to the learning individual and the individual's adaptation to fast-changing society.

4 Methods of active learning enhancing participation

The application of the learning methods is conditioned by the purpose of learning, programme target group, interests, preferences, experience, and aims of teaching and learning. Active participation in the learning process enhances the effectiveness of learning, cooperation among learners, and strengthens their trust in each other. The application of active methods helps the learners activate stored knowledge and strengths and to channel them in a useful direction for the whole group. Older people are often experts in diverse areas. They bring their life and work experience to the group, the focus in such situation lies on the spread of such expertise for the advantage of the whole group and not only for the learning needs of one individual.

The choice of the learning method affects the learning quality which may be interpreted as a degree of the learning outcome. Some methods directly influence the achievement of the learning outcomes, and some methods influence vicariously and as such do not produce a substantial impact upon learning outcomes. Active participation in the learning process boosts learning effectiveness [21]. Socializing and cooperation among participants of the process brings forth the atmosphere of trust and advances further learning. Yet, learning should not only be active, but also interesting, successful and pleasant. Indrašienė and Žibėnienė [21] state that among the best learning motives are success and pleasure of learning, which can be achieved through active learning, since learning through self-analysis, the opportunity to evaluate oneself and to make choices create an atmosphere of successful learning. According to Longworth [22], learning today is seen as a natural and unfolding process, which turns into a pleasant activity spanning the length of individual's life. It happens in all areas and allows individuals to augment personal abilities, competencies, and provides self-trust for an individual in the future.

Bonwell and Eison [23] stated that "active learning" occurs when learners are given the opportunity to take more interactive relationship with the subject matter of a course, encouraging them to generate rather than simply receive knowledge. On a practical level, active learning includes complex group exercises in which learners apply course material to "real life" situations and/or to new problems. Bonwell and Eison [23] defined "interactive learning" as the interactive means to acquire information through hands on. Teachers use interactive learning as the way to get their learners engaged and cause their interest in subject matters as they are not always successful at "getting their students on board" with a typical lecture format. Also, interactive learning often involves the use of computers and other tangible equipment. Scientists claim that active learning can support learners' development as:

- successful learners through using their imagination and creativity, tackling new experiences and learning from them, and developing important t skills;
- confident individuals through succeeding in their activities, having the satisfaction of the task accomplished, learning about bouncing back from setbacks, and dealing safely with risk;
- responsible citizens through encountering different ways of seeing the world, learning to respect themselves and others, and taking part in making decisions;

- efficient contributors through interacting together in leading or supporting roles, tackling problems, extending communication skills, taking part in sustained talking and thinking, and respecting the opinions of others [23].

Learning from experience is one of the most important and natural means of learning available to everyone. The great strength of experiential learning is that it provides an underpinning philosophy that acts as a thread, joining many learning theories together in a more unified whole. Dewey [24] posits that learners' learning needs had to be integrated with social demands. While he supported the notion that learners required some degree of freedom, a correspondent degree of structure was equally essential. He urged that freedom and structure (students' interests and needs and subject demands) must be integrated rather than made "antagonists." Dewey [24] thought that a potentially organic relationship existed between learners and their subjects. For this potential to be realized, Dewey formulated two educational principles: experience and interaction. Kolb [25] stressed that experiential learning theory offers the foundation for an approach to education and learning as a lifelong process. Kolb's experiential learning cycle involved concrete experience, observations and reflections, formation of abstract concepts and generalizations and testing implications of concepts in new situations. Silberman [26] stated that an interesting and disquieting aspect of experiential learning is that people do not learn from experience alone. To produce learning, experiential episodes with briefing, guidance, planning, feedback, reflection and sharing of insights should be combined. Here are three chronological contexts in which active experience education is integrated with conceptual and factual content, and deliberate, and collaborative reflection.

5 Research methodology

Qualitative research methodology was chosen for the study. The reasons for the selection of a qualitative research are the holistic approach to the object, obtaining of quality data, the openness of the research, the status of the researcher as a cognitive instrument, and the experience gained and used in describing the object in the study itself. According to Gibbs [27] and Flick [28], the holistic approach to the object is when the object is perceived as a solid, complex system. A researcher should not only explore the application of ICT in learning, but also to understand them as details of the educational process, the importance of the whole educational process, and most importantly, for the informant himself. The investigator as a cognitive tool interacts directly with the informant and other components of the educational situation. By performing the function of a scientific cognitive tool, the investigator tries to perceive the nature of the phenomena of education [29]. From this point of view, it is important for a researcher to conduct an investigation that not only records statistical data, but also observes the investigative reaction. Researcher's question should not provide the answer in disguise. It is very important to analyse the experience of research participants. The investigator seeks to uncover categories and dimensions, deepens the details, and interprets them in the context of the entire. The emphasis is on educational processes; individuals and culture are analysed, and the results are interpreted in the historical and social context.

The focus group method was chosen as a qualitative research form for detailed descriptive data collection in small groups, focusing on all topics of interest that were gathered. It is a form of research that highlights the experiences, interests, preferences, attitudes and opinions of group members [30]. Each member of the group is encouraged to express an opinion on the subject.

The research was conducted in the international project "Location-based games as a contemporary, original, and innovative method of seniors' teaching and learning" (project number: 2017-1-PL01-KA204-038869), Erasmus$^+$ programme key action 2: cooperation for innovation and the exchange of good practices. Two training stages, a pilot stage and a main stage, were conducted in the period of 2018 April 9 – 30 and 2018, September 24 – October 5. The training course lasted for 10 days with 10 same topics for each group. During each stage two groups of seniors were formed. The requirements for the participants were: seniors at the age 60 plus, possession of smartphone and basic knowledge of smartphone use, appropriate state of health – being able to walk about two kilometres per day. There were 10–12 participants in each group. Totally there were 49 participants: 9 males and 40 females of age 60–80 from Vilnius city. Participants had no possibility to transfer between groups. The research took place at Mykolas Romeris University in two periods: April 2018 and September-October 2018.

A narrative strategy was chosen. The method of data collection was a semi-structured focus group interview after the training course. This strategy meets the goal of the study. This is also appropriate for informants, since the conversation about seniors' experience is quite attractive to them. Four focus group interviews were conducted with an average duration of 40 minutes. The researchers used a sampling method, which means that the focus group interviews were attended by all training participants and split into groups according to how the training was held. Most of the participants were women. First of all, the participants of the study were analysed and coded. Questions were addressed in an attempt to respond to the purpose of the research, which is inseparable from the experience of the research. All questions are open-ended.

5.1 Research ethics

The participants in the study were given a written consent form to sign before they participated in the study. All participants voluntarily took part in the study. The data transcription and analysis of the research data did not reveal any personal information of any participant. All names and other personal data used in the study are changed at the discretion of the researcher. The data is processed ethically. The quotes, presented in the article with the consent of the participant, are completely unaltered, and are the authentic considerations of the participants (except identity data). Generalised answers are presented from all group interviews: P-pilot stage; M-main stage; Lgb – location- based games method group; Clm-classroom method-based group.

The data are encoded according to authentic phrases in Lithuanian and translated directly to English. The data are interpreted at the discretion of the investigator. The data analysis will be introduced to the study participants during the project events.

The validity of the research is related to naturalistic principles: natural environment as the main source of data; detailed description; the data are presented using the terms of the participants, not the researcher; data analysed inductively, without predetermined categories.

5.2 Results of research data analysis

The participants were asked about their expectations for coming to the training and if their expectations had been successfully met. The research results show that participants have fulfilled their expectations as they "think what they will show us here" (P-Clm), "In fact, expectations have proven" (M-Lbg), "Everything looks perfect to me" (M-Clm), "Really, really, really very good" (M-Lbg), "Everything looks perfect to me" (M-Clm). Some participants were disappointed because the course was short: "Time has passed like three weeks in 2–3 days, there could be more" (P-Clm), "I would like to have more" (M-Clm).

The participants were asked about the competences they have acquired during the training. The answers are categorised into a few pillars. First, the participants confirmed that they acquired new knowledge about applications and their capabilities: "I used to think about that application that I do not need it, and now I see that it is even very necessary" (P-Lgb), or "Still, there are lots of things you need on your phone" (M-Lbg), "It's useful to know all, some do not even know the applications that they are" (M-Clm), "But most importantly, there are a lot of features I did not know about, you need to look for, you need to try" (M-Clm). One of the most often answers is that "Still, there are a lot of things you need on your phone" (M-Lbg), "I did not know that such things exist" (M-Lbg).

Second, part of participants has enhanced their existing skills "But just like I expected, the topics were somehow heard or somewhat known, but we will learn how to use it very well" (M-Lbg), "And I'll add something <...> that's the kind of things you did not know, now we know more, deeper" (M-Lbg). They also give examples of what was known and new: "And my surprise is that there can be a translation of voices that everything is clear to the hand, but that's the voice" (M-Lbg). But at the same time, they agree that they have improved themselves.

Third, the participants were absolutely happy for the practical applicability of the acquired competences: "It's very good that everything was done right away, we try, we do, you advise how" (P-Clm), "And I immediately used [QR] in the store because they are everywhere" (P-Clm), "A lot of experience" (M-Clm).

The participants were asked a question about their experiences during the training. Their answers were categorised into two pillars, such as success experience and other emotions they felt during the training. There was full gamut of emotions, but he most highlighted were "fear" and "joy".

Fear is possible to be analysed from 2 different perspectives. One perspective is "Fear not to damage" the device "It's the fear of pushing something wrong, damaging what we have" (P-Lgb), "Just be afraid to push something because all sorts of funny things happen" (M-Lbg), "I did know much about it, but I was afraid to click" (M-Lbg), also fear to catch the virus "Fear is always. We are afraid of the virus" (M-Lbg), or fear be charged for unknown reasons "I'm afraid that I'll click – and then pay" (M-Lbg). This fear has disappeared because of the knowledge and courage that came after training "It helped me to get used to it" (P-Clm), "What if it does not work, it's nothing" (M-Clm), "Now it will be more courageous to click – it's clear how the children click, click and succeed" (M-Clm), "I know that I can press phone [buttons] and that [bad] happens" (P-Clm).

Another reason is the fear of asking and the fear to look ignorant "I am sensitive and I felt uncomfortable to ask" (P-Lbg), and anxiety "You may not know, you can ask ten times the same, we always get the answer, here are a lot of positive emotions" (P-Lbg). After the training the courage to ask appear "The feeling is that I'm absolutely not afraid not to know, you will not be considered as fool [laughing]" (P-Lbg), and gratitude to the teachers for attention and patience "And really thank you very much for the patience of the teachers that you put your energy into us" (P-Lbg), "No shame to ask you" (P-Lbg).

Also, joy was highlighted, the joy of cognition and joy of repeating school, "As in the eleventh class; something new; that joy of cognition, like a child" (P-Lbg). The participants were asked what they have enjoyed most and what was the easiest for them. The responses were different: "Most importantly, that was not boring" (M-Lbg), "There was nothing monotonous to be tired of and outraged" (M-Lbg).They did not highlight any learning method. They enjoyed everything "If I could choose, I would take this and that" (M-Lbg).

The local-based games methods caused tension in the beginning "I thought I came here not to play, but to learn the programs" (M-Lbg).

The participants were asked the question about why material that they have learned is valuable to them. The research data shows the practical application of knowledge and courage to help friends during and after classes: "I showed to one acquaintance how to install WhatsApp" (P-Lbg), "Yesterday, a friend came to show me something and shouting "do not break it", I answered "wait" and showed her what I know [laughing]" (P-Clm). The participants were demonstrating a desire to continue learning: "What courses are planned to improve our knowledge" (P-Lbg), "I want to study it further" (P-Lbg).

Summarizing the research, it can be stated that positive experiences, the seniors had during the training course, encourage their more active involvement in learning and provide more positive attitude to the surrounding environment.

6 Conclusions

The participation of older adults in educational activities reflects possibilities for personal development, socialization and the importance of learning. An older person's satisfaction with life is connected to his or her ability to keep their social roles, be active, get involved into voluntary and cultural activities, and learn new things. Therefore, active participation in educational activities helps older adults stay independent and valuable members of their communities.

This qualitative study has shown that it is important for seniors to have successful training, especially when learning how to use smart phones. It raises their self-esteem, encourages them to try something new, and also improves their quality of life.

References

[1] Telia (2016) `https://www.telia.lt/pranesimai-spaudai/siuolaikiniai-senjorai-griauna-mitus`

[2] S. Mikulioniene, *Socialinė gerontologija* (2011)

[3] V. Kanopienė, S. Mikulionienė, V. Česnuitytė, *Lietuvos šeima Europos kontekste* (2015)

[4] S. Cahill, J. Macijauskiene, A.M. Nygård, J.P. Faulkner, I. Hagen, T & D **19**, 55–60 (2007)

[5] N. Selwyn, S. Gorard, J. Furlong, L. Madden, *Ageing & Society* **23**, 561–582 (2003)

[6] V. Thalhammer, *Learning across generations in Europe: contemporary issues in older adult education* (2014)

[7] L. Dromantienė, I. Žemaitaitytė, International scientific conference: conference proceedings (2017) p. 49–60

[8] H. Boshier, *Towards a Learning Society* (1980)

[9] I. Žemaitaitytė, *Society, integration, education.* **2**, 260–268 (2014)

[10] G. Gray G. *Culturally Responsive Teaching. Theory, Research and Practice* (2000)

[11] J.M. Peters, J.L. Armstrong, *New Directions for Adult and Continuing Education* **79**, 75–85 (1998)

[12] Lisbon strategy (2000). Available at: `http://www.europarl.europa.eu/summits/lis1_en.htm`

[13] National programme for 2012 European Year for Active Ageing and Solidarity between Generations (2012). Available at: `https://www.e-tar.lt/portal/forms/legalAct.html?documentId=TAR.B0F07BD2FA4B`

[14] M. Formosa, *Learning across generations in Europe: contemporary issues in older adult education* (2014)
[15] F. Webster, *Informacinės Visuomenės Teorijos* (2006)
[16] L. Tomczyk, E-mentor **4**(36) (2010)
[17] D.A. Wagner, R. Kozma, UNESCO (2005)
[18] V. Dagienė, *Informacinės mokymo technologijos* (2008)
[19] H. Feistm, K. McDougall, *AP & MRCPB.* **1**(8) (2013)
[20] G. Tolutienė, S. Puskorius, S. TILTAI **3**, 123–140 (2014)
[21] V. Indrašienė, G. Žibėnienė, *Šiuolaikinėdidaktika* (2017)
[22] N. Longwoth, *LlinE,1,* 16–20 (1997)
[23] C.C. Bonwell, J.A. Eison, *Active learning: Creating excitement in the classroom* (1991)
[24] J. Dewey, *Democracy and Education* (1922)
[25] D.A. Kolb, *Experiential Learning experience as a source of learning and development* (1984)
[26] S. Silberman, *The Handbook of Experiential Learning* (2007)
[27] G.R. Gibbs, *Analyzing Qualitative Data* (2007)
[28] U. Flick, *An Introduction to Qualitative Research* (2014)
[29] B. Bitinas, Pedagogika **83**, 9–15 (2006)
[30] D. Wilkinson, P. Birmingham, *Using Research Instruments: A guide for Researcher* (2003)

AUTHOR INDEX

EDP Sciences – Web of Conferences
17, Avenue du Hoggar
Parc d'Activité de Courtabouf
BP 112
F-91944 Les Ulis Cedex A
France

ISBN 978-1-5108-9864-6

9 781510 898646

CURRAN ASSOCIATES INC.
proceedings.com